COSPAR COLLOQUIA SERIES

VOLUME 13

MULTI-WAVELENGTH OBSERVATIONS OF CORONAL STRUCTURE AND DYNAMICS

***

Yohkoh 10th Anniversary Meeting

PERGAMON

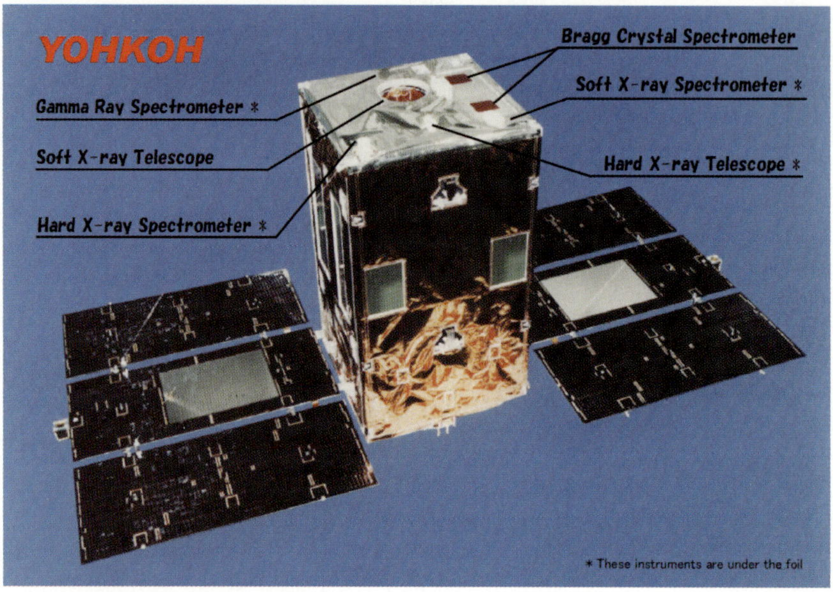

Yohkoh Spacecraft Illustration:

The *Yohkoh* spacecraft was named through a contest among Japanese schoolchildren. The name means "Sunbeam"

Cover Illustration:

This full disk 1991 Yohkoh-SXT image was selected by the readers of Sky an Telescope magazine as one of the 10 most inspiring astronomical images of the 20th century (January 2000 issue. Photo credit: Greg Slater).

MULTI-WAVELENGTH OBSERVATIONS OF CORONAL STRUCTURE AND DYNAMICS

***

Yohkoh 10th Anniversary Meeting

***

Proceedings of the COSPAR Colloquium
Held in Kona, Hawaii, USA
20-24 January, 2002

Edited by

Petrus C. H. Martens
Department of Physics, Montana State University
P.O.Box 3840
Bozeman, MT 59717 USA

and

David P. Cauffman
Lockheed Martin Advanced Technology Center
3251 Hanover Street, Palo Alto CA 94304 USA
now at:
8111 Possession Ridge Lane, Clinton WA 98236 USA

2002

PERGAMON
An Imprint of Elsevier Science

Amsterdam – Boston – London – New York – Oxford – Paris
San Diego – San Francisco – Singapore – Sydney – Tokyo

ELSEVIER SCIENCE Ltd
The Boulevard, Langford Lane
Kidlington, Oxford OX5 1GB, UK

First edition 2002

Library of Congress Cataloging in Publication Data
A catalog record from the Library of Congress has been applied for.

British Library Cataloguing in Publication Data
A catalogue record from the British Library has been applied for.

ISBN: 0-08-044060-6

♾ The paper used in this publication meets the requirements of ANSI/NISO Z39.48-1992 (Permanence of Paper).
Printed in The Netherlands.

PREFACE

These are the Proceedings of the *Yohkoh* 10th Anniversary Meeting, a COSPAR Colloquium held in Kona, Hawaii, USA, on January 20-24, 2002. There were 133 participants from 14 countries. The Colloquium was sponsored by (in no specific order) COSPAR, NASA, the MSU Solar Physics Group, LMSAL, ISAS, the Solar Physics Research Corporation, Elsevier Science, and the NSF. Thanks to the largesse of these sponsors the conference organizers were able to provide financial assistance in part or in whole to 30 participants.

The meeting had originally been planned for September 2001 at the same location, but had to be postponed due to the difficulty in travel following the terrible events of September 11, 2001. We appreciate the understanding and flexibility of the participants and of the conference hotel in rescheduling this meeting.

The title of the meeting was *Multi-Wavelength Observations of Coronal Structure and Dynamics.* We reviewed the many and varied advances in our understanding of the dynamic solar atmosphere in the past ten years of observations by *Yohkoh*, often in collaboration with *SOHO*, *TRACE*, *Ulysses*, and many ground-based observatories. The Scientific Organizing Committee made a concerted effort to invite younger scientists to present the Invited Reviews. This was a success and we have a large number of excellent review papers in this volume. This policy had the additional positive effect of freeing up several of the senior scientists to present exciting new results from their own research in contributed talks. The large number of different sessions in these Proceedings reflects the large variety in science topics that are being addressed with *Yohkoh* data. In all, these Proceedings reflect the quality, breadth, and depth of solar physics research inspired by the decade of *Yohkoh* observations.

The introductory talk for the meeting was given by Prof. Atsuhiro Nishida, former Director-General of ISAS, the lead institute for the *Yohkoh* mission. After that Prof. Takeo Kosugi (see the next paper) of ISAS gave us a succinct and very clear description of the events that had led to the unfortunate *Yohkoh* accident on December 14, 2001, and of the recovery efforts that were underway at the time of the meeting . Prof. Kosugi also chaired the final session in which Prof. Ogawara (also ISAS), the *Yohkoh* Project Manager, received a well deserved ovation for his unfailing efforts in leading the *Yohkoh* project from its conception to end, and the *Yohkoh* instrument builders were recognized.

The meeting was held at the lovely King Kamehamea Beach Hotel in Kona on the Big Island of Hawaii. This wonderful setting, and the friendly and helpful hotel staff made for a relaxed yet focused atmosphere. Social events included a welcome reception, a traditional Hawaiian Luau with a performance of dances from several of the Pacific island groups, and tours of the Keck and Subaru Observatories at Mauna Kea or the HAO Coronagraph Station at Mauna Loa. It is perhaps fitting that this milestone *Yohkoh* Anniversary meeting was held in the Hawaiian Isles, because one of the first *Yohkoh* international meetings was also held there, at the University of Hawaii in Oahu.

We are most grateful to the local organizing committee, in particular the professional work of Jana Halvorson, and the great efforts of MSU graduate students Rebecca McMullen and Elizabeth Noonan. The LOC was chaired by Profs. Loren Acton and Piet Martens of MSU. The computer setup worked flawlessly thanks to the efforts of Alisdair Davey. We are also most grateful for the efforts of the members of the Scientific Organizing Committee, chaired by the same persons as the LOC.

We owe a special word of thanks to Dr. Mark Weber, who designed the LaTex template for the Proceedings papers, and who has provided us continuous assistance with word processing problems, as well as with the final edits of the papers.

The *Yohkoh* 10th Anniversary Meeting was followed by a two day preparatory meeting at the same location for the next major Japanese solar physics mission, Solar-B. This mission builds upon the success of *Yohkoh* and promises to carry our understanding of the solar corona and photosphere even further. Again the US and UK are international partners in some of the instruments. With several solar missions being prepared by both NASA and ISAS for the current decade, and a growing recognition of the importance of space weather, the future of solar physics seems bright.

The Editors:
Piet Martens
Dave Cauffman

TEN YEARS OF YOHKOH AND ITS CURRENT STATUS: A BRIEF SUMMARY

T. Kosugi[1] and L. W. Acton[2]

[1]*Institute of Space and Astronautical Science, 3-1-1 Yoshinodai, Sagamihara, Kanagawa 229-8510, Japan*
[2]*Physics Department, Montana State University, P.O. Box 173840, Bozeman, MT 59717, USA*

ABSTRACT

The *Yohkoh* satellite was launched on August 30, 1991 by the Institute of Space and Astronautical Science (ISAS) from its Kagoshima Space Center, and operated until December 14, 2001, despite having a design goal of a lifetime of three years. Two of the onboard instruments were joint efforts with international partners, namely the United States and the United Kingdom. In the ten years of operations since, *Yohkoh* has acquired more than 6 million soft X-ray snapshot images and detected more than 2800 high energy X-ray solar flares. More than a thousand papers by authors worldwide have been published from *Yohkoh* in journals and proceedings, and a new generation of solar physicists has begun thesis research using *Yohkoh* data.

DISCUSSION

The *Yohkoh* satellite was launched on August 30, 1991 by the Institute of Space and Astronautical Science (ISAS) from its Kagoshima Space Center. It is a medium-sized satellite, 1 x 1 x 2 m in size and 400 kg in weight, and carries four advanced X-ray (and gamma-ray) instruments for studying the solar corona and the high-energy phenomena that take place there. They are the Soft X-ray Telescope (SXT), the Hard X-ray Telescope (HXT), the Bragg Crystal Spectrometer (BCS), and the Wide-Band Spectrometer (WBS). The SXT and the BCS were joint efforts with international partners, namely the United States and the United Kingdom (USA-Japan for SXT and UK-USA-Japan for BCS). Science operations and data analysis have been conducted by the same international framework. *Yohkoh* data have been made available to and been utilized by the worldwide solar physics and related science communities, however. In addition, they have been used by real-time, space-weather prediction services.

It was only after launch that we fully recognized the epoch-making excellence of the *Yohkoh* instruments, which all worked just as designed. The dynamically changing nature of the solar corona was for the first time vividly unveiled with high-quality, high-resolution, CCD images taken by SXT at high cadence. Acceleration (and transport) of energetic electrons in solar flares became traceable with HXT, thanks to its imaging capability in the purely nonthermal X-ray emission above 30 keV. In the subsequent years since then, we have accumulated more than 6 million SXT snapshot images and detected with HXT more than 2800 solar flares, together with their spectra from BCS and WBS.

More than a thousand papers have been published from *Yohkoh* in journals and proceedings. By the 10th Anniversary of *Yohkoh* (August 30, 2001), 53 graduate students around the world had completed their *Yohkoh*-related thesis work and received PhD doctorates. In Japan alone recipients of master of science degrees have totaled 47. These papers and theses cover a wide variety of topics.

The following *Yohkoh* observations have revolutionized our view of the solar corona:

- Dynamically changing coronal structure involving magnetic reconnection with various temporal and spatial scales, from microflares and jets to large-scale coronal restructuring events;

- Cusp-shaped soft X-ray arcades in long-duration event flares and above-the-loop-top hard X-ray sources in impulsive flares as evidence for on-going magnetic reconnection in solar flares;

- Arcade formation and coronal dimming identified as the soft X-ray counterpart of coronal mass ejections (CMEs), that are used for issuing alerts concerning the arrival of interplanetary disturbances at the Earth; and

- Sigmoidal soft X-ray structures in active regions, identified as signatures of the likely onset of CMEs.

The cooperative nature of *Yohkoh* flight operations and data sharing fostered the creation of a unique file structure and systems for data archive and analysis software. The *Yohkoh* data system has served as a model for subsequent missions. The system of analysis software has evolved into the widely used SolarSoft or SSW system that is increasingly becoming a standard for analysis of all kinds of solar observational data.

In addition to the science achievements mentioned above, we can be proud of *Yohkoh* for its successful public outreach programs. Many science museums in the world exhibit *Yohkoh* SXT movies, which contribute to making one of the modern science frontiers intimately available to the public. Similarly, internet outreach programs have been developed and are frequently visited by interested audiences.

When we designed the *Yohkoh* satellite, three years of mission lifetime was a target. In this sense it is far beyond our expectation that *Yohkoh* and its scientific instruments were perfectly operable for more than a decade until December 14, 2001, when an unlucky accident was triggered the by transit of *Yohkoh* in a solar near-total eclipse zone. *Yohkoh*'s encounter with the eclipse caused an attitude control anomaly, related to a defect in the software that had been installed when the control logic had been modified for overcoming deterioration of one of the attitude actuators on board, and made the satellite lose its attitude toward the Sun. As a consequence *Yohkoh* eventually lost its battery charge. None of the recovery efforts were successful. Nonetheless, we believe that *Yohkoh* science will continue to develop. *Yohkoh* data is full of many treasures to be analyzed, still hidden and still untouched.

REFERENCES

The *Yohkoh* instrument papers have been published in a special issue of *Solar Physics*, **136** (1991).

Initial results from *Yohkoh* have been published in a special issue of *Publications of the Astronomical Society of Japan*, **44**(5) (1992).

A comprehensive list of *Yohkoh* papers is posted at `http://solar.physics.montana.edu/sxt/`.

Sponsored by

THE COMMITTEE ON SPACE RESEARCH (COSPAR)
THE NATIONAL AERONAUTICS AND SPACE ADMINISTRATION (NASA)
THE NATIONAL SCIENCE FOUNDATION (NSF)
THE INSTITUTE FOR SPACE AND ASTRONAUTICAL SCIENCE (ISAS)
THE MONTANA STATE UNIVERSITY SOLAR PHYSICS GROUP
LOCKHEED SOLAR AND ASTROPHYSICS LABORATORY
SOLAR PHYSICS RESEARCH CORPORATION
and
ELSEVIER SCIENCE

Scientific Organizing Committee

Co-Chairpersons:

Loren Acton (Montana State University)
Piet Martens (Montana State University)

Other Members:

David Alexander (Lockheed Martin Solar and Astrophysics Lab)
Len Culhane (Mullard Space Science Laboratory)
Leon Golub (Center for Astrophysics)
Richard Harrison (Rutherford Appleton Laboratory)
Takeo Kosugi (Institute of Space and Astronautical Science)
Barry LaBonte (Institute for Astronomy)
Yoshi Ogawara (Institute of Space and Astronautical Science)
Toshifumi Shimizu (National Astronomical Observatory of Japan)
Keith Strong (Lockheed Martin Solar and Astrophysics Lab)
Yutaka Uchida (Science University of Tokyo)

Organizing Committee

Loren Acton (Montana State University)
Piet Martens (Montana State University)
Jana Halvorson (free-lance creative services)
Elizabeth Noonan (Montana State University)
Rebecca McMullen (Montana State University)

YOHKOH 10 CONFERENCE, KONA, HAWAII, JANUARY, 2002

1 Masuda
2 Nagata
3 Tanuma
4 Sakao
5 Ohyama
6 Kusano
7 Katsukawa
8 Sui
9 Panasenco
10 Morita
11 Kobayashi
12 Kliem
13 Parnell
14 Morimoto
15 Moore
16 Romashets
17 Aurass
18 Martens
19 Davey
20 Klimchuk
21 Altrock
22 Hagino
23 Cirtain
24 Canfield
25 Tarbell
26 Slater
27 Bartus
28 Nightingale
29 Farnik
30 Yokoyama
31 Sterling
32 Sersen
33 Sakurai
34 Miyagoshi
35 Tikhomolov
36 Fludra
37 Hansteen
38 Foley
39 Warren
40 Deluca
41 Karlicky
42 LaBonte
43 Banerjee
44 Miller
45 Hudson
46 Gary Nitta
47 J. Sylwester
48 Hirose
49 Shimojo
50 Matsuzaki
51 B. Sylwester
52 Emslie
53 Yashiro
54 Schmahl
55 Ryan
56 DeForest
57 Correia
58 Asai
59 Weber
60 Ichimoto
61 Lang
62 Virani
63 Takeda
64 Madjarska
65 Brosius
66 Reeves
67 Yoshimori
68 Ko
69 Isobe
70 Litvinenko
71 Saba
72 Aschwanden
73 Verma
74 Nariaki Nitta
75 van Driel
76 Falconer
77 Vats
78 Metcalf
79 Kundu
80 Mason
81 Wills-Davey
82 Youping Li
83 Jing Li
84 Cauffman
85 Yaji
86 McMullen
87 Wikstol
88 Larson
89 Harra
90 Khan
91 Winter
92 Doschek
93 Shin
94 Hassler
95 Hara
96 Benevolenskaya
97 Sato
98 Yoshimura
99 Shibata
100 Takeuchi
101 Kozu
102 Narukage
103 Akiyama
104 McKenzie
105 Noonan
106 Hanaoka
107 Pevtsov
108 Hori
109 Gburek
110 Magara
111 Shimizu
112 Watanabe
113 Kosugi
114 Acton
115 Tsuneta
116 Ogawara
117 Nishida
118 Uchida
119 Sawant
120 Sturrock
121 Bruner
122 Petrosian

CONTENTS

Section V. Reconnection in Flares

Section VI. MHD Simulations of Emergence and Eruptions

Section VII. Fine Structure in Flares

Section I.

Quiet Sun and Solar Wind

CORONAL HOLES AND THE SOLAR WIND

S. R. Cranmer

Smithsonian Astrophysical Observatory, 60 Garden Street, Cambridge, MA 02138

ABSTRACT

Coronal holes are the darkest regions of the ultraviolet and X-ray Sun, both on the disk and above the limb. Coronal holes are associated with rapidly expanding open magnetic fields and the acceleration of the high-speed solar wind. This paper reviews measurements of the plasma properties of coronal holes and how these measurements have been used to put constraints on theoretical models of coronal heating and solar wind acceleration. Heat deposition at the dense and collisional coronal base is of comparable importance (in determining, e.g., temperature gradients and asymptotic outflow speeds) as extended heating in the collisionless regions above 2 solar radii. Thus, a complete understanding of the physics requires both observations of the solar disk and inner corona (*Yohkoh,* EIT, CDS, SUMER) and coronagraphic observations of the wind's acceleration region (UVCS, LASCO). Although strong evidence has been found to suggest that the high-speed wind is driven mainly by proton pressure, the differences between proton, electron, and heavy ion velocity distributions are extremely valuable as probes of the dominant physical processes.

INTRODUCTION

The existence of coronal holes was first recognized by Waldmeier (1957, 1975), who noticed long-lived regions of negligible intensity in coronagraphic images of the 5303 Å green line. Waldmeier called the features that appeared more-or-less circular when projected onto the solar disk *Löcher* (holes), and the more elongated features were called *Rinne* (grooves) or *Kanal* (channels). The fact that coronal holes coincide with regions of open magnetic field that extend into interplanetary space was realized during the first decade of *in situ* solar wind observations (e.g. Wilcox, 1968). Coronal holes were effectively "re-discovered" in the early 1970s as discrete dark patches on the X-ray and ultraviolet solar disk, and their connection with the high-speed component of the solar wind soon became evident (Krieger *et al.* 1973, Zirker 1977). The term "coronal hole" thus has come to denote both the on-disk features and their open-field extensions off the solar limb. This paper provides a brief review of the physics of coronal holes and the acceleration of the high-speed solar wind.

Coronal holes become distinguishable from neighboring quiet and active regions several Mm above the photosphere, where the temperature exceeds $\sim 10^5$ K. At these low coronal heights, holes exhibit lower densities and temperatures than other regions (see, e.g., Esser & Habbal 1997). At larger heights, as the plasma becomes less collision-dominated, coronal hole densities remain relatively low but the temperatures of different plasma components begin to depart strongly from thermal equilibrium, with $T_e < T_p < T_{ion}$. Despite the large-scale identification of coronal holes with open magnetic field lines, they contain a wide variety of magnetic structures, from X-ray bright points and spicules on the smallest scales to plumes and jets on larger scales (see Figure 1). At the minimum of the Sun's 11-year activity cycle, large coronal holes exist at the north and south heliographic poles and extend into a large fraction of the volume of the heliosphere. At times other than solar minimum, smaller and more transient coronal holes appear at all

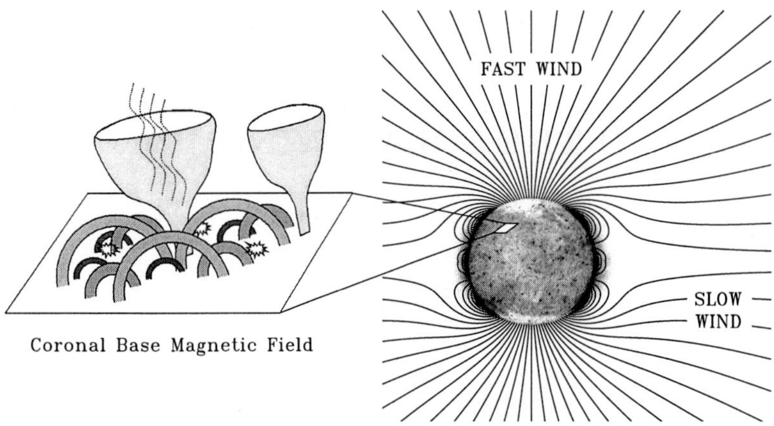

Fig. 1. Schematic view of the solar magnetic field at the minimum of the 11-year activity cycle. The stochastic distribution of small-scale loops and open flux tubes at the base (e.g. Dowdy *et al.* 1986) gives way to a more ordered set of field lines in the extended corona (Banaszkiewicz *et al.* 1998). The ultraviolet image of the solar disk (dark colors represent brighter regions) was taken by the EIT instrument on the *SOHO* spacecraft (Delaboudinière *et al.* 1995).

latitudes, with plasma properties intermediate between those of polar coronal holes and the higher-density portions of the corona.

UNANSWERED QUESTIONS

The energy that heats the corona and accelerates the solar wind originates in subphotospheric convective motions. However, even after a half-century of investigation, the physical processes that transport this energy to the corona and convert it into thermal, magnetic, and kinetic energy are still not known (e.g. Parker 1991, Marsch 1999, Gómez *et al.* 2000). Below are two lists of unanswered questions about the physics of coronal holes, separated by spatial scale into the coronal base and the extended corona.

The Coronal Base

1. *What physical processes are responsible for basal coronal heating?* To a certain extent, this major question cannot be answered until a more basic and phenomenological question is answered: "What is the time scale distribution of the mechanical energy pumped into coronal magnetic footpoints?" The traditional division into AC versus DC heating mechanisms (Ionson 1985; Narain & Ulmschneider 1990, 1996)—where AC [DC] denotes driving motions on time scales shorter [longer] than representative transit times—may give way to a more unified approach if the corona contains a continuous spectrum of time scales spanning both limits (e.g. Milano *et al.* 1997). It is likely that the ultimate heating processes all involve the dissipation of fluctuations on kinetic "microscales," and thus requires physics beyond ideal magnetohydrodynamics (MHD); see, e.g., Viñas *et al.* (2000), Leamon *et al.* (2000).

2. *Are coronal holes distinguished by different heating rates than neighboring quiet regions, or do they appear different only because of different relative fractions of closed and open magnetic flux?* The answer to this question may be "the latter," but it depends on the elucidation of the nature of coronal heating on the smallest scales (see, e.g., Hearn 1977, Axford *et al.* 1999, Falconer *et al.* 1999, Priest *et al.* 2000).

3. *How is the mass flux of the high-speed wind determined and regulated?* Leer & Marsch (1999) contrasted two proposed scenarios: (1) that because the wind is driven by the energy deposition in the low corona, the mass flux should be proportional to this mechanical energy flux, and (2) that the supply of plasma into open "funnels" is constrained and determined by rapid ionization processes. These concepts may be related to one another and it is not yet apparent to what extent each process contributes (see also Sandbæk *et al.* 1994, Peter & Marsch 1997, Chashei 1997).

The Extended Corona

1. *How much of the solar wind comes from coronal holes?* It is reasonably well established that the fast solar wind (i.e. with flow speed greater than ~500 km s^{-1} at 1 AU) is accelerated in coronal holes, but it is unclear how much of the slow component of the solar wind comes from: (1) the edges of coronal holes (Wang & Sheeley 1990), (2) transient reconnections in closed-field streamers (e.g. Wu *et al.* 2000), or (3) active regions (Hick *et al.* 1999). Conversely, there is also some controversy concerning how much of the fast wind may be associated with quiet regions on the solar disk (and thus *not* with the superradial expansion of coronal hole fields; see Habbal *et al.* 2001).

2. *How and where are the plasma fluctuations (i.e. waves, turbulence, and shocks) that are believed to drive extended heating and acceleration produced and damped?* Propagating fluctuations are believed to dominate the energy and momentum deposition in the extended corona because the ultimate source is presumably from the solar surface, and thus *propagation* of some kind is required to reach large heliocentric distances. The self-consistent determination of the radial evolution of both: (1) the wavenumber spectrum of all relevant fluctuation modes, and (2) the velocity distributions of electrons, protons, and minor ions was called "the Holy Grail of this line of inquiry" by Hollweg (1999). The section below titled *Proposed Physical Processes* contains a summary of recent work toward this goal.

3. *Does the acceleration of the high-speed wind require independent momentum deposition, or are pressure gradient forces sufficient?* Recent UVCS observations of proton temperatures perpendicular to the magnetic field ($T_{p\perp}$) as large as 3–4 million K (see below) suggest that a significant fraction of the driving of the high-speed wind comes from the anisotropic pressure gradient force (e.g. magnetic mirror force) on protons. Traditionally, however, wind speeds at 1 AU in excess of 600 to 700 km s^{-1} have been explained only as a result of additional momentum deposition from wave pressure (Jacques 1977), diamagnetic acceleration of plasmoids (Pneuman 1986), or other processes (see also Tziotziou *et al.* 1998). Cranmer (2002) produced a series of empirically based solar wind models which implied that a maximum $T_{p\perp}$ of 6 million K was required—in a model *without* additional momentum deposition—to produce a realistic fast wind. It thus seems likely that momentum deposition is required, but the uncertainties in the determination of $T_{p\perp}$ from H I Lyα line widths are still large enough so that it is not yet possible to answer this question definitively.

4. *To what degree do the observed filamentary inhomogeneities (e.g. polar plumes and jets) contribute to the mass, momentum, and energy budget of the fast wind?* Polar plumes contain denser (Ahmad & Withbroe 1977), cooler (e.g. Kohl *et al.* 1999), and slower (Giordano *et al.* 2000, Wilhelm *et al.* 2000) plasma than the "ambient" interplume corona. It is not known, though, whether the high-speed solar wind comes primarily from the interplume regions, or if it is a result of plume-interplume mixing somewhere between ~20 and ~60 $R_\odot$ (see also Reisenfeld *et al.* 1999, Parhi & Suess 2000, DeForest *et al.* 2001).

SUMMARY OF OBSERVATIONS

In order to understand how coronal holes are produced and maintained, one must have detailed empirical knowledge about the properties of the plasma. The two most useful means of measuring these properties have been *in situ* spacecraft detection and the remote sensing of coronal photons. Some key results of such measurements are summarized below. Other diagnostic techniques that cannot be discussed in detail in this brief review are the scintillation of radio waves passing through the corona (Bastian 2001), the analysis of backscattered solar radiation by interstellar atoms (Bertaux *et al.* 1996), and using comets as probes of the solar wind energy budget (Raymond *et al.* 1998).

Spacecraft have measured particle velocity distribution functions and electromagnetic fields as close to the Sun as 60 $R_\odot$ (*Helios 1* and *2*), and as far as 12,000 $R_\odot$ (*Voyager 2*). Departures from Maxwellian velocity distributions have been used as sensitive constraints on the kinetic physics on microscopic scales (see, e.g., Feldman & Marsch 1997). *In situ* instruments have also measured fluctuations in magnetic field strength, velocity, and density on time scales ranging from 0.1 second to months and years. Both propagating waves (mainly Alfvénic in nature) and nonpropagating, pressure-balanced structures advecting with the wind are observed. Nonlinear interactions between different oscillation modes create strong turbulent mixing, and Fourier spectra of the fluctuations show clear power-law behavior—indicative of inertial and dissipation ranges—in agreement with many predictions for fully developed MHD turbulence (Goldstein *et al.* 1995, Tu & Marsch 1995).

Because spacecraft measurements have not been able to probe the wind where its acceleration occurs (typically from the base of the corona to $\sim$10 $R_\odot$), we have relied on complementary observations of photons from the corona to study this key region. Instruments aboard the *Yohkoh, TRACE* (*Transition Region And Coronal Explorer*), and *SOHO* (*Solar and Heliospheric Observatory*) spacecraft—especially EIT, CDS, and MDI on the latter—have revealed strong variability and complexity at the coronal base on the smallest observable scales (100 to 1000 km; see, e.g., Watanabe *et al.* 1998, Engvold & Harvey 2000). The improved understanding of explosive, flarelike events from *Yohkoh* has led to many new ideas for the heating of the entire corona (e.g. Shimizu 1996, Moore *et al.* 1999, Priest *et al.* 2000). The SUMER instrument on *SOHO* has investigated the origins of the high-speed solar wind in coronal holes by mapping out blueshifts in coronal emission lines (Hassler *et al.* 1999). SUMER measurements have also shown that ion temperatures exceed electron temperatures at very low heights (Seely *et al.* 1997, Tu *et al.* 1998). Obtaining reliable electron temperatures above the limb ($\sim$1.1–1.4 $R_\odot$), though, has proved difficult. Relatively low values of T_e in the range 0.3–1.1 million K have been inferred by David *et al.* (1998) and Doschek *et al.* (2001) at these heights, which also agrees with the theoretical models of Hansteen *et al.* (1997). However, relatively high values of order 1.3–1.7 million K were inferred in coronal holes by Ko *et al.* (1997), Foley *et al.* (1997), and Aschwanden & Acton (2001). The reconciliation of this controversy may be the existence of non-Maxwellian electron distributions at low coronal heights (Esser & Edgar 2000), but there also may be selection effects due to different instrumental sensitivities in an intrinsically multi-thermal distribution of temperatures.

In the acceleration region of the wind, the ultraviolet emission from coronal holes is at least 5 orders of magnitude dimmer than the solar disk. Thus, the technique of occulting the disk in *coronagraph* telescopes—often combined with spectroscopy to isolate individual ion properties—has led to a dramatic increase in our knowledge about how the high speed wind is driven. The UVCS instrument aboard *SOHO* provided the first measurements of ion temperature anisotropies and differential outflow speeds in the acceleration region of the wind (Kohl *et al.* 1995, 1997, 1998). UVCS measured O^{5+} perpendicular temperatures exceeding 300 million K at a height of 2 $R_\odot$ (see Figure 2), with $T_\perp/T_\parallel$ of order 10–100. Temperatures for both O^{5+} and Mg^{9+} are significantly greater than mass-proportional when compared to hydrogen, and outflow speeds for O^{5+} may exceed those of hydrogen by as much as a factor of two (see also Cranmer *et al.* 1999a). These results are similar in character to the *in situ* data, but they imply more extreme departures from thermodynamic equilibrium in the corona. Because of the perpendicular nature of the heating, and because of the ordering $T_{\rm ion} \gg T_{\rm p} > T_{\rm e}$, UVCS observations have led to a resurgence of interest in models of coronal ion cyclotron resonance (see below).

Note from Figure 2 that the protons (as measured by proxy via the H I Lyα emission line) are heated more strongly than electrons, and thus provide the bulk of the pressure gradient force in coronal holes. The observed proton temperature gradient allows us to estimate the heating rate *per proton* to be of order $\sim$0.05

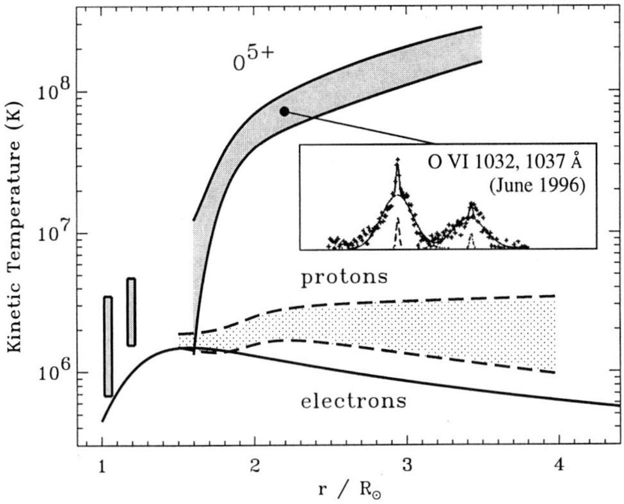

Fig. 2. Summary plot of coronal hole and high-speed wind temperature measurements. Perpendicular temperatures for protons and O^{5+} above 1.5 $R_\odot$ are from an empirical model that reproduced UVCS line widths (Kohl *et al.* 1998, Cranmer *et al.* 1999a). The two O^{5+} boxes at lower heights are representative of ion temperatures derived from SUMER line widths (e.g. Hassler *et al.* 1997), and the electron temperature is from Ko *et al.* (1997). Additional uncertainties, mainly due to differences between plumes and interplume regions, and differences between coronal holes at various latitudes, are not shown here.

to 0.1 eV s^{-1} at 2 $R_\odot$. Surprisingly, this is of the same order of magnitude as the heating rate per proton that is believed to exist at the coronal base, where an energy flux $F \approx 5 \times 10^5$ erg cm^{-2} s^{-1} (e.g. Parker, 1991) that is dissipated in a scale length ℓ of order 0.01 to 0.1 $R_\odot$, in a plasma with number density n of 10^8 to 10^{10} cm^{-3}, yields a heating rate per proton $F/(\ell n)$ in the range 0.01 to 1 eV s^{-1}. This result implies that both the base and the extended corona are of comparable importance in influencing particle velocity distributions in the high-speed wind.

PROPOSED PHYSICAL PROCESSES

Different physical mechanisms for heating the corona probably govern closed magnetic loops, active regions, and the open field lines that dominate coronal holes (e.g. Priest *et al.* 2000). There is also a growing realization that the coronal base ($r \lesssim 1.5\, R_\odot$) is probably heated by different processes than those that apply at larger heliocentric distances. This heuristic division into *creating* the lower corona versus *maintaining* and *evolving* the extended corona is supported by the drastic differences in Coulomb collision rates at the base (where all species seem to be collisionally coupled) and in the supersonic wind (which is nearly collisionless). The two regimes are also differentiated by the complexity and topology of the magnetic field (see Figure 1).

The remainder of this paper discusses the extended heating in the acceleration region of the high-speed wind, which as stated above is expected to be dominated by the dissipation of propagating fluctuations. It is not known, however, how or where the fluctuations responsible for extended heating are generated. Alfvén waves have received the most attention because they seem to be the least damped by collisional processes (i.e. viscosity, conductivity, resistivity) at the coronal base, but there have been recent observations that imply the presence of slow magnetosonic waves in various kinds of open flux tubes (Ofman *et al.* 1999). At heights greater than 2–3 $R_\odot$, wave dissipation should be dominated by collisionless processes. The most likely dissipation mechanism seems to be ion cyclotron resonance, since Landau damping mainly tends to heat electrons in a low-β plasma (Habbal & Leer 1982). Some have suggested that left-hand polarized ion cyclotron waves are generated impulsively at the base of the corona and propagate virtually unaltered to where they are damped (Axford *et al.* 1999). A related idea is that the same basal impulsive events generate fast shocks that fill the extended corona and convert some of their energy into anisotropic heating and ion acceleration (Lee & Wu 2000). Problems with these ideas include: (a) the neglect of minor ions that can easily absorb a basal fluctuation spectrum before any primary plasma constituents (protons or He^{2+}) can come into resonance (e.g. Cranmer 2000, 2001), and (b) a significant shortfall in observed density fluctuations, compared to model predictions consistent with basal wave generation models (Hollweg 2000).

More numerous are proposed scenarios of local wave generation; i.e. where "secondary" fluctuations arise throughout the extended corona as the result of either turbulent cascade, plasma instability, or mode conversion (e.g. Hollweg 1986, Matthaeus *et al.* 1999, Markovskii 2001). Ion cyclotron frequencies in the corona are typically 10 to 10,000 Hz, but the oscillation frequencies observed on the surface of the Sun (generated mainly by convection) are of order 0.01 Hz. Any wave generation mechanism must therefore bridge a gap of many orders of magnitude in frequency (or wavenumber). Most models of MHD turbulence favor the transfer of energy from small to large wavenumbers transverse to the background magnetic field ($\mathbf{k} \cdot \mathbf{B} \approx 0$); see, e.g., Shebalin *et al.* (1983), Goldreich & Sridhar (1997). However, ion cyclotron damping of Alfvénic fluctuations (believed to be the only mode that can survive into the solar wind) requires large *parallel* wavenumbers ($k_{\parallel} \approx \Omega_{\rm ion}/V_A$) that seemingly are not produced by MHD cascade. This inability to produce ion cyclotron waves locally in the corona is a major roadblock in our attempts to understand the origin of the observed anisotropic heating and preferential ion acceleration.

Despite our present lack of understanding about how ion cyclotron waves may be generated, there has been no shortage of attempts to "work backward" from the observational constraints to derive further details of the required wave properties and their kinetic effects. In addition to moment-based models assuming bi-Maxwellian distributions (e.g. Cranmer *et al.* 1999b, Hu *et al.* 1999, Tu & Marsch 2001), there has been a recent flurry of activity to understand kinetic departures from simple parameterized velocity distributions (Galinsky & Shevchenko 2000, Isenberg *et al.* 2001, Vocks & Marsch 2001, Cranmer 2001). The results from these investigations are still being digested, and it is not yet clear which aspects of the physics can be neglected and which ones are required for a basic understanding.

CONCLUSIONS

Considerable progress has been made in the last decade in characterizing the plasma state of coronal holes and their associated high-speed solar wind streams. The observations have guided theorists to a certain extent, but *ab initio* kinetic models are still required before we can claim a full understanding of the physics. Future spectroscopic measurements of the corona are expected to provide constraints on specific departures from bi-Maxwellian velocity distributions (Cranmer 2001), and NASA's Solar Probe (e.g. Möbius *et al.* 2000) should make valuable *in situ* measurements as close to the Sun as 4 $R_{\odot}$. Observations of the coronal base from X-ray and ultraviolet space-based telescopes are a key ingredient in determining the source regions and lower boundary conditions of the wind. To make further progress, the lines of communication must be kept open between theorists and observers, and also between the two traditionally separated observational communities of "solar physics" (i.e. near-Sun astronomy) and "space physics" (i.e. interplanetary plasma physics).

ACKNOWLEDGEMENTS

This work is supported by the National Aeronautics and Space Administration under grant NAG5-10093 to the Smithsonian Astrophysical Observatory, by Agenzia Spaziale Italiana, and by the Swiss contribution to the ESA PRODEX program.

REFERENCES

Ahmad, I. A., and G. L. Withbroe, EUV analysis of polar plumes, *Solar Phys.*, **53**, 397 (1977).

Aschwanden, M. J., and L. W. Acton, Temperature tomography of the soft X-ray corona, *Astrophys. J.*, **550**, 475 (2001).

Axford, W. I., J. F. McKenzie, G. V. Sukhorukova, M. Banaszkiewicz, A. Czechowski, and R. Ratkiewicz, Acceleration of the high speed solar wind in coronal holes, *Space Sci. Rev.*, **87**, 25 (1999).

Banaszkiewicz, M., W. I. Axford, and J. F. McKenzie, An analytic solar magnetic field model, *Astron. Astrophys.*, **337**, 940 (1998).

Bastian, T. S., Radio wave propagation in the corona and the interplanetary medium, *Astrophys. Space Sci.*, **277**, 107 (2001).

Bertaux, J. L., R. Lallement, and E. Quémerais, UV studies and the solar wind, *Space Sci. Rev.* **78**, 317 (1996).

Chashei, I. V., Energy balance of the quiescent solar corona and the problem of regulating the solar wind mass flux, *Astron. Reports*, **41**, 85 (1997).

Cranmer, S. R., Ion cyclotron wave dissipation in the solar corona: The summed effect of more than 2000 ion species, *Astrophys. J.*, **532**, 1197 (2000).

Cranmer, S. R., Ion cyclotron diffusion of velocity distributions in the extended solar corona, *J. Geophys. Res.*, **106**, 24,937 (2001).

Cranmer, S. R., Coronal holes and the high-speed solar wind, *Space Sci. Rev.*, submitted (2002).

Cranmer, S. R., G. B. Field, and J. L. Kohl, Spectroscopic constraints on models of ion cyclotron resonance heating in the polar solar corona and high speed solar wind, *Astrophys. J.*, **518**, 937 (1999b).

Cranmer, S. R., J. L. Kohl, G. Noci, E. Antonucci, G. Tondello, *et al.*, An empirical model of a polar coronal hole at solar minimum, *Astrophys. J.*, **511**, 481 (1999a).

David, C., A. H. Gabriel, F. Bely-Dubau, A. Fludra, P. Lemaire, and K. Wilhelm, Measurement of the electron temperature gradient in a solar coronal hole, *Astron. Astrophys.*, **336**, L90 (1998).

DeForest, C. E., S. P. Plunkett, and M. D. Andrews, Observation of polar plumes at high solar altitudes, *Astrophys. J.*, **546**, 569 (2001).

Delaboudinière, J.-P., G. E. Artzner, J. Brunaud, A. H. Gabriel, J. F. Hochedez, *et al.*, EIT: Extreme-ultraviolet imaging telescope for the SOHO mission, *Solar Phys.*, **162**, 291 (1995).

Doschek, G. A., U. Feldman, J. M. Laming, U. Schühle, and K. Wilhelm, Properties of solar polar coronal hole plasmas observed above the limb, *Astrophys. J.*, **546**, 559 (2001).

Dowdy, J. F., Jr., D. Rabin, and R. L. Moore, R. L., On the magnetic structure of the quiet transition region, *Solar Phys.*, **105**, 35 (1986).

Engvold, O., and J. W. Harvey (eds.), *The Physics of the Solar Corona and Transition Region*, Kluwer, Dordrecht (2000).

Esser, R., and R. J. Edgar, Reconciling spectroscopic electron temperature measurements in the solar corona with in situ charge state observations, *Astrophys. J.*, **532**, L71 (2000).

Esser, R., and S. R. Habbal, Coronal holes and the solar wind, in *Cosmic Winds and the Heliosphere*, ed. J. R. Jokipii, C. P. Sonett, and M. S. Giampapa, p. 297, U. Arizona Press, Tucson, AZ (1997).

Falconer, D. A., R. L. Moore, J. G. Porter, and D. H. Hathaway, Large-scale coronal heating, clustering of coronal bright points, and concentration of magnetic flux, *Space Sci. Rev.*, **87**, 181 (1999).

Feldman, W. C., and E. Marsch, Kinetic phenomena in the solar wind, in *Cosmic Winds and the Heliosphere*, ed. J. R. Jokipii, C. P. Sonett, and M. S. Giampapa, p. 617, U. Arizona Press, Tucson, AZ (1997).

Foley, C. R., J. L. Culhane, and L. W. Acton, Yohkoh soft X-ray determination of plasma parameters in a polar coronal hole, *Astrophys. J.*, **491**, 933 (1997).

Galinsky, V. L., and V. I. Shevchenko, Nonlinear cyclotron resonant wave-particle interaction in a nonuniform magnetic field, *Phys. Rev. Lett.*, **85**, 90 (2000).

Giordano, S., E. Antonucci, G. Noci, M. Romoli, and J. L. Kohl, Identification of the coronal sources of the fast solar wind, *Astrophys. J.,* **531**, L79 (2000).

Goldreich, P., and S. Sridhar, Magnetohydrodynamic turbulence revisited, *Astrophys. J.,* **485**, 680 (1997).

Goldstein, M. L., D. A. Roberts, and W. H. Matthaeus, Magnetohydrodynamic turbulence in the solar wind, *Ann. Rev. Astron. Astrophys.,* **33**, 283 (1995).

Gómez, D. O., P. A. Dmitruk, and L. J. Milano, Recent theoretical results on coronal heating, *Solar Phys.,* **195**, 299 (2000).

Habbal, S. R., and E. Leer, Electron heating by fast mode magnetohydrodynamic waves in the solar wind emanating from coronal holes, *Astrophys. J.,* **253**, 318 (1982).

Habbal, S. R., R. Woo, and J. Arnaud, On the predominance of the radial component of the magnetic field in the solar corona, *Astrophys. J.,* **558**, 852 (2001).

Hansteen, V. H., E. Leer, and T. E. Holzer, The role of helium in the outer solar atmosphere, *Astrophys. J.,* **482**, 498 (1997).

Hassler, D. M., I. E. Dammasch, P. Lemaire, P. Brekke, W. Curdt, H. E. Mason, and J.-C. Vial, Solar wind outflow and the chromospheric magnetic network, *Science,* **283**, 810 (1999).

Hassler, D. M., K. Wilhelm, P. Lemaire, and U. Schühle, Observations of polar plumes with the SUMER instrument on SOHO, *Solar Phys.,* **175**, 375 (1997).

Hearn, A. G., An explanation of the observed differences between coronal holes and quiet coronal regions, *Solar Phys.,* **51**, 159 (1977).

Hick, P., Z. Svestka, B. V. Jackson, F. Farnik, and H. Hudson, Quiet solar wind signatures above active regions observed in X-rays, in *Solar Wind Nine,* ed. S. Habbal, R. Esser, J. Hollweg, and P. Isenberg, p. 231, AIP Press, Woodbury, NY (1999).

Hollweg, J. V., Transition region, corona, and solar wind in coronal holes, *J. Geophys. Res.,* **91**, 4111 (1986).

Hollweg, J. V., The cyclotron resonance in coronal holes, 1, Heating and acceleration of protons, O^{5+}, and Mg^{9+}, *J. Geophys. Res.,* **104**, 24,781 (1999).

Hollweg, J. V., Compressibility of ion cyclotron and whistler waves, *J. Geophys. Res.,* **105**, 7573 (2000).

Hu, Y.-Q., S. R. Habbal, and X. Li, On the cascade processes of Alfvén waves in the fast solar wind, *J. Geophys. Res.,* **104**, 24,819 (1999).

Ionson, J. A., Coronal heating by resonant (A.C.) and nonresonant (D.C.) mechanisms, *Astron. Astrophys.,* **146**, 199 (1985).

Isenberg, P. A., M. A. Lee, and J. V. Hollweg, The kinetic shell model of coronal heating and acceleration by ion-cyclotron waves, 1, Outward-propagating waves, *J. Geophys. Res.,* **106**, 5649 (2001).

Jacques, S. A., Momentum and energy transport by waves in the solar atmosphere and solar wind, *Astrophys. J.,* **215**, 942 (1977).

Ko, Y.-K., L. A. Fisk, J. Geiss, G. Gloeckler, and M. Guhathakurta, An empirical study of the electron temperature and heavy ion velocities in the south polar coronal hole, *Solar Phys.,* **171**, 345 (1997).

Kohl, J. L., R. Esser, L. D. Gardner, S. Habbal, P. S. Daigneau, *et al.,* The Ultraviolet Coronagraph Spectrometer for the Solar and Heliospheric Observatory, *Solar Phys.,* **162**, 313 (1995).

Kohl, J. L., G. Noci, E. Antonucci, G. Tondello, M. C. E. Huber, *et al.,* First results from the SOHO Ultraviolet Coronagraph Spectrometer, *Solar Phys.,* **175**, 613 (1997).

Kohl, J. L., G. Noci, E. Antonucci, G. Tondello, M. C. E. Huber, *et al.,* UVCS/SOHO empirical determinations of anisotropic velocity distributions in the solar corona, *Astrophys. J.,* **501**, L127 (1998).

Kohl, J. L., R. Esser, S. R. Cranmer, S. Fineschi, L. D. Gardner, *et al.,* EUV line profiles in polar coronal holes from 1.3 to 3.0 $R_\odot$, *Astrophys. J.,* **510**, L59 (1999).

Krieger, A. S., A. F. Timothy, and E. C. Roelof, A coronal hole and its identification as the source of a high velocity solar wind stream, *Solar Phys.* **29**, 505 (1973).

Leamon, R. J., W. H. Matthaeus, C. W. Smith, G. P. Zank, D. J. Mullan, and S. Oughton, MHD-driven kinetic dissipation in the solar wind and corona, *Astrophys. J.,* **537**, 1054 (2000).

Lee, L. C., and B. H. Wu, Heating and acceleration of protons and minor ions by fast shocks in the solar corona, *Astrophys. J.*, **535**, 1014 (2000).

Leer, E., and E. Marsch, Working Group 1 report: Solar wind models from the Sun to 1 AU: Constraints by in situ and remote sensing measurements, *Space Sci. Rev.*, **87**, 67 (1999).

Markovskii, S. A., Generation of ion cyclotron waves in coronal holes by a global resonant magnetohydrodynamic wave, *Astrophys. J.*, **557**, 337 (2001).

Marsch, E., Solar wind models from the Sun to 1 AU: Constraints by in situ and remote sensing measurements, *Space Sci. Rev.*, **87**, 1 (1999).

Matthaeus, W. H., G. P. Zank, S. Oughton, D. J. Mullan, and P. Dmitruk, Coronal heating by magnetohydrodynamic turbulence driven by reflected low-frequency waves, *Astrophys. J.*, **523**, L93 (1999).

Milano, L. J., D. O. Gómez, and P. C. H. Martens, Solar coronal heating: AC versus DC, *Astrophys. J.*, **490**, 442 (1997).

Möbius, E., G. Gloeckler, B. Goldstein, S. Habbal, R. McNutt, J. Randolph, A. Title, and B. Tsurutani, B., Here comes Solar Probe, *Adv. Space Res.*, **25** (9), 1961 (2000).

Moore, R. L., D. A. Falconer, J. G. Porter, and S. T. Suess, On heating the Sun's corona by magnetic explosions, *Astrophys. J.*, **526**, 505 (1999).

Narain, U., and P. Ulmschneider, Chromospheric and coronal heating mechanisms, *Space Sci. Rev.*, **54**, 377 (1990).

Narain, U., and P. Ulmschneider, Chromospheric and coronal heating mechanisms II, *Space Sci. Rev.*, **75**, 453 (1996).

Ofman, L., V. M. Nakariakov, and C. E. DeForest, Slow magnetosonic waves in coronal plumes, *Astrophys. J.*, **514**, 441 (1999).

Parhi, S., and S. T. Suess, Alfvénicity of fluctuations associated with the Kelvin-Helmholtz instability, *Phys. Plasmas*, **7**, 2995 (2000).

Parker, E. N., Heating solar coronal holes, *Astrophys. J.*, **372**, 719 (1991).

Peter, H., and E. Marsch, Ionization layer of hydrogen in the solar chromosphere and the solar wind mass flux, in *The Corona and Solar Wind Near Minimum Activity*, Fifth SOHO Workshop, ed. A. Wilson, p. 591, ESA SP-404, Noordwijk, The Netherlands (1997).

Pneuman, G. W., Driving mechanisms for the solar wind, *Space Sci. Rev.*, **43**, 105 (1986).

Priest, E. R., C. R. Foley, J. Heyvaerts, T. D. Arber, D. Mackay, *et al.*, A method to determine the heating mechanisms of the solar corona, *Astrophys. J.*, **539**, 1002 (2000).

Raymond, J. C., S. Fineschi, P. L. Smith, L. Gardner, R. O'Neal, A. Ciaravella, *et al.*, Solar Wind at 6.8 solar radii from UVCS observation of Comet C/1996Y1, *Astrophys. J.*, **508**, 410 (1998).

Reisenfeld, D. B., D. J. McComas, and J. T. Steinberg, Evidence of a solar origin for pressure balance structures in the high-latitude solar wind, *Geophys. Res. Lett.*, **26**, 1805 (1999).

Sandbæk, Ø., E. Leer, and V. H. Hansteen, On the relation between coronal heating, flux tube divergence, and the solar wind proton flux and flow speed, *Astrophys. J.*, **436**, 390 (1994).

Seely, J. F., U. Feldman, U. Schühle, K. Wilhelm, W. Curdt, and P. Lemaire, Turbulent velocities and ion temperatures in the solar corona obtained from SUMER line widths, *Astrophys. J.*, **484**, L87 (1997).

Shebalin, J. V., W. H. Matthaeus, and D. Montgomery, Anisotropy in MHD turbulence due to a mean magnetic field, *J. Plasma Phys.*, **29**, 525 (1983).

Shimizu, T., Yohkoh observations related to coronal heating, *Magnetic Reconnection in the Solar Atmosphere,* ed. R. D. Bentley and J. T. Mariska, ASP Conf. Ser. 111, p. 59, Astron. Soc. Pacific, San Francisco (1996).

Tu, C.-Y., and E. Marsch, MHD structures, waves and turbulence in the solar wind: Observations and theories, *Space Sci. Rev.,* **73**, 1 (1995).

Tu, C.-Y., and E. Marsch, On cyclotron wave heating and acceleration of solar wind ions in the outer corona, *J. Geophys. Res.,* **106**, 8233 (2001).

Tu, C.-Y., E. Marsch, K. Wilhelm, and W. Curdt, Ion temperatures in a solar polar coronal hole observed by SUMER on SOHO, *Astrophys. J.,* **503**, 475 (1998).

Tziotziou, K., P. C. H. Martens, and A. G. Hearn, Energy and momentum deposition in coronal holes, *Astron. Astrophys.,* **340**, 203 (1998).

Viñas, A. F., H. K. Wong, and A. J. Klimas, Generation of electron suprathermal tails in the upper solar atmosphere: Implications for coronal heating, *Astrophys. J.,* **528**, 509 (2000).

Vocks, C., and E. Marsch, A semi-kinetic model of wave-ion interaction in the solar corona, *Geophys. Res. Lett.,* **28**, 1917 (2001).

Waldmeier, M., *Die Sonnenkorona* **2**, Verlag Birkhäuser, Basel (1957).

Waldmeier, M., The coronal hole at the 7 March 1970 solar eclipse, *Solar Phys.,* **40**, 351 (1975).

Wang, Y.-M., and N. R. Sheeley, Jr., Solar wind speed and coronal flux-tube expansion, *Astrophys. J.,* **355**, 726 (1990).

Watanabe, T., T. Kosugi, and A. C. Sterling (eds.), *Observational Plasma Astrophysics: Five Years of Yohkoh and Beyond,* Kluwer, Dordrecht (1998).

Wilcox, J. M., The interplanetary magnetic field, solar origin and terrestrial effects, *Space Sci. Rev.,* **8**, 258 (1968).

Wilhelm, K., I. E. Dammasch, E. Marsch, and D. M. Hassler, On the source regions of the fast solar wind in polar coronal holes, *Astron. Astrophys.,* **353**, 749 (2000).

Wu, S.-T., A. H. Wang, S. P. Plunkett, and D. J. Michels, Evolution of global-scale coronal magnetic field due to magnetic reconnection, *Astrophys. J.,* **545**, 1101 (2000).

Zirker, J. B., ed., *Coronal Holes and High-speed Wind Streams,* Colorado Assoc. Univ. Press, Boulder (1977).

SOLAR WIND ACCELERATION, CRITICAL POINTS AND MASS FLUX, AND CORONAL HEATING DUE TO SUPRA-THERMAL ELECTRON BEAMS

T. Hirayama

Meisei University, Ome-shi, Tokyo, 198-8655, Japan

ABSTRACT

We present a simple steady model of the solar wind, where its acceleration is governed by the gas pressure gradient resulting from heating at lower altitudes, and by the magnetic pressure gradient at larger distances. Both accelerations stem from the same Poynting magnetic energy flux. We present a theory of coronal heating, assuming twisting magnetic tubes as DC energy input (Paper I). Twisting inevitably produces charge imbalance: $\mathbf{div}\boldsymbol{E} = -\mathbf{div}(\boldsymbol{V} \times \boldsymbol{B}) = \sigma/\varepsilon_0 \neq 0$. The charges generate strong electric fields along the magnetic field, creating supra-thermal electron beams. Classical Coulomb collisions of these beams with ambient particles will heat the corona and start to accelerate the solar wind in accord with various observations. This is not Joule heating of any kind, but a co-spatial frictional heating. Thin sheaths and reconnections are not involved.

ACCELERATION BY THE HEAT DEPOSITED AND THE MAGNETIC PRESSURE GRADIENT

Simple Derivation of Terminal Velocity Here we try to understand, from empirical study, the basic physics of how the solar wind is accelerated. Vanishing interstellar pressures cannot be recognized by the Sun because the flow is supersonic and super Alfvénic, hence not usable as the restricting condition (e.g. Heyvaert 1996). Then what is the reason for the solar plasma to take the supersonic rather than the subsonic solution at the very location of the so-called critical point? Critical points do not appear explicitly in simulation studies (e.g. Lie-Svendsen *et al.* 2002, and a recent summary in Kohl & Cranmer 1999). We find below that critical points do not exist as physical entities 'in the solar wind', if purely mathematical ones, and that energy and mass flux in the innermost corona along with area expansion will fix the fate of the solar wind.

In a steady flow, the total energy which passes cross sectional area $S(z)$ of a coronal flux tube must be constant along distance z parallel to the magnetic field;

$$(F_m + F_c + F_r)S + \rho VS[(V^2 + V_\varphi^2)/2 - GM_\odot/z + 5R_gT] = \text{constant}. \tag{1}$$

V is the solar wind velocity along the magnetic field, V_φ is the velocity perpendicular to the field such as in twisting motions, F_m (W m^{-2}) is mechanical energy flux density along the magnetic tube, F_c is conduction flux, F_r is radiation loss, G = gravity constant, $M_\odot$ = solar mass, and R_g = gas constant. We apply Eq. 1 between the transition region of $T = 10^{4.8}$ K (subscript $\odot$) and 1 AU (subscript $\oplus$). Dividing by the constant mass flux, ρVS(kg s^{-1}), we obtain

$$\frac{1}{2}V_{\oplus}^2 \approx (F_m/\rho V)_{\odot} - GM_{\odot}/R_{\odot} \quad \text{or} \quad V_{\oplus}^2 \approx V_{throw}^2 - V_{esc}^2. \tag{2}$$

Here F_m in $(F_m/\rho V)_{\odot}$ can be anything if transmitted from the 10^5K layers and decayed appreciably before reaching e.g. 0.3 AU. Also we use $5R_g(T_{\odot} - T_{\oplus}) \ll GM_{\odot}/R_{\odot} \equiv \frac{1}{2}V_{esc}^2$, $V_{\odot} \ll V_{\oplus}$, $(V_{\varphi})_{\odot} \ll V_{\oplus}$, and negligible gravity at 1 AU = 215 $R_{\odot}$ or 0.3 AU. We define 'throw velocity' as $V_{throw} \equiv \sqrt{(2F_m/\rho V)_{\odot}}$, as if throwing a solid body from the Sun beyond escape velocity V_{esc} = 617 km s^{-1}. We adopt the base of the transition region to be where $F_m \gg |F_c + F_r| \approx 0$ or $dF_c/dz \approx -n_e^2\Gamma_{radiation}(T)$ (W m^{-3}, $n_e \equiv$ electron number density). Eq. 2 is also equivalent to the case of T = constant and no radiation, without relation to the transition region (Leer *et al.* 1982, Hansteen *et al.* 1995, Fisk *et al.* 1999). We 'only' need to assume $(F_m/\rho V)_{\odot} = 5.1 \times 10^{11}$ J kg^{-1} to obtain a fast wind velocity of $V_{\oplus}$ = 800 km s^{-1}, and $(F_m/\rho V)_{\odot}$ = 2.7×10^{11} gives a slow velocity of 400 km s^{-1}. If we use $(F_m)_{\odot}$ = 410 W m^{-2}(= 4.1×10^5 erg cm^{-2}s^{-1}), $(nV)_{\odot} = 4.9\times10^{17}$ m^{-2} s^{-1}and a non-radial expansion factor of S_{nr} = 4.2 (Ulysses), we can reproduce $V_{\oplus}$ = 800 km s^{-1} and $(nV)_{\oplus} = 2.5\times10^{12}$ at 1 AU. Here $S_{\oplus}/S_{\odot} = 215^2 S_{nr}$. These were derived without considering the so-called critical point.

Velocity and Energy Flux Distribution Figure 1 shows the fast wind velocity calculated from mass conservation $\rho VS = (\rho VS)_{\oplus}$ that matches the reliable observed average polar density in the minimum phases (Saito 1970). As the non-radial expansion function, we assume

$$f_{nr}(r) \equiv r^{-2}S(r)/S(1) = 1 + (S_{nr} - 1)\sin[\tfrac{1}{2}\pi(r-1)/(r_0 - 1)]$$

for $r \le r_0$ and $f_{nr}(r) = S_{nr}$ for $r > r_0$ ($r \equiv z/R_{\odot}$). We adopt $S_{\rm nr}$ = 4.2 from Ulysses and the area of the polar facular belt from the Nobeyama radio-map, r_0 = 3.3, $n_e = 3.1 \times 10^6(215/r)^2$ [m^{-3}] beyond r_2 = 4.5, and smooth transition from r_1 = 2.7 to r_2 (r_0, r_1 and r_2 are adjustable). From ρ, S, and V, we can calculate each term (> 0) of Eq. 1 in units of $S(r = 1) = 1$ m^2 as shown in right panel, neglecting conduction and radiation ($T = 10^6$ K for $5R_gT$ term). The constant total flux (dotted line) is taken as $(\rho SV^3/2)_{\oplus}$, corresponding to V = 2.9 km s^{-1} and $n_e = 1.7\times10^{14}$ at r = 1. The term $\rho VSV_{\varphi}^2/2$ shown as 'twist, kinetic' is first included in F_mS, and later separated out using the explicit form of F_m. We also show conductive flux of F_cS, enthalpy $5R_gT\rho VS$, and (integrated) radiation loss F_rS, by adopting observed temperatures [$T_{max} = 10^{5.95}$ K at $r \approx 1.15$, $10^{5.5}$ (0.3 AU), and $10^{5.4}$ (1 AU)]. We readily see that effects due to F_c, F_r and $5R_gT$ are small at least for determining the solar wind velocity, so we neglect them in the following.

To Look the Situation from a Different Angle We rewrite Eq. 1 divided by ρVS for $T \approx$ constant and $F_r \approx$ 0 as

$$V(r)^2 = (2F_m'/\rho V)_{\odot}\{1 - r^2 f_{nr}(r)\exp[-(r-1)/L_D(r)]\} - (1 - 1/r)V_{esc}^2 + V(1)^2. \tag{3}$$

Here we defined $F_m' = F_m + \rho VV_{\varphi}^2/2 \approx F_m$ and the decay length $L_D(r)$ by $F_m'(r) = F_m'(1)\exp[-(r-1)/L_D(r)]$, and '$\odot$' means r = 1 (coronal base). At large r, Eq. 3 is reduced to Eq. 2, regardless of the values of S_{nr} (typically 3-7), L_D (0.1-0.7), and r_0 ($\approx$ 2-8). Eq. 3 contains the major ingredients of the 'basic theory' of the wind 'velocity': $F_{m\odot}$, $(\rho V)_{\odot}$, f_{nr} and L_D. To reproduce the Saito density fairly well, we need increasing decay-lengths L_D from 0.3 at $r \approx 1$ (above T_{max}) to 0.5 beyond r = 4. In some cases a velocity maximum, though inconspicuous, appeared at around r_{max} = 4-7 where the velocity is already well beyond the sound speed. We calculate r_{max} using $r_{max} \approx (V_{esc}/V_{throw})^{1/2}[L_D/f_{nr}(r_{max})]^{1/4}\exp[(r_{max}-1)/4L_D]$. Eq. 3 is free from the critical point, yet it predicts the behavior of gas pressure P as $P \propto n_e \propto r^{-2}V^{-1} \rightarrow 0$ at infinity.

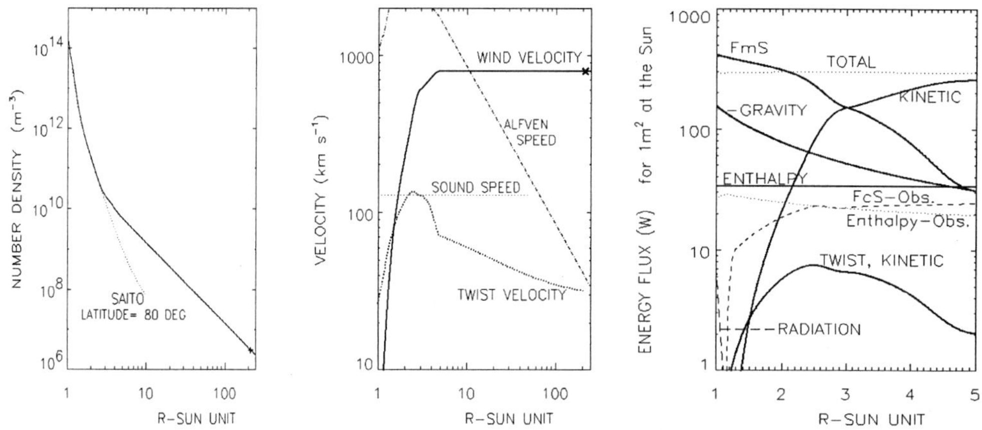

Fig. 1. Density (left) and velocity (center) versus distance. Right: energy flux (> 0) of Eq. 1.

This is because V stays finite as seen in Eq. 2. Since $L_D(r)$ is uncertain, Eq. 3 is, except for the lowest corona, accurate enough for 'the velocity', by far the most important parameter in the solar wind energetics. This implies that 'higher temperature' is not a necessary requirement of 'faster' flows. We regard the source velocity of the solar wind as an evaporating velocity due to excess heating of the transition region as in flares (see Paper I on closed loops).

Cause of the Fast Wind Acceleration The steady momentum equation is

$$\rho V\frac{dV}{dz}+\rho\frac{GM_\odot}{z^2}=-\frac{dP}{dz}+(\boldsymbol{j}\times\boldsymbol{B})_z. \tag{4}$$

Knowing the values of the left hand side from Figure 1 and $P = 2n_e kT$ ($T = 10^6$ K is assumed), we can infer $(\boldsymbol{j}\times\boldsymbol{B})_z$ at each distance z. This z-component of the Lorentz force, which is the magnetic pressure gradient, is found to be effective above $1.5R_\odot$, while the gas pressure gradient is effective below it. Thus we find that the acceleration of the solar wind is a two-step process, heating as in Eq. 5 and magneto-hydrodynamic expansion, coming from the same source. Therefore L_D beyond $1.5R_\odot$ is not a dissipation, but a decay length of the original mechanical energy flux by conversion to kinetic energy flux. (For a quick education, it might even be possible to neglect the gas pressure gradient: namely an atmosphere of $T = 0$ can reach e.g. 800 km s^{-1}.) Figure 1 shows that the solar wind velocity becomes the sound speed below $1.5R_\odot$, showing that the dP/dz term is larger there than the Lorentz force, yet dV/dz is positive. This indicates that there is no room for the critical point to appear. Slow winds may behave similarly.

To Recover Eq. 1 We first note that the steady thermal energy equation for T = constant is

$$-VdP/dz = H \quad (H \equiv \text{coronal heating rate, W m}^{-3}). \tag{5}$$

As asserted below, if the work done by Alfvénic twisting motions V_φ is the origin of coronal heating, then $H = -f_\varphi V_\varphi$, where f_φ is the effective friction against the twisting motion. The steady equation of twisting motions is thus

$$\rho V dV_\varphi/dz = (\boldsymbol{j} \times \boldsymbol{B})_\varphi + f_\varphi = (\boldsymbol{j} \times \boldsymbol{B})_\varphi + V dP/V_\varphi dz. \tag{6}$$

Using this and Eq. 4, we calculate $(\boldsymbol{j}\times\boldsymbol{B})_\varphi V_\varphi + (\boldsymbol{j}\times\boldsymbol{B})_z V$. With the use of the Poynting theorem, $d(SF_m)/Sdz \equiv \mathbf{div}(\boldsymbol{E} \times \boldsymbol{B})/\mu_0 = -\boldsymbol{E}\cdot\boldsymbol{j} = (\boldsymbol{V} \times \boldsymbol{B})\cdot\boldsymbol{j} = -(\boldsymbol{j} \times \boldsymbol{B})\cdot\boldsymbol{V}$, we recover Eq. 1 after multiplying by S and integrating with z.

Effect of Co-spatial Friction The energy flux density used is, assuming $1 + V/V_A \approx 1$, $F_m \equiv (\boldsymbol{E}\times\boldsymbol{B})_z/\mu_0 = -B_\varphi V_\varphi B_z/\mu_0$ in cylindrical coordinates $(r,\ \varphi,\ z)$ along the magnetic tube (now 'r' is different). If there is friction as advocated below, the twist velocity is smaller for a given Lorentz force. Defining α by $V_\varphi = -\alpha B_\varphi V_A/B_z$, we found that $\alpha \approx 0.5$ (assumed constant) matches better with $(\boldsymbol{j}\times \boldsymbol{B})_z = d(B_\varphi^2/2\mu_0)/dz$ in Eq. 4 than $\alpha = 1$ (no friction). Here $V_A = B_z/(\ \mu_0\rho)^{1/2}$ is the Alfvén velocity, and $B_z S = (B_z)_\odot S_\odot$ where $(B_z)_\odot = 6 \times 10^{-4}$ T = 6G is fixed from Ulysses and S_{nr}. (If Joule dissipation were important, $\alpha > 1$ would be expected, since conversely for a given twist velocity the change of B_φ would be smaller than with no dissipation.) Hence we actually adopted $F_m = 2\rho V_\varphi^2 V_A$. Smaller twists (factor 4) correspond to the same pressure $B_\varphi^2/2\mu_0$ as compared to $\alpha = 1$. This V_φ, used in Figure 1, reached a maximum of 140 km s^{-1} at $1.5R_\odot$, the increase being mainly from the rapid density decrease in F_m as compared to the decay of F_m.

The large V_φ and the large solar wind velocity at lower heights, required by the Saito density, might explain the coronal large line-widths of ions. Also $V_\varphi \geq 30$ km s^{-1}at 1 AU in Figure 1 is close to observations. We expect this as the remnant of photospheric twisting magnetic knots of >1 km radius (Paper I), and hence its temporal frequency at the Sun and 1 AU may be 0.1Hz or less.

CORONAL HEATING

Scenario of Coronal Heating We assume that magnetic twisting originating from 0.1 T faculae (10^3 Gauss) is the source of the mechanical energy flux. In cylindrical coordinates we assume $B = (0,\ B_\varphi,\ B_z)$ and $V = (0,\ V_\varphi,\ 0)$. The twisting motion V_φ 'in the corona' is nearly perpendicular to the main magnetic field B_z. Hence V_φ produces almost-radial electric fields $E_r = -(\boldsymbol{V}\times\boldsymbol{B})_r$. This inevitably produces a non-zero divergence by 'definition' and a charge separation:

$$\sigma/\varepsilon_0 = \mathbf{div}\boldsymbol{E} \approx \partial r E_r/r\partial r = -B_z \partial r V_\varphi/r\partial r. \tag{7}$$

The σ cannot be compensated unless the twist stops (V_φ would have to be zero all the time). The charge necessarily produces electric field E_z parallel to the magnetic tube. This E_z accelerates electrons. Beams of accelerated electrons n_b are thermalized by the classical Coulomb collisions with protons and background electrons n_0 ($n_b + n_0 \equiv n_e$, $n_b \ll n_0 \approx n_e$, and n_e = number density of electrons).

Slow counter flows of background electrons will keep plasmas neutral except for the small charges of $\sigma/e \equiv n_i - n_e \approx 3 \times 10^{-10} n_e$, however large E_z may be ($E_z \approx 20\times$Dreicer field $\approx 10^{-3} E_r$ as in Remarks paragraph (A); $B_z = 1\times10^{-3}$ T = 10 G, $V_\varphi = 25$ km s^{-1}, $n_e = 10^{14}$ m^{-3} and tube radius $R_c = 100$ km as in Paper I). Thus:

$$n_b V_b + n_0 V_0 = n_e V_d \equiv -j_z/e \approx 0 \tag{8}$$

(V_b = beam velocity $\approx 10^4$ km s^{-1} >> $V_0 \approx 10$ km s^{-1} >> $V_d = 0.01$ km s^{-1}). Hence no extra electric currents arise other than $j_z = \mathbf{rot}_z \boldsymbol{B}/\mu_0$. For a large E_z, we cannot use Ohm's law parallel to B_z, which is effectively the same as a 'fluid' momentum equation for electrons, because beam electrons carry substantial momentum and energy. Hence a 'fluid' treatment is irrelevant. As a substitute we use $m_e n_b V_b + m_e n_0 V_0 \approx 0$ within a constraint of a 'finite' energy input rate F_m, where beam energy is finite (Paper I). Acceleration and damping occur everywhere as in electric wires in the laboratory. In this situation, we might imagine a coronal tube in which infinite numbers of tiny batteries are buried, and infinitely small wires lead beam and bulk electrons to circulate co-spatially.

Damping Length of the Mechanical Energy Flux and Heating Rate In this section we adopt $\alpha = 1$ for simplicity (unlike in the solar wind section). The damping length of the beam, i.e. damping distance of the twisting energy, is expected to be on the order of

$$L_D = V_b/(2\nu_b). \tag{9}$$

Here ν_b is the collision frequency of beam electrons with protons plus bulk electrons: $\nu_b\ (s^{-1}) \approx 3 \times 10^{-4}(V_{Te}/V_b)^3 n_e/T_e^{1.5}$ (Spitzer, 1962, Eq. 5-28; concisely in Paper I). T_e is electron temperature and $V_{Te} = \surd(2kT_e/m_e)$ the electron thermal speed. This L_D ($\propto n_e^{-1}$ which is the only rapidly changing quantity with z) is inversely proportional to the electron pressure in an isothermal corona, consistent with Withbroe (1988). To match with $L_D \approx 0.4R_\odot$ from above and Withbroe, V_b should be $\approx 2V_{Te}$. In closed loops, a favorable value of $L_D \approx$ scale height requires $V_b \approx 4V_{Te}$. Because electron acceleration is almost instantaneous (Paper I), we should expect $\langle\frac{1}{2}m_e n_b|V_b^3|\rangle_z \approx \langle\rho V_\varphi^2 V_A\rangle_z$ (both 'rates' are equal : $\langle\frac{1}{2}m_e n_b V_b^2\rangle_z \approx 0.2\langle\frac{1}{2}\rho V_\varphi^2\rangle_z$). Using $\rho V_\varphi^2 V_A$ ($V_\varphi \approx 25$ kms^{-1}) and $V_b \approx 3V_{Te}$, we infer, thus empirically, $n_b \approx 10^{-3} n_e$. (This empirically inferred relation of $V_b \approx 3V_{Te}$ might be related to the fact that if V_b were e.g., $10V_{Te}$, the total kinetic energy of beam electrons would become far larger for a given total number of beam electrons.) The heating rate H_{beam} is given by [viscous force due to the beam $m_e n_b V_b \nu_b$] $\times$ [beam velocity V_b], namely

$$H_{beam} = m_e n_b V_b^2 \nu_b = \rho V_\varphi^2 V_A / L_D \approx 3 \times 10^{-5}\text{W m}^{-3} = 3 \times 10^{-4}\text{erg cm}^{-3}\text{ s}^{-1} \propto \frac{n_e^2}{\sqrt{T_e}}\frac{n_b/n_e}{V_b/V_{T_e}}. \tag{10}$$

This is sufficient to heat the corona and 'to start' to accelerate the solar wind. Since the loop length L and gas pressure P are related 'observationally' to T_e as $T_e \propto (LP)^{1/3}$ (RTV scaling law), the heating rate H, in general, should be proportional to $n_e^2/\sqrt{T_e}$, quite satisfactorily the same as our H_{beam} (Paper I). Apart from the line-widths problem, ion temperature may be raised since bulk ions are everywhere heated by 1/2 of bulk electrons from the electron beams even if there are no other mechanisms of ion heating (Paper I). This is because heavy and hence almost-zero-speed ions collide with the beam electrons only at a 1/2 rate as compared to fast electrons. Excitation of ion cyclotron waves may be possible due to supra-thermal electron and ion beams (on decay) (see Cranmer *et al.* 1999), since ions are also accelerated due to the same Dreicer fields, although probably not vital in total heating. This is an interesting area of further research.

Remarks The three points below, (A)-(C), are remarks omitted from Paper I.

(A) Here we demonstrate E_z (from charges) $\neq 0$. If we assume $\partial B_\varphi/\partial t = 0$, then $\mathbf{rot}\boldsymbol{E} = 0$, so that $\boldsymbol{E} = -\nabla\Phi$ holds. Thus static potential Φ, for given charges, i.e. given twist velocity, leads to $E_z \neq 0$. Also $\partial B_\varphi/\partial t = 0$ needs to be satisfied only locally, temporarily, and roughly. Conversely, to keep $\partial B_\varphi/\partial t \neq 0$ all the time is impossible since $|B_\varphi|$ goes to ∞ as time elapses. For $\partial/\partial t \approx 0$, $\partial B_\varphi/\partial t = \partial E_r/\partial z - \partial E_z/\partial r = 0$ leads to $E_z \approx R_c E_r/L_z \neq 0$ in accord with Paper I (L_z = distance of V_φ-change, e.g. due to change of photospheric facular motions, and R_c = loop radius). In closed loops, the equation of the twisting motion is $\rho\partial V_\varphi/\partial t = -j_r B_z + f_\varphi$. Here the 'internal friction force' f_φ due to beam electrons is evaluated from $f_\varphi = -H_{beam}/V_\varphi$ (heating rate = force $\times$ velocity). If $\rho\partial V_\varphi/\partial t \approx 0$ is satisfied locally, H_{beam} should be proportional to $\partial V_\varphi^2/\partial z$. The steady Boltzmann equation for large E_z may be satisfied by a nearly flat electron distribution function in both beam and bulk parts near the Sun, keeping $T_{e,mean} = 10^6$ K. In our scenario this is the most important task in the future.

(B) One might ask why all the electrons do not move in the same direction as given by the parallel electric field. As stated before, if all moved in the same direction, there would arise enormous charges and thus electric field, and hence the back flow of the background majority is the only possibility to keep plasmas almost neutral everywhere. For this question, the following circumferential knowledge might help. First, it is well known that even in small electric fields where the Spitzer resistivity η is applicable such as $E_z = \eta j_z$, small numbers of supra-thermal electrons carry most of the electric current (Spitzer 1962, Figure 5.3, line b). That is, in non-zero electric fields the majority are not moving, and the so-called electron drift velocity $V_d \equiv -j_z/en_e$ (≈ 0.01 km s^{-1}) is only an averaged quantity [cf. Eq. 8]. We must accept this apparent contradiction to the Newton's equation of motion for each particle as the nature of collective behavior of plasmas (once accelerated, less collisions). Spitzer cautions that his resistivity starts deviating already in very small field-aligned electric fields. Secondly, in the classical theory of runaway electrons the same situation applies. This was once expressed as an "all-or-nothing" picture by Kruskal & Bernstein (1964, bottom of p. 417), meaning that runaway electrons carry "all" the electric current and the majority carry "nothing", i.e. the majority do not move. Of course Ohm's law cannot be applied here, even though the field strength treated in the past is at most $0.2E_D$. Here E_D is the Dreicer field defined by $E_D = 6 \times 10^{-12} n_e/T_e$ [V m^{-1}], where a test electron increases its velocity by one electron thermal speed during one collision time (e.g. Parail & Pogutze 1982). In our case if we would take a coordinate moving with the bulk plasma velocity of $V_0 \approx 10$ km s^{-1}($\ll V_b$), then the situation would become similar. The charge neutrality condition in the actual corona of finite length may well prohibit possible electrostatic micro-instabilities.

(C) The edge of a coronal loop may not be twisting: $V_\varphi(R_c) = 0 = B_\varphi(R_c)$. This is because the magnetic Reynolds number in a thin edge of 'photospheric' magnetic knots is small, hence not frozen, or slipping. This leads to a favorable situation in the corona, where $\int 2\pi r j_z dr \propto \int[\partial(rB_\varphi)/\partial r]dr = [rB_\varphi]_0^{R_c} = 0$ and $\int 2\pi r \sigma dr \propto \int[\partial(rV_\varphi)/\partial r]dr = [rV_\varphi]_0^{R_c} = 0$. The former means e.g. $j_z > 0$ in $r < 0.9R_c$ and $j_z < 0$ in $0.9R_c < r < R_c$ at any height z. The latter means that charges σ appear by separation among the inner (e.g. $\sigma > 0$) and outer ($\sigma < 0$) parts of the cross section at each height. Numerical integrations of Coulomb's law using σ from prescribed $V_\varphi(r,z)$ show that whenever $\partial V_\varphi/\partial z \neq 0$, large E_z persists, since e.g. at $r \approx 0$ the inner-σ predominates. Note that values of E_z are not critical. It is easily shown that the damping of the energy flux occurs whether $E_z > 0$ or $E_z < 0$.

I thank Drs. T. Watanabe and S. R. Cranmer for discussions, particularly on the solar wind.

REFERENCES

Cranmer, S.R., Field, G.B. & Kohl, J.L., *Astrophy.J.*, **518**, 937 (1999).

Fisk, L.A., Scwadron, N.A., & Zurbuchen, T.H., *J.Geophys.Res.*, **104**, 19766 (1999).

Hansteen, V.H. & Leer, E., *J.Geophys.Res.*, **100**, 21677 (1995).

Heyvaert, J., in *Plasma Astrophysics*, C. Chiuderi and G. Einaudi (eds.), p. 31, Springer (1996).

Hirayama, T., *Internat. Astron.Union Symp.*, **203**, 495 (2001), referred to as Paper I.

Kohl, J.L. & Cranmer, S.R. (eds.), *Space Sci.Rev.*, **87**, Nos.1-2 (1999).

Kruskal, M.D. & Bernstein, I.B., *Phys. Fluids*, **7**, 407 (1964).

Leer, E., Holzer, T.E., & Flå, T., *Space Sci. Rev.*, **33**, 161 (1982).

Li-Svendsen, Ø., Hansteen, V.H., Leer, E., & Holzer, T.E., *Astrophys.J.*, **566**, 562 (2002).

Parail, V.V. & Pogutse, O.P. in *Rev. Plasma Phys.*, M.A. Leontovich (ed.), Consultants Bureau, **11**, 1 (1986).

Saito, K., *Ann.Tokyo Astron.Obs.*, 2nd Ser., **12**, 53 (1970).

Spitzer, L., *Physics of Fully Ionized Gases*, Interscience Publ. (1962).

Withbroe, G., *Astrophys.J.*, **325**, 442 (1988).

LONG–PERIOD OSCILLATIONS IN POLAR CORONAL HOLES AS OBSERVED BY CDS ON SOHO

D. Banerjee[1], E. O'Shea[2] J. G. Doyle[3], and M. Goossens[1]

[1]*Centre for Plasma Astrophysics, Katholieke Universiteit Leuven, Celestijnenlaan 200B, 3001 Heverlee, Belgium*
[2]*Instituto de Astrofisica de Canarias, 38200 La Laguna, Tenerife, Spain*
[3]*Armagh Observatory, Armagh, BT61 9DG, N. Ireland*

ABSTRACT

In this work we examine spectral time series of two lines; a transition region line due to O V 629Å and the chromospheric line He I 584Å, which were observed simultaneously in a coronal hole with the Coronal Diagnostic Spectrometer (CDS) onboard the *SOHO* spacecraft. Using wavelet methods, we perform a time series analysis on several small individual regions which allow us to derive the duration as well as the period of the oscillations. The statistical significance of the oscillations was estimated by using a randomisation method. In this short contribution we will present results from one dataset, s18778r00, taken on March 7, 2000. Our observations indicate the presence of compressional waves with periods of 20-30 minutes or longer. These slow magneto-acoustic waves may provide enough energy flux for the acceleration of the fast solar wind.

INTRODUCTION

It is now generally accepted that the fast solar wind originates from coronal holes. Often they are "peppered" with vertical radial structures, "plumes", which sometimes reveal flare-like dynamics at their footpoints. Wang *et al.* (1998) detected 27 correlated white light and EUV jets in a polar coronal hole and interpreted them in terms of magnetic reconnection between magnetic bipoles and neighbouring unipolar flux. There have been several attempts to study the structure and dynamics of these plumes. Banerjee *et al.* (2000, 2001a) have found the signatures of long-period magneto-acoustic waves in the plumes and inter-plume regions respectively. It was conjectured that these long-period waves originate from the network boundaries of the coronal hole. In this paper we study the dynamics of the coronal hole regions and try to trace back the origin of these long-period waves to the solar disk part of the coronal hole. We report on the temporal behaviour of the polar coronal hole as observed by the CDS/*SOHO* instrument. Observations were performed with the chromospheric (He I 584Å) line, and series of oxygen lines, O III 599Å, O IV 554Å, O V 629Å, formed in the transition region. Here, we only use the He I and O V lines. Detailed results from other lines have been presented in Banerjee *et al.* (2001b).

OBSERVATIONS AND DATA REDUCTION

To obtain these observations we used the normal incidence spectrometer (NIS) (Harrison *et al.* 1995), which is one of the components of the Coronal Diagnostic Spectrometer (CDS) on-board the Solar Heliospheric

Observatory (*SOHO*). The temporal series dataset, s18778r00, was obtained on 2000 March 7, in four lines: He I 584Å (log T = 4.3 K), O III 599Å, (log T=5.0 K), O IV 554Å (log T = 5.2 K) and O V 629Å (log T = 5.4 K), with exposure times of 60 sec, using the 2 × 240 arcsec slit. Fitting of both the He I and O V was done using a single Gaussian. Details on the CDS reduction procedure, plus the wavelet analysis, may be found in O'Shea *et al.* (2001).

Figure 1 shows an image of the north polar coronal hole region taken with Extreme ultraviolet Imaging Telescope (EIT) on *SOHO* in Fe IX/X 171Å at 19:00 UT on 2000 March 7, with the slit position superimposed (for dataset s18778r00). This figure confirms that our observation was pointed in a coronal hole. In order to get good time resolution the rotational compensation was switched off (sit-and-stare mode), so it is important to calculate the lowest possible frequency we can detect from this long time sequence (see Doyle *et al.* 1998 for details). For our dataset, s18778r00, we can work out the solar rotation expected at the coordinates $x = 127, y = 781$. We find, by using the routine ROT_XY in the *SOHO* software tree that over one hour the Sun should have rotated by ~3 arcsec. For our 2 arcsec slit width, the lowest resolution possible will be 3 arcsec/hour/2 arcsec = 0.42 mHz. So one should note that we do not have any confidence in power below 0.42 mHz.

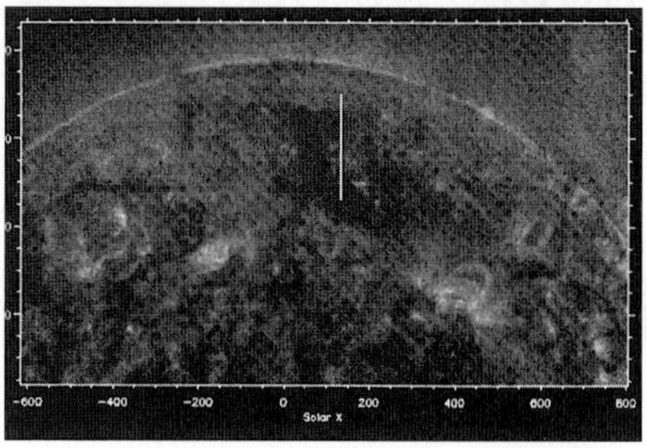

Fig. 1. Position of the observing slit for the s18778r00 dataset (2000 March 7) on an EIT/*SOHO* image of the north polar coronal hole in Fe IX/X 171Å taken during the run of the temporal series (courtesy of the EIT consortium).

The statistical significance of the observed oscillations was estimated by using a Monte Carlo or randomisation method. The advantage of using a randomisation test is that it is independent of noise distribution or nonparametric, i.e. it is not limited or constrained by any specific noise models, such as Poisson, Gaussian etc. We follow the method of Fisher randomisation as outlined in Nemec & Nemec (1985) (details can be found in O'Shea *et al.* 2001). We performed 250 random permutations to calculate the probability levels, choosing a value of 95% as the lowest acceptable probability level. To improve the signal-to-noise ratio of this data we binned by three pixels along the slit (i.e. 3 × 1.68 arcsec), in effect creating new pixels of ~5×2 arcsec^2. The velocity values presented in this paper are relative velocities, that is they are calculated relative to an averaged profile that was obtained by summing over all pixels along the slit and all time frames.

RESULTS

We first show the space time behaviour as observed by the He I 584Å and O V 629Å lines in the form of X-T slices (left panels of Figures 2 & 3). The left top panels show the original intensity map. To bring out the details of the original intensity map (X-T slice) we have filtered out the bright components in the image. The intensity map $I(y,t)$ is convolved in the time direction with a Gaussian $G(t)$. This results in a smoothed image $S(y,t) = I * G$ which contains no high frequencies. Then dividing the original intensity map by the smoothed map results in the contrast-enhanced map, i.e. $C(y,t) = I(y,t)/S(y,t)$ (see Doyle *et al.* 1999 for details). The grey scale coding has the most intense regions as white. From the contrast-enhanced images, fluctuations in the bright features seem to have a periodicity ~ 25 minutes at several locations across the slit. To study the statistical behaviour of individual pixels across the slit we show the spatial behaviour of the oscillation frequencies measured from the He I and O V intensity and velocity time series in the middle panels of Figures 2 & 3. These figures show the measured frequencies as a function of position along the slit (X-F slice). The primary and the secondary maxima from the global wavelet power spectra, which have a probability of more than the 95%, are indicated by crosses and plus symbols respectively. The total number

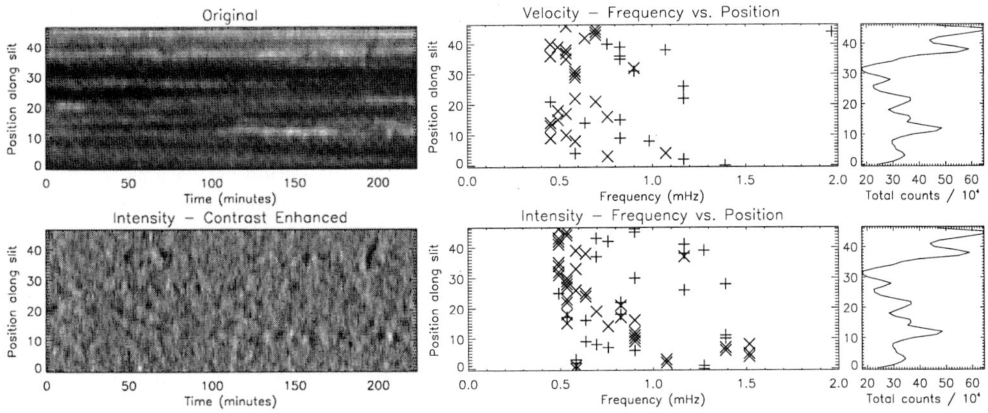

Fig. 2. Space-time behaviour of the intensity in the He I 584Å line corresponding to the s18778r00 dataset. The left panels show the intensity maps (X-T slice). The gray scale coding has the most intense regions as white. The middle panels show the frequencies measured in the intensity and velocity fluctuations of the He I 584Å line, as a function of spatial position along the slit. The crosses represents frequencies corresponding to the maximum power and the pluses the secondary power maxima, both measured above the 95% confidence level, after the randomisation test, in the global wavelet spectra. The right panels show the total number of counts in a pixel (summed counts) over the observation time.

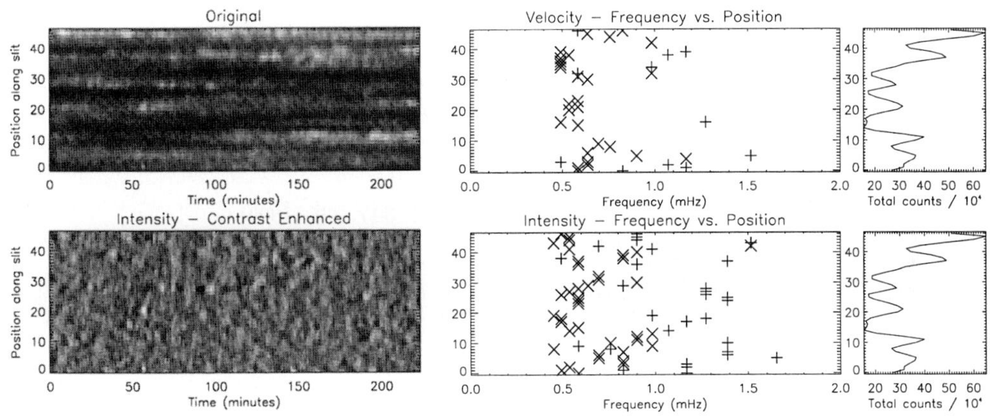

Fig. 3. Same as Fig. 2 but for the O V 629Å line

of counts in a pixel (summed counts) during the observation is shown in the right column, and is useful in identifying the network brightening. The intensity and velocity results both show that the primary maxima in the global wavelet spectra lie primarily in the range 0.5-0.8 mHz. The appearance of more crosses in the intensity X-F slices as compared to the velocity also indicates that the intensity oscillations are stronger and more reliable. Both He I and O V behave more or less in a similar way. The other two oxygen lines formed in the transition region also show the existence of these low-frequency oscillations, particularly in the intensity but with slightly less reliability in the velocity. A detailed plot of wavelet spectra for individual pixels is beyond the scope of this presentation, but will be presented in a forthcoming paper. An interesting point to note here is that these long-period oscillations are present both in the bright pixels (representing network) and also in the darker pixels (internetwork).

CONCLUSION

Compressional modes reveal themselves in the form of intensity oscillations, through variations in the emission measure, and also as velocity oscillations through fluctuations in the plasma density. This fact allows us to interpret the measured oscillations in this work as being due to slow magneto-acoustic waves. The important point to note here is that we find the existence of these long-period oscillations in the bright network pixels and also in the darker internetwork regions. This implies that these slow waves are present all over the coronal hole region, which raises the question: how and where are these waves generated? The energy carried by the slow magneto-acoustic waves can be estimated as $\rho[(\delta v)^2/2]v_s$, where δv is the wave velocity amplitude, and $v_s = c_s = 150$ km s^{-1} in the low β coronal plasma. The non-thermal velocity of the O V 629Å line is $\xi \sim 29$ km s^{-1} in the 'quiet Sun' (Teriaca *et al.* 1999), where ξ is related to the wave amplitude as $\xi^2 = (\delta v)^2/2$. Using $\rho = 1.67 \times 10^{-15}$ gm cm^{-3} , we get a wave energy flux of $\sim 2.1 \times 10^5$ ergs cm^{-2} s^{-1}. These slow waves are rather difficult to dissipate, so they can carry enough energy flux to larger distances for the acceleration of the fast solar wind. It is likely that the waves detected at 1.9 $R_\odot$ by Ofman *et al.* (2000) using UVCS/*SOHO* and the waves detected around 1.2 $R_\odot$ by DeForest & Gurman (1998) using EIT/*SOHO* are the same as those reported here.

ACKNOWLEDGEMENTS

DB wishes to thank the organisers of the meeting for partial support. CDS and EIT are part of *SOHO*, the *Solar and Heliospheric Observatory*, which is a mission of international cooperation between ESA and NASA. EOS is a member of the European Solar Magnetometry Network (`www.astro.su.se/~dorch/esmn/`).

REFERENCES

Banerjee, D., E. O'Shea, J.G. Doyle, & M. Goossens, *A&A*, **377**, 691 (2001a)
Banerjee, D., E. O'Shea, J.G. Doyle, & M. Goossens, *A&A*, **380**, L39 (2001b)
Banerjee, D., E. O'Shea, & J.G. Doyle, *Solar Physics*, **196**, 63 (2000)
DeForest, C.E., & J.B. Gurman, *ApJ*, **501**, L217 (1998)
Doyle, J.G., G.H.J. van den Oord, E. O'Shea, D. Banerjee, *A&A*, **347**, 335 (1999)
Harrison, R. *et al. Solar Physics*, **162**, 233 (1995)
Nemec, A. F., & J.M. Nemec, *AJ*, **90**, 2317 (1985)
Ofman, L., M. Romali, G. Poletto, G. Noci, & J.L. Kohl, *ApJ*, **529**, 592 (2000)
O'Shea, E., D. Banerjee, J.G. Doyle, B. Fleck, & F. Murtagh, *A&A*, **368**, 1095 (2001)
Teriaca L., D. Banerjee, & J.G. Doyle, *A&A*, **349**, 636 (1999)
Wang, Y.-M. *ApJ*, **501**, L145 (1998)

UV AND SOFT X–RAY POLAR CORONAL JETS

D. Dobrzycka[1] J. C. Raymond[1], S. R. Cranmer[1], and J. Li[2]

[1]*Harvard—Smithsonian Center for Astrophysics, 60 Garden Street, Cambridge, MA 02138, USA*
[2]*Institute for Astronomy, Honolulu, HI 96822, USA*

ABSTRACT

We present results of simultaneous SXT and UVCS observations of jets from polar coronal holes. Although we did not identify the UV counterparts of the X–ray jets, on one occasion UVCS recorded H I Lyα profile variations consistent with a UV jet at the position of the X–ray jet but prior to that event. We discuss the possible relation between the UV and X–ray events and consider the magnetic reconnection models developed for X–ray jets, as a model for UV jet formation. The rough estimates of the total energies of the X–ray and UV jets show the energies to be comparable.

INTRODUCTION

Coronal jets are spectacular dynamic events originating from different structures in the solar corona. Various jet–like phenomena were observed by *Yohkoh*'s Soft X–ray Telescope (SXT) as well as instruments aboard Solar and Heliospheric Observatory (*SOHO*). The relation among the different types of jets is not yet clear. The Ultraviolet Coronagraph Spectrometer (UVCS/*SOHO*) provided us with spectroscopy of polar coronal jets that were recorded as a significant enhancement in the integrated intensities of the strongest coronal emission lines: mainly H I Lyα and O VI $\lambda\lambda 1032, 1037$ (see e.g. Dobrzycka *et al.* 2001). Most of the detected jets were correlated with *SOHO*'s Extreme ultraviolet Imaging Telescope (EIT) Fe XII 195 Å and *SOHO*'s Large Angle and Spectrometric Coronagraph (LASCO) C2 white-light events. Polar coronal jets are short lived bursts of material presumably triggered by magnetic reconnection near flaring UV bright points within polar coronal holes. They are ejected with broad distributions of initial velocities – the leading edge moves with speeds of $400 - 1100$ km s^{-1} and the "centroid" with average initial speed of 500 km s^{-1}. As the jets travel they decelerate and are incorporated into the ambient solar wind within the C2 field of view (Wang *et al.* 1998). In this paper we present results of simultaneous SXT and UVCS observations of polar jets in December 1996.

OBSERVATIONS AND RESULTS

On 1996 December 16, 09:42 UT, SXT observed a jet originating in the north polar coronal hole at position angle (measured counterclockwise from the north) P.A.= 27°. UVCS/*SOHO* was obtaining synoptic observations at that time. The radial scan centered at P.A.= 0° began at 08:51 UT and covered heights: $1.4, 1.7, 2.0, 2.25, 2.5\ R_\odot$. The scan centered at P.A.= 45° began at 12:06 UT. The strongest lines observed were H I Lyα, O VI $\lambda\lambda 1032, 1037$, H I Ly$\beta$, Si XII and we did not see any obvious intensity variations. Another X–ray jet in a north polar coronal hole was observed by SXT at P.A.$\approx$ 12° between 00:53 UT and 01:01 UT on 1996 December 11 (Shibata 1998). UVCS was executing sit-and-stare observations at 1.98 $R_\odot$, P.A.= 0° from 16:35 UT of the previous day to 02:29 UT. The exposure time was 300 s. The H I Lyα line showed obvious temporal intensity variations consistent with UV polar jets at the approximate position of the X–ray jet but about 2 hours before the X–ray jet was observed, between 23:00 UT and 23:15 UT. The profiles indicated ~ 70 km s^{-1} blueshift.

General comparison of the EIT and SXT observations revealed that the bright EUV points – footpoints of the UV polar coronal jets (Dobrzycka *et al.* 2000) – were also very bright in soft X–rays. That indicates that the initial temperature of the UV jets is at least 2×10^6 K, which is consistent with our model predictions (Dobrzycka *et al.* 2000). Presented simultaneous SXT and UVCS observations did not identify the UV counterparts of the X–ray jets. However, the UVCS instrument configuration was not ideal at the time of jet observations (short time coverage, coarse binning in the wavelength and spatial directions, etc.) and some signal could have been missed. UVCS observations of H I Lyα brightening at the position of the X–ray jet but prior to that event on December 11, 1996 may suggest that the X–ray and UV jets are ejected from the same bright points. For the average masses, velocities and temperatures (5×10^{12} g, 5×10^{12} g; 200 km s^{-1}, 500 km s^{-1}; 5.6×10^6 K, 2×10^6 K) for the X–ray and UV jets respectively, we estimate that the jet's kinetic energies are $E_{k,X} \sim 10^{27}$ erg, $E_{k,UV} \sim 6 \times 10^{27}$ erg and thermal energies are $E_{t,X} \sim 5 \times 10^{27}$ erg, $E_{t,UV} \sim 1.5 \times 10^{27}$ erg. Thus, the total energies of the X–ray and UV jets are comparable. Shimojo & Shibata (2000) and Shimojo *et al.* (2001) concluded that most X–ray jets are *evaporation flows* resulting from flare heating due to reconnection. In this scenario the acceleration mechanism is the gas pressure force and the jets are ejected with velocities comparable to the sound speed. UV polar jets' initial speeds are considerably higher than the sound speed ($C_{s,UV} \approx 215$ km s^{-1}) suggesting that they are accelerated by the magnetic force. Two magnetic force acceleration scenarios have been considered in literature and both are supported by observations of UV jets. In the first, *magnetic twist jet* scenario, acceleration is due to magnetic pressure in relaxing magnetic twist resulting from reconnection between twisted and untwisted loops. The EIT observations of several UV jets confirmed the presence of the magnetic twist in the early stages of ejection. In the second, *reconnection jet* scenario, acceleration is by the magnetic tension force in the reconnection process. Yokoyama & Shibata (1996) performed numerical simulations of such a reconnection jet. Polar UV jets could be produced as their *anemone* type of jet in the vertical coronal field. The simulations predict formation of a cool jet together with a hot one. EIT 195 Å observations from 1996 July 13 revealed a macrospicular dark surge that occurred at the end of the UV jet sequence, at the base of the jet.

ACKNOWLEDGEMENTS

This work is supported by NASA under grant NAG5–10093 to the Smithsonian Astrophysical Observatory, by Agenzia Spaziale Italiana, and by the ESA PRODEX program (Swiss contribution).

REFERENCES

Dobrzycka, D., J.C. Raymond, & S.R. Cranmer, *ApJ,* **538**, 922 (2000).

Shibata, K., X–ray Jets and X–ray Plasmoids, in *Solar Jets and Coronal Plumes*, ed. T.-D. Guyenne, p. 142, ESA Publication Division SP-421; ESTEC, Noordwijk, The Netherlands (1998).

Shimojo, M., & K. Shibata, *ApJ,* **542**, 1100 (2000).

Shimojo, M., K. Shibata, T. Yokoyama, K. Hori, *ApJ,* **550**, 1051 (2001).

Wang, Y.-M., *et al.*, *ApJ,* **508**, 899 (1998).

Wood, B.E., M. Karovska, J.W. Cook, R.A. Howard, & G.E. Brueckner, *ApJ,* **523**, 444 (1999).

Yokoyama, T., & K. Shibata, *PASJ,* **48**, 353 (1996).

SPECTROSCOPIC OBSERVATION OF CORONAL OSCILLATIONS

T. Sakurai[1,2] K. Ichimoto[2], K. P. Raju[3], and J. Singh[3]

[1]*Department of Astronomical Science, The Graduate University for Advanced Studies, Mitaka, Tokyo 181-8588, Japan*
[2]*National Astronomical Observatory, Mitaka, Tokyo 181-8588, Japan*
[3]*Indian Institute of Astrophysics, Bangalore 560 034, India*

ABSTRACT

A time sequence of coronal green line spectra was obtained with a coronagraph at the Norikura Solar Observatory. The Fourier analysis shows oscillatory powers in the Doppler velocity of the line in the 1–3 mHz range. The phase relationship between the Doppler velocity and the line intensity indicates that the oscillatory signal is due to propagating (rather than standing) sound waves.

INTRODUCTION

The current idea is that the corona is heated either by microflares or by some kind of waves. The detection and mode identification of oscillations in the corona are of vital importance in understanding the coronal heating mechanism. For this purpose, a high-dispersion spectroscopic observation of coronal emission lines by using a coronagraph is the most promising approach.

OBSERVATION AND RESULTS

The observation was carried out on October 28, 1998 using the 25 cm coronagraph at Norikura Solar Observatory. The time sequence of one-dimensional spectroscopic data of the coronal green line [Fe XIV] 5303 Å was obtained for about 80 minutes with a cadence of 25 s. The slit covered the height range of $30''$–$210''$ above the limb. To this spectral data we have applied a Gaussian fitting program and obtained the Doppler velocities, line intensities, and line widths (FWHM) as a function of time and position along the slit.

Properties of Doppler Oscillations

The Doppler velocity shows an rms amplitude of 0.2 – 0.6 km s^{-1}, generally growing with height. This amplitude is smaller than the turbulent broadening of 10 – 20 km s^{-1} of the line (Singh *et al.* 1999). The power spectrum of the Doppler velocity at each height shows peaks in the range of 1 – 3 mHz, but a k-ω diagram (the distribution of power in the wavenumber (k) and frequency (ω) space, often used in helioseismology) shows no global oscillations (oscillations coherent over the full slit height of $180''$).

By cross-correlating the Doppler velocities observed at two different heights, we found that the coherence is lost for separations exceeding roughly 20″. Therefore, we cross-correlated the two height levels separated by 18″, and derived the time lag between them. We found both positive and negative lags (with roughly equal populations), indicating that both upward and downward propagating waves are present. The histogram of lags peaks at zero lag, and implies that there are signals faster than 500 km s^{-1}. These may indicate Alfvén waves (whose speed in the corona is typically 1000 km s^{-1}), but the interpretation is not unique because we only observed the one-dimensional cut of the corona at the position of the spectrograph slit. The histogram is broad and extends to time lags of 100 s or more, indicating slow waves whose propagation speed is of the order of 100 km s^{-1}. These waves could be slow-mode or sound waves.

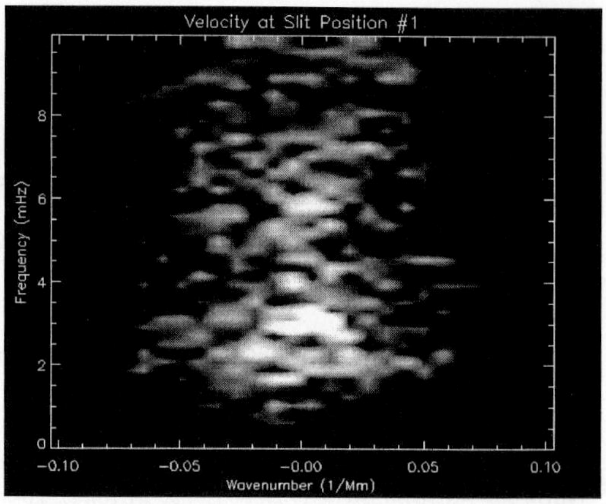

Fig. 1. The wavenumber-frequency ($k - \omega$) diagram of Doppler velocity.

Properties of Line Intensity Oscillations

The line intensity shows an rms variation of $\delta I/I = 1 - 2$ %. The Fourier analysis shows no distinct frequencies in power spectra. However, the correlation between the Doppler velocity and the line intensity is found to maximize at zero time lag. Therefore, the waves are propagating: Standing waves will give zero correlation at zero time lag. The Doppler amplitude of 0.3 km s^{-1} and the line intensity variation $\delta I/I =$ 1 % are consistent if the waves are sound waves. This also implies that the Alfvén waves are not the major contributor to the Doppler signal. This agrees with Hara & Ichimoto (1999) who found that the anisotropy in line widths that is expected for transverse oscillations (like Alfvén waves) is actually small ($\simeq$ 20%).

CONCLUSIONS

(1) The in-phase relationship between the Doppler velocity and the line intensity suggests the existence of propagating slow-mode or sound waves.
(2) The Alfvén waves are not the dominant (but could be an equally important) contributor to the observed Doppler signal.
(3) In any case, the observed coherent waves do not carry enough energy to heat the corona, although unresolved motions observed as turbulent line broadening may.

The full account on this topic will be submitted to *Solar Physics.*

REFERENCES

Hara, H., & K. Ichimoto, *Astrophys. J.*, **513**, 969 (1999).

Singh, J., K. Ichimoto, H. Imai, T. Sakurai, & A. Takeda, *Publ. Astron. Soc. Japan*, **51**, 269 (1999).

Section II.

Active Region and Bright Point Studies

CONNECTION BETWEEN PHOTOSPHERIC MAGNETIC FIELDS AND CORONAL STRUCTURE/DYNAMICS

T. Shimizu

National Astronomical Observatory of Japan, Mitaka, Tokyo 181-8588, Japan

ABSTRACT

Yohkoh has provided new observations of the X-ray corona for over 10 years and has made several advances in understanding magnetic origins of the heating and energy releases in the corona. We have, especially, learned much about weak transient activities, such as jets and transient brightenings (microflares) from *Yohkoh* observations and also from EUV observations by the *SOHO* and *TRACE* spacecraft. Some observations have been published, clearly showing a key role of magnetic flux just after newly emerging from below the surface. These observations can be explained by an emerging flux model.

INTRODUCTION

For more than 10 years while the *Yohkoh* satellite was in operation on orbit, a large number of observations of photospheric magnetic fields that were simultaneous with *Yohkoh* X-ray observations were conducted at various ground-based observatories and by the Michelson Doppler Imager (MDI) on *SOHO*. These observations have provided new views for understanding magnetic origins of the heating of transient and steady structures in the corona.

It has been well known that the location and strength of identifiable X-ray features in the corona are associated with strong magnetic fields at the surface. Bright features seen in soft X-rays are located above sunspot groups observed in visible light. *Yohkoh* observations have quantitatively confirmed that the thermal properties of the "active-region" corona are well correlated to the integrated and averaged magnetic properties derived from magnetogram observations. The total thermal energy involved in an active region is well related to its total magnetic flux from a tiny active region ($\sim 3 \times 10^{20}$ Mx) to a large active region ($\sim 7 \times 10^{22}$ Mx) (Fisher *et al.* 1998). The coronal gas pressure averaged over an active region is also correlated to the average of the magnetic flux density (Yashiro & Shibata 2001) from a diffuse active region ($\sim$ 40 gauss) to a well-confined active region ($\sim$ 300 gauss).

The active-region corona consists of a variety of coronal loops, which trace out magnetic field structures filled with hot plasma in the corona (Figure 1). Bright coronal loops connect the leading sunspot area to the following sunspot area, and these loops appear to be rooted beside and/or in the penumbra of sunspots. It is interesting that the corona is in general dark above the umbra of sunspots, although sunspots are the cross-section of strong magnetic field bundles at the surface. Therefore, for better understanding of the heating of the active-region corona, it is important to look more deeply into how the thermal properties of coronal structures in active regions are related to magnetic fields at the sites where the coronal structures are rooted.

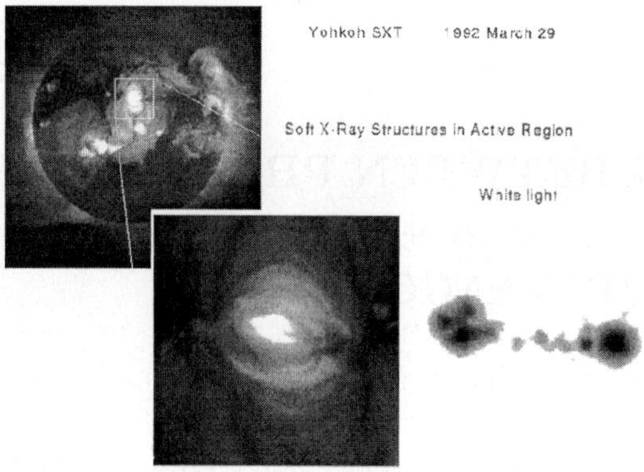

Fig. 1. An active region observed in soft X-rays (left) and visible light (right) by *Yohkoh* on 29 March, 1992

One of the remarkable discoveries made by *Yohkoh* is that the corona is more dynamic than thought before. This view is brought to us by the soft X-ray telescope (SXT) onboard *Yohkoh*, which provides continuous sequences of soft X-ray coronal (> 3 MK plasma) images with high temporal/spatial resolution and high sensitivity, making it possible to detect weak transient activities that have not been well observed so far, such as X-ray jets and active-region transient brightenings (e.g. Shibata *et al.* 1992; Shimizu *et al.* 1992). Shimizu (1993) showed that X-ray transient brightenings (microflares) have preferred locations in active regions for their occurrence (Figure 2). Transient brightenings are well observed around the outer boundary of the penumbra of well-developed spots in emerging flux regions. Magnetic fields around the penumbra are one of the key features for understanding magnetic origins of weak transient activities. Also, since bright coronal loops are rooted beside and/or in the penumbrae, magnetic fields around the penumbrae are important for understanding the heating of the active-region corona.

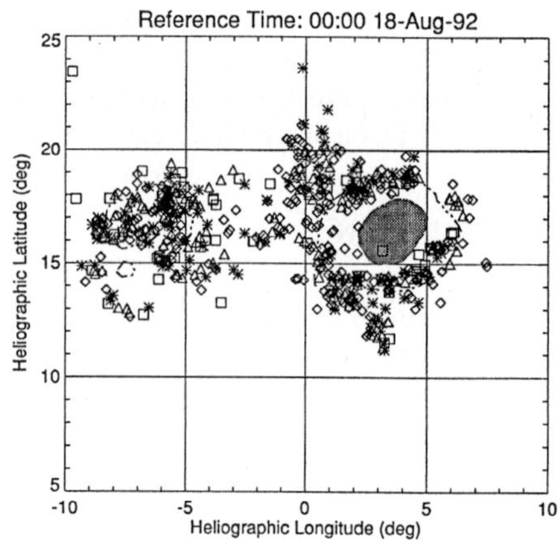

Fig. 2. Spatial distribution of X-ray transient brightenings (microflares) in active region NOAA 7260 (Shimizu 1993). This map shows 639 transient brightenings observed from 15 through 20 August, 1992.

This paper first reviews what kinds of evolution of magnetic fields are well observed in active regions at the surface, and their possible associations with the heating of transient and steady coronal structures. Coronal observations with simultaneous magnetogram observations have been obtained to investigate how magnetic fields at the surface are responsible for weak transient activities. The next section briefly reviews weak transient activities and also some observational examples of jets and microflares that clearly show magnetic connection between the coronal and photospheric magnetic fields. Using unique simultaneous observations by *Yohkoh*/SXT and La Palma, magnetic origins of point-like transient brightenings (microflares) were studied by Shimizu *et al.* (2002). Some observations presented in Shimizu *et al.* (2002) are reviewed and discussed.

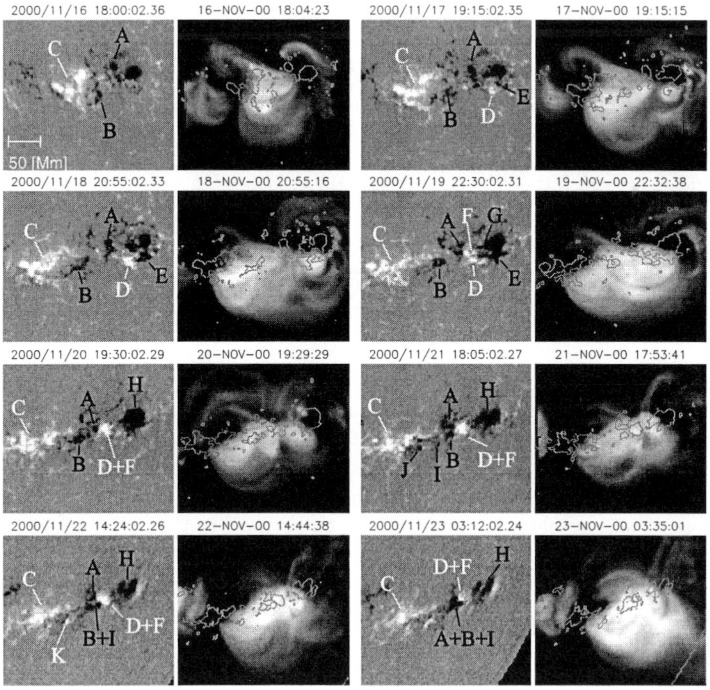

Fig. 3. Magnetic field and coronal evolution of active region NOAA 9231 from 16 November to 23 November 2000. Longitudinal magnetograms are taken with *SOHO* MDI, and coronal images are obtained with *Yohkoh* SXT. Contours in coronal images are +/-200 gauss levels in magnetograms.

MAGNETIC EVOLUTION IN ACTIVE REGIONS

Figure 3 shows the day-by-day evolution of magnetic fields at the photosphere and magnetic structures in the corona of an active region. Typical magnetic activities have been well observed in series of magnetograms: newly emerging magnetic flux, small patches around well-developed sunspots, merging of the same polarity flux, flux canceling with opposite polarity flux, disappearing magnetic flux maybe due to fragmentation and diffusion, shearing magnetic bipoles, and so on.

The large-scale emergences of magnetic flux from below the photosphere are observed as a magnetic bipole labeled D and E and another magnetic bipole labeled F and G. As magnetic flux successively emerges, the bundle of bright loops is newly developed above the newly emerging magnetic flux and frequent occurrence of transient brightenings (microflares) and X-ray jets is observed (Kawai *et al.* 1992, Yoshimura & Kurokawa 1999). The positive-polarity patches labeled D and F approach negative-polarity patches labeled A, B, and I after 19 November, and it appears that they are partially canceled. Associated with the cancellation, a new loop system connecting the positive-polarity area (D and F) to the negative-polarity area (A and B) is observed after 20 November. A magnetic bipole labeled B and C shows shearing motion; the negative patch B slowly separates from the positive patch C, and the direction of the line across B and C slightly rotates counterclockwise. The loops connecting B to C maintain their brightness during the observation.

In Figure 3, compact X-ray sources and faint loops extending from the compact X-ray sources are seen around the leading large sunspot, where a large number of X-ray transient brightenings are observed with

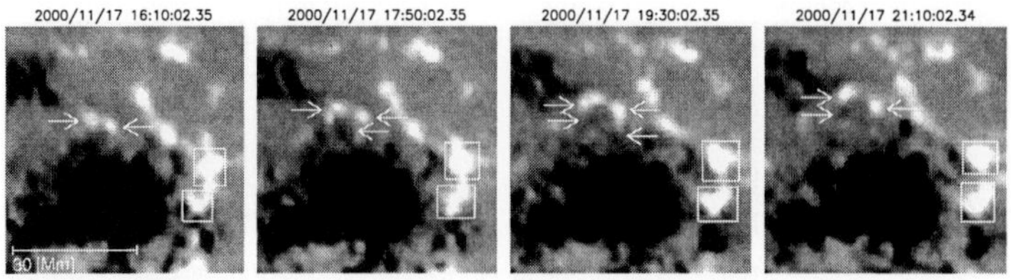

Fig. 4. Satellite spots, enclosed by squares, and moving magnetic features, marked by arrows. Longitudinal magnetograms are taken with *SOHO* MDI on 17 November 2000.

Yohkoh. They are associated with the formation of satellite spots, the polarity of which is opposite to the leading spot (examples enclosed by squares in Figure 4, Leka *et al.* 1994, Shimizu 1993, Shimojo *et al.* 1998). Moreover, numerous moving magnetic features (MMFs) are commonly observed around well-developed spots (examples shown by arrows in Figure 4). The MMFs are small magnetic bipoles which are born at the outer edge of the penumbra of well-developed spots and then go outward in the radial direction from the spot. It appears that compact X-ray sources are not associated with MMFs.

WEAK TRANSIENT ACTIVITIES

Variety of Weak Transient Activities

Yohkoh SXT has revealed that X-ray transient brightenings (microflares) occur in the bright corona (Shimizu *et al.* 1992, 1994). SXT also discovered X-ray jets as transitory X-ray enhancements with apparent collimated motion (Shibata *et al.* 1992, Shimojo *et al.* 1996). They are associated with small flares in XBPs, transient brightenings or small flares in active regions or emerging flux regions. Since then, several kinds of weak transient activities have been reported from the *Yohkoh*, *SOHO*, and *TRACE* observations. They are named using different terminology, because some differences can be found from weak transient activities previously reported, or observations are made with different instruments. In Figure 5, weak transient activities observed in the corona are summarized as a function of involved energy in the vertical axis and the location of occurrences in the horizontal axis. Note that weak transient activities found in transition-region EUV lines, such as blinkers (Harrison 1997), explosive events and EUV jets (Brueckner & Bartoe 1983, Innes *et al.* 1997) are not included in this figure.

Newly observed coronal weak activities are distributed between 10^{29} and 10^{24} ergs. Weak transient activities are more easily found in quiet regions because of the low quasi-steady X-ray background level. No significant differences except for the involved energy and occurrence location may be found among X-ray transient brightenings, XBP flares (Strong *et al.* 1992, Kundu *et al.* 1994), network flares (Krucker *et al.* 1997), and EUV transient brightenings (Berghmans *et al.* 1998, Krucker & Benz 1998, Benz & Krucker 1998). The durations of these activities are all less than roughly 10 min, and they show soft X-ray light curves with a sudden increase at the beginning and a slow decrease in the late phase, which are temporal behaviors similar to those of standard flares. The coronal loops showing these activities are compact, and smaller energy activities appear to be confined into more compact loops.

Coronal jets are observed in a wide range of weak transient activities, as illustrated by a hatched region in Figure 5. Because of a lack of observations, it is currently uncertain whether small variations in quasi-steady long loops are similar to the other weak transient activities. However, they appear to be small variations at limited parts within coronal loops, whereas the other weak activities are X-ray brightenings of entire compact loops (Shimizu & Tsuneta 1997, Katsukawa & Tsuneta 2001).

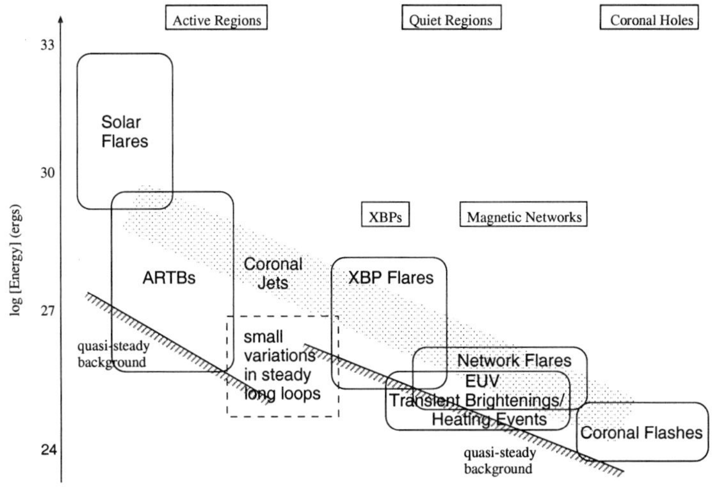

Fig. 5. Coronal weak transient activities as a function of energy in vertical axis and occurrence location in horizontal axis (Shimizu 1999). The abbreviation ARTB means X-ray transient brightenings (microflares).

Implication for Heating the Corona

One of the attractive concepts for the heating of the corona is that numerous small energy releases (microflares, nanoflares, or picoflares) may be a possible source for heating the corona (Parker 1988). The frequency distribution of energy releases by weak transient activities as a function of the magnitude has been well studied to examine the concept of heating by microflares and nanoflares. Shimizu (1995) studied the frequency distribution in the microflare energy range with *Yohkoh* observations, and then Krucker & Benz (1998), Parnell & Jupp (2000) and Aschwanden *et al.* (2000) estimated the frequency distribution in nanoflare energy range with EUV observations by *SOHO*/EIT and *TRACE*. They are all well represented by a power-law function with the slope similar to that of standard flares (e.g. Crosby, Aschwanden, & Dennis 1993), meaning that the flare power-law distribution is maintained over almost eight orders of magnitude in energy ($10^{24} \sim 10^{32}$ ergs). The total thermal energy supplied is estimated to be at most a factor of 5 smaller than the heating rate required for the active-region corona (Shimizu 1995, Benz & Krucker 2002), and the total energy released by weak transient activities observed is not sufficient to explain the entire heating of the corona. The energy released by weak transient activities, however, plays a key role in generating > 5 MK hot plasma in the corona (Watanabe *et al.* 1995, Yoshida & Tsuneta 1996).

Photosphere-Corona Connection

A lot of observational studies have shown that major solar flares frequently occur in sheared magnetic regions (e.g. Sakurai *et al.* 1992), in emerging flux regions (e.g. Hanaoka 1996), and with complicated magnetic topology in active regions. However, since the photospheric magnetic field configurations of major flares are generally too complicated to completely understand, small-scale activities in the corona as seen in X-ray and EUV wavelengths can provide a better opportunity to understand fundamental physical mechanisms of energy build up and triggering. It is expected that the photospheric magnetic field dynamics responsible for small-scale activities is smaller than for major flares, but recent visible light data with high spatial and temporal resolution make the detailed study of small-scale activities in the corona possible and reliable.

By comparing coronal data with magnetograms, some studies were recently made to understand the energy build up and trigger of these small-scale activities. Shimojo *et al.* (1998) provided a statistical result on photospheric magnetic-field patterns favorable to the occurrence of X-ray jets by studying longitudinal magnetograms at the footpoints of X-ray jets. Zhang *et al.* (2000) observed simultaneous occurrence of an X-ray jet and a surge in Hβ at the site where the pre-existing magnetic flux was "canceled" by newly emerging flux of opposite polarity. Chae *et al.* (1999) found several EUV jets that repeatedly occurred where pre-existing magnetic flux was "canceled" by newly emerging flux of opposite polarity. Yoshimura *et al.* (2002) reported that surge activities were observed in Hα where the pre-existing magnetic flux was "canceled" by newly emerging flux of opposite polarity, although no enhanced X-ray emissions were found. On the other hand, Tang *et al.* (2000) found a soft X-ray microflare for which the impulsive enhancement of the emerging flux in magnetograms occurred about 20 minutes before the the peaks of the soft X-ray brightening. Shimizu *et al.* (2002) have found several X-ray transient brightenings (microflares) showing close relationships with emergence of magnetic flux, as described in detail in the next section.

These observations indicate that newly emerging flux and/or magnetic cancellation with newly emerging flux play a vital role in causing transient energy releases in the upper solar atmosphere.

YOHKOH/SXT-LA PALMA OBSERVATIONS

By combining *Yohkoh* soft X-ray images with high resolution magnetograms simultaneously obtained at La Palma, Shimizu *et al.* (2002) studied photospheric magnetic signatures responsible for soft X-ray transient brightenings (microflares).

Identification of Associated Magnetic Activities

In order to have a reliable correspondence between the photosphere and the corona, 16 point-like transient brightenings with X-ray source size less than 10 arcsec occurring during periods when the seeing is excellent at La Palma have been studied, although a lot of transient brightenings are in the form of multiple or single loop structures. In half of the studied events, small-scale emergences of magnetic flux loops are found in the vicinity of the transient brightenings (Figure 6). Six events of the half show that a small-scale flux emergence occurs 5 $\sim$ 30 minutes prior to the onset of the X-ray brightening (Figure 7). In the other half of the studied events, no apparent evolutionary change of magnetic flux elements is found associated with the transient brightenings. Many of these events are found in rather strong magnetic fields, such as sunspots and pores, implying that small-scale changes of magnetic flux are obscured or suppressed by strong magnetic fields.

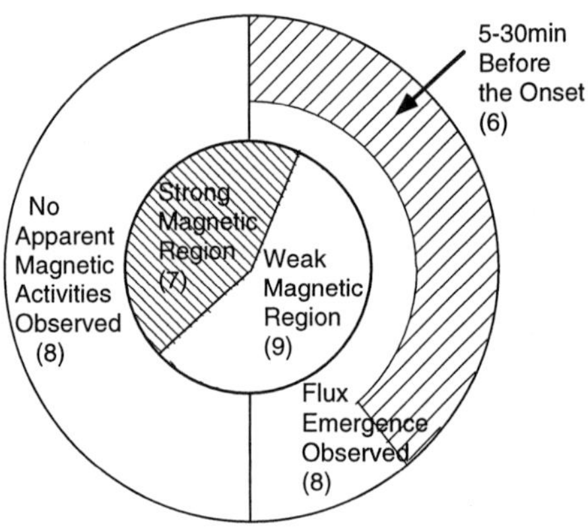

Fig. 6. Summary of photospheric magnetic activities associated with transient brightenings (Shimizu *et al.* 2002). The outer circle shows what kinds of magnetic activities are observed in the vicinity of 16 pointlike transient brightenings. The inner circle indicates whether transient brightenings are observed in strong magnetic regions or weak magnetic regions.

Detailed Spatial Relationship

The spatial relationship among newly emerging flux, soft X-ray source, and tiny brightenings observed in Hα provides information on the configuration of magnetic fields involved in the energy release. The location of a

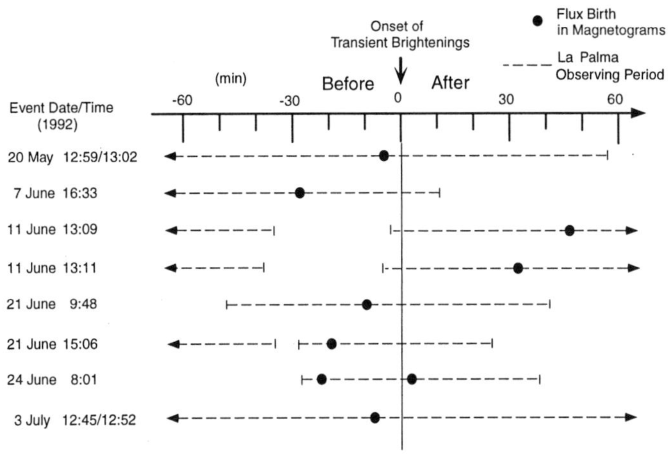

Fig. 7. Temporal relationships between the first appearance of small-scale magnetic flux in magnetograms and the onset of transient brightenings (Shimizu *et al.* 2002). For 8 pointlike transient brightenings associated with the emergence of small-scale magnetic flux, each line shows the timing of flux emergence relative to the onset of transient brightening, with the period of magnetogram observation at La Palma.

small emerging magnetic bipole detected in magnetograms tells where the newly emerging flux appears at the photosphere from the interior. The soft X-ray source is the signature of the energy release, giving the place where magnetic fields involved in the energy release are located in the corona. Tiny brightenings observed in Hα can be used to infer the location of footpoints of the heated soft X-ray loops, because Hα brightenings are probably heated by thermal conduction from the soft X-ray source. Since the La Palma images are co-aligned with SXT soft X-ray images with an accuracy limited by the SXT pixel size (2.46 arcsec), the spatial distribution with larger than this accuracy gives meaningful information on the spatial relationship. Figure 8 shows the spatial relationship among emerging flux, soft X-ray source, and Hα brightenings for the 6 transient brightenings in which newly emerging flux is detected in the magnetograms, showing that the center position of the soft X-ray source core is not spatially coincident with the newly emerging magnetic flux. In 5 of the 6 events, one of the tiny brightenings in Hα is observed at one end of an emerging magnetic bipole and the other brightenings are located apart from the emerging bipole.

DISCUSSION

Observed Temporal Delay

The driving force for the emergence of magnetic loops is the enhanced magnetic buoyancy of flux tubes. The first appearance of emerging flux in the photosphere is an anomalously dark intergranular lane observed in white light granules. The dark intergranular structures last about 10 minutes. The dark lanes indicate an emerging magnetic flux loop crossing the photospheric layer (Strous *et al.* 1996). At this time, the signature of the emerging flux may not be observable in longitudinal magnetograms because the loop is nearly horizontal (Lites, Skumanich, & Martinez Pillet 1998). Then, at the ends of the elongated dark structures, bright elements appear in G-band images and the elements separate from each other. At this time, the signature of the emerging flux is observable in longitudinal magnetograms, because the ends of the emerging loop are no longer horizontal at the photospheric level. In successive emergences of magnetic flux in ephemeral active regions, the rate of expansion is the order of 5 km s^{-1} in the first few minutes after

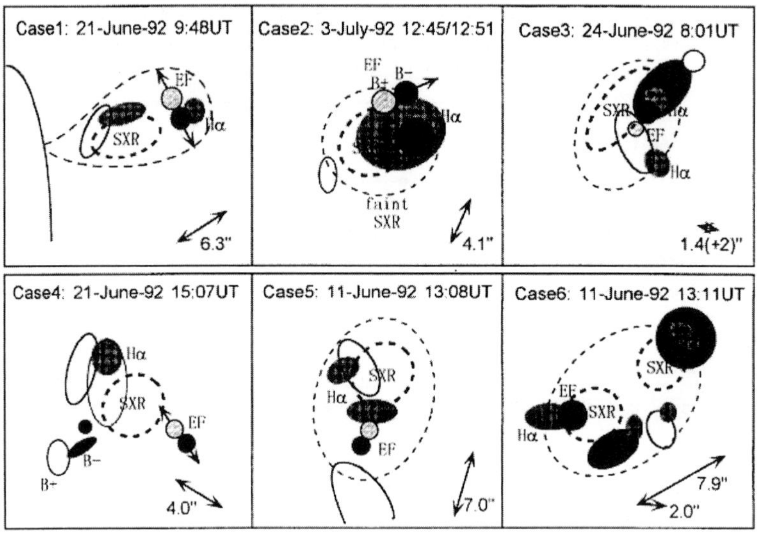

Fig. 8. Spatial relationship among newly emerging flux, soft X-ray source, and tiny brightenings in Hα for 6 transient brightenings. The distance between the center position of the soft X-ray source core and the newly emerging flux is given at the lower right corner in each frame.

the emergence, then drops to values between 1.3 and 0.7 km s^{-1} during the next several hours (Harvey & Martin 1973). We have found that the speed of moving magnetic elements is 2.8 km s^{-1} in the 9:48 UT 21 June 1992 transient brightening, which is in good agreement with previous observations. In this transient brightening, the first appearance of the new magnetic flux in magnetograms is about 10 minutes prior to the onset of the X-ray brightening. Assuming the vertical speed of the emergence to be approximately equal to the measured horizontal speed, the emerging magnetic flux loop would reach $\sim$ 1700 km height above the photospheric level, which may be at the mid-upper chromosphere.

After the emergence, the chromospheric response to emerging flux is observed in Hα. An arch filament system is observed to form, connecting the plages of opposite polarity. The rise velocity of filaments is 10 $\sim$ 15 km s^{-1}. The rise velocity of emerging flux loops is accelerated due to magnetic buoyancy from less than a few km s^{-1} at the photospheric level to 10 $\sim$ 15 km s^{-1} at the chromospheric level, although there is lack of observational information on the acceleration. However, the dynamical behavior of emerging flux loops is well demonstrated by numerical simulations. The time scale for the emergence of magnetic flux from the photosphere to the coronal level is about 20 minutes (Shibata *et al.* 1989). This is comparable to the observed time difference between the first appearance of the flux emergences in magnetograms and the onset of coronal X-ray transient brightenings. This observation suggests that high spatial observations with temporal resolution of less than a few minutes are essential for investigating in detail the dynamical response of the coronal magnetic fields to magnetic emergences.

Spatial Relationship and Emerging Flux Model

The observations mentioned in this paper show that flux emergence is involved in the occurrence of X-ray transient brightenings (microflares) and X-ray/EUV jets, strongly suggesting that the magnetic fields just emerged from below the photosphere play a key role in the transient release of magnetic energy in the corona. The emerging flux model (Heywaerts *et al.* 1977, Yokoyama & Shibata 1996) has been considered as an important process for converting magnetic energy into thermal and kinetic energy in the corona (Figure 9). In this model, a new magnetic flux loop rises and collides with pre-existing magnetic fields, creating a current sheet between them. Recent numerical simulations show that X-ray emitting hot plasma can be created by a magnetic reconnection in a neutral sheet between emerging and pre-existing coronal magnetic fields, and the hot plasma is ejected upwards with a compact micro-flaring loop (X-ray jet).

The emerging flux model explains the observed spatial and temporal relationships. No X-ray jet is, however, observed in the events examined in the previous section, although the model predicts the existence of an X-ray jet. Instead, the observations show tiny ejections from the brightening site in Hα in 3 cases. Whether an X-ray jet exists or not may depend on the pre-existing magnetic field environment. When the pre-existing field is rather strong, strong magnetic pressure may force the reconnection site to the lower atmosphere. In this case, an X-ray jet may be produced with a micro-flaring compact loop. It may be also observed at the photospheric level that one polarity of the emerging magnetic bipole is canceled with the pre-existing magnetic flux, because of the magnetic reconnection in the lower atmosphere. When the pre-existing field is rather weak, weak magnetic pressure may force the reconnection site to the higher atmosphere. In this case, a micro-flaring loop may be produced with no apparent appearance of X-ray plasma ejection, because of low plasma density and low magnetic tension.

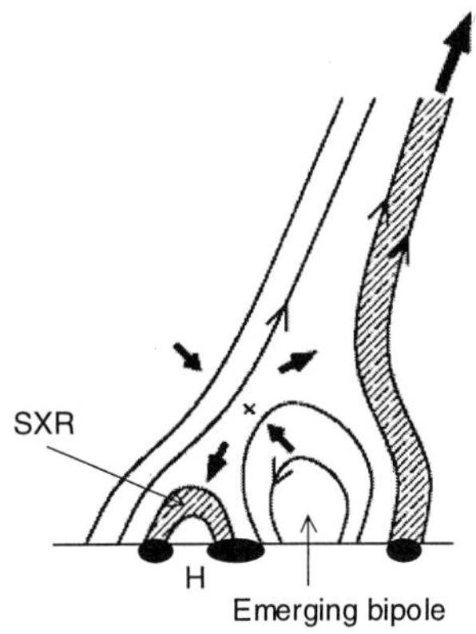

Fig. 9. An emerging flux model for explaining the simultaneous occurrence of X-ray jet and microflare. Hα brightenings are added to the picture from Yokoyama (1996).

ACKNOWLEDGEMENTS

The author would like to express his thanks to the Scientific Organizing Comittee members of the symposium. *Yohkoh* observations have made it possible for the author to study magnetic connection between the corona and the photosphere for soft X-ray transient brightenings, and the author thanks the *Yohkoh* project personnel and all the people who have made contributions to the *Yohkoh* observations. The author also thanks M. Kubo for providing Figure 3 and Figure 4 from his master thesis.

REFERENCES

Aschwanden, M.J., Tarbell, T.D, Nightingale, R.W., Schrijver, C.J., Title, A, Kankelborg, C.C., Martens, P., and Warren, H.P. 2000, *ApJ* **535**, 1047

Benz, A.O. and Krucker, S. 1999, *A&A* **341**, 286

Benz, A.O. and Krucker, S. 2002, *ApJ* in press

Berghmans, D., Clette, F., Moses, D. 1998, *A&A* **336**, 1039

Brueckner, G.E. and Bartoe, J.-D.F. 1983, *ApJ* **272**, 329

Chae, J., Qiu, J., Wang, H., and Goode, P.R. 1999, *ApJ* **513**, L75

Crosby, N.B., Aschwanden, M.J., Dennis, B.R. 1993, *Sol. Phys.* **143**, 275

Fisher, G.H., Longcope, D.W., Metcalf, T.R., and Pevtsov, A.A. 1998, *ApJ* **508**, 995

Hanaoka, Y. 1996, *Sol. Phys.* **173**, 319

Heyvaerts, J., Priest, E.R., and Rust, D.M. 1977, *ApJ* **216**, 123

Harrison, R.A. 1997, *Sol. Phys.* **175**, 467

Harvey, K.L. and Martin, S.F. 1983, *Sol. Phys.* **32**, 389

Innes, D.E., Inhester, B., Axford, W.I. and Wilhelm, K. 1997, *Nature* **386**, 811-813

Katsukawa, Y. and Tsuneta, S. 2001, *ApJ* **557**, 343

Kawai, G., Kurokawa, H., Tsuneta, S., Shimizu, T., Shibata, K., Acton, L.W., Strong, K.T., and Nitta, N. 1992, *PASJ* **44**, L193

Krucker, S., Benz, A.O., Bastian, T.S., Acton, L.W. 1997, *ApJ* **488**, 499

Krucker, S., Benz, A.O. 1998, *ApJL* **501**, L213

Kundu, M.R., Strong, K.T., Pick, M., White, S.M., Hudson, H.S., Harvey, K.L., Kane, S.R. 1994, *ApJ* **427**, L59

Leka, K.D., van Driel-Gesztelyi, L., Nitta, N., Canfield, R.C., Mickey, D.L., Sakurai, T., and Ichimoto, K. 1994, *Sol. Phys.* **155**, 301

Lites, B.W., Skumanich, A. and Martinez Pillet, V. 1998, *A&A* **333**, 1053

Parker, E.N. 1988, *ApJ* **330**, 474

Parnell, C.E. and Jupp, P.E. 2000, *ApJ* **529**, 554

Sakurai, T., Shibata, K., Ichimoto, K., Tsuneta, S., Action, L.W. 1992, *PASJ* **44**, L123

Shibata, K., Tajima, T., Steinolfson, R.S., and Matsumoto, R. 1989, *ApJ* **345**, 584

Shibata, K., Ishido, Y., Acton, L.W., Strong, K.T., Hirayama, T., Uchida, Y., McAllister, A.H., Matsumoto, R., Tsuneta, S., Shimizu, T., Hara, H., Sakurai, T., Ichimoto, K., Nishino, Y., and Ogawara, Y. 1992, *PASJ* **44**, L173

Shimizu, T., Tsuneta, S., Acton, L.W., Lemen, J.R., and Uchida, Y. 1992, *PASJ* **44**, L147

Shimizu, T. 1993, in *Proceedings of the 2nd Japan-China Seminar on Solar Physics*, eds. T. Sakurai, T. Hirayama, & G. Ai (Tokyo: National Astronomical Observatory of Japan) p. 193

Shimizu, T., Tsuneta, S., Acton, L.W., Lemen, J.R., Ogawara, Y., and Uchida, Y. 1994, *ApJ* **422**, 906

Shimizu, T. 1995, *PASJ* **47**, 251

Shimizu, T., Tsuneta, S. 1997, *ApJ* **486**, 1045

Shimizu, T. 1999, in *Proceedings of Nobeyama Symposium on Solar Physics with Radio Observations*, eds. T. Bastian, N. Gopalswamy, & K. Shibasaki (NRO report No. 479) p. 123

Shimizu, T. Shine, R.A., Title, A.M., Tarbell, T.D., and Frank, Z. 2002, *ApJ* **574**, August 1 issue, in press

Shimojo, M., Hashimoto, S. Shibata, K., Hirayama, T., Hudson, H.S., and Acton, L.W. 1996, *PASJ* **48**, 123

Shimojo, M., Shibata, K., and Harvey, K.L. 1998, *Sol. Phys.* **178**, 611

Strong, K. T., Harvey, K. L., Hirayama, T., Nitta, N., Shimizu, T., Tsuneta, S. 1992, *PASJ* **44**, L161

Strous, L.H., Scharmer, G., Tarbell, T.D., Title, A.M., and Zwaan, C. 1996, *A&A* **306**, 947

Tang, Y.H., Li, Y.N., Fang, C., and Sakurai, T. 2000, *ApJ* **534**, 482

Watanabe, T., Hara, H., Shimizu, T., Hiei, E. *et al.* 1995, *Sol. Phys.* **157**, 169

Yoshida, T., Tsuneta, S. 1996, *ApJ* **459**, 342

Yashiro, S. and Shibata, K. 2001, *ApJ* **550**, L113

Yokoyama, T. and Shibata, K. 1996, *PASJ* **48**, 353

Yoshimura, K. and Kurokawa, H. 1999, *ApJ* **517**, 964

Yoshimura, K., Kurokawa, H., Shimojo, M., and Shine, R. 2002, *these proceedings*

Zhang, J., Wang, J., and Liu, Y. 2000 *A&A* **361**, 759

CONTAGIOUS CORONAL HEATING FROM RECURRING EMERGENCE OF MAGNETIC FLUX

R. L. Moore, D. A. Falconer, and A. C. Sterling

SD 50, NASA/MSFC, Huntsville Al, 35812, USA

ANALYSIS AND DISCUSSION

There are two ways by which the body of the corona (the magnetically closed part not rooted in coronal holes) is heated via its magnetic field. One is by dissipation of nonpotential magnetic energy built up in the field by convection driving of its feet and/or by interaction of the extant magnetic field with emerging fields (e.g. Parker 1983; Longcope 1996; Schrijver *et al.* 1998). The other way is by absorption of MHD waves, energetic particles, and/or hot plasma that enter from below, generated by photospheric turbulence or fine-scale magnetic activity (microflaring) in the chromosphere and transition region (e.g. Porter *et al.* 1987; Krucker & Benz 1998; Moore *et al.* 1999; Aschwanden, Schrijver, & Alexander 2001). While it is accepted that these two avenues each in principle contributes some heating, and together deliver the total heating of the bulk of the corona, it is not known whether both contributions are substantial or one is negligible. We present evidence that some obvious gradual large-scale coronal heating is by internal magnetic energy release (the first way stated above).

For each of four old bipolar active regions, we followed in *Yohkoh*/SXT full-frame images and *SOHO*/MDI full-disk magnetograms the development, over several days, of enhanced coronal heating in and around the old bipole in response to new magnetic flux emergence within the old bipole. In each active region, new flux emerged in the equator-ward/leading end of the old bipole, around a lone remaining sunspot (as in Figures 1 and 2) and/or on the neutral line of the old bipole (as in Figures 2 and 3). As usual, the emerging field was continually marked by intense coronal heating (brightness) (as in each Figure), and often caused enhanced heating in extended loops stemming from the emergence site (as in the first panel of Figure 3). In addition, a "rooster tail" of coronal loops in the poleward/trailing extent of the old bipole underwent brightening and became more extensive in response to the flux emergence (as in each Figure). There were gradual (hours-long) episodes of enhanced heating in the far end of the rooster tail and/or in external fields, in loops that were either part of or contiguous with the old bipole but were not directly connected to the emerging field (as in each Figure).

Apparently, the accommodation of localized newly emerged field within an old active region entails far-reaching adjustments in the 3D magnetic field throughout the active region and in surrounding fields in which the active region is embedded. An episode of contagious heating in indirectly impacted fields is like a confined flare (e.g. the confined flares shown by Moore *et al.* 2001), only weaker and slower, waxing and waning over many hours. We infer that the heating in these events comes from the body of the loops rather than from the feet because the events are flare-like and apparently occur in response to the remote emerging flux. Pevtsov & Acton (2001) have reported the long-range positive effect of active regions on coronal heating in remote quiet regions. Our contagious heating events are apparently examples of this effect in action, and support the conclusion of Pevtsov & Acton that the stimulated remote coronal heating comes from the body of the corona and not from activity at its feet.

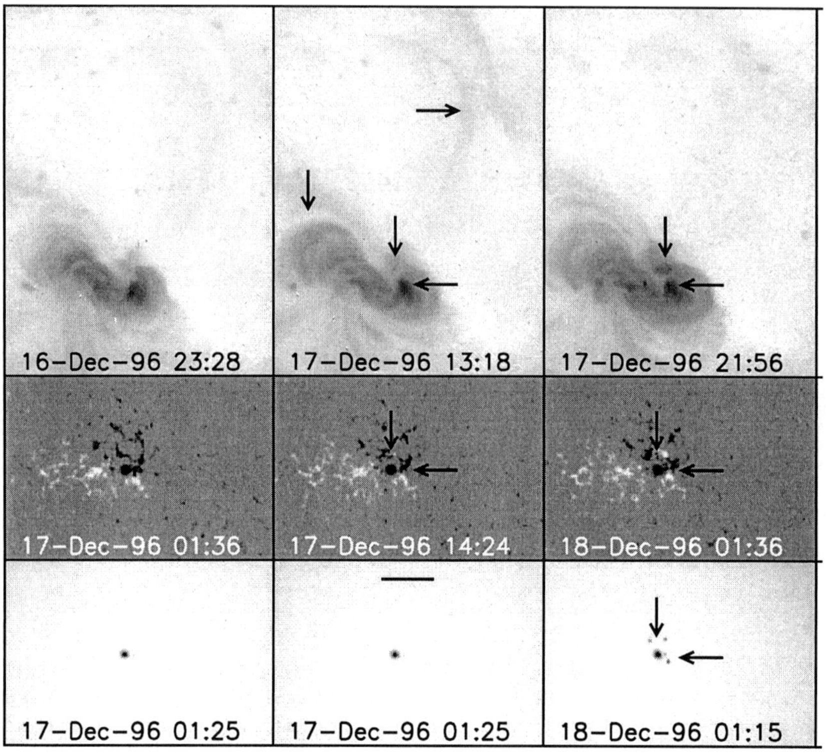

Fig. 1. Rooster-tail brightening and external brightening in response to new flux emergence in active region NOAA AR 8004. Top: SXT coronal images; middle: MDI magnetograms; bottom: MDI photospheric brightness images. All images are co-registered to span the same longitudes and have the same southern edge; north is up, west right. The horizontal bar spans 100,000 km. In the middle SXT panel, from south to north, the first two arrows point to heating in the emerging field, the third points to contagious heating in the far end of the rooster tail, and the fourth points to contagious heating in adjacent old coronal loops external to the active region. In the magnetograms and photospheric images, the arrows point to new magnetic flux and new sunspots.

This work was funded by NASA's Solar & Heliospheric Physics Program.

REFERENCES

Aschwanden, M. J., Schrijver, C. J., & Alexander, D. 2001, *ApJ*, **550**, 1036.

Krucker, S. & Benz, A. O 1998, *ApJ*, **501**, L213.

Longcope, D. W. 1996, *Solar Phys.*, **169**, 91.

Moore, R. L., Falconer, D. A., Porter, J. G., & Suess, S. T. 1999, *ApJ*, **525**, 505.

Moore, R. L., Sterling, A. C., Hudson, H. S., & Lemen, J. R. 2001, *ApJ*, **552**, 833.

Pevtsov, A. A. & Acton, L. W. 2001, *ApJ*, **554**, 416.

Parker, E. N. 1983, *ApJ*, **264**, 642.

Porter, J. G., Moore, R. L., Reichmann, E. J., Engvold, O., & Harvey, K. L. 1987, *ApJ*, **323**, 775.

Schrijver, C. J. *et al.* 1998, *Nature*, **394**, 152.

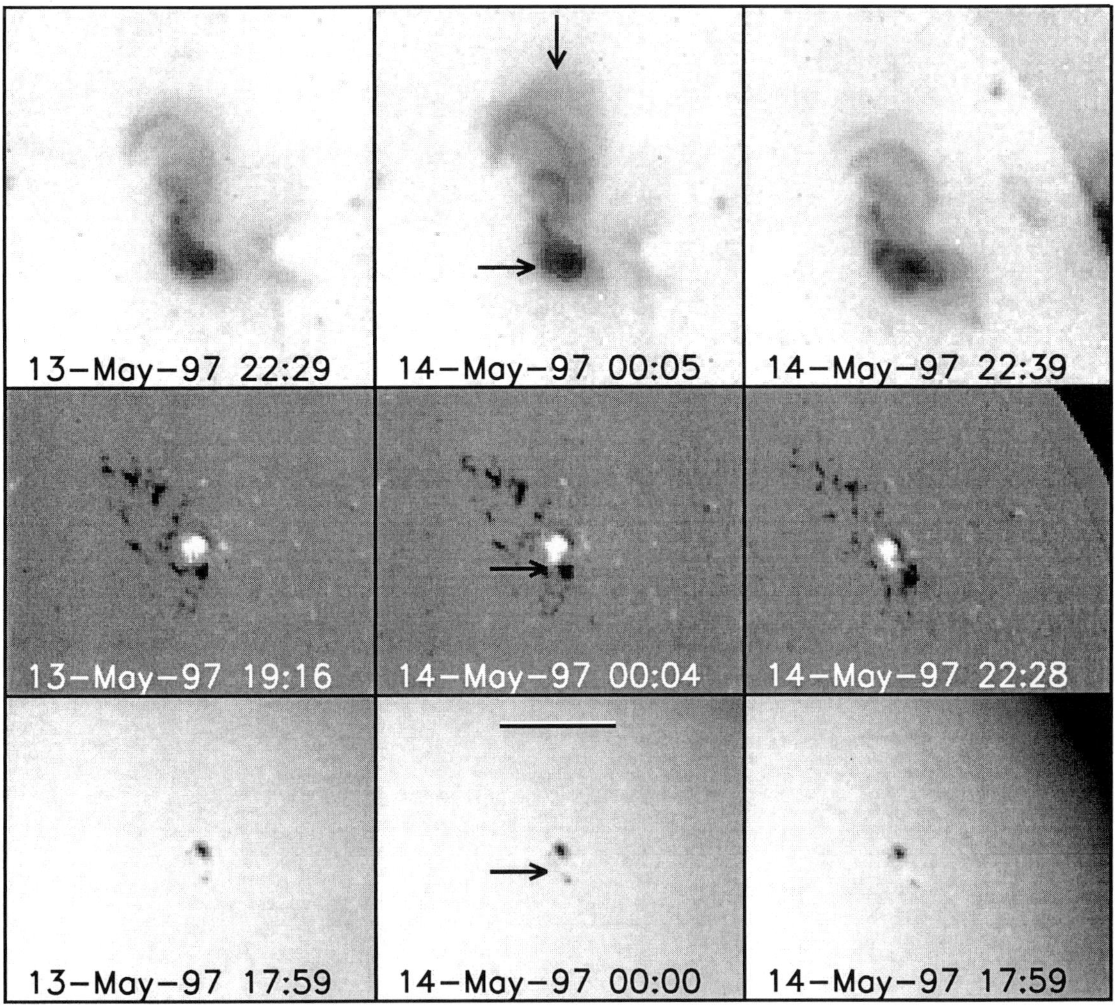

Fig. 2. Contagious heating in far end of rooster tail as new flux emerges at edge of old sunspot at the equatorward end of the neutral line of old bipolar active region NOAA AR 8038. Same format as Figure 1.

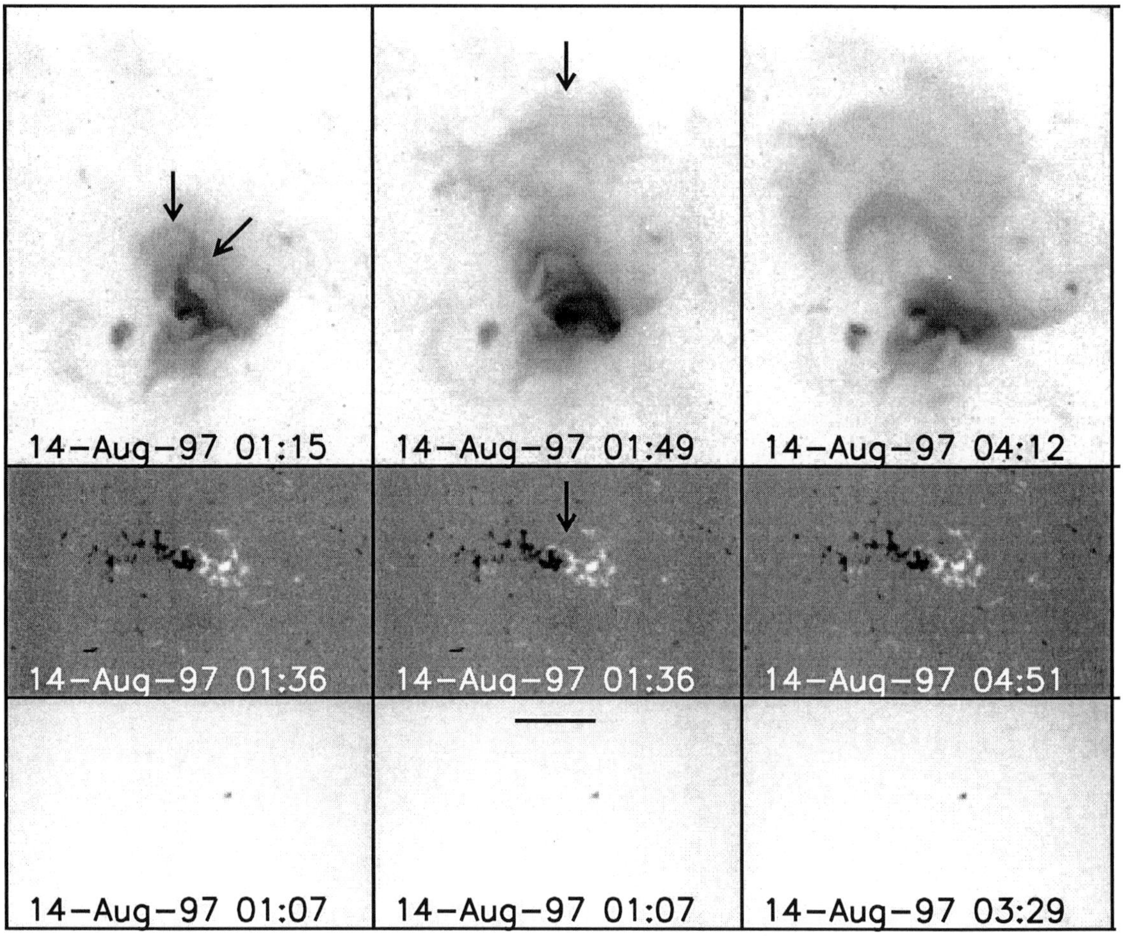

Fig. 3. Large rooster-tail far-end heating event as new field spreads and melds with the old bipole of NOAA AR 8071. Same format as Figure 1. The arrows in the leftmost SXT panel point to examples of extended loops, bright old-field loops that are rooted in contact with the new field. The new field emerged during 11 & 12 August as a small bipole centered on the neutral line of the old bipole and aligned east-west with positive polarity leading. By 14 August (shown here), the new field arched from in and around the sunspot pore to most of the strongest patch of following negative polarity. The strongest heating seen in the SXT panels is in this new field and intermixed old field in the interior of the old bipole.

HEATING RATE OF CORONAL ACTIVE REGIONS

S. Yashiro[1] and K. Shibata[2]

[1] *Center for Solar Physics and Space Weather, The Catholic University of America, Washington DC 20064, USA*
[2] *Kwasan Observatory, Kyoto University, Yamashina, Kyoto, 607-8471, Japan*

ABSTRACT

Scaling laws between thermal and magnetic properties have been reported to investigate the coronal heating mechanism. Yashiro and Shibata (2001) examined the properties of entire active regions observed with *Yohkoh* and *SOHO*, and found the following empirical scaling laws:

$$P \propto B^{0.78} L^{-0.16},$$

where P, B, and L are the gas pressure, magnetic flux density, and region dimension, respectively, of active regions. We have made detailed comparisons with previous studies from the *Skylab* results to the steady loop scaling laws found by Klimchuk & Porter (1995). We conclude that the scaling law of entire active regions is consistent with that for steady coronal loops.

INTRODUCTION

One approach to understanding coronal heating mechanisms is to investigate correlations between the heating rate and other physical parameters. Most coronal heating models require the following relationship for the heating flux F (erg cm^{-2} s^{-1}), magnetic flux density B, and loop length l:

$$F \propto B^{\alpha} l^{\beta}, \tag{1}$$

where α and β are exponents which are different for different heating mechanisms. Rosner *et al.* (1978) found the following scaling laws for the observed gas pressure p, loop length l, and loop temperature T:

$$T \approx 1.4 \times 10^3 (p \cdot l)^{1/3}, \quad E_H \approx 10^5 p^{7/6} l^{-5/6}, \tag{2}$$

where E_H is the volumetric heating rate (erg cm^{-3} s^{-1}). These relations are called RTV (Rosner-Tucker-Vaiana) scaling laws. From Eq. 1 and 2, and noting that $E_H = F\, l^{-1}$, we can obtain the following relation:

$$p \propto B^{6\alpha/7} l^{(6\beta-1)/7}. \tag{3}$$

Therefore establishing the relationship between the magnetic and thermal properties of active regions may play a key role in understanding the coronal heating process.

THERMAL AND MAGNETIC PROPERTIES OBSERVED WITH YOHKOH

In a previous paper (Yashiro & Shibata 2001), we examined the thermal and magnetic properties of entire active regions by analyzing *Yohkoh* (Ogawara *et al.* 1991) and *SOHO* (Domingo *et al.* 1995) data. We carried out the temperature analysis for 64 active regions observed with the Soft X-ray Telescope (Tsuneta *et al.* 1991). Figures 1 and 2 show the scatter plots among the region size, temperature, and pressure. We found that (i) the temperature of the quasi-steady component ranges from 1.7 to 3.8 MK, (ii) large active regions ($\simeq 10^{10}$ cm) have higher temperatures ($\simeq$ 3 MK), and small active regions ($\simeq 4 \times 10^9$ cm) have lower temperatures ($\simeq$ 2 MK), and (iii) the pressure of active regions has a weak dependence on the region size. Using the least-squares method, we obtained the following relationship between the region size L, temperature T, and pressure P:

$$T \propto L^{0.28\pm0.08}, \quad P \propto L^{-0.16\pm0.21}, \tag{4}$$

where the 3 standard deviation error of the power-law index is indicated. Note that these relations are for the *entire active regions.*

We investigated the magnetic field properties of active regions using the *SOHO*/Michelson Doppler Imager (MDI) (Scherrer *et al.* 1995). We analyzed 31 active regions located within 0.7 $R_\odot$ from disk center, and examined the total magnetic flux. Figure 3 demonstrates the relationship between the mean magnetic flux density and the coronal gas pressure. We estimated the mean magnetic flux density B as $B = \Phi L^{-2}$, where L is the region size of the *coronal* active region. Using the least-squares method, we obtained the following scaling law:

$$P \propto B^{0.78\pm0.23}, \tag{5}$$

where the error shows the 3 standard deviation of the power-law index. There is no clear relation between B and L in our data set. We have concluded that

$$P \propto B^{0.78\pm0.23} L^{0.28\pm0.08}. \tag{6}$$

From Eq. 3 and 6, we found $\alpha = 0.91 \pm 0.27$ and $\beta = -0.02 \pm 0.25$ with 3 sigma errors. This is an observational requirement for the coronal heating mechanism.

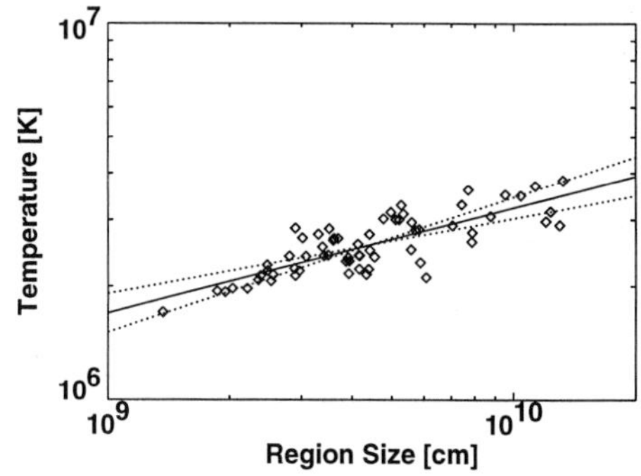

Fig. 1. Relation between the region size L and temperature T. Solid line shows $T \propto L^{0.28}$, and the two dotted lines show the 3 sigma error of the slope (from Yashiro & Shibata 2001).

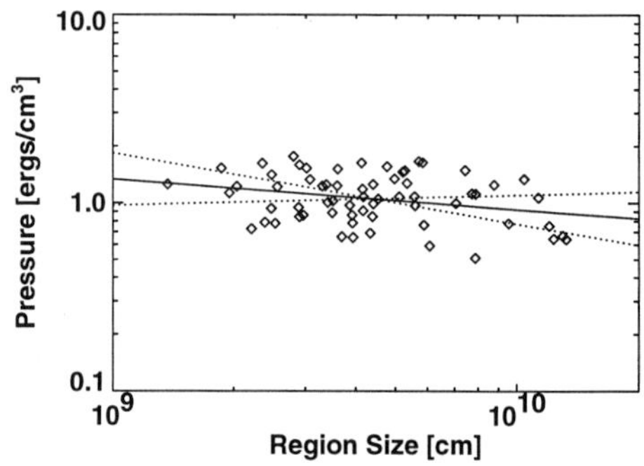

Fig. 2. Relation between the region size L and coronal gas pressure P. Solid line shows $P \propto L^{-0.16}$, and the two dotted lines show the 3 sigma error of the slope (from Yashiro & Shibata 2001).

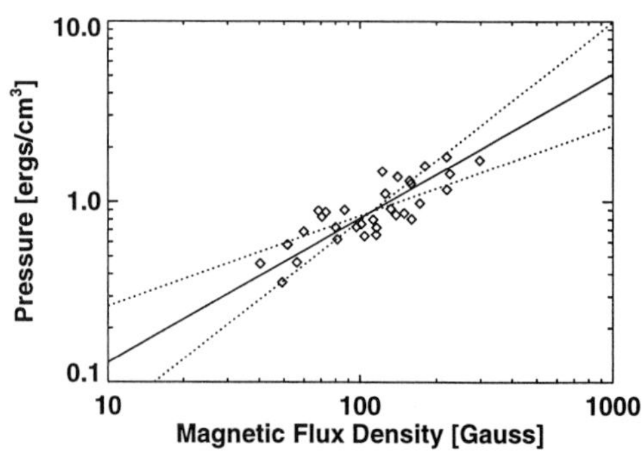

Fig. 3. Relation between the magnetic flux density B and coronal gas pressure P. Solid line shows $P \propto B^{0.78}$, and the two dotted lines show the 3 sigma error of the slope (from Yashiro & Shibata 2001).

COMPARISON WITH SKYLAB

Let us compare the *Yohkoh* scaling law with *Skylab*. Golub *et al.* (1980) found $P \propto B^{1.6}$ for entire active regions. This is significantly different from *Yohkoh* results, $P \propto B^{0.78 \pm 0.23}$. Sturrock & Uchida (1981) pointed out that the Golub *et al.* results are biased by a single data point that comes from another study (marked by LSS in Figure 2 of Golub *et al.*) We re-examine the best-fit relation for the data of Golub *et al.* excluding LSS. Using the least square method and assuming $\sigma_{\log p} = \sigma_{\log B} = 0.1$, we obtained $P \propto B^{1.11 \pm 0.64}$, where the error is the 3 standard deviation of the power-law index again (Figure 4). This estimate is closer to our results.

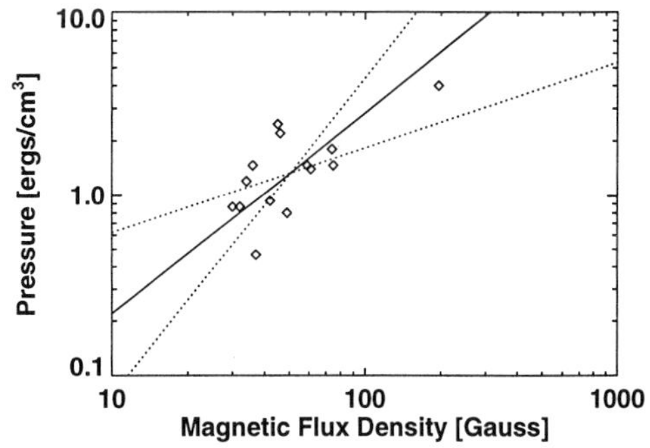

Fig. 4. Re-examination of *Skylab* results. Solid line shows $P \propto B^{1.11}$, and the two dotted lines show the 3 sigma error of the slope.

COMPARISON WITH OTHER SCALING LAWS

Many researchers have investigated observable scaling laws of the form: $P \propto B^a$, $P \propto L^b$, $T \propto L^c$, and $B \propto L^d$. We found $a = 0.78 \pm 0.23$, $b = -0.16 \pm 0.21$, $c = 0.28 \pm 0.08$, and $d \sim 0$ with 3 sigma errors (Yashiro & Shibata 2001). Since an active region is a cluster of several loops, it of interest to compare the scaling law of entire active regions with that of *coronal loops.*

Klimchuk & Porter (1995) studied steady coronal loops using *Yohkoh* data, and found that the temperature is independent of loop size, suggesting $c \sim 0$ (see also Porter & Klimchuk 1995). They also found a power-law index of $-1.82 \leq b \leq -0.22$ (90% confidence range). Their results are consistent with the RTV scaling laws, suggesting that the radiative loss is comparable with conductive loss for SXR loops. They did not examine the magnetic properties of each loop at that time, but Mandrini *et al.* (2000) studied the relation between magnetic field strength and loop length on the same active regions analyzed by Klimchuk & Porter. Using photospheric magnetograms, they computed the magnetic filed line in the corona, and made scatter plots to find the correlation between them. For the intermediate length loop, $50 < L < 300$ Mm, they found $d = -0.97 \pm 0.25$.

Fludra & Ireland (2002) studied the relationships between the photospheric magnetic flux and EUV lines of active regions observed by the *SOHO* Coronal Diagnostic Spectrometer, and found that the intensity of the Fe XVI line is proportional to B^2. They assumed that the active region loops are in static equilibrium, and obtained an Fe XVI intensity proportional to P^2 suggesting $a = 1$. This is similar to our results.

Properties of coronal loops observed with the *SOHO* Extreme ultraviolet Imaging Telescope (EIT) were investigated by Aschwanden *et al.* (1999). They found $b = -0.41 \pm 0.12$ (1 σ standard deviation) and $c \sim 0$. They also examined magnetic field strengths of each EIT loop, and found $d = -1.02 \pm 0.25$ with 1 sigma error. This is the same as the result of Mandrini *et al.* (2000). In contrast with SXR observations, EIT loops are distinctly different from RTV scaling laws. Aschwanden *et al.* argued that radiative loss is dominant compared with conductive loss because EUV loops have lower temperatures than SXR loops.

Table 1 summarizes the power-law index for X-ray active regions, X-ray loops, EUV active regions, and EUV loops. Note that the $B - L$ relation for SXR loops is from Mandrini *et al.* (2000). The observed scaling laws for entire active regions are different from those for coronal loops. However there are many physical parameters that are related to each other, although it is not easy to find which relations are fundamental.

Table 1. Summary of Empirical Scaling Laws

	a	b	c	d	RTV	Heating Rate
X-ray Active Regions [1] (2 $\sim$ 4 MK)	0.78	−0.16	0.28	~ 0	yes	$E_H \propto B^1 L^{-1}$
X-ray Loops [2] (2 $\sim$ 4 MK)		−0.96	~ 0	$(-0.97^{\dagger})$	yes	$E_H \propto L^{-2-d}$
EUV Active Regions [3] (2 $\sim$ 2.5 MK)	~ 1					$E_H \propto B^{1.17}$
EUV Loops [4] (1.2 MK)		−0.41	~ 0	−1.02	no	

Scaling laws are of the form: $P \propto B^a$; $P \propto L^b$; $T \propto L^c$; and $B \propto L^d$.
The 5th column shows whether these parameters consistent with the RTV scaling laws.
References.– (1) Yashiro & Shibata 2001; (2) Klimchuk & Porter 1995, Porter & Klimchuk 1995; (3) Fludra & Ireland 2002; (4) Aschwanden *et al.* 1999. † Mandrini *et al.* 2000.

We found the following scaling law:

$$P \propto B^{0.78} L^{-0.16}. \tag{7}$$

Using Mandrini *et al.*'s result ($B \propto L^{0.97}$), we obtain:

$$P \propto (L^{-0.97})^{0.78} L^{-0.16} \propto L^{-0.92}. \tag{8}$$

This is quite similar to the result obtained by Klimchuk & Porter. We conclude that Eq. 7 is the fundamental scaling law not only for entire active regions but also for coronal loops observed in soft X-rays (except for EUV loops). Note that Eq. 7 suggests $F \propto B^1 L^0$ ($E_H \propto B^1 L^{-1}$). We are able to use the relation to test different models, since the different coronal heating models predict different power-law indices.

ACKNOWLEDGEMENTS

The authors would like to thank all of the *Yohkoh* and *SOHO* teams for providing the good data and software, and also thank N. Gopalswamy and J. A. Klimchuk for fruitful comments.

REFERENCES

Aschwanden, M. J. *et al.*, *Astrophysical Journal*, **515**, 842 (1999).
Domingo, V., Fleck, B., & Poland, A. I., *Solar Physics*, **162**, 1 (1995).
Fludra, A. & Ireland, J., in Proceedings of the 12th Cambridge Workshop on Cool Stars, Stellar Systems and the Sun, (2002).
Golub, L., Maxson, C., Rosner, R., Vaiana, G. S. & Serio, S., *Astrophysical Journal*, **238**, 343 (1980).
Klimchuk, J. A. & Porter, L. J., *Nature*, **377**, 131 (1995).
Mandrini, C. H., Démoulin, P., & Klimchuk, J. A., *Astrophysical Journal*, **530**, 999 (2000).
Ogawara, Y. *et al.*, *Solar Physics*, **136**, 1 (1991).
Porter, L. J. & Klimchuk, J. A., *Astrophysical Journal*, **454**, 499 (1995).
Rosner, R., Tucker, W.H., & Vaiana, G.S., *Astrophysical Journal*, **220**, 643 (1978).
Scherrer, P. H. *et al.*, *Solar Physics*, **162**, 129 (1995).
Sturrock, P. A. & Uchida, Y., *Astrophysical Journal*, **246**, 331 (1981).
Tsuneta, S. *et al.*, *Solar Physics*, **136**, 37 (1991).
Yashiro, S. & Shibata, K., *Astrophysical Journal*, **550**, L113 (2001).

X-RAY BRIGHT POINTS AND OTHER QUIET SUN TRANSIENT PHENOMENA

C. E. Parnell

Department of Mathematics and Statistics, University of St Andrews, St Andrews, Fife, KY16 9SS, Scotland

ABSTRACT

Over the last decade, the unprecedented, uninterrupted, high resolution, coverage of the Sun has led to the discovery of a number of new types of small-scale phenomena, as well as a better understanding of known phenomena such as X-ray bright points and explosive events. This paper reviews our current understanding of X-ray bright points and various phenomena in the corona (X-ray jets and nanoflares), transition region (blinkers and explosive events) and photosphere (ephemeral regions and cancelling magnetic features), that may be related to X-ray bright points. The relations that are known to exist between these phenomena are discussed, as are the potential relationships that warrant further investigation.

INTRODUCTION

The quiet Sun is not quiet. It plays host to many small-scale phenomena. The coronal phenomena must be responsible for maintaining temperatures of more than a million degrees Kelvin in the corona, due to the constraints of thermal conduction across field lines. Examples of quiet Sun coronal phenomena are X-ray bright points (XBPs), X-ray jets and nanoflares/microflares/network flares. In the transition region, events such as blinkers/network brightenings/cell brightenings/EUV brightenings, explosive events and unit brightenings (non-velocity, or weak velocity, intensity enhancements observed by SUMER) are observed. Although a number of these phenomena have been known about for several decades, relatively little is known about the connections between them and their relation to the underlying photospheric magnetic concentrations. This is mainly because the different phenomena are observed by different instruments at different temperatures. Due to the last decade's fleet of solar spacecraft, we now have a wealth of data with the potential to reveal these connections. In this review article, I briefly discuss the observations of phenomena in the photosphere, corona and transition region and also indicate the possible mechanisms that have been suggested to explain them. Then the connections that are known to exist between these phenomena and connections that seem likely, but as yet unexplored, are discussed.

PHOTOSPHERIC PHENOMENA

The quiet Sun's magnetic field is made up of many small-scale concentrations of flux that are continuously moving around, breaking up into smaller concentrations, interacting with each other, merging into larger concentrations, if of the same polarity, or cancelling and wiping each other out, when opposite polarities meet. This seething mass of magnetic blobs, known as the magnetic carpet, is continuously being fed through the emergence of new pairs or clumps of opposite polarity flux. Indeed, so much flux is cancelling and emerging that it is estimated that the total flux in the quiet Sun is replaced every 14 hours (Hagenaar 2001). This turnover of flux and, hence, the massive topological changes in magnetic connectivity that are the natural consequence of such behaviour, are clearly going to give rise to wide spread energy release in many parts of the quiet Sun's atmosphere. Understanding the connections between the behaviour of the magnetic activity in the photosphere and the upper atmosphere is essential to determine how the different phenomena in the atmosphere are created. The two most easily identifiable types of behaviour in the magnetic carpet are emergence and cancellation, which are discussed in more detail below.

Ephemeral Regions

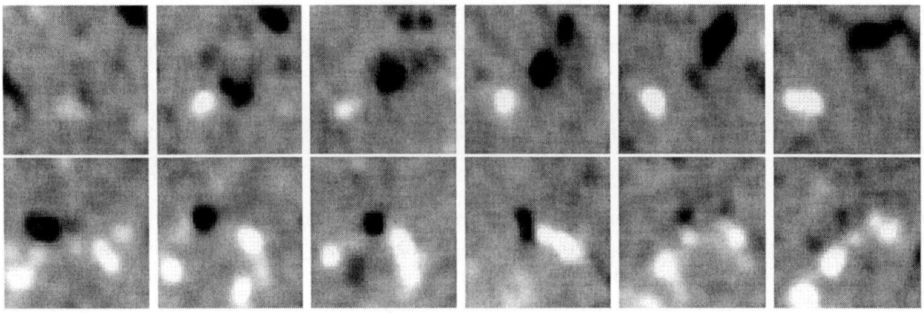

Fig. 1. Sections cut from MDI high resolution images showing (top) the emergence of an ephemeral region and (bottom) a cancelling magnetic feature.

The pairs or, more often than not, clumps of opposite polarity magnetic concentrations that emerge in the magnetic carpet are called ephemeral regions (ERs) (Figure 1). They were first described by Harvey and Martin (1973) and over the years, with increasing resolution, their characteristics have been refined. A recent study of ERs using MDI magnetograms has been made by Hagenaar (2001). The author finds a total absolute flux in a single ER of approximately 1.1×10^{19} Mx with equal amounts of positive and negative flux. The newly emerged concentrations appear at the edges of supergranular cells, then grow in flux as they move apart. Their rate of divergence is estimated at 2.3 km s^{-1} and their rate of flux growth is approximately 1.6×10^{15} Mx s^{-1}. Due to the ceaseless activity of the magnetic carpet, ERs are hard to track for any more than 3 hours. It is estimated that at solar minimum, when there are no active regions, some 5×10^{23} Mx of flux is injected through ERs per day (Hagenaar 2001).

Cancelling Magnetic Features

New flux is injected into the photosphere from the convection zone below. It can also leave the photosphere through events called cancelling magnetic features(CMF) (Martin 1984). In this process, almost the inverse of emergence, initially unconnected, opposite polarity, magnetic concentrations converge and mutually lose flux (Figure 1). Although these features were studied during the 1980's, good estimates of the numbers of events and the typical flux per cancellation are not known. Clearly, since the total absolute flux on the Sun remains approximately constant the rate of cancellation must be equal to the rate of emergence. However, it has been estimated that there are around 3 to 5 times as many cancellation events as emergence events, which clearly means that the typical flux lost in each event must be about $3 - 5 \times 10^{18}$ Mx (Parnell 2001).

When cancellation takes place, it could be the result of a small loop lifting off into the atmosphere or due to subduction of flux back down into the convection zone. Since the concentrations are observed to be unconnected at the start of each cancellation, the former seems the most likely. However, a study by Harvey *et al.* (2000) looking at magnetograms taken at varying depths, showed that the concentrations disappeared from the highest magnetograms first suggesting that the flux was actually submerging. If, as is most likely since equal amounts of flux of both polarities is being lost, the flux is being lost through the subduction of a loop, reconnection must take place before every cancellation and therefore CMFs also inject energy into the quiet Sun's atmosphere. Indeed, they are observed to be the source of some of the largest events in the quiet corona.

X-RAY BRIGHT POINTS

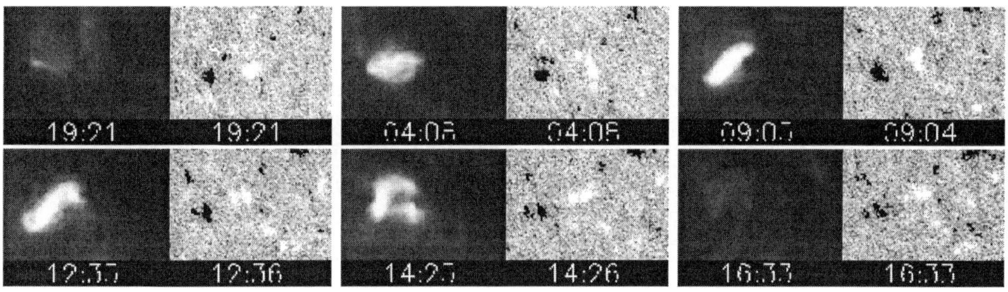

Fig. 2. Pairs of images cut from *TRACE* Fe XII and MDI high resolution images showing an XBP taken on the 13th June 1998 and the magnetic carpet below.

XBPs are small, compact bright features observed in X-ray and EUV images. They were discovered using rocket based X-ray telescopes in the late 1960's and their general characteristics were determined from Skylab data (Golub *et al.* 1974, 1976a, 1976b, 1977). XBPs have a typical size of 5.4×10^7 km^2, an average lifetime of 8 hours and release a total of between 10^{26} and 10^{28} ergs during their life. From the *Skylab* data XBPs were found to be out-of-phase with the solar cycle (Golub *et al.* 1979). The first observations of the magnetic field below XBPs revealed bipoles (Figure 2), which were assumed to be ERs. More recent results, however, show that in actual fact 66% of XBPs lie above cancelling magnetic features and about 33% lie above ERs (Harvey 1984, 1985, 1996) Moreover, detailed studies using *Yohkoh* that take into account the scatter due to high intensity events such as active regions, suggest that XBPs are actually uncorrelated with the solar cycle (Nakakubo & Harra 2000).

Due to the evidence of cancellation and emergence of magnetic flux below XBPs, the most likely mechanism for powering these events is driven reconnection of opposite polarity magnetic concentrations. The first model for XBPs and CMFs was the Converging Flux Model by Priest *et al.* (1994). Initially unconnected opposite polarity concentrations converge until they begin to interact magnetically, forming a neutral point in the photosphere. As the concentrations continue to converge, reconnection takes place at the neutral point, releasing energy and raising the neutral point up into the corona. The energy released injects hot dense plasma along field lines, giving rise to the XBP which will last until almost all the flux between the two concentrations has been reconnected and the neutral point drops back to the photosphere. Then reconnection takes place in the photosphere producing the CMF. This basic idea has been developed further and extended into three-dimensions by a number of authors (Parnell *et al.* 1994b, 1995; Longcope 1996, 1998; Longcope & Kankelborg 1999, 2001). This idea has also been applied to observed XBPs (Parnell *et al.* 1994a, van Driel Gesztelyi *et al.* 1996).

Reconnection resulting in coronal energy release can occur not just from the emerging of new flux or the cancellation of existing flux, but also from the simple movement of one concentration past another. Numerical simulations of energy release in these types of events are shown in Figure 3 (Galsgaard *et al.* 2000).

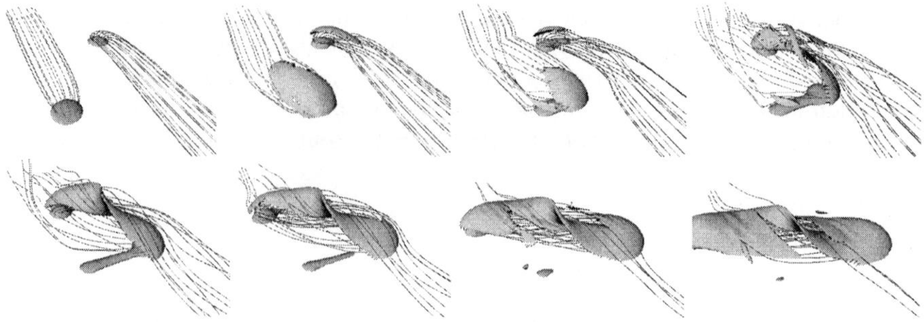

Fig. 3. Frames from an MHD numerical simulation showing the interaction of two opposite polarity magnetic fragments in an over-lying field as they are driven past one another (Galsgaard *et al.* 2000). Iso-surfaces of flux (circular blobs on the base) and of current are shown. The lines represent field lines from the fragments.

CORONAL PHENOMENA

X-ray Jets

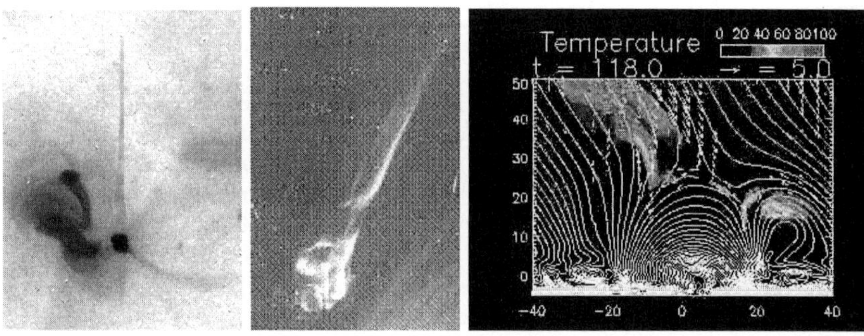

Fig. 4. (left) Bi-directed X-ray jet seen by SXT on 7th July 1998. (middle) X-ray jet seen in the *TRACE* Fe XII line on 3 November 2000. (right) MHD numerical simulation of magnetic reconnection model for X-ray jets (from Yokoyama & Shibata 1995, 1996).

One of the major discoveries made by *Yohkoh* was X-ray jets (Shibata *et al.* 1992, Strong *et al.* 1992). These small-scale features appear above XBPs and small active regions (Figure 4) and are often associated with microflaring at their footpoints. Over a hundred jets were studied by Shimojo *et al.* (1996) who found that they typically have lengths of $10^4 - 4 \times 10^5$ km, widths of $5 \times 10^3 - 10^5$ km and velocities of 10-1000 km s^{-1}. They have lifetimes of up to 10 hours. The temperature of the jets is about $3 - 6 \times 10^6$ K and they are estimated to have kinetic energies of $10^{25} - 10^{28}$ ergs. Like XBPs, X-ray jets are seen not only in X-ray images, but also EUV images, e.g., such as the jet observed in the *TRACE* Fe XII image in Figure 4.

Since X-ray jets are associated with XBPs or small active regions and therefore, with the cancellation or emergence of flux, it is natural to assume that the mechanism that explains how they are created must be similar to that for X-ray bright points. At the reconnection site of an XBP hot dense plasma can be injected along newly reconnected field lines. If this injection is sufficiently impulsive and if enough plasma is injected in one go a jet can occur. The emergence of new flux into a slanted over-lying magnetic field was suggested by Heyvaerts and Priest (1984) as a mechanism for solar flares, although it works equally well as a mechanism for X-ray jets. Simulations of emerging flux interacting with a slanted over-lying field have been performed by Yokoyama and Shibata (1995, 1996) and compare well with observations for X-ray jets. It is not essential that the over-lying field be slanted such that one end is open. X-ray jets can have both single or bi-directed jets (Figure 4). Bi-directed X-ray jets typically occur above XBPs where the magnetic field is likely to be more closed than above an active region. Simulations of these types of jets have been made by Birk *et al.* (1996).

Nanoflares

The corona is also home to much smaller events than either XBPs and X-ray jets. These events have areas of just 10^7 km^2 and release $100 - 10^4$ times less energy than XBPs. They occur throughout the quiet Sun and are called nanoflares or microflares. Their typical lifetime is just 10 minutes. The existence of these small, short-lived events was postulated by Levine (1974). He suggested that the quiet corona may be populated by millions of little events whose total combined energy was enough to maintain the corona at a million degrees Kelvin. The first evidence for these events came from analysis of EIT images by Krucker & Benz (1998), but they have also been found in *TRACE* (Parnell & Jupp 2000, Aschwanden *et al.* 2000).

The distribution of nanoflare/microflare energies is observed to follow the form of a power-law over several orders of magnitude. If the power-law index is less than -2, then small-scale events determines the total energy, however, if the index is greater than -2, the energy is determined by the large-scale flares. Thus, there are two key questions: (i) what is the power-law index of the distribution of flare energies; and (ii) are there sufficient nanoflares/microflares to explain the energy losses from the corona? Answering these questions has been a hot topic over the last few years and a range of indicies between -2.6 and -1.3 have been derived. Why is there such a range of indicies? There are many factors that effect the derived index:

- Detection algorithm - e.g. use of macro pixels, use of selection effects, detecting jump in intensity or emission measure, size of minimum jump, synchrony - size of time window used to determine events.
- Instrument passband (restricted temperature coverage) and sensitivity
- Determination of line-of-sight depth - e.g. constant or dependent on area ($A^{1/2}$, $A^{1/3}$ or fractal).
- Energy estimate - total energy is thermal+kinetic+accelerated particles. All these components can not be calculated from the observations, so observed energies are likely to be too low. Methods of estimating energy are either emission measure increase or radiative loss.
- Line fitting algorithm - Fitting a line to a histogram of energies is the usual approach, although this is not robust and depends on the bin size of the data and also whether you weight the bins or not when fitting a line. A more robust method preferred by statisticians is to use maximum likelihood. Here, the index of the power-law, δ, equals

  $$\delta = 1/(mean(log(E/E_0))) + 1 ,$$

  where E is event energy and E_0 is the smallest event energy.
- Finally, since XBPs can be affected by scatter, it is highly likely that nanoflares will also be obscured if there is a large amount of activity on the disk. Therefore, the time of the solar cycle could also affect the derived indicies.

As a result of all of these problems the errors on any derived index are large, at least ± 0.5, if not greater.

There are many possible mechanisms for nanoflares and microflares. Parker (1983b, 1988) suggested that they are the results of turbulent reconnection. Flux braiding has also been suggested (Parker 1983a, Berger 1984, Galsgaard & Nordlund 1996). In this scenario, magnetic footpoints are continuously driven past each other, causing the field from these footpoints to tangle or braid. Current sheets can form where field lines have become highly sheared and reconnection releases energy in the form of a nanoflare. Even if the driving is continuous and rhythmic the resulting energy release can be sporadic in time and space, leading to an approximate uniform heating in space. A similar explanation has been proposed recently by Priest, Heyvaerts & Title (2002) and is called 'coronal tectonics'. Here, the main difference is the realisation that one so-called footpoint in the photosphere will have many loops extending out of it that will connect to a set of different footpoints. This results in an atmosphere being full of myriads of current sheets due to the multiple loop structure of the magnetic field.

TRANSITION REGION PHENOMENA

Blinkers

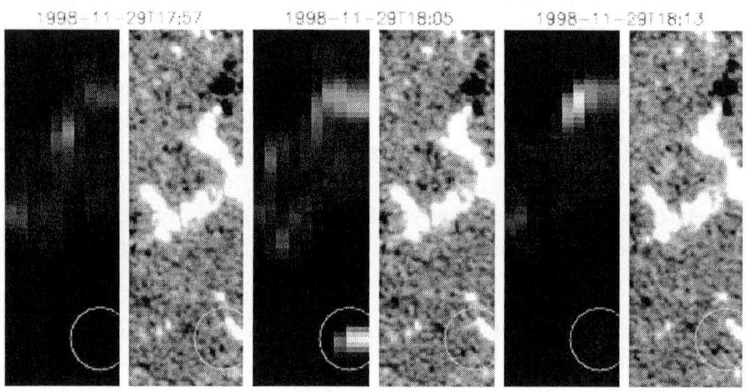

Fig. 5. Pairs of CDS rasters and MDI high resolution images taken on the 19th November 1998 showing a blinker (circled) and the underlying magnetic field.

In transition region lines, small bright intensity enhancements known as blinkers are observed (Figure 5). They were discovered by Harrison (1997) using CDS. They can be observed in a range of EUV lines with temperatures between 2×10^4 and 2.5×10^5 K, but are seen best in O V, O IV and He II, since these lines are strong. By considering ratios of oxygen lines, it has been found that blinkers are enhancements in density or filling factor, not temperature. They have a mean lifetime of about 16 minutes, a mean area of about 3×10^7 km^2 and a global frequency of between 5-20 s^{-1} (Bewsher *et al.* 2002). Surprisingly, blinkers are observed not only in the quiet Sun, but also in active regions and above sunspots (Parnell *et al.* 2002).

There have been varying reports of magnetic field below blinkers. Harrison *et al.* (1999) suggested they occurred above bipolar regions. Bewsher *et al.* (2002), however, suggest that they are closely correlated with regions of strong magnetic field, with 52% of blinkers observed above a single polarity region, 36% above a region with a dominant polarity and just 12% above regions of mixed polarity. Blinkers are known to occur mostly above network regions, but can also occur in cell centres. It is therefore likely that blinkers, network brightenings, cell brightenings, and EUV brightenings are all the same phenomena.

One strange observation is the absence of any real velocity signature in blinkers. This is either because the plasma is moving at less than 5 km s^{-1}, which is too slow for CDS to observe with confidence, or that they have velocities of more than 100 km s^{-1}. Again, this is rather unlikely since we would then expect to see the enhancements actually moving over their 15 minute lifetimes.

Harrison *et al.* (1999) suggested that blinkers were the result of driven reconnection in the same way as bright points and nanoflares. However, since they are not temperature events, but enhancements in either density or filling factor, this is unlikely. Possible alternative mechanisms have been suggested by Priest *et al.* (2002). These are the compression of

- spicule material on re-entry - unlikely, since there are significantly more spicules than blinkers.
- cool low-lying loops - short loops could remain filled with plasma at temperatures of approximately 10^5 K for several minutes.
- the coronal base of hot loops.
- material that has been heated and evaporated or material that has cooled and is draining.

Explosive Events

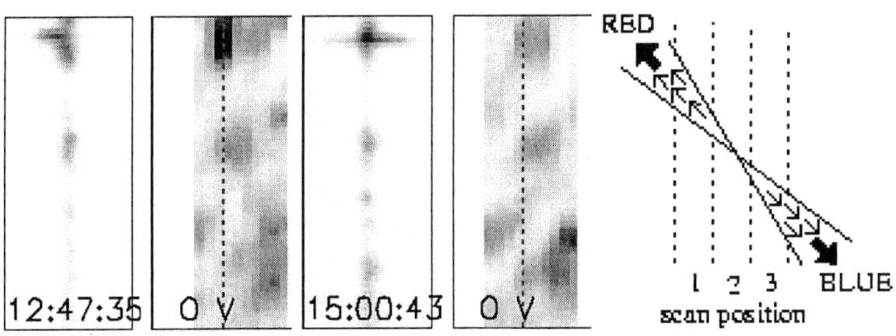

Fig. 6. SUMER Si IV slit and CDS OV rasters taken on 19th June 1998. The left pair show an explosive event simultaneous and co-spatial with a blinker, whereas the middle two show an explosive event without a blinker in CDS (courtesy of Davina Innes). The right-hand sketch shows the jets from a reconnection event. In the first explosive event the SUMER slit would be sited at scan position 1 and in the second at scan position 2.

Explosive events are the other main quiet transition-region phenomena. They were first observed from rocket flights (Bruecker & Bartoe 1983, Dere *et al.* 1989). They are UV velocity events that are detected above the network at transition-region temperatures and are seen well with the SUMER instrument (Innes *et al.* 1997, Innes 2001). They are very small, short-lived events, having areas of just 2×10^6 km^2 and lifetimes of only 60 s. They typically have velocities of 150 km s^{-1}. These velocities can have both blue and red shifts or either a blue or red shifts (Figure 6). They are very numerous with a global rate of 600 s^{-1} and they do not appear to be well correlated with coronal emission, just like blinkers.

Clearly, it is interesting to know whether the explosive events observed by SUMER are the same as the blinkers observed in CDS. Unfortunately, there are few good CDS blinker data sets and SUMER explosive events data sets that overlap. One such data set was taken on the 19th June 1998. In Figure 6 two sets of SUMER Si IV image/CDS OV rasters are shown. The dashed lines on the CDS rasters indicate the position of the SUMER slit. In the first pair, a red shifted explosive event can be seen in the SUMER image, corresponding with a blinker in the CDS. In the second pair, a bi-directed explosive event is seen but this time there is no blinker observed in the CDS raster. It is, therefore, not clear what the relationship is between explosive events and blinkers.

Reconnection again features as the most likely mechanism to explain explosive events (Dere *et al.* 1991, Innes *et al.* 1999, Roussev *et al.* 2001), since it can result in the rapid outflow of oppositely directed plasma jets. The rate of outflow of these plasma jets will vary greatly, depending on stratification, magnetic topology and projection affects. It is thought that these bi-directed jets are responsible for the velocity events seen by SUMER (Figure 6).

DISCUSSION

All the different types of phenomena described above occur in the quiet Sun and it is likely that a number of them are related. We already know that XBPs occur above ERs and CMFs. Indeed, it is the global restructuring of the magnetic field due to footpoint motions that gives rise to the bright point. We also know that X-ray jets are connected with XBPs, as again they are driven by reconnection. What about the other phenomena? Recently, XBPs have been found to be made up of many little nanoflares (Parnell 2002). The total numbers of these nanoflares per bright point are unknown, but will of course be dependent on the size and lifetime of the XBP. Furthermore, it is not clear whether nanoflares can explain the total energy of XBPs. As for the transition region phenomena, neither blinkers nor explosive events have a strong correlation with coronal emission. They are observed to occur with or without a coronal signature. As yet no one has looked to see if all XBPs can occur without blinkers and explosive events. XBPs and nanoflares are all thought to be powered by reconnection and, therefore, one might suppose that explosive events, which are also thought to be associated with reconnection, should occur below XBPs.

Over the last decade, our understanding of the individual phenomena in the quiet Sun has increased considerably. We know more about each phenomenon and have also discovered additional phenomena. What is lacking though is an understanding of how all of these phenomena are connected to one another. *Yohkoh*, *SOHO* and *TRACE* have provided us with a wealth of data, which should continue to be analysed to determine these correlations. It is important that this existing data is analysed such that when the SolarB and SDO missions are launched, we can take the best advantage of their increased spatial and temporal resolution to focus our efforts at answering more specific questions.

ACKNOWLEDGEMENTS

The author would like to thank the Royal Astronomical Society for the Sir Norman Lockyer Fellowship which she has held for the past three years. She would also like to thank the conference organizers for the invitation to give this review and for their partial financial support.

REFERENCES

Aschwanden, M. J., R. W. Nightingale, T. D. Tarbell and C. J. Wolfson, Time Variability of the "Quiet" Sun Observed with TRACE. I. Instrumental Effects, Event Detection, and Discrimination of Extreme-Ultraviolet Microflares, *Astrophys. J.*, **535**, 1047 (2000)

Berger, M.A., Rigorous New Limits on Magnetic Helicity Dissipation in the Solar Corona, *Geophys. Astrophys. Fluid Dynamics*, **30**, 79 (1984)

Bewsher, D., C.E. Parnell and R.A. Harrison, Transition Region Blinkers I: Quiet-Sun Properties, *Solar Phys.*, in press (2002)

Birk, G.T., J. Dreher and T. Neukirch, Three-Dimensional Numerical Studies on Coronal Heating of X-ray Bright Points, *Magnetic Reconnection in the Solar Atmosphere, Proc. of a Yohkoh Conference, Bath, England*, eds R.D. Bentley and J.T. Mariska, **111**, 89 (1996)

Bruecker, G. E. and J.-D. F. Bartoe, Observations of High-Energy Jets in the Corona above the Quiet Sun, the Heating of the Corona, and the Acceleration of the Solar Wind, *Astrophys. J.*, **272**, 329 (1983)

Dere, K.P., J.-D. F. Bartoe and G. E. Bruecker, Explosive Events in the Solar Transition Zone, *Solar Phys.*, **123**, 41 (1989)

Dere, K.P., J.-D. F. Bartoe, G. E. Bruecker, J. Ewing and P. Lund, Explosive Events and Magnetic Reconnection in the Solar Atmosphere, *J. Geophys. Res.*, **96**, 9399 (1991)

Galsgaard, K and A. Nordlund, Heating and Activity of the Solar Corona 1. Boundary Shearing of an Initially Homogeneous Magnetic Field, *J. Geophys. Res.*, **315**, 312, (1996)

Galsgaard, K., C.E. Parnell and J. Blaizot, Elementary Heating Events - Magnetic Interactions between Two Flux Sources, *A&A*, **362**, 395 (2000)

Golub, L., J. Davis and A.S. Krieger, Anti-Correlation of X-ray Bright Points with Sunspot Number, *Astrophys. J.*, **229**, L145 (1979)

Golub, L., A.S. Krieger, J.W. Harvey and G.S. Vaiana, Magnetic Properties of X-ray Bright Points, *Solar Phys.*, **53**, 111 (1977)

Golub, L., A.S. Krieger, J.K. Silk, A.F. Timothy and G.S. Vaiana, Solar X-ray Bright Points, *Astrophys. J.*, **189**, L93 (1974)

Golub, L., A.S. Krieger and G.S. Vaiana, Distribution of Lifetimes for Coronal Soft X-ray Bright Points, *Solar Phys.*, **49**, 79 (1976a)

Golub, L., A.S. Krieger and G.S. Vaiana, Observations of Spatial and Temporal Variation in X-ray Bright Point Emergence Patterns, *Solar Phys.*, **50**, 311 (1976b)

Hagenaar, H.J., Ephemeral Regions on a Sequence of Full-Disk Michelson Doppler Imager Magnetograms, *Astrophys. J.*, **555**, 448 (2001)

Harrison, R.A., EUV Blinkers: The Significance of Variations in the Extreme Ultraviolet Quiet Sun, *Solar Phys.*, **175**, 467 (1997)

Harrison, R.A., J. Lang, D. H. Brooks and D. E. Innes, A Study of Extreme Ultraviolet Blinker Activity, *Astron. Astrophys.*, **351**, 1115 (1999)

Harvey, K.L., Solar Cycle Variation of Ephemeral Active Regions, Proc. 4th European Meeting on Solar Physics. ESA SP, **220**, 335 (1984)

Harvey, K.L., The Relation between Coronal Bright Points as seen in He 10830 and the Evolution of Photospheric Magnetic Network Fields, *Aust. J. Phys.*, **38**, 875 (1985)

Harvey, K.L., Observations of X-ray bright points, *Magnetic Reconnection in the Solar Atmosphere, Proc. of a Yohkoh Conference, Bath, England*, eds R.D. Bentley and J.T. Mariska, **111**, 9 (1996)

Harvey K.L., H.P. Jones, C.J. Schrijver and M.J. Penn, Does Magnetic Flux Submerge at Flux Cancelation Sites?, *Solar Phys.*, **190**, 35 (2000)

Harvey K.L. and S.F. Martin, Ephemeral Active Regions, *Solar Phys.* **32**, 389 (1973)

Heyvaerts, J. and E.R. Priest, Coronal Heating by Reconnection in DC Current Systems - A Theory Based on Taylor's Hypothesis, *A&A*, **137**, 63 (1984)

Innes, D. E., Coordinated Observations of the Quiet Sun Transition Region using SUMER Spectra, TRACE Images and MDI Magnetograms, *A&A*, **378**, 1067 (2001)

Innes, D. E., P. Brekke, D. Germerott and K. Wilhelm, Bursts of Explosive Events in the Solar Network, *Solar Phys.* **175**, 341 (1997)

Innes, D. E. and G. Toth, Simulations of Small-Scale Explosive Events on the Sun, *Solar Phys.* **185**, 467 (1999)

Krucker, S. and A. O. Benz, Energy Distribution of Heating Processes in the Quiet Solar Corona, *Astrophys. J.*, **501**, 213 (1998)

Levine, R.H., A New Theory of Coronal Heating, *Astrophys. J.*, **190**, 447 (1974)

Longcope, D. W., Topology and Current Ribbons: A Model for Current, Reconnection and Flaring in a Complex, Evolving Corona, *Solar Phys.*, **169**, 91, (1996)

Longcope, D. W., A Model for Current Sheets and Reconnection in X-ray Bright Points, *Astrophys. J.*, **507**, 417 (1998)

Longcope, D. W. and C. C. Kankelborg, Coronal Heating by Collision and Cancellation of Magnetic Elements, *Astrophys. J.*, **524**, 483 (1999)

Longcope, D. W. and C. C. Kankelborg, J. L. Nelson and A. A. Pevtsov, Evidence of Separator Reconnection in a Survey of X-ray Bright Points, *Astrophys. J.*, **553**, 429 (2001)

Martin, S.F., Dynamic Signatures of Quiet Sun Magnetic Fields, *Proc. Symp. on Small-scale Dynamical Processes in Quiet Stellar Atmospheres* (ed. S.L. Keil), 30 (1984)

Nakakubo, K. and H. Hara, Variation of X-ray Bright Point Number over the Solar Activity Cycle, *Advances in Space Research*, **25**, 9, 1905 (2000)

Parker, E.N., Magnetic Neutral Sheets in Evolving Fields I - General Theory *Astrophys. J.*, **264**, 635 (1983a)

Parker, E.N., Magnetic Neutral Sheets in Evolving Fields II - Formation of the Solar Corona *Astrophys. J.*, **264**, 642 (1983b)

Parker, E.N., Nanoflares and the Solar X-ray Corona *Astrophys. J.*, **330**, 474 (1988)

Parnell, C.E., A Model of the Solar Magnetic Carpet, *Solar Phys.*, **200**, 23 (2001)

Parnell, C.E., On the Relation between X-ray Bright Points and Nanoflares, *Astron. Astrophys.*, in preparation (2002)

Parnell, C.E., D. Bewsher and R.A. Harrison, Transition Region Blinkers II: Active-Region Properties, *Solar Phys.*, in press (2002)

Parnell, C.E., E.R. Priest and L. Golub, The Three-Dimensional Structures of X-ray Bright Points, *Solar Phys.*, **151**, 57 (1994a)

Parnell, C.E., E.R. Priest and V.S. Titov, A Model for X-ray Bright Points due to Unequal Cancelling Magnetic Sources, *Solar Phys.*, **153**, 217 (1994b)

Parnell, C.E. and E.R. Priest, A Converging Flux Model for the Formation of an X-ray Bright Point above a Supergranule Cell, *Geophys. Astrophys. Fluid Dynamics*, **80**, 255 (1995)

Parnell, C.E. and P.E. Jupp, Statistical Analysis of the Energy Distribution of Nanoflares in the Quiet Sun, *Astrophys. J.*, **529**, 554 (2000)

Priest, E.R., C.E. Parnell and S.F. Martin, A Converging Flux Model for an X-ray Bright Point and an Associated Cancelling Magnetic Feature, *Astrophys. J.*, **427**, 459 (1994)

Priest, E.R., A. Hood and D. Bewsher, The Nature of Blinkers and the Solar Transition Region, *Solar Phys.*, **205**, 249 (2002)

Priest, E.R., J. Heyvaerts and A. Title, A Flux Tube Tectonics Model for Solar Coronal Heating Driven by the Magnetic Carpet, *Astrophys. J.*, in press (2002)

Roussev, I., K. Galsgaard, R. Erdelyi and J. G. Doyle, Modelling of Explosive Events in the Solar Transition Region in a 2D Environment. I. General Reconnection Jet Dynamics, *Astron. Astrophys.*, **370**, 298 (2001)

Shibata, K *et al.*, Observations of X-ray Jets with the Yohkoh Soft X-ray Telescope, *PASJ*, **44**, L173 (1992)

Shimojo, M., S. Hashimoto, K. Shibata, T. Hirayama, H.S. Hudson and L.W. Acton, Statistical Study of Solar X-ray Jets Observed with the Yohkoh Soft X-ray Telescope, *PASJ*, **48**, 123 (1996)

Strong, K., K. Harvey, T. Hirayama, N. Nitta, T. Shimizu and S. Tsuneta, Observations of the Variability of Coronal Bright Points by the Soft X-ray Telescope on Yohkoh, *PASJ*, **44**, L161 (1992)

van Driel-Gesztelyi, L., B. Schmieder, G. Cauzzi, N. Mein, A. Hofmann, N. Nitta, H. Kurokawa, P. Mein and J. Staiger, X-ray Bright Point Flares Due to Magnetic Reconnection, *Solar Phys.*, **163**, 145 (1996)

Yokoyama, T and K. Shibata, Magnetic Reconnection as the Origin of X-ray Jets and H-Alpha Surges on the Sun, *Nature*, **375**, 42 (1995)

Yokoyama, T and K. Shibata, Numerical Simulation of Solar Coronal X-ray Jets Based on the Magnetic Reconnection Model, *PASJ*, **48**, 353 (1996)

RECONCILIATION OF THE CORONAL HEATING FUNCTION BETWEEN YOHKOH AND TRACE

M. J. Aschwanden

Lockheed-Martin Advanced Technology Center, Solar and Astrophysics Lab., Bldg. 252 Org. L9-41, 3251 Hanover Street, Palo Alto, CA 94304

ABSTRACT

We model the geometry and hydrostatic steady-state solution of a trans-equatorial loop system observed with *Yohkoh*/SXT at disk center. From this we determine the heating scale height s_H and find a value of $s_H = 8.4 \pm 2.5$ Mm. This is comparable with the value ($s_H = 13 \pm 1$ Mm) found in another *Yohkoh*-observed loop system above the limb, and with the values ($s_H = 12 \pm 5$ Mm) found from 40 other loops observed with *TRACE*. These results demonstrate that the heating scale heights s_H determined from *Yohkoh* and *TRACE* can be reconciled with forward-fitting of hydrostatic solutions to multi-wavelength fluxes, as opposed to utilizing inversions of filter-ratio temperatures which fail in multi-temperature plasmas.

INTRODUCTION

Progress in the *coronal heating problem* can only be achieved by quantitative measurements from observations, i.e. from volumetric heating rates, spatial heating scales (e.g. characterized with an exponential *heating scale height* s_H), and its temporal variations. Here we concentrate on the heating scale height s_H. Mixed results are quoted in the literature. There seems to be a discrepancy between *Yohkoh* and *TRACE* results. Essentially, heating functions $E_H(s)$ derived from the temperature-broadband instrument SXT/*Yohkoh* yield best fits for uniform or looptop heating (Priest *et al.* 1998, 2000; Wheatland, Sturrock, & Acton 1997), while the same heating function derived from temperature-narrowband instruments like *TRACE* (or *SoHO*/EIT) yield best fits for footpoint heating with scale heights of $s_H \lesssim 15$ Mm. The problem seems to be rooted in an inadequate treatment of the multi-temperature distribution that is encountered along every line-of-sight, which cannot be characterized by a filter-ratio (FR) temperature T_{FR}, implicitly making the assumption of a single temperature (for a given line-of-sight)! The filter-ratio temperature is particularly problematic for instruments with a broadband response such as *Yohkoh*/SXT, which is sensitive to emission measure-weighted temperatures of the entire $dEM(T)/dT$ distribution $T \gtrsim 1.5$ MK.

We present the results of hydrostatic modeling and determinations of the loop heating scale height s_H for two different loops observed with *Yohkoh*. One has been observed at the limb and was analyzed previously (Priest *et al.* 1998, 2000) and re-analyzed recently (Aschwanden 2001). Another one is observed at disk center and is analyzed here for the first time. The results demonstrate that both the *Yohkoh*/SXT and *TRACE* data yield comparable values for the heating scale height, in the order of $s_H \approx 5 - 20$ Mm. We shall attempt to shed light on the effect of filter-ratio assumptions in this study. We will see that both cool loops ($T \approx 1$ MK) seen in EUV as well as hot loops ($T \approx 2 - 3$ MK) seen in soft X-rays are subject to a similar heating function, concentrated near the footpoints.

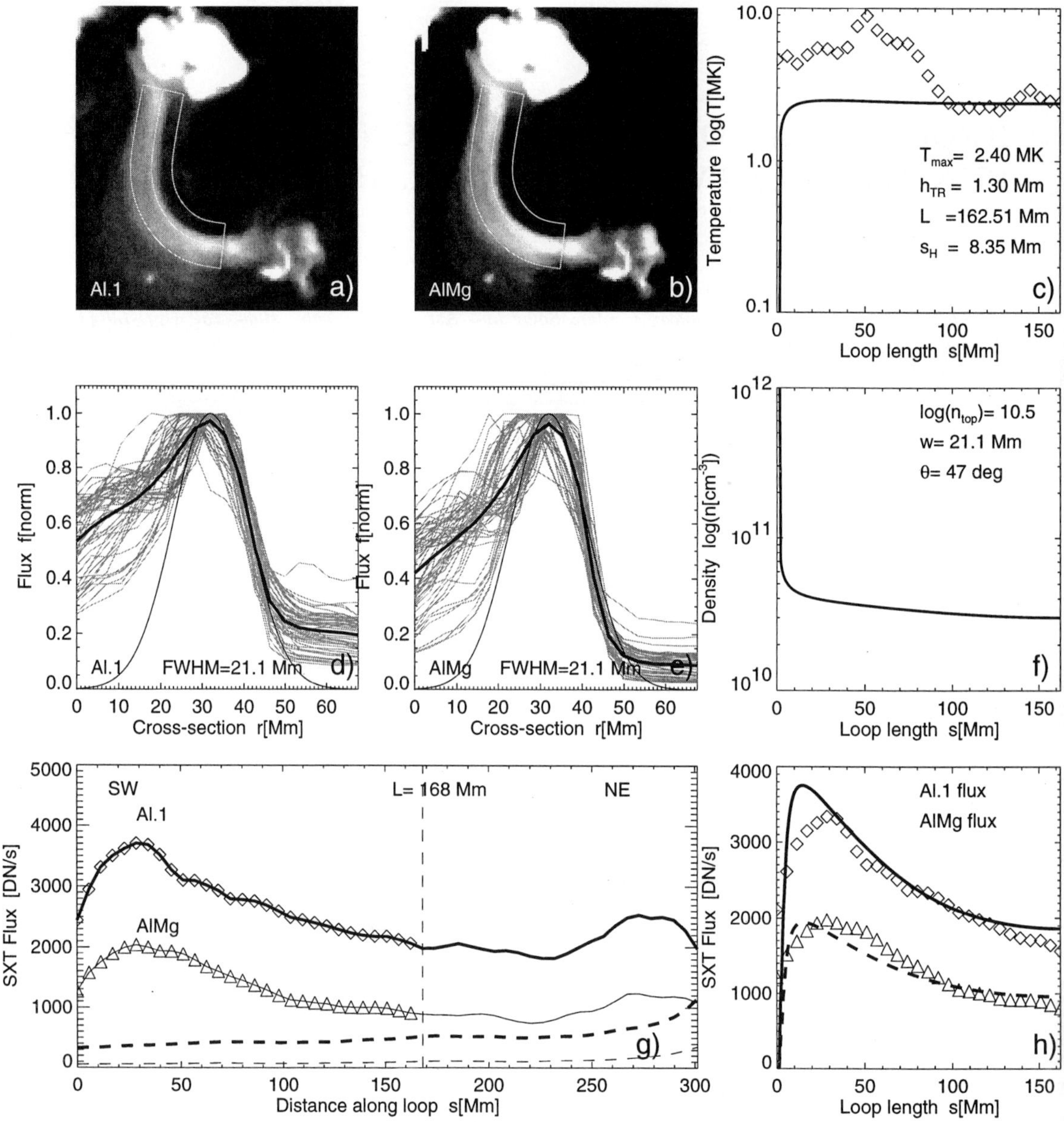

Fig. 1. a) *Yohkoh*/SXT Al.1 image, 1999 April 8, 00:51:45 UT, 1.9 ms exposure time, 227 Mm field of view; b) *Yohkoh*/SXT AlMg image, 1999 Apr 08, 00:46:09 UT, 5.4 ms exposure time; c) filter ratio temperature along southern half loop length (diamonds) and hydrostatic temperature solution $T(s)$; d) loop cross-sections from Al.1 image; e) loop cross-sections from AlMg image; f) hydrostatic density solution $n_e(s)$; g) Al.1 (diamonds) and AlMg fluxes (triangles) along axis and lowest background flux at edge of stripes (dashed lines); h) forward-fit of fluxes from hydrostatic solution (thick solid and dashed lines) to background-subtracted observed Al.1 (diamonds) and AlMg fluxes (triangles).

DATA ANALYSIS

Analysis of *Yohkoh* loops above the limb

The first determination of a coronal heating function $E_H(s) = E_{H0} \exp(-s/s_H)$, which balances the radiative losses $E_R(s) \approx -n_e(s)^2\Lambda[T(s)]$ and the conductive losses $-\nabla F_C(s)$ along the loop coordinate s (see Serio *et al.* 1981 for definitions) for loops observed with *Yohkoh* was published by Priest *et al.* (1998, 2000). The method applied therein is based on modeling of the filter-ratio temperatures $T_{FR}(s)$ by the energy balance equation between heating rate and conductive loss rate, finding best fits for loop-top and uniform heating. This solution was criticized for unphysical heights of the footpoints, if hydrostatic solutions are applied that extend all the way down to the transition region boundary (Mackay *et al.* 2000), as well as for unphysical solutions of the column depths, if one relates the hydrostatic scaling laws to the observed fluxes (Aschwanden 2001). This loop system has been re-analyzed recently and it was found that a combination of hot loops ($T \approx 2.6$ MK) embedded in a cooler background corona ($T \approx 1.0$ MK) fit the fluxes $F(s)$ in both filters, as well as the filter ratio temperatures $T_{FR}(s)$. It yields a heating scale height ($s_H \approx 13 \pm 1$ Mm) that is much smaller than the loop half length ($L \approx 380$ Mm), and thus supports the conclusion of footpoint heating (Aschwanden 2001).

Analysis of *Yohkoh* loops near disk center

We select a relatively bright loop observed at disk center, which minimizes the contamination by cooler background plasma, and thus, by proper subtraction of the background, enables a cleaner single-loop modeling than is possible for loops at the limb, which always are embedded in a haze of background loops. The analyzed loop has been described as a *trans-equatorial loop* (courtesy of Nariaki Nitta, *Yohkoh* Science Nugget of April 09, 1999). First we determine the inclination of the loop plane by a best-fit of a symmetric circular geometry and find an inclination angle of $\vartheta = 47^0$ with respect to the vertical, a loop half length of $L = 168$ Mm, and mean loop cross-section width of $w = 21$ Mm. Half-resolution (pixel size of 4.9") images in the Al.1 and Al.Mg filters are shown in Figure 1a and 1b. We trace the "backbone" of the loop, interpolate its coordinates with a spline fit, and extract a curvi-linear stripe with a width of 20 pixels (Figure 1a and 1b). The cross-sections along the loop axis are shown in Figure 1d and 1e, revealing an average width of $w \approx 21$ Mm and a background flux of $\approx 10 - 20$ %. The fluxes $F(s)$ extracted along the loop axis s are shown in Figure 1g, along with the background fluxes $F_B(s)$ (dashed lines in Figure 1g).

We fit the background-subtracted fluxes $f(s) = F(s) - F_B(s)$ (Figure 1h) in the two filters and the filter-ratio temperature $T_{FR}(s)$ (Figure 1c) simultaneously with analytical approximations of hydrostatic solutions for the temperature $T(s)$ and density $n_e(s)$, derived in Aschwanden & Schrijver (2001). The hydrostatic solutions are specified by three independent parameters: The loop top temperature T_{max}, the loop half length L, and the heating scale height s_H. Since we measure the loop half length $L = 168$ Mm directly, while the filter-ratio temperature constrains $T_{max} = 2.4$ MK at the looptop (Figure 1c), the only variable we have to optimize is the heating scale height, for which we find a best fit of $s_H = 8.35 \pm 0.05$ Mm. The inclination of the loop plane reduces the effective gravity along the loop and is considered in the hydrostatic solutions by introducing an effective gravity $g = g_\odot \cos\vartheta$. The best-fit solution yields a looptop density of $log(n_e) = 10.5$ (Figure 1f). Repeating the procedure of loop tracing and background selection, we can estimate the systematic errors from the range of slightly different solutions. We find approximate solutions in a range of $T_{max} \approx 1.8 - 2.6$ MK, $L \approx 150 - 170$ Mm, and $s_H \approx 5 - 10$ Mm. Thus, the solutions are quite robust and exclude uniform heating $s_H = \infty$ or looptop heating ($s_H < 0$). The best fit to the flux data has a standard deviation of $\approx 10\%$ (Figure 1h). We have to be aware, that such small deviations can easily be explained by cross-sectional variations, dynamical changes (e.g. siphon flows) that deviate from hydrostatic equilibrium, or by multi-temperature bundles of loops. Our fit of a single-loop model represents then only a best fit to the average of the hydrostatic solutions of a multi-temperature loop bundle. Nevertheless, it shows that forward-fitting to the two filter fluxes yield more robust solutions than fits to filter-ratio temperatures.

Table 1. Summary of measurements of heating scale heights s_H in coronal loops

Data set	Instrument	Method	Fit	Heating scale height	Refs.
40 AR loops	*TRACE*	SL	$F_1(s), F_2(s)$	$s_H = 17 \pm 6$ Mm	Aschwanden *et al.* 2000
40 AR loops	*TRACE*	HS	$F_1(s), F_2(s)$	$s_H = 12 \pm 5$ Mm	Aschwanden *et al.* 2001
1 limb loop	*Yohkoh*	HS	$T(s)$	$s_H \approx \infty$	Priest *et al.* 1998, 2000
		HS	$T(s)$	$s_H \ll L$	Mackay *et al.* 2000
		HS	$F_1(s), F_2(s)$	$s_H \approx 13 \pm 1$ Mm	Aschwanden 2001
1 disk loop	*Yohkoh*	HS	$F_1(s), F_2(s)$	$s_H \approx 8.4 \pm 2.5$	This work

Method: SL = Scaling laws, HS = Hydrostatic solutions
Fit: $F_1(s), F_2(s)$ = fluxes of two filters, $T(s)$ = filter-ratio temperature

The example shown in Figure 1 illustrates that a local deviation of 10-20% of the model to the observed fluxes (Figure 1h) cause changes of more than 100% in the filter-ratio temperatures (Figure 1c, in lower half of the loop, at $s < 100$ Mm), making filter-ratio inversions highly unstable.

CONCLUSIONS

The determinations of heating scale heights s_H are compiled in Table 1. For every analyzed loop system a solution is found in the range of $s_H \approx 5-20$ Mm, for loop lengths of $L \approx 10-300$ Mm. We conclude therefore that coronal loops are generally heated near footpoints, in active regions as well as in Quiet Sun regions. Previous results that concluded uniform or looptop heating seem to be based on unreliable methods, flawed by inadequate modeling of the filter-ratio temperatures in the multi-temperature corona (e.g. neglecting the hydrostatic weighting bias). There seems to be no fundamental difference in the spatial heating function of cool EUV loops ($T \approx 1$ MK) and hotter soft X-ray loops ($T \approx 2-3$ MK), nor between data from different spacecraft (*TRACE* versus *Yohkoh*).

ACKNOWLEDGEMENTS

This work has been supported by *Yohkoh* contract (NAS8-40108).

REFERENCES

Aschwanden, M. J., Nightingale, R. W., and Alexander, D., 2000, *ApJ* **541**, 1059.

Aschwanden, M. J., Schrijver, C. J., and Alexander, D. 2001, *ApJ* **550**, 1036

Aschwanden, M. J., Schrijver, C. J., 2001, *ApJ*, subm.

Aschwanden, M. J. 2001, *ApJ* **559**, L171.

Mackay, D. H., Galsgaard, K., Priest, E. R., and Foley, C. R. 2000, *Solar Phys.* **193**, 93

Priest, E. R., Foley, C. R., Heyvaerts, J., Arber, T. D., Culhane, J. L., & Acton, L. W. 1998, *Nature* **393**, 545

Priest, E. R., Foley, C. R., Heyvaerts, J., *et al.* 2000, *ApJ* **539**, 1002

Serio, S., Peres, G., Vaiana, G. S., and Rosner, R. 1981, *ApJ* **243**, 288

Wheatland, M. S., Sturrock, P. A., and Acton, L. W., 1997, *ApJ* **482**, 510

SMALL FLUCTUATIONS OF CORONAL X-RAY INTENSITY: A SIGNATURE OF NANOFLARES

Y. Katsukawa[1] and S. Tsuneta[2]

[1]*Department of Astronomy, School of Science, University of Tokyo, Bunkyo-ku, Tokyo 113-0033, Japan*
[2]*Solar Physics Division, National Astronomical Observatory, 2-21-1, Osawa, Mitaka, Tokyo 181-8588, Japan*

ABSTRACT

We analyze fluctuations of X-ray intensity observed with the Soft X-Ray Telescope (SXT) aboard the *Yohkoh* satellite to detect the tiny events which may occur in the solar corona but cannot be observed as discrete events. We create simple histograms of the X-ray intensity fluctuation around the mean intensity, and measure the width of the histograms. The width of this component becomes broader with increasing intensity, and is larger than the width of the predicted noise distribution. We suggest that nanoflares produce the observed fluctuation of the X-ray intensity. The estimated energy of nanoflares that causes the fluctuation of X-ray intensity is 10^{20} – 10^{22} erg. We also discuss the relationship between the nanoflare energy and magnetic fields.

INTRODUCTION

If the ensemble of small energy events (dubbed "nanoflares" by Parker 1988) heats the solar corona, how can we investigate such small energy events? A direct way is to identify small events. *Yohkoh*, *SOHO*, and *TRACE* revealed that many small events occur not only in active regions (e.g. Shimizu 1995) but also in quiet regions (e.g. Krucker *et al.* 1998, Parnell *et al.* 2000, Aschwanden *et al.* 2000). The thermal energy deposited by each event and the occurrence rate are investigated, and the energy distribution function has been examined by many authors. Although the distribution functions are slightly different among different authors, all the distribution functions appear to be roughly expressed by a single power-law distribution in the energy range of 10^{24} to 10^{33} erg.

These observational works showed that flares and microflares that can be identified as discrete events do not supply enough energy to heat the solar corona. So, the events that mainly contribute to heating of the solar corona should have much smaller energy, and may not be identifiable as discrete events. Thus, we suggest a new method to diagnose such small events (Katsukawa & Tsuneta 2001). If the persistently heated corona is maintained by many small events ("nanoflares"), the X-ray intensity from the corona cannot be constant but is expected to fluctuate around the mean intensity. When we simply create a histogram of the X-ray intensity fluctuation, the peak of the histogram is located around the mean intensity, and is broadened by the fluctuation due to nanoflares. The width of the histogram is the fundamental quantity that contains information about nanoflares.

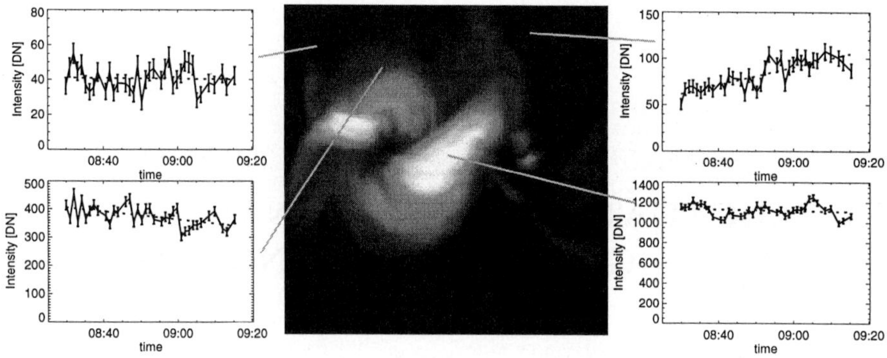

Fig. 1. Sample light-curves of the X-ray intensity observed with *Yohkoh*/SXT. Each observing sequence lasted about 40 minutes, and there are about 20 data points in each sequence.

OBSERVATIONS OF SMALL FLUCTUATIONS

We analyzed the time variation of the X-ray intensity observed with *Yohkoh*/SXT (see Figure 1). We made histograms of the X-ray intensity, and compared the histograms with predicted noise distributions. The observed histograms of X-ray intensity are found to be broader than the predicted noise distribution. This shows that some fluctuating sources exist in the solar corona (see Katsukawa & Tsuneta 2001). We suggest that nanoflares produce the observed fluctuation of the X-ray intensity. In addition to the amplitude of the fluctuation, we investigated the spatial and temporal extent of the fluctuation (Katsukawa 2002), and found the following properties: (1) The spatial extent of the fluctuation is smaller than the spatial resolution of SXT. (2) The temporal extent of the fluctuation is smaller than the time resolution.

To derive the information on nanoflare energy from the observed X-ray fluctuation, we derive the analytical expression for the intensity fluctuation. The number of nanoflares during each exposure and in each pixel N is assumed to be time dependent. The mean value of N is N_0, and N fluctuates around N_0. If nanoflares occur randomly, N has the fluctuation $\sqrt{N}$ around the mean N_0. We assume that the observable X-ray intensity I is proportional to the heat flux into the corona, that is, $I = CNE$, where NE is the heat input by nanoflares of energy E and C is the coefficient to convert the heat input into the X-ray intensity. The fluctuation of $\sqrt{N}$ causes the intensity fluctuation $\sigma_I = CE\sqrt{N} = \sqrt{CI_0E}$, where I_0 is the mean intensity and is expressed by $I_0 = CN_0E$. The intensity fluctuation σ_I and the mean intensity I_0 are obtained from observation, and then we can estimate the nanoflare energy E. To be exact, the duration of each event also affects the fluctuation of X-ray intensity. The detailed description of the analytical expression for the intensity fluctuation is in Katsukawa (2002).

ENERGY SCALE OF NANOFLARES

We overlay the energy of individual nanoflares derived in this manner and the occurrence rate upon the previously observed frequency distribution in Figure 2. Our derived energy (10^{22} erg) is much smaller than the value previously observed (greater than 10^{24} erg). And the occurrence rate of nanoflares is much larger than that obtained by extrapolating the distribution of $E^{-1.5}$.

The physical origin of nanoflares is believed to be magnetic reconnection. Energy available from magnetic reconnection may be generated by photospheric convective motion. When the magnetic free energy reaches

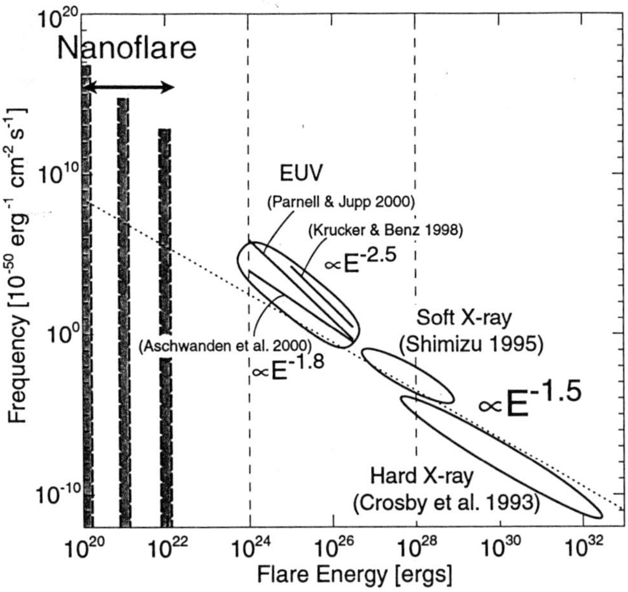

Fig. 2. Previously observed frequency distribution of flares, and the distribution of nanoflares estimated in our analysis.

a critical value, magnetic reconnection occurs, and releases the free energy. Here we briefly discuss the relationship between magnetic field strength and energy input by such sporadic energy release. Parker (1994) gives a very simple expression for energy release through nanoflares. We assume that magnetic reconnection occurs when the acute angle of the magnetic fields reaches a critical angle θ_c. At the critical angle θ_c, the free magnetic energy per unit area is

$$E_c = \frac{B^2}{8\pi} L \tan^2 \theta_c \,, \tag{1}$$

where B is the magnetic field strength and L the loop length. Magnetic reconnection reduces the magnetic free energy from E_c to some fraction $E_\gamma = \gamma E_c$ $(0 < \gamma < 1)$. The energy of $(1-\gamma)E_c$ is released per unit area in this process. Then storage of the free magnetic energy goes on until another reconnection occurs, and the time interval of reconnection is

$$\Delta t = \frac{L}{v} \tan \theta_c (1 - \gamma^{1/2}) \,. \tag{2}$$

The mean heat input rate q_0 is the energy released through single reconnection divided by the time interval Δt, and is given by

$$q_0 = \frac{(1-\gamma)E_c}{\Delta t} = \frac{B^2}{8\pi} v \tan \theta_c (1 + \gamma^{1/2}) \,. \tag{3}$$

The energy and the occurrence rate of individual events are characterized by γ. When $\gamma \simeq 1$, the energy release through a single nanoflare $(1-\gamma)E_c$ is very small, and the occurrence rate is very high $(\Delta d \simeq 0)$.

On the other hand, the free magnetic energy stored in the corona is almost completely released with only a single reconnection when $\gamma \simeq 0$. However the mean heat input rate q_0 is almost independent of γ, because $1 < (1 + \gamma^{1/2}) < 2$. Equation (3) shows that the mean heat input rate q_0 mainly depends on the magnetic field strength B.

The mean X-ray intensity in the solar corona varies over 1 or 2 orders of magnitude. This implies that the heat input rate is position-dependent. In an active region core, the heat input rate is about 10^7 erg $\text{cm}^{-2}\text{s}^{-1}$, while that in a dark (quiet) region is about 10^5 erg $\text{cm}^{-2}\text{s}^{-1}$. In the context of the nanoflare heating, the heat input rate is determined either by the individual energy of nanoflares or by the occurrence rate. In our analysis of the fluctuation of X-ray intensity, we suggested that the heat input rate is not determined by the occurrence rate of nanoflares, but by the energy of individual nanoflares.

The interval of the energy release Δt is determined by the photospheric convective motion (v in equation (2)) and the reconnection process (γ and θ_c) according to equation (2), and is independent from magnetic field strength B. Thus, the occurrence rate may be almost constant everywhere on the solar surface. On the other hand, the mean heat input rate q_0 (see equation (3)) and the energy released by a single event $(1 - \gamma)E_c$ (see equation (1)) strongly depend on the magnetic field strength B. The magnetic field strength depends on the locations on the Sun, and strong magnetic fields are observed in active regions while there are weak magnetic fields in quiet regions. The strong magnetic fields may cause nanoflares with larger energies while the occurrence rate is almost constant, and result in large mean heat input rate.

REFERENCES

Aschwanden, M. J., Nightingale, R. W., Tarbell, T. D. and Wolfson, C. J. , 2000, *ApJ*, **535**, 1027

Katsukawa, Y. & Tsuneta, S. 2001, *ApJ*, **557**, 343

Katsukawa, Y. , 2002, *PASJ*, submitted

Krucker,S. & Benz, A. O. , 1998, *ApJ*, **501**, L213

Parker, E. N. , 1988, *ApJ*, **330**, 474

Parker, E. N. , 1994, *Spontaneous Current Sheets in Magnetic Fields* (Oxford University Press)

Parnell, C. E. & Jupp, P. E. 2000, *ApJ*, **529**, 554

Shimizu, T. , 1995, *PASJ*, **47**, 251

OBSERVATION AND THEORY OF CORONAL LOOP STRUCTURE

J. A. Klimchuk

Space Science Division, Naval Research Laboratory, Washington, DC 20375, USA

ABSTRACT

Following up on an initial study of 10 soft X-ray loops observed by *Yohkoh* (Klimchuk *et al.* 1992), we have carefully examined 43 additional *Yohkoh* loops and 24 EUV loops observed by *TRACE*, and we confirm our original finding that most coronal loops have a nearly uniform thickness. This implies that: 1. the magnetic field in these loops expands with height much less than standard coronal models would predict; and 2. the shape of the loop cross section is approximately circular. We have investigated whether these surprising results can be explained by locally enhanced twist in the field, so that observed loops correspond to twisted coronal flux tubes. Our approach is to construct numerical models of fully three-dimensional force-free magnetic fields. To resolve the internal structure of an individual loop embedded within a much larger dipole configuration, we use a nonuniform numerical grid of size 609×513×593, the largest ever applied to a solar problem, to our knowledge. Our models indicate that twist does indeed promote circular cross sections in the corona, even when the footpoint cross section is irregular. However, twist does not seem to be a likely explanation for the observed minimal expansion with height.

INTRODUCTION

Early in the *Yohkoh* mission, we published a study finding that coronal loops observed by the Soft X-ray Telescope (SXT) have a nearly uniform thickness (Klimchuk *et al.* 1992). This was a surprising result, since loops are believed to coincide with magnetic field lines, and the field must on average diverge with height above the solar surface. This suggests that a majority of loops should be wider at their tops than at their footpoints. Some expansion is in fact observed, but much less than predicted by extrapolation models of idealized or observed photospheric magnetic flux distributions (see Klimchuk 2000). Longcope (1996, 1998) has suggested that coronal loops lie in the immediate vicinity of magnetic separators, but the expansion properties of such field lines have not been fully investigated (point magnetic charge models seem to indicate minimal expansion, but these models may not accurately represent the real solar field).

Our initial study was rather limited, since it involved only 10 loops observed in half resolution mode (4.9 arcsec pixels). We have recently completed a more extensive follow-up study involving 43 loops observed at both half and full (2.45 arcsec) resolution (Klimchuk 2000). In addition, we have completed a companion study of 24 loops observed by the *Transition Region and Coronal Explorer* (*TRACE*), which has much higher spatial resolution (0.5 arcsec pixels) (Watko & Klimchuk 2000). The results of these new studies are reviewed here.

In our original paper, we suggested that the uniform thickness might be explained by localized twist in the magnetic field. It is well known that the central portions of straight axisymmetric flux tubes become constricted as the tubes are twisted (e.g. Parker 1977, Zweibel & Boozer 1985, Lothian & Hood 1989).

Because the constriction is greater for weak magnetic fields than for strong fields, and because no constriction is possible at the photospheric footpoints, where line tying applies, one might imagine that an expanding loop within a potential magnetic field configuration would become more and more uniform as the loop is twisted. We have recently computed 3D force-free magnetic field models to investigate this idea (Klimchuk *et al.* 2000), and those results are also reviewed here.

OBSERVED LOOP EXPANSION FACTORS

Details of our data analysis procedure can be found in the full papers, but two important points are worth emphasizing. First, we carefully subtract the background emission from the loops before making the width measurements. Second, we attempt to account for spatial resolution effects by correcting the measurements for the point spread function of the telescope and the finite size of the CCD pixels. To characterize the footpoint-to-apex expansion of the loops, we define a parameter called the "expansion factor":

$$\Gamma_r \equiv \frac{r_{mid}}{r_{foot}}\,, \qquad (1)$$

where r_{mid} and r_{foot} are the widths measured at the loop midpoint and footpoint, respectively. Each loop has two expansion factors, one for each leg. Figure 1 shows Γ_r plotted against loop length for the loops of the new SXT study. Stars and diamonds represent measurements made from full- and half-resolution observations, respectively. The median Γ_r is 1.30, meaning that the loops are typically 30% wider at their midpoints than at their footpoints. This is much less than predicted by standard magnetic field models (Klimchuk 2000). Standard models also predict that Γ_r should increase with loop length, which is not seen in the data. Finally, the full- and half-resolution observations give similar results, suggesting that spatial resolution effects are not important.

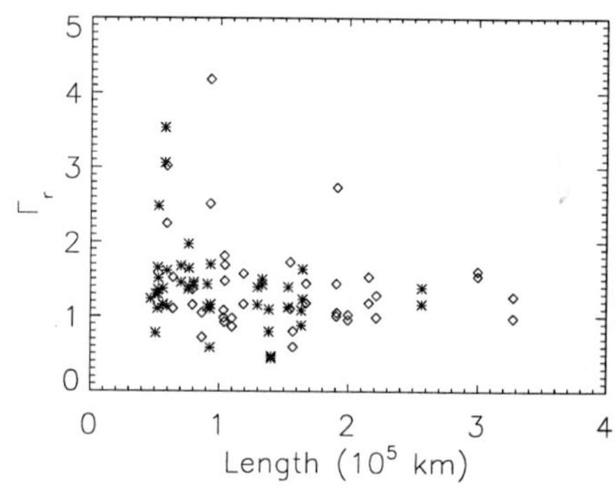

Fig. 1. Expansion factor versus loop length. Stars and diamonds are for full- and half-resolution SXT observations, respectively (from Klimchuk 2000).

The loops of our *TRACE* study were observed mostly in the 171 Å and 195 Å bands, and, unlike the SXT loops, several were clearly associated with a flare. The 15 non-flare cases have a median Γ_r of 0.99 (actually narrower at the midpoints!), and the 9 post-flare cases have a median Γ_r of 1.13. These are listed in Table 1 together with the median values from both the old and new SXT studies. Estimated uncertainties are given in parentheses. It is rather striking how consistent the results are from the different studies. It is also interesting that the expansion factors measured from the higher resolution *TRACE* observations are, if anything, smaller than the SXT values. This is further indication that spatial resolution is not a concern.

TWISTED FLUX TUBE MODELS

To test the idea that the observed thickness uniformity can be explained by localized twist in the magnetic field, we have used the magnetofrictional method to model a twisted coronal flux tube embedded within a much larger dipole potential field configuration. Other modelers have examined active-region-scale twist in the field, which is appropriate to active region evolution (e.g. Sakurai 1979, Van Hoven *et al.* 1995, Amari

Table 1. Median Expansion Factors

Study	Γ_r	# Loops
SXT (old)	1.13 (0.10)	10
SXT (new)	1.30 (0.12)	43
TRACE (non-flare)	0.99 (0.04)	15
TRACE (post-flare)	1.13 (0.34)	9

et al. 1996), but our models are the first to have a truly localized twist. In order to numerically resolve the internal structure of the loop and at the same time include a large volume of surrounding field, it was necessary to adopt a very large nonuniform grid of size $609 \times 513 \times 593$.

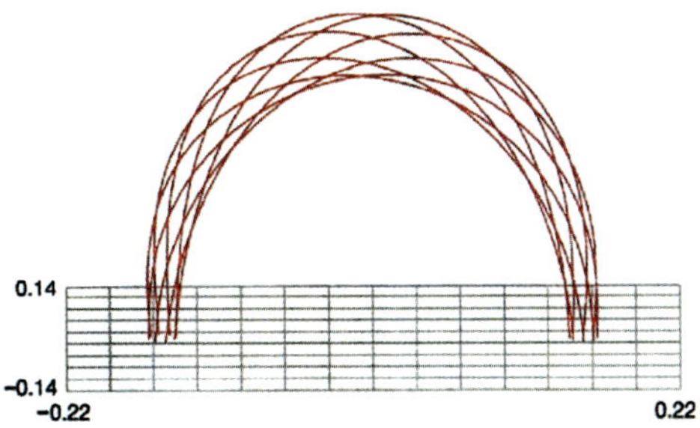

Fig. 2. Flux tube of 2π twist as viewed from the side at a 15° angle to horizontal. Every twentieth grid line is shown (from Klimchuk *et al.* 2000).

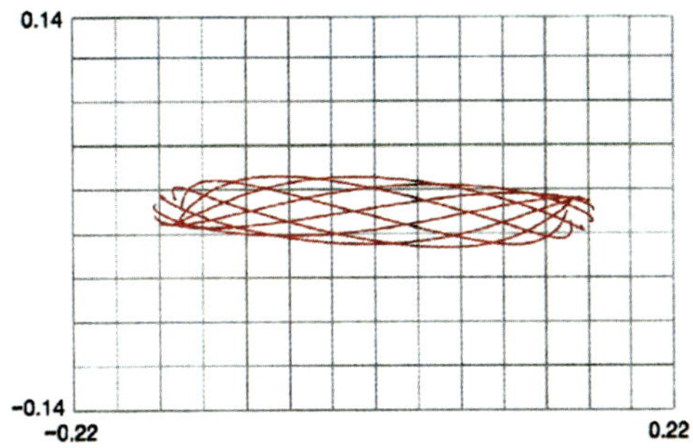

Fig. 3. Flux tube of 2π twist as viewed from directly above. Every twentieth grid line is shown (from Klimchuk *et al.* 2000).

Figure 2 shows a side view of the inner part of the flux tube after an end-to-end twist of 2π has been applied. Compared to the untwisted potential state, the expansion factor of the tube has decreased from 2.8 to 2.3. This is still much larger than the values given in Table 1. Figure 3 shows the same flux tube, only this time viewed from directly above. From this perspective, the twist has an opposite affect and causes the expansion factor to increase rather than decrease, from 2.0 to 2.4! Based on these results, it seems doubtful that the thickness uniformity of observed loops can be explained by magnetic twist. However, there remains an issue of line-of-sight overlap which prevents us from ruling out the twist explanation with absolute certainty. This can be seen in Figure 3, where the footpoints of the loop overlap with the lower portions of the loop leg.

There is another very interesting aspect to the simulations. Each flux tube in the original potential field configuration has a cross section that varies in shape as well as area along the tube. Our twisted flux tube has a circular cross section in the photosphere, by definition, since we rotate a circular patch at the footpoints. Before the twist is applied, the circle maps to an oval at the loop apex (elongated in the vertical direction), but after a twist of 2π, the oval has changed into a near circle. Thus, the primary affect of twist is to circularize the cross section while maintaining a nearly constant area. The physical reason for this can be understood in terms of the magnetic tension associated with the azimuthal field component that is introduced by the twist. Magnetic field lines trace out closed paths when viewed projected onto a cross section, and the tension in these lines will act to make the paths circular. The stronger the twist, the greater the force that organizes the flux into an axisymmetric bundle.

This result is quite significant because it offers a natural explanation for why coronal loops are observed to have approximately circular cross sections (Klimchuk 2000). Although circular cross sections are often assumed, there has until now been no obvious reason to expect them. The magnetic field has a very clumpy distribution in the real solar photosphere, so the footpoints of loops are likely to have highly irregular shapes (unlike our model). In the absence of twist, we expect the irregular footpoints to map to similarly irregular cross sections in the corona. Twisted loops, on the other hand, should tend to have circular cross sections irrespective of the shape of their footpoints.

In closing, we note that observed loops are actually plasma structures, and we have assumed that they coincide with magnetic flux tubes. In principle, they could be different, though this seems highly unlikely given the efficiency with which thermal energy and plasma flow along but not across the field lines. It has been suggested that the scale height of the plasma might be greater at the axis of a flux tube than at the outer edge. This could produce a plasma loop with a uniform cross section even if the magnetic flux tube is expanding. For this to feasible, however, the scale height at the outer edge would need to be smaller than both the scale height at the axis *and* the geometric height of the loop. This requires radial temperature stratification, which is not observed, and temperatures in the outer layers that are far too cool for *Yohkoh* to detect. The puzzle remains!

REFERENCES

Amari, T., Luciani, J.F., Aly, J.J., & Tagger, M., Very Fast Opening of a Three-dimensional Twisted Magnetic Flux Tube, *ApJ,* **466**, L39 (1996).

Klimchuk, J.A., Cross-Sectional Properties of Coronal Loops, *Solar Phys.,* **193**, 53 (2000).

Klimchuk, J.A., Lemen, J.R., Feldman, U., Tsuneta, S., & Uchida, Y., Thickness Variations Along Coronal Loops Observed by the Soft X-Ray Telescope on Yohkoh, *PASJ,* **44**, L181 (1992).

Klimchuk, J.A., Antiochos, S.K., & Norton, D., Twisted Coronal Magnetic Loops, *ApJ,* **542**, 540 (2000).

Longcope, D.W., Topology and Current Ribbons: A Model for Current, Reconnection and Flaring in a Complex, Evolving Corona, *Solar Phys.,* **169**, 91 (1996).

Longcope, D.W., A Model for Current Sheets and Reconnection in X-ray Bright Points, *ApJ,* **507**, 433 (1998).

Lothian, R.M., & Hood, A.W., Twisted Magnetic Flux Tubes: Effect of Small Twist, *Solar Phys.,* **122**, 227 (1989).

Parker, E.N., The Origin of Solar Activity, *ARA&A,* **15**, 45 (1977).

Sakurai, T., A New Approach to the Force-Free Field and Its Application to the Magnetic Field of Solar Active Regions, *PASJ,* **31**, 209 (1979).

Van Hoven, G., Mok, Y., & Mikić, Z., Coronal Loop Formation Resulting From Photospheric Convection, *ApJ,* **440**, L105 (1995).

Watko, J.A., & Klimchuk, J.A., Width Variations Along Coronal Loops Observed By TRACE, *Solar Phys.,* **193**, 77 (2000).

Zweibel, E.G., & Boozer, A.H., Evolution of Twisted Magnetic Fields, *ApJ,* **295**, 642 (1985).

DYNAMICS AND DIAGNOSTICS OF EXPLOSIVE EVENTS AND BLINKERS

M. S. Madjarska[1], J. G. Doyle,[1] and L. Teriaca[2]

[1]*Armagh Observatory, College Hill, Armagh BT1 9DG, N. Ireland*
[2]*Osservatorio Astrofisico di Arcetri, Largo E. Fermi 5, 50125 Firenze, Italy*

ABSTRACT

The knowledge of the main physical parameters of UV explosive events and blinkers, such as electron density and temperature, is of great interest for a better understanding of the true nature of these transient phenomena. In this context, electron density and temperature diagnostics based on lines belonging to O IV and O III multiplets (SUMER/*SOHO*) are presented for both kind of events. The dynamics of the solar transient phenomena is revealed through their temporal and spatial evolution as observed in spectral lines covering a wide temperature range. The events are also studied, searching for their chromospheric origin and further propagation higher in the solar corona.

INTRODUCTION

Observations have revealed the existence of high velocity small-scale events seen in lines from ions formed at temperatures from 2 10^4 K (C II) to 2 10^5 K (N V) with no signature in chromospheric lines (Brueckner & Bartoe 1983). They are characterized by non-Gaussian profiles due to an enhancement in the blue and red wings. Most of the events are predominantly blueshifted, showing velocities up to 250 km s^{-1}. The explosive events are located in the network lanes at the boundaries of the super-granulation cells. They appear preferably in regions with weak fluxes of mixed polarity or on the border of regions with a large concentration of magnetic flux and are observed in regions undergoing magnetic cancellation. The average lifetime of explosive events ranges from $\sim$ 60 to 350 s with spatial dimensions of $\sim$ 2500 km. Their first identification was followed by several studies that extended the description of their general characteristics naming them by the term 'explosive events' (Madjarska & Doyle 2002, and references therein).

A new transient phenomenon observed by CDS/*SOHO* as enhancements in the flux of transition region lines at network junctions was recently introduced by Harrison (1997) who named them 'blinkers'. Blinkers are mainly observed in lines of O III (10^5 K), O IV (1.6 10^5 K) and O V (2.5 10^5 K) while lines formed at lower and higher temperatures (such as He I and Mg X) show only a modest increase in intensity. They show a typical lifetime of $\sim$ 17 minutes over an area of $\sim$ 5 10^7 km^2, with an average intensity increase in O V of $\sim$ 1.5 which, in extreme cases, can reach values as high as five times the pre-event level (Harrison *et al.* 1999).

OBSERVATIONS OF TRANSIENT PHENOMENA

The analysed observational material is presented in Table 1. Details on line blends and methods of analysis can be found in Teriaca *et al.* (2001), Madjarska & Doyle (2002), and Teriaca *et al.* (2002).

Table 1. Details on the SUMER observational material

Date	Detector	Slit	Exp. time	Solar X, Y	Spectral lines (Å)
1996 July 10	A	1″×120″	20	+ 3, + 0	O IV 1399, 1401, 1404, 1407
1997 May 31	B	0.3″×120″	12	+600, +576	O IV 1401, 1404, Si IV 1393, 1402
1997 June 5	B	1″×120″	30	+696,+158	Full spectrum (from 903 to 943 Å)
2001 Oct 23	A	1″×120″	20	+170,+0	N V 1238, Mg X 625

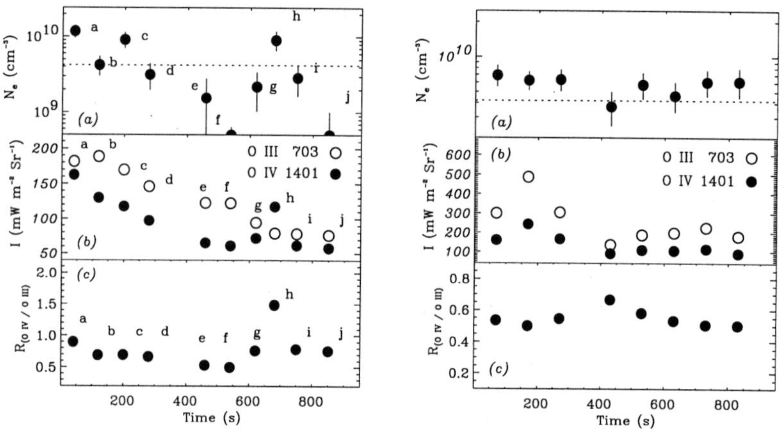

Fig. 1. *Left panel:* Electron density obtained during an explosive event on July 10, 1996. *Right panel:* Electron density obtained during a blinker event on July 10, 1996. The dashed line indicates the average density over the observed region. For details see the text.

DENSITY AND TEMPERATURE VARIATIONS DURING TRANSIENT PHENOMENA

Despite the large number of observational works on UV explosive events and blinkers, large uncertainties about their basic physical parameters, such as electron density and temperature, still exist. In order to derive the electron density of transition region plasma, the most reliable method involves the use of density-sensitive line ratios of lines belonging to the same ion. In the present contribution we studied the behaviour of the O IV 1401.16/1404.81 line ratio during the appearance of explosive events and a UV blinker. The ratio of allowed lines from different ionization stages of the same element (O IV 1401.16/O III 703) were used in order to verify whether the observed solar plasma undergoes temperature variations.

Figure 1 (left panel) presents the temporal variation of the electron density during the explosive event identified in the dataset from 1996 July 10 (a). Figure 1(b) shows the intensities of the O IV 1401 (filled circles) and O III 703 (open circles) lines, while the ratio between the intensities of the former and the latter line is shown in (c). During the last 400 seconds (**e** – **j**) the intensity of both lines is more or less constant with the remarkable exception of **h**. This is the point where a strong explosive event was observed in the blue wing of the Si IV 1402 line and in O IV 1401.16 Å, with an *electron density enhancement of a factor* ~ 3 with respect to the pre (**g**) and post-event (**i**) values. The intensity ratio O IV/O III increase in **h** (~ 30 % more than **g** and **i**) suggests *a temperature increase during the explosive event.*

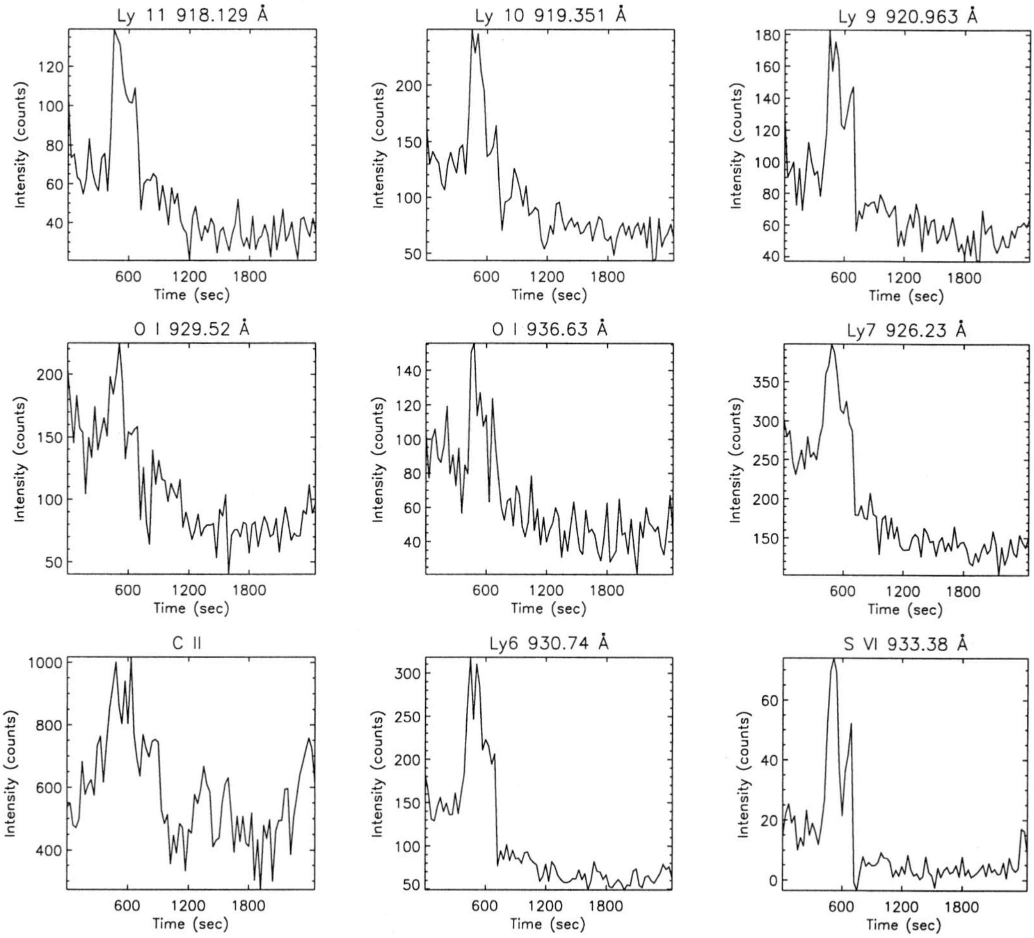

Fig. 2. Integrated intensity in the blue wing of Ly 11, Ly 10, Ly 9, Ly 7, Ly 6 and S VI in the region of the explosive event as a function of time. We also show the total intensity of the O I and C II (includes four blended C II lines) lines (Curdt *et al.*, 1997, 2001).

In Figure 1(a) (right panel) the temporal variation of N_e during a blinker phenomenon, identified in the dataset in July 10, 1996, is presented. Figure 1(b) shows the intensities of the O IV 1401 (filled circles) and O III 703 (open circles) lines, while the ratio between the intensities of the former and the latter line is shown in (c). *No appreciable N_e and T_e variations were detected.*

CHROMOSPHERIC ORIGIN AND CORONAL COUNTERPART OF EXPLOSIVE EVENTS

Figure 2 shows the integrated intensity in the blue wings of Ly 11, Ly 10, Ly 9, Ly 7, Ly 6 and S VI obtained as described in Madjarska & Doyle (2002). We were especially interested in the response of the O I (15 000 K) lines. The visual inspection of these lines suggested some intensity increase during the explosive event, but this is uncertain because of the low emission of these lines. Therefore we used the total intensity in the two selected unblended oxygen lines O I 929.52 and 936.63 Å. During explosive events the central

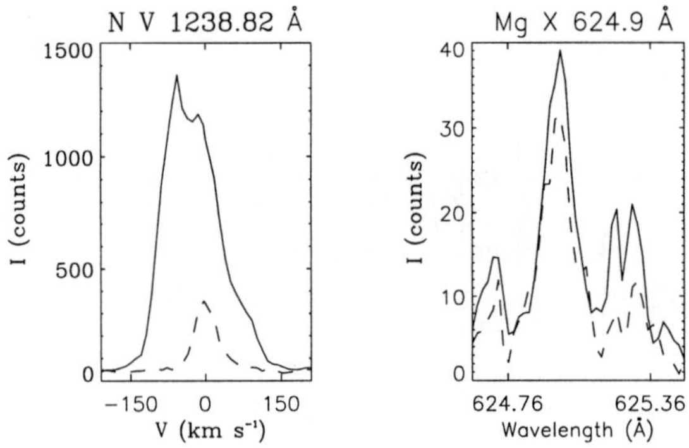

Fig. 3. The N v 1238 Å and Mg x 625 Å line profiles before (dashed line) and during (solid line) the explosive event. The small increase in the Mg IX is due to first order blends (see Teriaca *et al.*, 2002).

intensity of optically thin lines increases by at least 1.6 and in optically thick lines by less than 1.2 times (Madjarska & Doyle 2002). Therefore, when it is impossible to detect blueshifted and redshifted emission as in the case of faint lines such as oxygen, and the event is already registered by other simultaneously recorded lines, the total intensity in these lines is a good indicator for the presence of the event in this spectral line. The plots in Figure 2 show that during an explosive event, plasma with a temperature starting from 15 000 is registered *suggesting a chromospheric origin of the explosive events.*

Simultaneous observations in the N v and Mg x lines were aimed at finding out whether during explosive events a plasma at coronal temperature exists. The strongest event was selected and the line profile before (dashed line) and during (solid line) the event in N v and Mg x are shown on Figure 3. The Mg x 625 Å line does not show any significant variations during the explosive event while the S II 1250.58 Å line (the second line on the right side of the Mg line) increase of a factor $\sim$ 3. These observations confirm once again the chromospheric origin of the explosive events and show no signature of the phenomena at coronal temperatures.

ACKNOWLEDGEMENTS

Research at Armagh Observatory is grant-aided by the N. Ireland Dept. of Culture, Arts and Leisure, while partial support for software and hardware is provided by the STARLINK Project which is funded by the UK PPARC. MM was supported by PPARC grant PPA/GIS/1999/00055. The SUMER project is financially supported by DLR, CNES, NASA, and PRODEX.

REFERENCES

Brueckner, G. E. & Bartoe, J.-D. F., *ApJ*, **272**, 329 (1983).

Curdt, W., Feldman, U., Laming, J. M. et al., *A&AS*, **126**, 281 (1997).

Curdt, W., Brekke, P., Feldman et al., *A&A*, **375**, 591 (2001).

Harrison, R. A., it Solar Phys., **175**, 467 (1997).

Madjarska, M. S. & Doyle, J. G., *A&A*, **382**, 319 (2002).

Teriaca, L., Madjarska, M. S. & Doyle, J. G., *Solar Phys.*, **200**, 91 (2001).

Teriaca, L., Madjarska, M. S. & Doyle, J. G., *A&A* (2002) (submitted).

A HIGH TEMPERATURE CORONA ABOVE AN ACTIVE REGION COMPLEX

Y.-K. Ko[1], J. C. Raymond[1], J. Li[2], A. Ciaravella[1,3], J. Michels[1,4], S. Fineschi[1,5], and R. Wu[1]

[1]*Harvard-Smithsonian Center for Astrophysics, 60 Garden St., Cambridge, MA 02138, USA*
[2]*Institute for Astronomy, University of Hawaii, 2680 Woodlawn Dr., Honolulu, HI 96822, USA*
[3]*now at Osservatorio Astronomico di Palermo, "G.S. Vaiana", P.za Parlamento 1, 90134 Palermo, Italy*
[4]*now at Princeton Materials Institute, Princeton University, Princeton, NJ 08544, USA*
[5]*now at Osservatorio Astronomico di Torino, Strada Osservatorio 20, I-10025, Pino Torinese, Italy*

ABSTRACT

We present the results of *SOHO*/UVCS and *Yohkoh*/SXT observations above an active region complex (AR8194/8195/8198) at the southeast limb on April 6-7, 1998. The electron temperature analysis indicates a two-temperature structure, one with $\sim 1.5 \times 10^6$ K which is similar to that observed in quiet Sun streamers, the other with a high temperature $\sim 3.0 \times 10^6$ K. We compare the electron temperature and emission measure from the *SOHO*/UVCS data with those from the *Yohkoh*/SXT data. The absolute elemental abundances show a general first ionization potential effect (FIP effect) and decrease with height for all the elements. We discuss mechanisms that may explain the observed abundances.

OBSERVATION AND RESULTS

The target of the observation is the corona above an active region complex at the southeast limb. AR8194 and AR8195 appeared at the east limb on April 5 followed by AR8198 1.5 days later. The heliographic latitude is S18, S27 and S28 for AR 8194, 8195, and 8198, respectively. UVCS observed for nine heliocentric heights centered at position angle (PA) of 120° from 13:27 UT, April 6 to 14:20 UT, April 7, 1998. Figure 1 shows the lowest and highest OVI Channel slit positions (1.22 $R_\odot$ to 1.60 $R_\odot$) on the EIT-UVCS composite image, and the *Yohkoh*/SXT image plotted with the regions extracted for SXT data analysis.

UVCS data show that besides the usual coronal lines, some high-ionization lines, such as Fe XVII λ1153, [Fe XVIII] λ974, Ne IX λ1248, [Ca XIV] λ943, are particularly bright compared to the quiet Sun corona. These high-ionization lines are still visible at heights up to 1.6 $R_\odot$. This indicates that this region is unusually hot compared with the 'average' solar corona observed at these heights. The UVCS slit is 40 arcmin ($\sim 2.5 R_\odot$) long as seen in the plane of the sky (see Figure 1). We average UVCS data over 280 arcsec of the spatial extent of the slit centered at PA=120° (corresponding to PA range from $\sim$ 114° to $\sim$ 126°). This is where the emission of those high-ionization lines is concentrated. The data have been wavelength and radiometrically calibrated, and corrected for stray light and flat field. The radiative and collisional components for the hydrogen Lyman lines and OVI λ1032/λ1037 doublets are separated. The collisional excitation rates are mostly adopted from the CHIANTI database version 3.01 (Dere *et al.*, 2001). The ionization equilibrium of Mazzotta *et al.* (1998) were adopted. For details of the UVCS data analysis, see Ko *et al.* (2002).

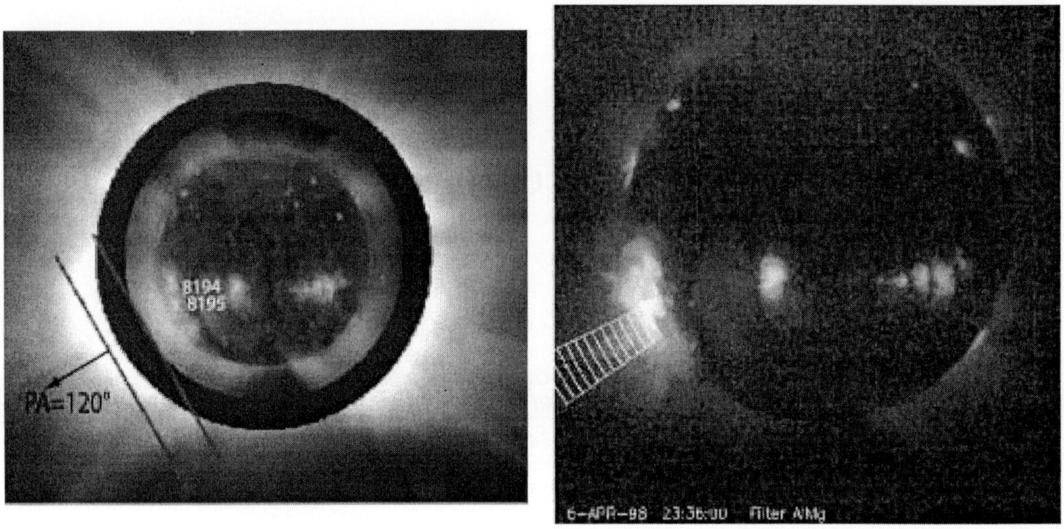

Fig. 1. Left panel: The pointing of UVCS observations on the composite image of EIT 284 (19:06 UT, April 6) and UVCS (synoptic image in OVI λ1032, taken from 22:06 UT, April 5 to 11:50 UT, April 6, 1998). Right panel: *Yohkoh*/SXT image at 23:36 UT, April 6 plotted with the regions extracted for data analysis.

Figure 2 plots the electron temperature derived from various line ratios of Si and Fe lines at all heights. It can be seen that the temperature distribution seems mainly to be clustered into two temperatures, one around 1.5×10^6 K, and the other around 3×10^6 K.

If we assume that the corona above this active region complex has a two-temperature structure along the line of sight, and that the elemental abundances in the two T_e regions are the same, the line intensity (photon $s^{-1}cm^{-2}sr^{-1}$) can then be expressed as:

$$I_{line} = \frac{1}{4\pi}\frac{n_{el}}{n_H}B_{line}[\frac{n_{ion}}{n_{el}}(T_{hi})q_{line}(T_{hi})(\int n_e n_H dl)_{hi} + \frac{n_{ion}}{n_{el}}(T_{lo})q_{line}(T_{lo})(\int n_e n_H dl)_{lo}] \quad (1)$$

where n_{el}/n_H is the elemental abundance relative to hydrogen (absolute abundance), n_{ion}/n_{el} is the ionic fraction, B_{line} is the branching ratio, q_{line} is the electron excitation rate, and $\int n_e n_H dl$ is the emission measure at a given T_e. T_{hi} and T_{lo} are the average of the ratio temperatures (see Figure 2) from the high-T_e gas (FeXV/FeXVII, FeXVIII/FeXVII, FeXVIII/FeXV) and the low-T_e gas (rest of the ratios), respectively. If we use [Fe X] λ1028, [Fe XII] λ1242 as the proxy for the low-T_e gas and Fe XVII λ1153, [Fe XVIII] λ974 for the high-T_e gas, the ratio of the emission measures at the two T_e's can be determined. The 'high-T_e' and the 'low-T_e' components for the lines can then be calculated analytically using Eq. 1 along with the absolute elemental abundance and the emission measure. Figure 3 plots the abundances relative to their photospheric values versus their first ionization potentials (FIP). The lines that are used to determine the adopted abundances are: N V λ1238, O VI λ1032, Ne IX λ1248, Si XII λ499, [S X] λ1196, [Ar XII] λ1018, [K XIII] λ994, [Ca XIV] λ944, the average of [Fe X] λ1028, [Fe XII] λ1242, [Fe XIII] λ510, Fe XVII λ1153 and [Fe XVIII] λ974 for Fe, and [Ni XIV] λ1034. We can see that the FIP effect, in which the abundances of the low-FIP elements (FIP smaller than $\sim$10 eV, such as Fe) are enhanced relative to those of the high-FIP elements (FIP larger than $\sim$10 eV, such as Ar) when compared with their photospheric values, is present at all heights. Furthermore, the abundance generally decreases with height in a systematic way.

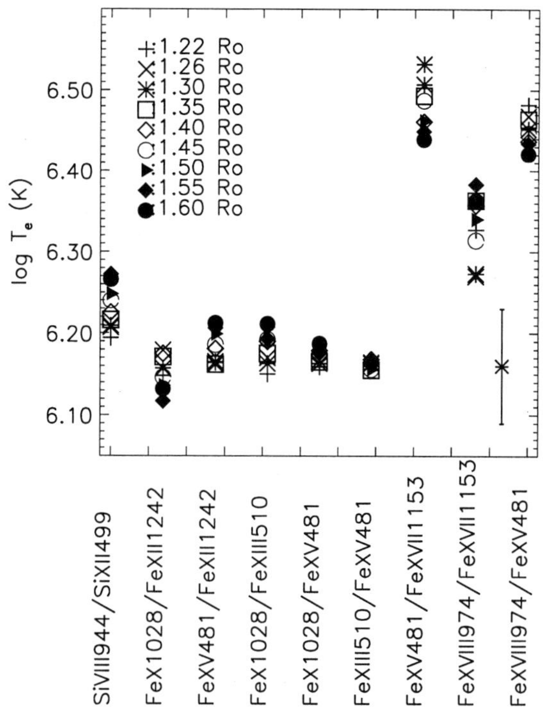

Fig. 2. Electron temperature derived from various line ratios of Si and Fe at all nine heights.

Figure 4 and Figure 5 compare the electron temperature and the emission measure, respectively, derived from the UVCS data with those from the SXT data which are taken from 09:09 UT, April 6 to 11:09 UT, April 7, 1998. The data were averaged over PA=120 $\pm$ 5° with increment of 0.05 $R_\odot$ (cp. Figure 1). The SXT Science Composite (SSC) images with filter pairs Al.1 and AlMg were used. In order to compare the SXT and UVCS measurements, we calculate theoretical X-ray spectra developed by Raymond & Smith (1977) ('RS code') using the elemental abundances measured by UVCS ('ab_UVCS'). The SXT response function is obtained by combining the theoretical X-ray spectra with the SXT effective area. The SXT temperature and emission measure are then derived from the SXT response function. The comparison shows that the electron temperatures of the 'high-T_e' region from UVCS are consistent with those from SXT. The temperature derived from SXT band ratios is expected to be an average of the high and low temperatures, strongly weighted toward the high temperature by the higher emissivity of the hotter plasma in the *Yohkoh* bandpasses. The emission measures derived from SXT are also consistent with the UVCS results. Figures 4 and 5 also show the SXT results using two other approaches: 1) SXT standard routine ('SXT code') which uses Meyer (1985) coronal abundances ('ab_Meyer'), and 2) Raymond & Smith (1977) code using Meyer (1985) coronal abundances. We can see that it is important to use consistent elemental abundances and plasma emission codes to calculate the electron temperature and the emission measure from broad band data such as SXT acquires. Using different sets of abundances (e.g. photospheric vs. coronal, coronal with FIP effect with enhanced vs. depleted low FIP elements) may give significantly different results.

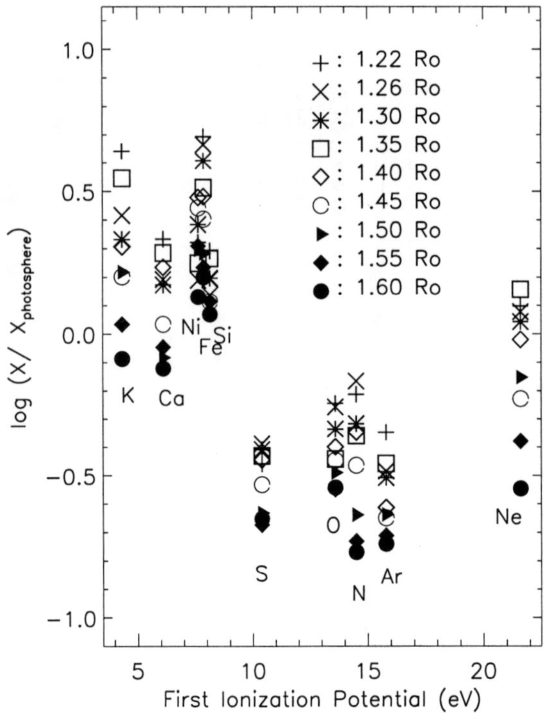

Fig. 3. Elemental abundances relative to the photospheric values vs. their first ionization potential at all nine heights.

DISCUSSION AND SUMMARY

We have shown that the corona above this active region complex can be characterized by electron temperature higher than that in the usual quiet Sun corona and streamers. Detailed analysis indicates that the temperature distribution is mainly clumped around two values, one at $\sim 1.5 \times 10^6$ K and the other at $\sim 3 \times 10^6$ K. The lower value is similar to that found in the quiescent streamers. The higher value is most likely associated with these active regions.

Our analysis shows that the FIP effect is present at all heights and the abundance decreases with height for all the elements. The FIP bias of about a factor of 4 is typical of the slow solar wind. Schwadron *et al.* (1999) modeled the elemental fractionation at the foot of large coronal loops and found that MHD wave heating is able to provide both mass-independent fractionation and low-FIP bias in coronal loops. The materials stored in the closed loops are then released by reconnection with adjacent open field lines and form the slow wind carrying FIP bias with them. A plausible explanation for the decreasing abundances at larger heights is gravitational settling of the heavier elements in closed magnetic loops (Raymond *et al.* 1997). We have shown that the absolute abundance declines by factors of 2-4 between 1.2 and 1.6 $R_\odot$, suggesting a scale height of about a few 10^{10} cm. There is no obvious dependence upon mass. Gravitational settling, however, cannot be totally responsible for the abundance variations we see here. Active regions usually evolve on a time scale of a few weeks. The settling times of the ions are roughly one day (Lenz *et al.*, 1998). Therefore

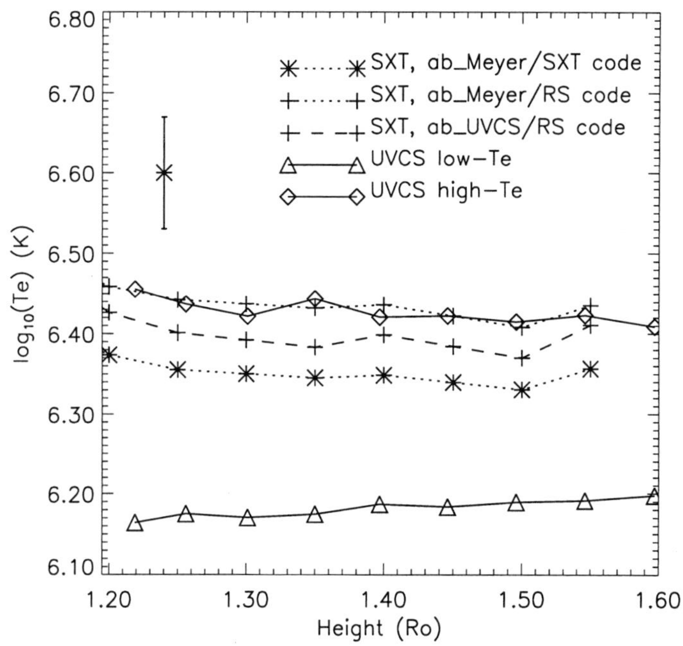

Fig. 4. The electron temperature derived from UVCS compared with those from SXT using three different approaches.

we would expect much smaller abundances than we observed here. One possibility to counteract the effect of gravitational settling is that plasma may cycle through the magnetic loops with a time scale of a few days. Another possibility is that the apparently static gas is flowing outwards, and the heavy ions are pulled along by ion drag of the protons (e.g. Ofman 2000). This latter option would suggest that very highly ionized plasma would be observable by *ACE* at times corresponding to the passage of these active regions. The decline in abundance with height would also suggest an increase with height in the ratio of outflow speed of the elements to the speed of hydrogen.

Elemental abundance is a powerful tool in understanding the coronal origin of the solar wind. This could be accomplished by comparing the elemental abundances in the corona and those in the solar wind measured *in-situ.* Previous and present work have shown that FIP effect (which is based on relative abundances of low-FIP to high-FIP elements) exists in the streamers and active region loops. However, the absolute abundances change with height and across structures. Therefore, absolute abundances, not the FIP effect solely, should be the more relevant parameter in understanding the coronal origin of the solar wind.

ACKNOWLEDGEMENTS

This work is supported by NASA grant NAG5-10093.

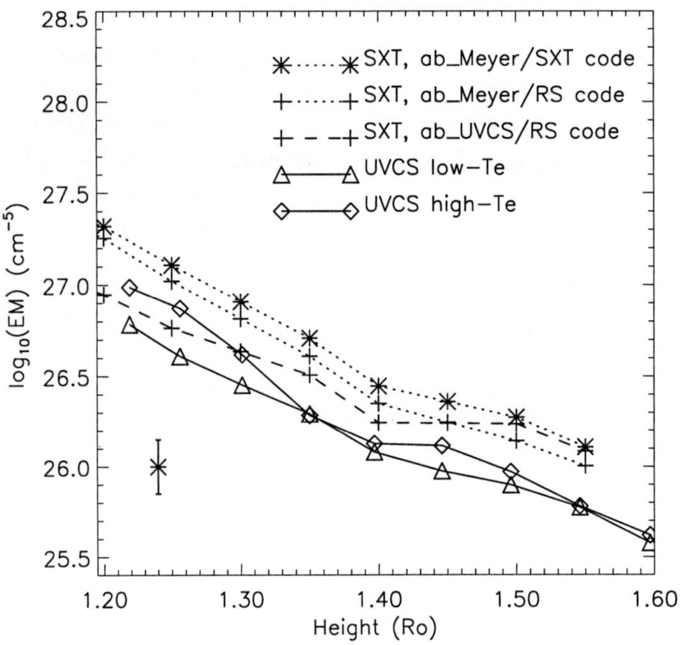

Fig. 5. The emission measure derived from UVCS compared with those from SXT using three different approaches (see text).

REFERENCES

Dere, K. P., Landi, E., Young, P. R., and Del Zanna, G., *Astrophys. J. Suppl.*, **134**, 331 (2001)

Ko, Y.-K., Raymond. J. C., Li, J., Ciaravella, A., Michels, J. *et al.*, *Astrophys. J.*, in press (2002)

Lenz, D. D., Lou, Y.-Q., and Rosner R., *Astrophys. J.*, **504**, 1020 (1998)

Mazzotta, P., Mazzitelli, G., Colafrancesco, S., and Vittorio, N., *Astron. Astrophys. Suppl.*, **133**, 403 (1998)

Meyer, J.-P., *Astrophys. J. Suppl.*, **57**, 173 (1985)

Ofman, L., *Geophys. Res. Lett.*, **27**, 2885 (2000)

Raymond, J. C., *et al.*, *Sol. Phys*, **175**, 645 (1997)

Raymond, J. C., & Smith, B. W., *Astrophys. J. Suppl.*, **35**, 419 (1977)

Schwadron, N. A., Fisk, L. A. and Zurbuchen, T. H., *Astrophys. J.*, **521**, 859 (1999)

ISOTHERMAL APPROXIMATION VS. DIFFERENTIAL EMISSION MEASURE ANALYSIS: HOW HOT ARE HOT LOOPS

J. W. Cirtain[1] and J. T. Schmelz[2]

[1]*Montana State University, P.O. Box 173840, Bozeman, MT 59717, USA*
[2]*University of Memphis, Physics Department, Memphis, TN 38152, USA*

ABSTRACT

Analysis of EUV data from both EIT/*SOHO* and *TRACE* suggests that active region loops may be isothermal. These results are in sharp contrast to the multi-thermal loops obtained from the analysis of X-ray data from the SXT instrument on the *Yohkoh* satellite. The analysis of all these observations uses an isothermal approximation, but the EUV results are derived from narrow-band filter ratios while the X-ray results use a broadband ratio. We have incorporated data from the CDS/*SOHO* instrument into the mix in two different ways: (a) we have used an isothermal approximation with different iron line ratios to determine temperatures at various pixels along a (relatively) isolated coronal loop on the limb; and (b) we have used multiple spectral lines from the same data set to produce differential emission measure distributions at these pixels. The data set was obtained from observations taken on 13 Nov 1997 by CDS, EIT and SXT. We find that different instruments and/or different methods of analysis give different results. In some sense, this is not surprising since the limitations of the isothermal approximation are well understood.

INTRODUCTION, DATA ANALYSIS AND RESULTS

The determination of the *actual* temperature of the plasma within a coronal loop has been the focus of much research for many years. With the recent deployment of many fine EUV and X-ray telescopes, much data has been collected on these structures. The information gathered was analyzed with methods prescribed by the physics of the device in use. It was our intention to determine an accurate temperature measurement for coronal loops and, as such, we used multiple methods to generate temperature and emission measure for loops with data from various instruments. Through the use of several techniques it was our hope that a correlation between the method of data analysis and the results could be shown, and furthermore, that a determination of the best way to use the current array of instruments for real temperature measurements could be found.

Spectral and imaging data of a distinct and relatively isolated loop visible on the northwest limb of the solar disk were collected on 13 November 1997 by three different instruments: spectral data in 11 different emission lines from CDS on *SOHO*; narrow-band filter data at 171, 195 and 284 Å from EIT on *SOHO*; and broadband images in the Al.1 and AlMg by SXT on *Yohkoh*. These data were co-aligned, calibrated, and analyzed using standard routines provided in SolarSoft. The atomic physics contained in the CHIANTI package was used throughout. Care was taken to insure that there was a spatial and temporal overlap of the loop features from one instrument to the next. Multiple points along the length of the loop were selected for analysis. Although impossible to present all line and filter pairs in this paper, we found the following results to be true for all pairs available.

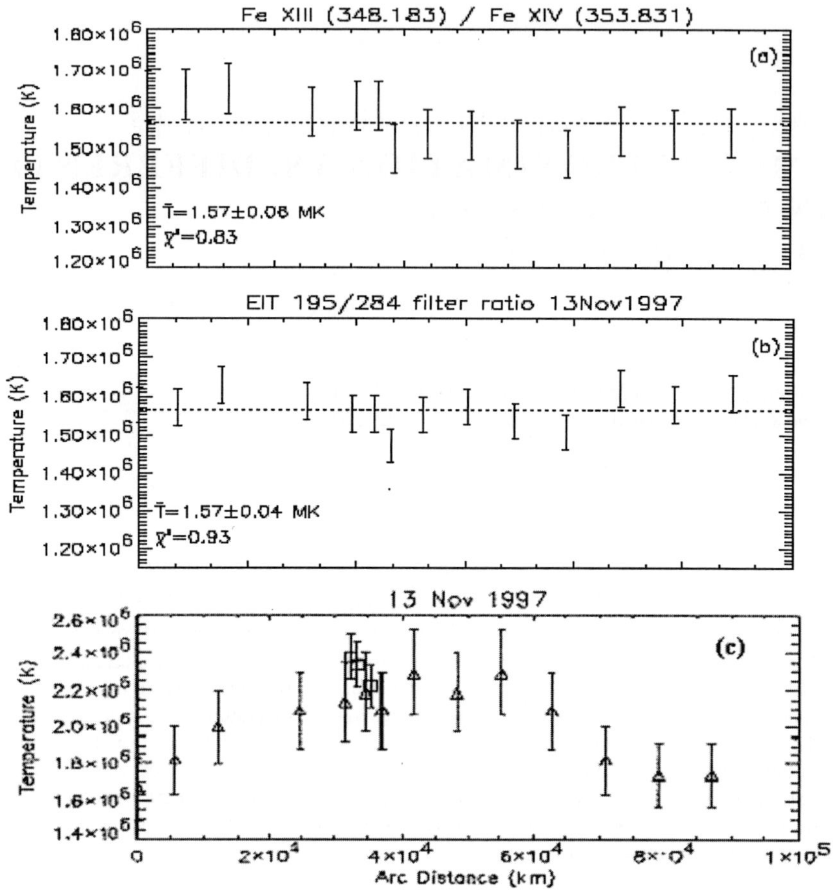

Fig. 1. Temperature as a function of arc distance along the loop.

The ratios of several CDS iron line pairs were used to determine the plasma temperature via a standard isothermal analysis; one of these results is plotted in Figure 1a, which shows the temperature plotted as a function of arc distance along the loop with the foot points of the loop at 0 and ~9e4 km. This analysis suggests that a single temperature can characterize the loop plasma. In Figure 1b the figure shows a similar result for the ratio of two of the EIT narrow band filters. It is apparent from Figures 1a and 1b that the chosen line and filter pairs produce statistically identical results. It is interesting to note, however, that other line and filter ratios produce statistically different temperatures for the same pixels.

The 11 CDS lines were also used to calculate the differential emission measure (DEM) at each of the chosen pixels. A weighted average temperature per DEM for each of these temperature distributions was calculated, and the results are plotted as the triangles in figure 1c. Contrary to the results shown in figures 1a and 1b, this CDS analysis indicates that an isothermal loop would be a poor model for the loop plasma as the distribution of temperatures is nearly 4e5K. We then plotted the temperatures determined from the SXT broadband filter ratio for several pixels near the top of the loop, which were the only points we felt could be accurately co-registered. These results are plotted as boxes in figure 1c and it is clear that they agree with those of the CDS multi-thermal analysis. An investigation into the reasons for the discrepancy among these different methods of temperature analysis is currently underway.

X-RAY JETS IN INTERCONNECTING LOOPS

F. Fárník[1] and Z. Švestka[2]

[1]*Astronomical Institute of the Czech Academy of Sciences, Ondřejov, Czech Republic*
[2]*CASS, UCSD, La Jolla, CA 92093-0424, U.S.A. and SRON Utrecht, The Netherlands*

ABSTRACT

We present examples of X-ray jets, observed by *Yohkoh*/SXT, that followed trajectories of transequatorial interconnecting loops (TILs). All these TILs were preexisting, but mostly invisible at the time of the onset of the jet that often made them bright along their total length. Most of these TIL-associated jets have properties very similar to those of other jets ejected inside active regions or along open field lines (properties such as footpoints in X-ray bright points, recurrence, strong collimation, and average speeds close to 350 km s^{-1}). However, they may reach larger lengths, up to 450,000 km. Quite a high number of X-ray jets may be of this TIL-associated kind.

OBSERVATIONS

One of the discoveries made by the Soft X-ray Telescope (SXT) on *Yohkoh* was the detection of X-ray jets by Shibata *et al.* (1992). Basic characteristics of these jets were summarized by Shibata *et al.* (1994). They are collimated features that are ejected from (micro)flare-like brightenings in active regions, emerging flux regions, or X-ray bright points. All jets described so far seemed to follow open field lines or short loop structures in active regions. We report here another kind of X-ray jets that are injected into loops that connect distant active regions, often across the equator (transequatorial interconnecting loops – TILs).

As an example Figures 1a–c show the development of a jet that originated in a large X-ray bright point to the southwest of AR 8602 on 26 June 1999 shortly before 14:20 UT. While in Figure 1a it looks like an ordinary jet similar to those described by other authors in the past, the following images reveal that the jet was injected into a TIL that connected in the south to an enhanced magnetic field region. The average jet speed was 383 km s^{-1}. The whole TIL became visible after the jet propagated through it, and stayed visible for more than 10 hours thereafter.

Observations of systems of TILs both on *Skylab* and *Yohkoh* showed that these transequatorial connections often last for many days, although their visibility in X-rays during this period is very variable. The same is true for the TIL in which the jet of 26 June appeared. It must have existed already on 25 June, as one could see another jet at 12:29 UT on that day (Figure 1d) that clearly followed the same trajectory as the jet in Figures 1a–c. This jet event on 25 June demonstrates the recurrence of TIL-associated jets, but its observation also has another important consequence: there are no other SXT images of this jet available; therefore, if we did not see the time development of the jet on 26 June, we would have no idea that the jet on 25 June was injected into a TIL. Thus, indeed, there may be many other jets observed at other times that were injected into a TIL, and we have no evidence for it.

Very similar events were observed on 26 April 2000 when, after 06:15 UT, a jet was injected into a TIL that

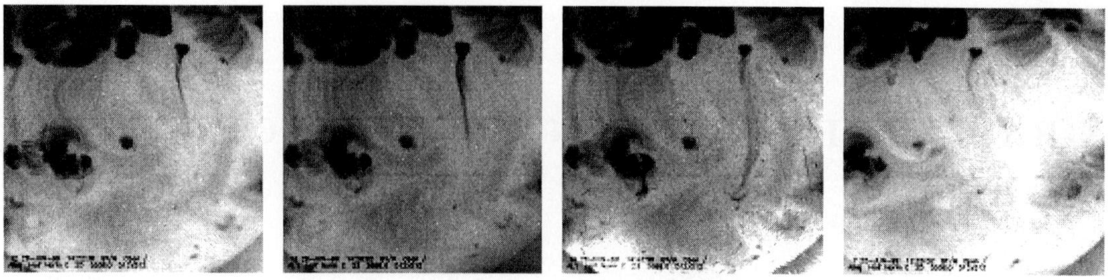

Fig. 1. From the left: (a–c) X-ray jet observed by *Yohkoh*/SXT on 26 June 1999: (a) 14:22:38 UT, AlMg filter, jet length 175 000 km. (b) 14:29:02 UT, Al.1 filter, jet length 308 000 km. (c) 14:43:58 UT, Al.1 filter, jet (TIL) length 455 000 km. (d) Image of another X-ray jet one day earlier, at 12:28:32 UT on 25 June 1999.

connected AR 8970 in the south with AR 8971 in the north. The total length of the jet (TIL) was 336,000 km. A recurrent jet was ejected along the same TIL eight hours later, with a speed of 420 km s^{-1} during the first (eight minute) period and >210 km s^{-1} during the second (less than eight minute) period of the jet propagation. On 9 April 1999 a TIL connected AR 8511 in the south with AR 8507 in the north. A series of recurrent jets appeared during that day near the southern footpoint, but this time the jets did not propagate through the whole length of the TIL. The speed of propagation was found to be 340 km s^{-1}. The jet's speed is a rough estimate only, because of the small number of SXT images available.

CONCLUSIONS

Injections of jets into loops that connect active regions (in all our examples across the equator - TILs) seem to be phenomena that occur frequently and, obviously, are one of the ways pre-existing magnetic interconnections become visible in soft X-rays. As we demonstrated in Figure 1d, there are jets injected into such loops that we would never include in this category. With the lifetime of jets in the range of minutes, and the lifetime of TIL brightenings in the range of hours, it is likely that we see many TIL brightenings caused by injected jets without noticing the jets themselves. Therefore, jets injected into interconnecting loops may represent a very significant fraction of all jets observed.

ACKNOWLEDGEMENTS

This work was supported by the Grant Agency of the Academy of Sciences of the Czech Republic Nos. 3003802, 3003003 and by the project No. S1003006 under the Key Project of the Astronomical Institute K2043105.

REFERENCES

Shibata, K., Ishido, Y., Acton, L.W., Strong, K.T., Hirayama, T., Uchida, Y., McAllister, A.H., Matsumoto, R., Tsuneta, S., Shimizu, T., Hara, H., Sakurai, T., Ichimoto, K., Nishino, Y., and Ogawara, Y. in *Publ. Astron. Soc. Japan* **44**, L173 (1992).

Shibata, K., Yokoyama, T., and Shimojo, M., in S. Enome and T. Hirayama (eds.), *Proceedings Kofu Symposium*, Nobeyama Radio Observatory, p. 75 (1994).

CONVECTIVE STRUCTURE IN AN EMERGING FLUX REGION

H. Kozu[1] and R. Kitai[2]

[1]*Kwasan Observatory, Kyoto University, Yamashina, Kyoto, Japan*
[2]*Hida Observatory, Kyoto University, Kurabashira, Gihu, Japan*

ABSTRACT

The magnetic field in the Sun is created near the bottom of the convective layer and emerges through the convective zone as emerging flux loops. Convection can play an important role at this stage. In this paper we analyzed NOAA 8582 with a Local Correlation Tracking Method (LCTM) in order to study the convective structure in the Emerging Flux Region (EFR). Besides the varying convective structure we found temporary continuous upflow structures. Such a structure existed under the loop top of each flux loop continually for about 100 minutes.

INTRODUCTION

Vertical velocity fields in the EFR have been studied with spectroscopic methods for several decades. Many authors reported downward flow at the footpoints of emerging flux in the early stage of pore formation (Kawaguchi & Kitai 1976, Zwaan *et al.* 1985, etc.). In contrast, upward velocity structures around pores are not observed so often (Brants 1985, Lites *et al.* 1998) probably because their speed is not so large. One efficient method to understand such velocity structures is to study horizontal flow patterns.

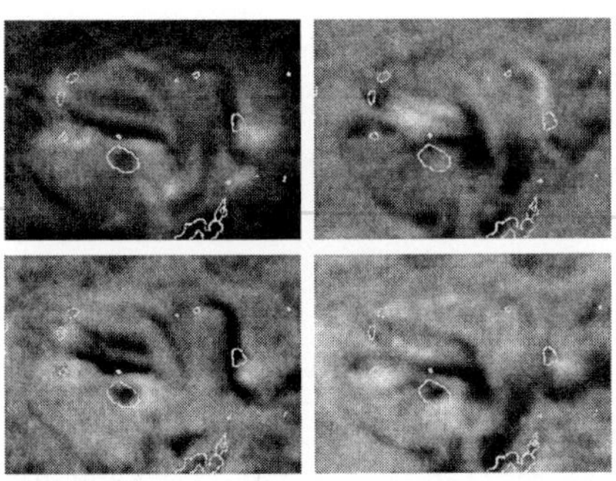

Fig. 1. Hα images with G-band contours. *left-upper*: line center; *left-lower*: Hα-0.6A; *right-lower*: Hα+0.6A; *right-upper*: red - blue. Each pane is 46.80Mm$\times$33.43Mm.

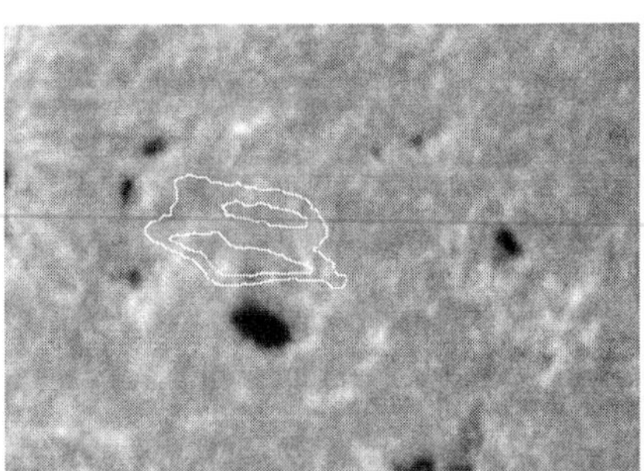

Fig. 2. G-band image and EF contour in Hα blue wing. Image size is 46.80Mm$\times$33.43Mm.

OBSERVATIONS AND ANALYSIS

We observed a newly born EFR NOAA8582 (N26E30) on 11th June 1999 from 21:24UT to 23:04UT with the Domeless Solar Telescope in Hida Observatory. We used the Real-Time Frame Selector2 for G-band (Figure 2) and a Lyot-Filter for Hα observations (Figure 1). No flare occurred but many surge activities were seen. The relation between pores and emerging fluxes is shown in Figure 2. We can see two distinct positions of emerging flux loops beside the largest pore. Two small pores at the upper left correspond to their footpoints. We applied LCTM to the series of G-band images. LCTM uses the correlation of intensity distribution in a small box in order to trace the motion of granules in the box. With this method we can derive the horizontal velocity field at the photosphere. We applied the following parameters for LCTM: the size of box is 1.92 arcsec; and the time separation between two images compared is 3 minutes. The result changes depending upon these parameters so we tested our set of parameters with a reverse test, which requires that the time-reversed LCTM result should be the reverse of the normal one.

RESULTS AND DISCUSSION

Horizontal velocity (arrows) and its divergence image averaged over the whole period are shown in Figure 3. The average speed is 0.145 km s^{-1} and the speed of the longest arrow corresponds to 0.487 km s^{-1}. The error ratio estimated by comparing results from other sets of LCTM parameters is about 50-100% for these values. Even with this large error the morphological divergence structure is reliable because its pattern doesn't change much around the best set of parameters. In Figure 3 we can see three strong divergent (white) structures and two of them locate beneath the loop tops of emerging flux loops. In the divergence movie we can see that these two divergence structures are rather stable. In interpreting the divergence structure as upward flow, this result is consistent with early works with spectroscopic methods (Brants 1985, Lites *et al.* 1998) but our result indicates that the upward flow is maintained even after it becomes too weak for detection with spectroscopic methods. Our result is also consistent with one reported by Strous *et al.* (1996), who used Feature Tracking (FT) and derived similar divergence structures in the horizontal field.

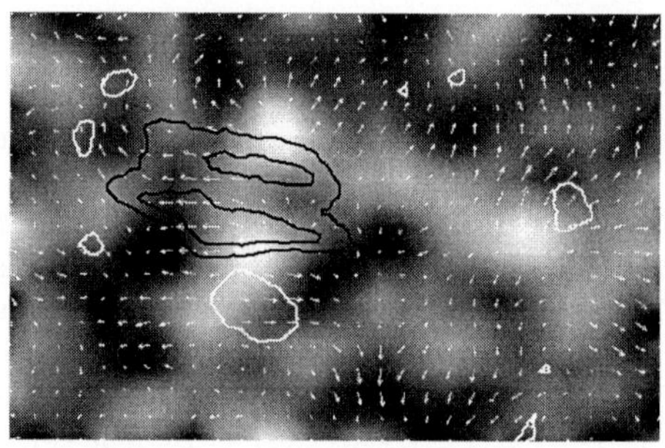

Fig. 3. Image shows flow velocity (arrows) and its divergence. White and black gray scale indicates divergence and convergence respectively. White and black contours indicate positions of pores and EF loops respectively. Image size is 37.63Mm×24.19Mm.

This work was supported in part by a Research Fellowship of the Japan Society for the Promotion of Science for Young Scientists (3583).

REFERENCES

Brants, J.J., *Solar Phys.*, **98**, 197 (1985).

Kawaguchi, I. & Kitai, R., *Solar Phys.*, **46**, 125 (1976).

Lites, B.W., Skumanich, A. & Pillet, V.M., *Astron. and Astrophys.*, **333**, 1053 (1998).

Strous, L. H., Scharmer, G., Tarbell, T. D., Title, A. M., and Zwaan, C., *Astron. and Astrophys.*, **306**, 947 (1996).

Zwaan, C., Brants, J.J. & Cram, L.E., *Solar Physics*, **95**, 3 (1985).

FREQUENCY DRIFT RATE MEASUREMENTS OF CORONAL TEMPERATURES

V. Krishan*, F. C. R. Fernandes, and H. S. Sawant

Divisão de Astrofísica - INPE, Cx. Postal 515, 12201-970, São José dos Campos - SP

ABSTRACT

The frequency drift rates of radio emission are traditionally used to determine the velocity of the exciting agency for a chosen coronal density model. The speed of the exciting agency, say an electron beam, is assumed to remain constant during its propagation through the radio emission region. Here, we allow the electron beam to decelerate either due to its collisions with the ambient coronal particles or due to any of the diffusion and plasma transport mechanisms. The deceleration is related to the time derivative of the frequency drift rate. Thus, assuming the plasma mechanism for the radio emission combined with the slowing down of the electron beam enables us to self consistently determine the plasma density profile and the temperature of the radio emitting region. Conversely, the frequency dependence of the drift rate can be determined for a given temperature of the emission region. A comparison with the observed drift rate can then tell us about the validity of the beam slow-down model.

INTRODUCTION

The bursty radio emission originating in the solar corona often shows a drift in its frequency of emission. The frequency drift arises due to the motion of the radio source (energetic electron beam) through a plasma of varying density such as the solar corona and the fact that emission frequency is a function of the coronal density. From the measured drift rate, the speed of the electron beam, assumed to remain a constant, is determined for a known density model. However, the beam may suffer collisional and or diffusional losses (Takakura & Shibahashi, 1976) during its propagation. The resulting deceleration can modify the frequency drift rate. Thus, it is shown here that a self consistent model of the radio emitting region, particularly its temperature, can be derived by including the physics of the deceleration of the beam.

TEMPERATURE AND THE BEAM DECELERATION

In the fluid model, the evolution of an electron beam of velocity u, undergoing collisions with a stationary plasma (the corona) can be described by $du/dt = -\nu u$, (Tanenbaum 1967), where the collision frequency ν is, in general, a function of the beam and the coronal parameters. For the plasma emission model, the radio frequency f (MHz) is related to the coronal density n as $f = f_o(n/n_o)^{1/2}$. This provides a relation between ν and f . The frequency drift rate df/dt of the radio emission is given by $df/dt = fu/2H_n$, where $H_n^{-1} = (1/n)\ (dn/ds)$, whith H_n^{-1} the characteristic spatial scale of the density variation and dn/ds denoting

*On sabbatical leave from the Indian Institute of Astrophysics, Bangalore, India

the density gradient along the propagation path of the radio source. From the above equations we find:

$$\nu = f^{-1}(df/dt) - (d^2f/dt^2)(df/dt)^{-1} . \tag{1}$$

Thus, the drift rate gets related to the coronal and beam parameters through the collision frequency. For the sake of an illustration, we choose an isotropic velocity distribution for the radio source since such a distribution has been shown to produce strong plasma emission (Melrose 1985). Thus, we model the Coulomb collision frequency as $\nu = \nu_o(f/f_o)^2$ with $\nu_o = 5 \times 10^{-4} f_o^2/T_{c6}^{1.5}\ g(x)\ \mathrm{s}^{-1}$, where x is the ratio of the beam energy and coronal thermal energy and g is a dimensionless function. Here we make the well-accepted assumption that the coronal temperature remains nearly constant in the emission region of a specified frequency band such as the decimeter, the meter and the decameter bands. All the zero subscripted quantities refer to the region of the starting emission frequency f_o. With this model, from the solution of Eq. 3 we find under the assumption $x \simeq 1$ for which $g \simeq 1$,

$$\begin{aligned} & df/dt = -\nu_o f^3/2f_o^2\ ; \quad u/u_o = (f/f_o)^2 = n/n_o = exp(-s + s_o)/|H_{no}|\ ; \\ & f/f_o = exp(\nu_o t)[1 - 0.5(1 - exp(2\nu_o t))]^{-0.5} \quad \text{and} \quad T_{c6} = [2.5 \times 10^{-4} f^3 (df/dt)^{-1}]^{2/3} . \end{aligned} \tag{2}$$

Thus, the knowledge of the drift rate enables us to determine the temperature of the radio emitting region for a given radio source.

DISCUSSION AND CONCLUSIONS

In order to determine the coronal temperature from the frequency drift rate, we need to know ν for which we need to know the distribution function of the radio source. The radio source could be a monoenergetic electron beam, a drifting Maxwellian with a large or small thermal spread or an isotropic thermal distribution. The latter produces strong plasma emission for $x > 1$. However, for a relatively weak radio source, $x \geq 1$, and $g \simeq 1$. Using the typical observed values of the drift rates, we find $T_{c6} \simeq 1$ for $|df/dt| = 250$ MHz s^{-1} at $f = 100$ MHz and $T_{c6} \simeq 40$ for $|df/dt| = 1000$ MHz s^{-1}. Further, it is seen that df/dt should vary as f^3. This, of course, is a consequence of our choice of the collision and plasma emission models valid only for an isotropic distribution of a weak radio source. It remains to be seen how the other distributions change it. In addition, the collision process may be other than the Coulomb. Anomalous collisions such as due to plasma instabilities may be operating. In this case the collision frequency ν will have a different density and temperature dependence and consequently lead to a different estimation of the temperature. The beam may propagate in a mode other than the free streaming mode. It may be undergoing one of the anisotropic diffusion processes with different parallel and perpendicular diffusion coefficients and slowing down as (Huba 1994):

$$d/dt(u - \langle u \rangle)_\perp^2 = \nu_\perp u^2 \quad \text{or} \quad d/dt(u - \langle u \rangle)_\parallel^2 = \nu_\parallel u^2 , \tag{3}$$

where the perpendicular and parallel ν depend on the temperature as $T^{-1/2}$ instead of $T^{-3/2}$. This will produce completely different estimates of the temperature along with different density and drift rate profiles. This investigation along with other issues such as beam distribution function and variation of the beam and coronal temperatures is in progress.

REFERENCES

Huba, J.D., *NRL Plasma Formulary*, 31 (1994).

Melrose, D.B., in *Solar Radiophysics*, Eds. D.J. McLean and N.R. Labrum (1985).

Takakura, T. and Shibahashi, H., *Solar Phys.*, **46**, 323 (1976).

Tanenbaum, B. S., *Plasma Physics*, p. 109, 251, McGraw-Hill Book Company, New York (1967).

MAGNETIC FIELDS IN THE PHOTOSPHERE ARE NOT FORCE-FREE

B. J. LaBonte

Johns Hopkins University, Applied Physics Laboratory, 11100 Johns Hopkins Road, Laurel MD 20723-6099, USA

ABSTRACT

Coronal magnetic fields are often inferred from the extrapolation of photospheric magnetic observations. The assumptions that the fields are potential or force-free are not correct in the photosphere. Therefore the extrapolation should be done by a method that accounts for this problem.

ANALYSIS AND INTERPRETATION

The magnetic field in the corona can be inferred using magnetic fields observed in the photosphere as a boundary condition if the nature of electric currents in the solar atmosphere is also known. The degree to which the field in the photosphere may be represented as force-free is a topic of debate (Metcalf *et al.* 1995; Moon *et al.* 2002). I show here that the predictions of field extrapolation are in gross disagreement with observations in the photosphere, for an obvious reason: the high electrical conductivity of the solar atmosphere.

The Imaging Vector Magnetograph (IVM) at Mees Solar Observatory, Haleakala, observes Stokes spectra over an active region field of view repeatedly through each day. I use an observation of AR8592 on 22 June 1999 taken from 17:41-1745 U.T. in the 630.3 nm line, with the region at North 21°, East 37°. The magnetic field was inferred using an inversion procedure based on the work of Landolfi & Landi degl'Innocenti (1982). Figure 1 shows that, over most of the plage, the transverse component of the magnetic field is well measured above the noise level of 52 G. The line-of-sight component of the field was used to compute potential and force-free fields (with spatially constant ratio of current to field, α) for the height of observation, based on the formulae of Alissandrakis (1981) and Gary (1989). The field azimuths were resolved using several algorithms to check for consistent results. The observed fields were rotated into coordinates oriented to the solar surface for comparison with the extrapolations. Figure 2 shows that the fields are actually radial in the plage. By contrast, both extrapolations predict substantial horizontal fields over most of the field of view. The median transverse field is 60 G in each extrapolation. Fields of such strength are not observed; the extrapolations do not describe the Sun.

The reason for the strict confinement of the observed fields is well known. The high electrical conductivity excludes the fields from the nonmagnetized plasma (Cowling 1946). Buffeting of fluxtubes on timescales of a few minutes implies that there must be a sheath of order 300 m surrounding each strong field, with a current that cancels the external field. The sheath must be continuous, turning horizontal as the fluxtubes expand, to continue to exclude the fields from the photosphere in the regions between strong radial fields. The difference of the extrapolated horizontal field from the observed contains information on the strength and locations of that sheath current. The sheath current doubles the strength of the field above it, given that it must cancel the field below it.

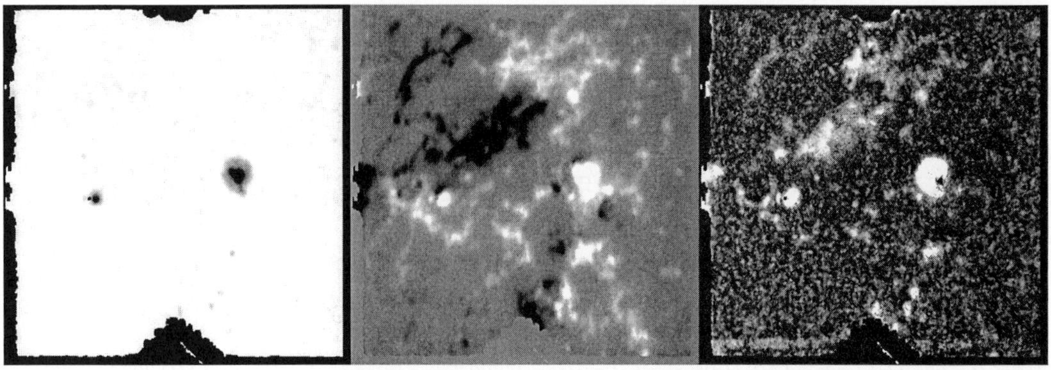

Fig. 1. AR8592 in the continuum, line-of-sight magnetic field scaled to ±300 G, transverse field magnitude scaled to 300 G.

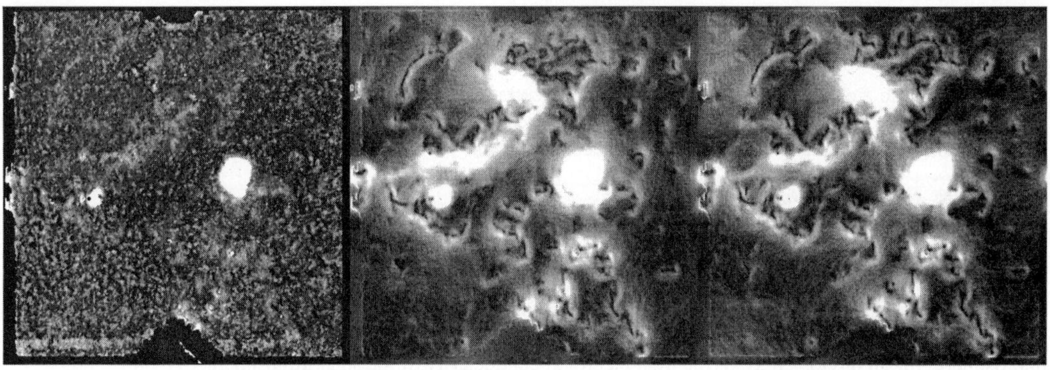

Fig. 2. Horizontal magnetic field magnitude in the solar surface all scaled to 300G. Left to right: observed field; potential extrapolation; force-free extrapolation for $\alpha = 0.01$, close to sunspot values.

The average filling factor of magnetic field in the photosphere is 8.5% over this active region. As the magnetic fields expand with height the filling factor becomes 100%. Spreading the observed flux over the region area yields the average field strength of 65 G, comparable with the extrapolations. The magnetic pressure of the overlying field then corresponds to the total pressure of an unmagnetized atmosphere at an altitude of 600 km, about 350 km above the height of observation (Maltby *et al.* 1986). Gas pressure in the magnetized layers will increase the total pressure. The magnetic structure from the photosphere into the lower corona will be distorted from that predicted by extrapolation. I conclude that electric currents in the photosphere are neither absent nor exclusively field-aligned, and the presence of currents affects magnetic structure and its extrapolation. Only above the chromosphere can the present methods be used to extrapolate the magnetic field into the corona. I thank K. D. Leka and T. R. Metcalf for major elements of the software development that was needed to analyze the IVM data. Observations at Mees Solar Observatory are supported by NASA Grant NAG5-4941.

REFERENCES

Alissandrakis, C. E.: *Astron. Astrophys.*, **100**, 197 (1981).
Cowling, T. G., *M.N.R.A.S.*, **106**, 218 (1946).
Gary, G. A.: *Ap. J. Supl.*, **69**, 323 (1989).
Landolfi, M., and Landi degl'Innocenti, E.: *Solar Phys.*, **78**, 355 (1982).
Maltby, P., Avrett, E. H., Carlsson, M, Kjeldseth-Moe, O., Kurucz, R. L., and Loeser, R.: *Ap. J.*, **306**, 284 (1986).
Metcalf, T. R., Jiao, L., McClymont, A. N., and Canfield, R. C.: *Ap. J.*, **439**, 474 (1995).
Moon, Y., Choe, G. S., Yun, H. S., Park, Y. D., and Mickey, D. L. *Ap. J.*, **568**, 422 (2002).

SOHO-CDS OBSERVATIONS OF AN X2 FLARE SPRAY INJECTION

H. E. Mason[1] and C. D. Pike[2]

[1]*Department of Applied Mathematics and Theoretical Physics, Silver Street, Cambridge CB3 9EW, UK*
[2]*Rutherford Appleton Laboratory, Chilton, Didcot, OXON OX11 0QX, UK*

ABSTRACT

An X-class flare was reported on 10 April 2001 in AR 9415, with a halo Coronal Mass Ejection (CME) associated. The Coronal Diagnostic Spectrometer (CDS) on board the Solar and Heliospheric Observatory (*SOHO*) recorded a very high velocity ejection of plasma. The spatial scanning and spectral capabilities of CDS allow the measurement of both transverse and line-of-sight velocities. Components of the plasma, seen in emission from OV at around 2.5 x 10^5 K, reached transverse velocities exceeding 600-800 km/s. The nature of the spectral line profiles suggests a rotational motion superimposed upon the general outward expansion. The ejection detected using CDS is thought to be a spray of plasma with a helical structure driven by the magnetic topology.

CDS OBSERVATIONS

The nature of plasma ejected during the initial stages of a flare is a subject of intense debate. Rarely has this plasma been observed with an EUV spectrometer, which allows measurement of velocities both along the line-of-sight and in the transverse direction. The *SOHO*-CDS instrument on *SOHO* allows just such an analysis for plasma covering a wide range of temperatures.

On 10 April, 2001, there was a class X2.3 flare in AR 9415. The soft X-ray emission began to increase at around 05:00 UT. The CDS observations consist of a 4' ×4' raster using the $2'' \times 240''$ slit. Here we consider only the O V (629.73 Å) emission. The CDS slit moves across the solar image from W to E (right to left), with the time between successive exposures of about 21s. The first indication of the high velocity event in the CDS data occurs at 05:07:26 UT when the OV profile at the northern limit of the CDS field began to show a bias to positive (red-shifted) velocities. The strong blue-shifted component of OV is first seen at 05:10:41 UT. This time corresponds with the sharp rise in X-ray emission from the flare, which was happening to the north-west. The soft X-ray emission did not reach its peak until about 05:25 UT. The X-ray images prior to the flare show the type of 'sigmoidal' structure frequently associated with eruptive phenomena.

Most of the OV emission associated with the ejection is seen in the form of *blobs* observed at different Y values for each exposure. From their nature, we conclude that these are spatially small, dynamic features, which are being *strobed* by the CDS raster. Analysis of these blobs indicates a transverse velocity of around 600-800 km/s.

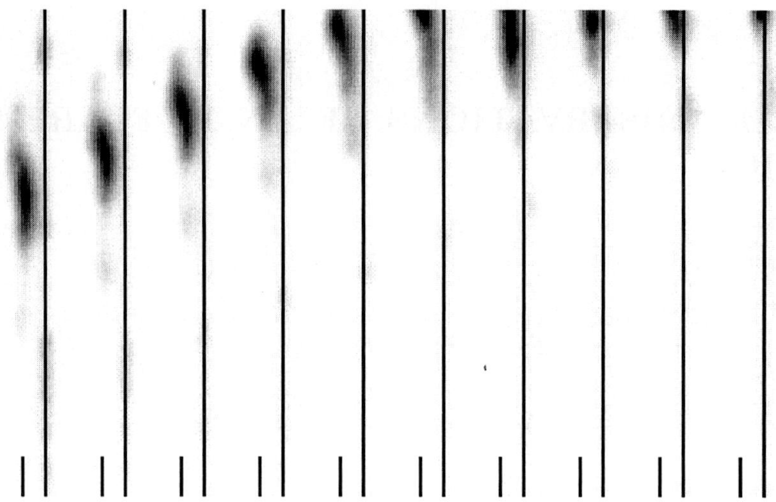

Fig. 1. CDS scans of O V. The individual images have dimensions of y axis: Solar-Y and x-axis: wavelength. The long vertical black bar marks the location of the zero-velocity component and a shorter tick mark indicates a relative velocity of -400km/s (blue-shifted). The time sequence of images is from right to left.

Figure 1 shows the spray ejection in the CDS OV spectral window as it crosses (and CDS scans) the field of view. The CDS slit images are shown at ten spatial locations. The axes of each slit image are y-axis: solar Y (along the stigmatic slit, which is oriented N-S) and x-axis: wavelength. The full line indicates the rest wavelength for O V and the short tick mark indicates a blue-shift in the spectral line profile of 400 km/s. The time dimension goes from right to left. The blue-shifted feature is clearly visible, moving down along the Y-axis (southward) with time. The feature is large enough to remain in the field of view as CDS scans (20 arcseconds) in solar longitude during these observations. The matter imaged in the northward end of the feature is rising more rapidly (blueshifted more) than the southernmost part.

In a more detailed analysis of the CDS data (Pike & Mason, 2002), we identified three component structures for the ejection. We saw the red-shifted feature first, followed by the extreme blue-shifted feature. At around 5:12 UT, the main part of the ejection, the brightest feature, is observed. The OV spectral line is both blue-shifted and broadened. The CDS data are consistent with a large rotating structure with filamentary structures. This is also the impression given by the *TRACE* movie.

ACKNOWLEDGEMENTS

H.E.Mason acknowledges the support of PPARC and the Royal Society. *SOHO* is a project of international co-operation between NASA and ESA.

REFERENCES

Pike, C.D. and Mason, H.E., *Solar Phys.*, in press (2002).

MULTI-TEMPERATURE STRUCTURE OF THE SOLAR CORONA OBSERVED BY YOHKOH AND SOHO

S. Nagata

Institute for Space and Astronautical Science, Yoshinodai, Sagamihara, Kanagawa, Japan

ABSTRACT

The relationship between higher-temperature ($T > 2$ MK) coronal loops seen with *Yohkoh* SXT and lower-temperature (~ 1 MK) loops seen with *SOHO* EIT is discussed. Hot and cool loops are not co-spatial, and their observed lifetime is much longer then the estimated cooling timescales. This suggests that each loop has its own destined heating rate, and that there are high heating-rate loops and low heating-rate (dormant) loops.

INTRODUCTION

In order to reveal the heating mechanism of the Solar corona, we have to know the thermal evolution of the magnetic loops of which the corona is composed. In other words, it is necessary to investigate the relationship between higher-temperature structures that can be observed with *Yohkoh* SXT (Yoshida & Tsuneta 1996, Kano & Tsuneta 1995), and lower-temperature structure that can be observed in *SOHO* EIT, *SOHO* CDS, and *TRACE* (Aschwanden *et al.* 1999, 2000; Flaud *et al.* 1997). In this poster paper, we present the multi-temperature structure and its evolution revealed by the simultaneous observations by the *Yohkoh* Soft X-ray Telescope (SXT), the Extreme-ultraviolet Imaging Telescope (EIT) aboard Solar and Heliospheric Observatory (*SOHO*) and the XUV Doppler Telescope (XDT) aboard a sounding rocket.

OBSERVATION AND ANALYSIS

We analyzed the data taken by the three telescopes on January 31, 1998 for the XDT flight and investigated the time evolution of Active region NOAA8143. We found that coronal loops with different temperatures are not co-spatial. We then tracked the evolution of eight cool loops (~ 1 MK) and eight hot loops ($T > 2$ MK) with *Yohkoh* and *SOHO* and found that the hot and cool patterns are conserved more than 6 hours. We examined whether the cool loops are remnants of hot loops or not by employing an analytical model of a coronal loop. If the cool loops are remnants of hot loops, the duration during which the loops can be observed with EIT is found to be ~ 1.5 hour; this is too short as compared with the observations. From this observation and an analytical calculation on loop evolution, we concluded that the cool loops are a different entity from the hot loops.

The observed lengths of the cool and hot loops are almost the same ($\sim 10^{10}$ cm), and their densities are also almost the same (1–3 $\times 10^9$ cm^{-3}). The temperature-column density distribution of these loops are shown in Figure 1. Hot loops seen with SXT are shown as S1–S8, and cool loops seen with EIT are shown as E1–E8. The SXT loops analyzed by Kano & Tsuneta (1996), and the EIT loop analyzed by Aschwanden *et al.* (1999) are also shown. In this figure two examples of the generalized scaling law with heating function

$H = H_0 \exp(-T/\beta)$ are shown (Landini & Monsignori Fossi 1978). The lower line corresponds to uniform heating over the loop ($\beta = 0$), and the upper line corresponds to a loop where the heating rate decreases with temperature ($\beta = 2.4$). SXT loops are located in the conduction dominant regime, and these loops and our results are located around the scaling law given by power index $\beta = 0$. On the other hand, EIT loops are located in the radiation dominant regime.

The difference between the cool loops ($\sim$ 1MK) and the hot loops ($T > 2$MK) might be explained by the difference in the heating function. The amount of energy required to sustain a hot loop ($\sim 10^7$ erg cm^{-2} s^{-1}) is larger than that of a cool loop ($\sim 2 \times 10^6$ erg cm^{-2} s^{-1}). Our conclusion thus suggests that each loop has its own destined heating rate; the heating rate is determined by the local conditions of each loop.

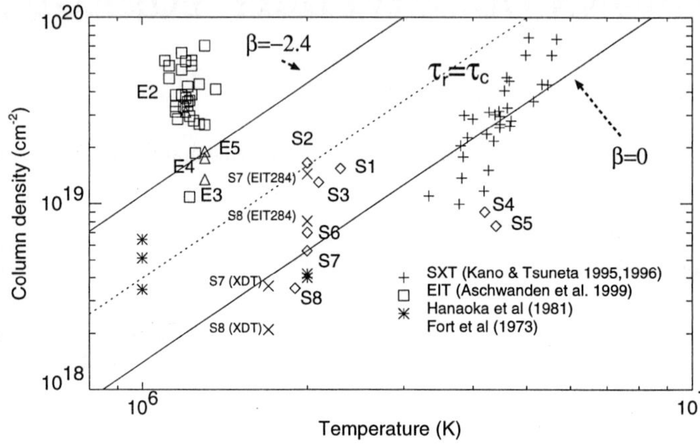

Fig. 1. Temperature-column density distribution of hot and cool loops.

DISCUSSION AND CONCLUSION

Based on our findings, we analyzed two coronal heating models: (a) a continuous heating model such as waves and (b) a nanoflare heating model that assumes the loops are heated by numerous small impulsive events. In the continuous heating scenario, the heating mechanism of the loops is examined through the scaling law for plasma loops; we found that the energy release in a hot loop is enhanced at the upper portion of the loop, whereas that of the cool loops is enhanced at the footpoints. In the nanoflare heating scenario, the amount of the elemental energy release in hot loops is larger than in cool loops.

REFERENCES

Aschwanden, M. *et al.* Three-dimensional Stereoscopic Analysis of Solar Active Region Loops. I. SOHO/EIT Observations at Temperatures of $(1.0 - 1.5) \times 10^6$ K *ApJ*, **515**, 842, (1999).

Aschwanden, M. *et al.* Evidence for Nonuniform Heating of Coronal Loops Inferred from Multithread Modeling of TRACE Data, *ApJ*, **541**, 1059, (2000).

Flaud, A. *et al.* Active Region Observed in Extreme Ultraviolet Light By The Coronal Diagnostic Spectrometer on SOHO, *Sol.Phys.*, **175**,487, (1997).

Kano, R., & Tsuneta, S., Scaling Law of Solar Coronal Loops Obtained with Yohkoh, *ApJ*, **454**, 934, (1995).

Landini, M., & Monsignori Fossi, B. C., Coronal Loops in the Sun and in the Stars, *Astron. Astrophys.*, **102**, 391, (1981).

Yoshida, T. and Tsuneta, S., Temperature Structure of Solar Active Regions, *ApJ*, **459**, 342, (1996).

COMBINING SXT AND CDS OBSERVATIONS TO INVESTIGATE CORONAL ABUNDANCES

H. D. Winter III and J. W. Cirtain

Montana State University, P.O. Box 173840, MT 59717-3840, USA

ABSTRACT

It has been shown that elemental abundances vary up to a factor of about four from the photosphere to the corona and that this variation is based on the first ionization potential of each element (Fludra *et al.*, 1999). Also, these abundances can vary greatly from feature to feature in the corona and can vary within a coronal structure such as a streamer or loop. Accurate knowledge of the elemental abundances is vital for characterizing solar corona plasma and observing how elemental abundances vary may provide insight into the underlying mechanisms causing the variation. There has not yet been a satisfactory theory as to the causes of these abundance variations that can be constrained by observations and none predict variations to occur within the corona (Fludra, *et al.* 1999). There are currently three different empirical models describing the average coronal elemental abundance (Fludra & Schmelz 1999) and these are averages over whole sun and do not take into account variations between coronal features. In order to study the abundances of a coronal loop, observations by the *Yohkoh* soft x-ray telescope (SXT) and the *SOHO* coronal diagnostic spectrometer (CDS) were combined using a multi-thermal differential emission measure (DEM) technique. This method of combining SXT broadband data with spectral data has proven to be very beneficial in the characterization of corona plasma parameters (Schmelz *et al.* 2001). A preliminary analysis of this loop suggests not only that the abundances values may differ from the Meyer (1985) values, which are used as the input for standard SXT flux calculations, but also that there may also be a variation of elemental abundances along the coronal loop structure.

DATA AND METHODOLOGY

On 18 September 2001 a coronal loop was observed by both CDS and SXT as part of *SOHO* joint observing program 146. The loop structure is well resolved by CDS in all waveband windows and co-temporal data for the region was obtained with SXT's AlMg filter. Data was acquired for three points along the loop corresponding to CDS pixels. The loop base is identified as Region A, Region B is midway between the base and the apex of the loop and Region C is at the loop top. Each CDS pixel was then co-registered with SXT pixels with 10% of the observed value being the assigned error of each measurement. The emissivities for the spectral lines under study were calculated using the CHIANTI Version 3.03 database using the abundance values of Meyer (1985), the ionization balance equations of Arnaud & Rothenflug (1985) with the Arnaud & Raymond (1992) adjustments for iron, and an estimated mean electron density of 5.0×10^9 cm^{-3} which was obtained by the method of Schmelz & Winter (1999). An input model for the DEM curve was then folded through the spectral line emissivities and the response function for SXT's AlMg filter, using Arnaud & Raymond (1992) ionization balances, in order to generate predicted intensity values. The DEM curve was then iteratively adjusted until the best agreement between predicted and observed data is reached, which is quantified by a reduced chi-square value. The number of ratios used in the analysis provides an important constraint with the SXT AlMg filter constraining the high temperature portion of the DEM curve. The abundances for each element were then adjusted and the predicted intensities were recomputed. This iterative process continued until the reduced chi-square reached a minimum value. The abundances were adjusted only for elements with more than one observed spectral line in a pixel. This was to prevent the data from being skewed by any one line, which may have been misidentified or have errors in the atomic physics calculations. Iron abundances were also left static due to the high response of SXT to Fe lines, although iron lines were used in the establishment of the DEM curve.

RESULTS

The first ionization potential of each element used in this analysis is listed in Table 1. Also listed in this table are the final percentages of the abundance for each element used with respect to the original Meyer value, with dashed lines representing abundances that were not adjusted for reasons mentioned previously. As can be seen, oxygen was the only element that seemed to maintain a stable abundance over the entire loop, while other elements required a great deal of adjustment in order to best fit the data. The initial reduced chi-square values for each region using Meyer abundance values are listed in Table 2 along with the reduced chi-square values for each region after abundance adjustment. As can be seen the fits between predicted and observed intensities for each region were greatly improved by the variation of abundance values. The greatest improvement was in Region C, which is at the loop top. This is primarily due to the Si X 347.403 line, which had a high initial predicted to observed ratio.

Table 1. Elemental adjustment with respect to Meyer abundance values.

Element	**First Ionization Potential (Fludra et al., 1999)**		**% with respect to Meyer abundances**	
		Region A	**Region B**	**Region C**
Mg	7.61	200%	150%	130%
Si	8.12	80 %	75%	55%
O	13.55	100%	90%	100%
Ne	21.47	------	100%	375%

Table 2. Reduced Chi-Square Analysis.

	Region A	**Region B**	**Region C**
Reduced Chi-Square Before Elemental Adjustment	16.1	10.1	35.6
Reduced Chi-Square After Elemental Adjustment	8.67	5.79	8.15

Due the way the CDS instrument builds rasters across the field of view, it is not possible to tell whether the abundance variations shown are a temporal or spatial feature of the loop analyzed. However, this very preliminary analysis of abundance variation along loop structures suggests that elemental abundances may not be constant across a loop structure and should not be ignored in future analysis. Further abundance analysis of loop structures utilizing modified SXT response files that reflect the changes in assumed abundances is suggested.

REFERENCES

Arnaud, M., Raymond, J. 1992, *ApJ*, **398**, 394

Arnaud, M., and Rothenflug, R., 1986, *A&AS*, **60**, 425

Fludra, A., Schmelz, J. T. 1999, *Astronomy & Astrophysics*, **348**, 286

Fludra, A. *et al.* 1999, *Many Faces of the Sun* (eds. Strong, K. T., Saba, J. L. R., Haisch, B. M., and Schmelz, J. T.,) New York: Springer pp. 89-141

Meyer, J. P. 1985, *ApJS*, **57**, 173-204

Schmelz, J. T., and Winter, H. D. 1999, *8th SOHO workshop*, ESA SP-446, 593-597

Schmelz, J. T., Scopes, R. T., Cirtain, J. W., Winter, H. D., Allen, J. D. 2001, *ApJ*, **556**, 896-904

NANOFLARE MODELING OF AN X-RAY BRIGHT POINT CORONAL LOOP

R. A. McMullen, D. W. Longcope, and C. C. Kankelborg

Montana State University, P.O. Box 173840, Bozeman, MT 59717, USA

ABSTRACT

We study the spatial structure and temporal evolution of an X-ray bright point loop in order to understand the role of magnetic energy dissipation. We use a time-dependent hydrodynamic model to simulate the corona and transition region in the x-ray bright point's coronal loop. For this work we model a bright point observed by *TRACE* and *SOHO* on June 17, 1998, where the magnetic field geometry is derived from an extrapolation of magnetograms. We study the effects of temporal distributions of heat deposition within the loop, as a series of long and short duration "nanoflares". The quantity of energy deposited and the location of the energy release is constrained by a model equilibrium magnetic field.

OBSERVATIONS

On June 17, 1998 the transient brightening of an x-ray bright point (XBP) was observed at approximately sun center by *TRACE*. The brightening lasted from about 09:45 to 10:45 UT, and appeared to be coincident with a flux cancellation observed in the same region by *SOHO*/MDI. Kankelborg & Longcope (1999) postulate that a bipole submergence model of reconnection along a current sheet applied to this particular loop will liberate $1.6 \cdot 10^{26}$ *erg*. While the mechanism of absorption of this energy by the plasma contained in the magnetic field is unknown, the question can be addressed indirectly through modeling nanoflares as either impulsive (short duration) events, or longer duration events.

The *TRACE* light curves integrated over the background-subtracted area of the bright point (about 100 pixels) are shown in Figure 1, along with the integrated 171Å/195Å filter ratio. A successful model of this transient brightening event must emulate a) the smoothness of the *TRACE* integrated light curves, b) the restriction of the filter ratio 171Å/195Å to values between 1.0 and 2.0, and c) the five minute time lag between the filter's maxima. We expect that the gradual rise in the *TRACE* data correlates with a gradual rise in heating event amplitude, frequency or duration, all of which are outside the scope of this study.

MODEL RESULTS

Extending the work of Kankelborg & Longcope (1999) to longer time heating functions, we use a modified version of their hydrodynamic code which differs mainly in the incorporation of the chromosphere. From the morphology of the *TRACE* observations, we derive the loop geometry including a half-length of $10^9 cm$ above the photosphere, and a cross-sectional area that varies from $1.8 \cdot 10^{15}$ cm^2 at the transition region, to $1.3 \cdot 10^{16}$ cm^2 at the loop apex. The method uses explicit finite volume advection with implicit thermal conduction. The numerical grid is static, but refined in the transition region where the temperature gradients are large. At the photospheric boundary the model continues into a gravitationally stratified chromosphere

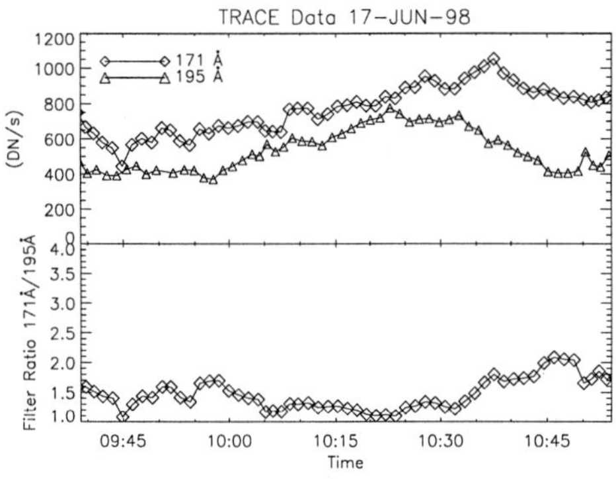

Fig. 1. *TRACE* integrated lightcurves and filter ratio.

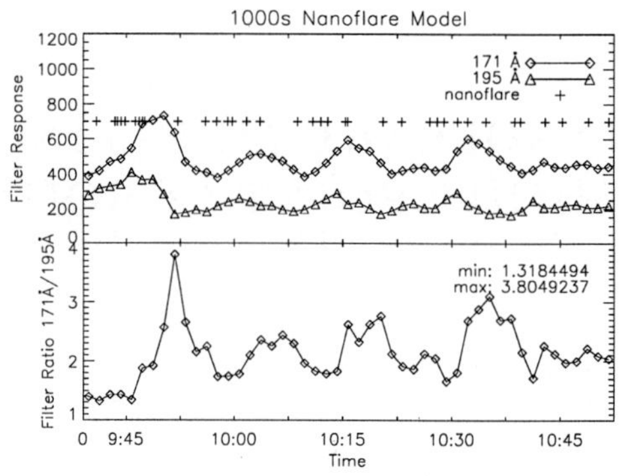

Fig. 2. Best fit model lightcurves and filter ratio.

with temperature $T = 10^4 K$. Heating of this chromosphere is not considered in the energetics of the transient brightening.

Initial equilibrium was established through a background heating of $7 \cdot 10^{-4}\ erg\ cm^{-3}\ s^{-1}$, applied uniformly in the loop, so that it settled into a steady state with a temperature maximum of $8 \cdot 10^5\ K$ and density minimum of $5.4 \cdot 10^8 cm^{-3}$ at the loop apex. From this equilibrium, spatially uniform volumetric heating functions were applied, stepping the heating up discontinuously to a constant level such that $1.6 \cdot 10^{26} erg$ of energy was added after 145s or 1000s, respectively for the two test cases. Each model was then returned to equilibrium with the original background heating. The resulting models were rescaled by a factor of 110 in amplitude, replicated, and added in a staggered manner such that on average there was a 110s time delay between "nanoflares".

The best fit to the overall amplitude and smoothness of the data was that of the 1000s (long duration) heating events, shown in Figure 2. The smoothness of the modeled *TRACE* filter responses is a clear improvement over previously published work, as are the amplitude and smoothness of the model filter ratio $171\mathring{A}/195\mathring{A}$, a new analysis tool. Problems with this model include the overall amplitude of the integrated signal, insufficient time lag between peaks, and inconsistency in the filter ratio range. None of the models considered thus far optimize time spent in *TRACE*'s peak sensitivity range. In future work we intend to include a study of regular versus random nanoflares, and the dependence of the heating function magnitude on the time lag between events.

ACKNOWLEDGEMENTS

This work is supported through NASA Grant NAG5-10489.

REFERENCES

Avrett, E.H., New Models of the Chromosphere and Transition Region, in *Mechanisms of Chromospheric and Coronal Heating*, Springer-Verlag, Berlin (1991).

Fontenla, J.M., E.H. Avrett, & R. Loeser, Energy Balance in the Solar Transition Region. I. Hydrostatic Thermal Models With Ambipolar Diffusion, *Astrophys. J.*, **355**, 700 (1990).

Kankelborg, C.C., & D.W. Longcope, Forward Modeling of the Coronal Response to Reconnection in an X-Ray Bright Point, *Solar Physics*, **190**, 59 (1999).

SOLAR CYCLE DEPENDENCY OF X-RAY BRIGHT POINTS AND PHOTOSPHERIC BIPOLES

I. Sattarov[1], A. A. Pevtsov[2], A. S. Hojaev[3], and C. T. Sherdonov[3]

[1] *Tashkent State Pedagogical University, Department of Physics and Mathematics, 103 Yusuf Xos-Hojib str., Tashkent 700100, Uzbekistan*
[2] *National Solar Observatory, Sacramento Peak, Sunspot, NM 88349, USA*
[3] *Ulugh Beg Astronomical Institute of Uzbek Academy of Sciences, Astronomicheskaya 33, Tashkent, 700052, Uzbekistan*

ABSTRACT

We employ daily *Yohkoh* SXT full disk images and National Solar Observatory/Kitt Peak (NSO/KP) photospheric magnetograms from 1993–1999 to study the solar cycle variation of X-ray bright points (XBP) and magnetic bipoles (BP). The XBP number follows a well-known anti-cycle variation, but the BP number does not. We see this as a clear indication that the anti-cycle variation of XBP number is an apparent, not a real effect.

XBPS AND MAGNETIC BIPOLES DURING 1993–1999

The number of X-ray bright points (XBPs) varies inversely with the sunspot cycle (e.g. Davis *et al.* 1977). To explain this, Golub *et al.* (1979) proposed the existence of a secondary cycle, operating in anti-phase with the active region cycle. On the other hand, Nakakubo & Hara (1999) showed that "haze" from bright active regions may alter XBP identification and result in an apparent anti-cycle dependency. The majority of XBPs can be identified with photospheric bipoles (Longcope *et al.* 2001). The "haze" does not affect the photosphere, and hence, the BPs number can be used to distinguish between a real and apparent cyclic variation of XBPs (Sattarov *et al.* 2002). For the present study XBPs were manually identified using full disk images from the *Yohkoh* soft X-ray telescope (SXT). The magnetic bipoles were found using National Solar Observatory/Kitt Peak (NSO/KP) full disk longitudinal magnetograms. The automatic procedure identified magnetic poles of negative and positive polarity with magnetic flux above 20 G, and a cross-section between 5"–55.2". The bipoles were established as the closest pairs of opposite polarity poles with separation between 5.5" and 48.3". The numerical values were adopted from Longcope *et al.* (2001) XBPs surveys. To minimize the influence of active regions we restricted our data to a narrow ±5° equatorial zone.

Figure 1 shows the variation of XBP and BP numbers observed in the equatorial region and averaged magnetic and X-ray fluxes at the heliographic center (Pevtsov & Acton 2001). There is no correlation between XBP and BP numbers (Spearman's rank correlation coefficient $r_s = -0.09$). There is strong negative correlation between XBP number and X-ray flux at heliographic center ($r_s = -0.59$). The magnetic flux and BP number remain constant through the declining phase of cycle 22, solar minimum and the rising phase of cycle 23. In contrast, the number of XBPs increases and reaches maximum at the end of 1996. We expect that the fraction of photospheric bipoles that exhibit XBPs is cycle independent. Hence, the cyclic variation in the XBP number is not real. X-ray flux at heliographic center (Figure 1d) varies with sunspot cycle, but the average magnetic flux at the same area does not change (Figure 1c). Pevtsov & Acton (2001) have argued that X-ray brightness of the quiet Sun corona in the SXT bandpass is determined mostly by the active regions, not the underlying magnetic field. Thus, we believe that the cyclic variation of XBPs number (Figure 1a) is due to "haze" from distant active regions in agreement with Nakakubo & Hara (1999).

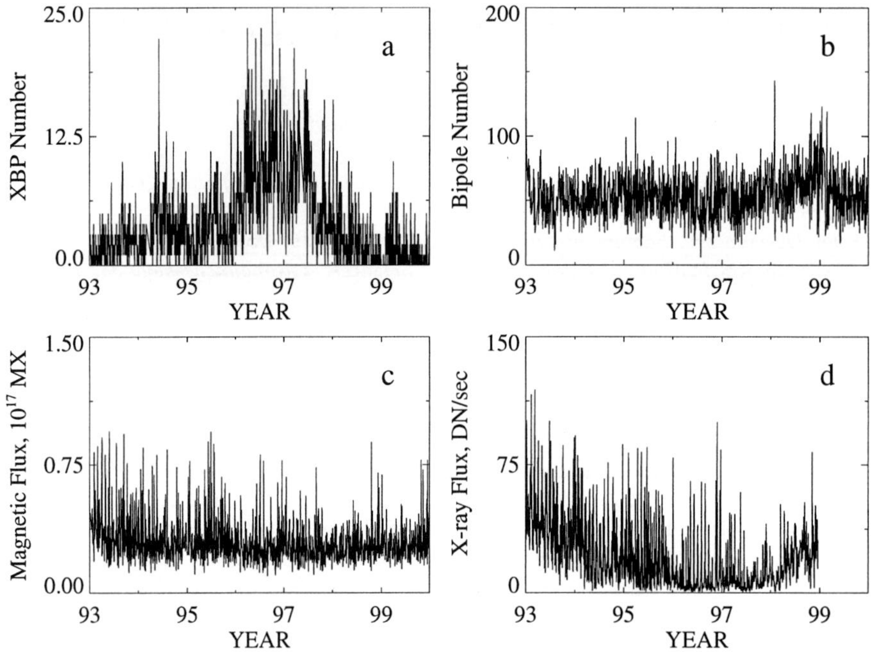

Fig. 1. Number of X-ray bright points (a), photospheric bipoles (b), average magnetic (c) and X-ray (d) flux at heliographic center. Pevtsov & Acton (2001) data, – panel (d) – extends only to the end of 1998.

REFERENCES

Davis, J. M., L. Golub, & A. S. Krieger, Solar Cycle Variation of Magnetic Flux Emergence, *Astrophysical Journal,* **214**, L141, (1977).

Golub, L., J. Davis, & A. Krieger, Anticorrelation of X-ray Bright Points With Sunspot Number, 1970–1978, *Astrophysical Journal,* **229**, L145, (1979).

Longcope, D. W., C. C. Kankelborg, J. L. Nelson, & A. A. Pevtsov, Evidence of Separator Reconnection in a Survey of X-ray Bright Points, *Astrophysical Journal,* **553**, 429, (2001).

Nakakubo, K. & H. Hara, Variation of X-ray Bright Points Number Over the Solar Activity Cycle, *Advances in Space Research,* **25**, 1905, (1999).

Pevtsov, A. A. & L. W. Acton, Soft X-ray Luminosity and Photospheric Magnetic Field in Quiet Sun, *Astrophysical Journal,* **554**, 416, (2001).

Sattarov, I. S., A. A. Pevtsov, A. S. Hojaev, & C. T. Sherdonov, X-ray Bright Points and Photospheric Bipoles During Cycles 22 and 23, *Astrophysical Journal,* **564**, 1042, (2002).

SURGES, MAGNETIC FLUX CANCELLATIONS, AND UV BRIGHTENINGS AROUND AN EMERGING FLUX REGION

K. Yoshimura[1], H. Kurokawa[2], M. Shimojo[3], and R. Shine[4]

[1]*Institute of Space and Astronautical Science, Sagamihara, Japan*
[2]*Kwasan and Hida Observatory, Kyoto University, Japan*
[3]*Nobeyama Radio Observatory, Japan*
[4]*Lockheed-Martin Solar and Astrophysics Laboratory, Palo Alto, CA, USA*

ABSTRACT

We show that the surge activities correlate well in time with the cancellation of magnetic fluxes around an emerging flux region. In particular, the highest surge in Hα and the brightest emission in *TRACE* 1600 Å images was related to the most pronounced cancellation of magnetic flux in one small area. These facts are consistent with the magnetic reconnection model. We found a spatial displacement between the area of magnetic cancellation and the bright area in UV images, which was about 8000 km in plane of sky. We could not find any bright structures in soft x-ray (SXR) and EUV images during this surge active phase. These may be clues for detailed modeling of surge activities.

INTRODUCTION

Surges are collimated ejections of chromospheric matter into coronal heights. They can be seen as dark features in Hα images. Sano (1996) showed that many surge activities were related to emergences of magnetic fluxes. These activities are thought to be caused by a magnetic reconnection (Shibata *et al.* 1992, Kurokawa & Kawai 1993). But the detailed relationships between emerging magnetic fluxes and surge activities are still unclear.

RESULTS

The surge activities, which lasted for 2 hours, were observed around an emerging flux region near the disk center (N19 W04 in heliocentric coordinates) on 10 June 1999. Though many data in various wavelengths of this region were available, we could not find any bright features during this surge active phase in SXR (*Yohkoh*/SXT) and EUV (171/195 Å, *TRACE*) images. So in this paper we analyzed Hα images (La Palma), magnetogram data (*SOHO*/MDI) and UV images (1600Å, *TRACE*). In the sequential magnetogram data we can see cancellations between the positive fluxes, which had already existed at the beginning of the observation, and the emerging negative ones in the area enclosed by the larger rectangle in Figure 1. Since few positive magnetic fluxes seemed to have emerged in this area, the decrease of positive fluxes must be a good index for the magnetic cancellation.

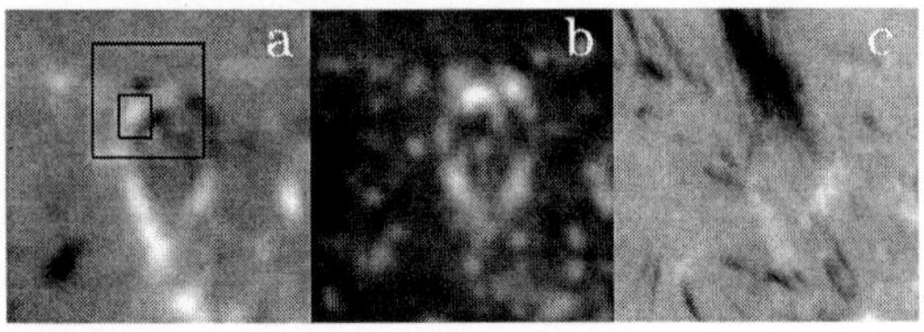

Fig. 1. The emerging flux region observed in (a) magnetogram, (b) UV and (c) Hα–0.7 Å.

Figure 2 shows temporal variations of the lengths of surges, total counts of 1600 Å flux and positive magnetic flux. All surge activities were observed during the declining phase of positive magnetic fluxes. The fastest cancellation occurred in the small area, which is enclosed by the small rectangle in Figure 1a. It lasted for 25 minutes (gray belt in Figure 2). The highest surge and the brightest feature in 1600 Å images appeared in this phase.

The total kinetic energy of these surges are less than 10^{27} erg, though it is difficult to estimate it accurately. The total energy expected to be released by magnetic field dissipation is estimated to 10^{30} erg. Hence the magnetic field dissipation can provide enough energy to produce these surge activities. The observational facts described above are consistent with the magnetic reconnection model.

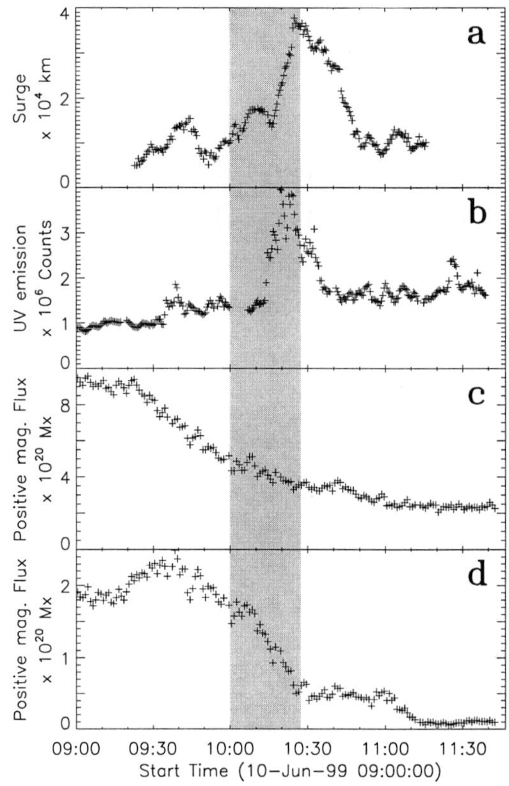

Fig. 2. Temporal variations of (a) the surges lengths, (b) total counts of 1600 Å fluxes and total positive magnetic fluxes in the larger rectangle (c) and in the small one (d) in Figure 1.

The area where the largest magnetic cancellation was observed is located close to the brightest region in 1600 Å images. But we can see nontrivial displacement ($\sim$ 8000 km) between them. The distance is too large to be a projection effect.

The observational fact that no bright feature was found in coronal heights may provide us with some clues to understand the detailed structures of the surge activities. They may suggest that the reconnection points were lying in the chromosphere.

REFERENCES

Kurokawa, H., & G. Kawai, *ASP Conference Series*, **46**, 507 (1993).

Sano, S., *Master Thesis in Kyoto University*, (1996).

Shibata, K., S. Nozawa, & R. Matsumoto, *Publ. Astron. Soc. Japan*, **44**, 256 (1992).

Section III.

Education and Public Outreach

YOHKOH: A DECADE OF DISCOVERY

D. Alexander and T. R. Metcalf

Lockheed Martin Solar and Astrophysics Lab., Org. L9-41 B/252, 3251 Hanover St., Palo Alto, CA 94304

ABSTRACT

Data from the *Yohkoh* satellite have led the way in the dissemination of solar science to the general public over the last decade. This paper will discuss the role that solar physics in general, and *Yohkoh* in particular, has played in stimulating a public interest in science and the effect they have had in the teaching of science in the classroom. To highlight *Yohkoh* we will discuss the extremely successful *Yohkoh* Public Outreach Project and its daughter-site Solar Week with particular emphasis on the impact of scientist involvement in the effective dissemination of the solar science.

The *Yohkoh* Public Outreach Project (YPOP) has been the recipient of a number of acknowledgements, examples of which are a Griffiths Observatory Star Award, a NASA Cool Picks award, inclusion of our activities in the SciLinks Program of the National Science Teachers Association, and the identification of YPOP as an *Exemplary Resource* by the Sun Earth Connection Education Forum. YPOP can be found at `http://www.lmsal.com/YPOP/`.

INTRODUCTION

In recent years the agencies that fund solar and space physics across the world have pressed for a greater accountability to the real people who fund our endeavors: the taxpayers. Typically, this has grown from simple requests for some basic information on any given scientific mission or project (a descriptive paragraph, an image etc.), to the requirement of an education and public outreach (E/PO) effort integrated with the funded scientific program. This increase in accountability, with the consequent increased involvement of the scientific community, has led to a great deal of creativity, cooperation and ingenuity with the end result that scientists, educators, teachers and students now have instant access to some of the best solar data we have ever produced.

When coupled to the rapid development of the world wide web, the dissemination of scientific data and content has become easier, more effective and, as a result, has created more of an impact on the primary goals of scientific E/PO, namely to increase the scientific literacy of the non-science community, to enhance pre-college science education for all students and, as a result, develop the scientific workforce of the future. It is far from perfect, however, and requires the continued and direct involvement of the scientific community at all levels of the dissemination process. In this paper, we will highlight some of the approaches taken within the field of solar physics, and our sister field of space physics, all within the auspices of the *Sun Earth Connection*, and will try to encourage more of you to participate and to share in the enjoyment of communicating our science to young people and the public in general. Due to the nature of this particular proceedings, we will concentrate on the contributions that *Yohkoh* has made to popularizing our science. These efforts will be placed in context with broader E/PO goals but will also furnish some valuable lessons learned that can help facilitate future and current E/PO efforts.

EDUCATION AND PUBLIC OUTREACH RESOURCES

There is no unique approach to the dissemination of solar science and, in fact, specific means often need to be developed to address specific targets. These range from curriculum development at the state and national level, through classroom activities and scientist visits, to the distribution of posters, CDs, videos and web-sites. Scientists can contribute in a variety of ways at all of these levels. A defining feature of many of the most successful space science products has been the direct "hands-on" involvement of scientists, either in the production of the materials or in their dissemination.

The available resources for solar and space science education are far too numerous to mention let alone do justice to. The Sun Earth Connection Education Forum (SECEF: see paper by Diane Kisich, Isabel Hawkins and Rich Vondrak in this proceedings), in collaboration with the Origins Education Forum, has produced a searchable online *Resource Directory* that provides an excellent interface between the available resources, most of which undergo a rigorous "product review" for scientific accuracy, age-specific content and educational effectiveness, and the teachers who wish to use them. This resource can be found at `http://teachspacescience.org/`. Scientists who have developed educational or outreach materials should make use of this directory sponsored and maintained by NASA Space Science to increase the access points to their resources. Equivalent directories in other countries should also be utilized to help develop a wider audience for your E/PO activities.

A number of *Yohkoh-centered* activities are included in this directory and are a direct result of a NASA-funded effort called *The Yohkoh Public Outreach Project* or YPOP, that will be discussed below. YPOP provides an excellent example of incorporating the differing aspects of education and public outreach and emphasizes the strong appeal of our space science data. The key is the provision of a diversity of entry points catering to both formal (e.g. lesson plans, classroom activities, games, interaction with scientists) and informal education (e.g. daily images and movies, virtual solar tour, introductory materials).

Since the advent of YPOP we have been treated to the incredibly effective and exciting outreach activities of both the *SOHO* and *TRACE* missions. The discussion that follows, while emphasizing the contributions of *Yohkoh*, has benefitted greatly from the experiences of the *SOHO* and *TRACE* outreach programs.

SCIENCE IN THE CLASSROOM

Ultimately, the aim of most of us who participate in education and public outreach is to instil an interest in and a respect for the work that we do and to share the excitement and enthusiasm we feel for our research and our quest to understand the workings of the Sun . If, along the way, we increase scientific literacy, help a budding scientist with her science project or replace a poster of Britney Spears with a poster depicting the Solar Cycle, then all the better. While it is difficult to measure any long-term impact from our E/PO efforts, it is clear to anyone who has talked to school children or adult classes that there is a willing and intellectually curious audience out there that is thrilled by the science of the Sun. The difficulty is not in overcoming any intrinsic lack of interest but providing the information in a way that has a lasting impact and that, depending upon the format, satisfies the needs of the teachers, who have to deal with busy schedules, limited class times, national and state standards, and a general lack of resources.

Formal and Informal Education

The dissemination of E/PO products is generally categorized as formal or informal education; the former is identified loosely as anything that is incorporated into the classroom, while the latter encompasses all non-classroom activities (e.g. science museum exhibits, astronomy club presentations, etc.). While the materials for each setting may be the same the constraints placed on their dissemination are very different with formal education requiring compliance with, for example, national standards, state standards, or specific pedagogical approaches. The transition between informal and formal education benefits greatly from a collaboration between scientists and educators at all levels to provide a direct path from the science to the intended audience.

Some examples of ideas and products include:

- Simple eye-catching representations (posters, lithographs, bookmarks etc.)
- Basic content and information provided through various media (web-site, videos, CD/DVDs)
- Hands-on involvement (games, activities)
- Detailed lesson plans
- Standards-based curricula

Special Events for Formal and Informal Education

Recently, there has been a focus within the solar E/PO community on special events targeting specific solar phenomena (e.g. eclipses) or dates (e.g. equinox). Often centering on museum exhibits, web-casts, live interaction or direct online participation, these events have the bonus of receiving wide publicity and can, in some cases, command audiences in the tens to hundreds of thousands of participants including students, teachers, and members of the general public. Two of the most successful events in recent years have been the live eclipse broadcasts co-produced by the San Francisco Exploratorium (`http://www.exploratorium.org/eclipse`) and SECEF, and the wealth of activities associated with Sun Earth Day (`http://sunearth.gsfc.nasa.gov/sunearthday`) produced by SECEF. The latter is a recently developed annual event that provides scientists with the tools and materials to assist them when presenting to K-12 audiences (see paper by Kisich, Hawkins, & Vondrak, these proceedings).

Additional events that have taken part in association with the Sun Earth Day activities, although developed and run independently, include the NASA Quest webcasts facilitated by the Stanford Solar Center (`http://solar-center.stanford.edu/webcasts.html`) and Solar Week (`http://www.lmsal.com/YPOP/solarweek/`) which brings scientists and classrooms together on the web in a week long series of content, games and activities. Solar Week is a daughter site to YPOP geared towards girls and young women. Solar Week will be discussed in more detail below.

Active dissemination of solar science via these kinds of special events make it extremely easy for scientists to participate by creating an audience primed and eager to learn about the Sun.

YOHKOH PUBLIC OUTREACH PROJECT

The *Yohkoh* Public Outreach Project (`http://www.lmsal.com/YPOP/`) is a NASA-funded education and public outreach site aimed at providing public access to high quality scientific data and is a product of a fruitful collaboration between Montana State University and Lockheed Martin Solar and Astrophysics Laboratory (see paper by Larson *et al.* in this proceedings). YPOP was the first project of its kind in the dissemination of solar science and was facilitated by NASA's interest in providing a public interface to the everyday discoveries of a current science mission. The generation of continuous full-disk images of the Sun at X-ray wavelengths by the Soft X-ray Telescope on board *Yohkoh* provided an ideal medium for bringing together the science and the public.

A key feature of YPOP was the early definition of what we called a Creative Design and Definition Team, or CDDT. The formation of this team was at the heart of the philosophy behind the YPOP project. The

Table 1. The *Yohkoh* Classroom

Activity	Description
Solar Cycle	Investigate the cycles of the Sun with 250 years of data! Learn to recognize common features and match x-ray images of the sun with visible light images from the same day.
Rotating Sun	Study the rotation rate of the Sun and the planets. This lesson includes several hands-on activities to familiarize the student with the concept of rotation. It then uses movies made with actual solar images to study the rotation rate of the Sun.
Sunspots	Learn image processing skills that will help to measure the size of sunspots. These skills can be used to measure features on any electronic image.
Digital Images	Magnify images of your school or yourself and see what happens. Learn about the pixels that make up all images. Use pixels to create secret messages or uncover a mystery picture!
Filter Wheels	Build your own inexpensive, color filter wheel and use it to study an image of the Crab Nebula! Discover why scientists use different filters to study astronomical images. View several images of the Sun as seen through different solar filters.
Satellite Orbits	In this lesson you obtain the period of the *Yohkoh* satellite from an image of its orbital path that is updated regularly. With this information you will calculate the height of the satellite above the Earth's surface.
Yohkoh Model	Build a model of the *Yohkoh* solar satellite. With a cardboard box and some paper you can make a model of the *Yohkoh* satellite. Instrument blueprints are provided along with detailed instructions for construction.
Build a Sundial	Learn how to make a portable, inexpensive sundial of your own. Wear it as a necklace or make a keychain!
Earth's Orbit	Investigate the shape of the Earth's orbit based on the apparent changing size of the Sun. Compare the difference in the Sun's diameter measured from two points in the Earth's orbit and see what you discover!

CDDT was comprised of scientists, science education specialists, and teachers, specifically to provide the broadest range of expertise and experience to the design of what was to become the YPOP site. The biggest impact of the creation of this diverse team was the enhancement of the original charge of producing a public outreach product to include educational products via our online classroom of activities. The nine activities in our classroom have proved to be one of the biggest attractions in YPOP (see Table 1).

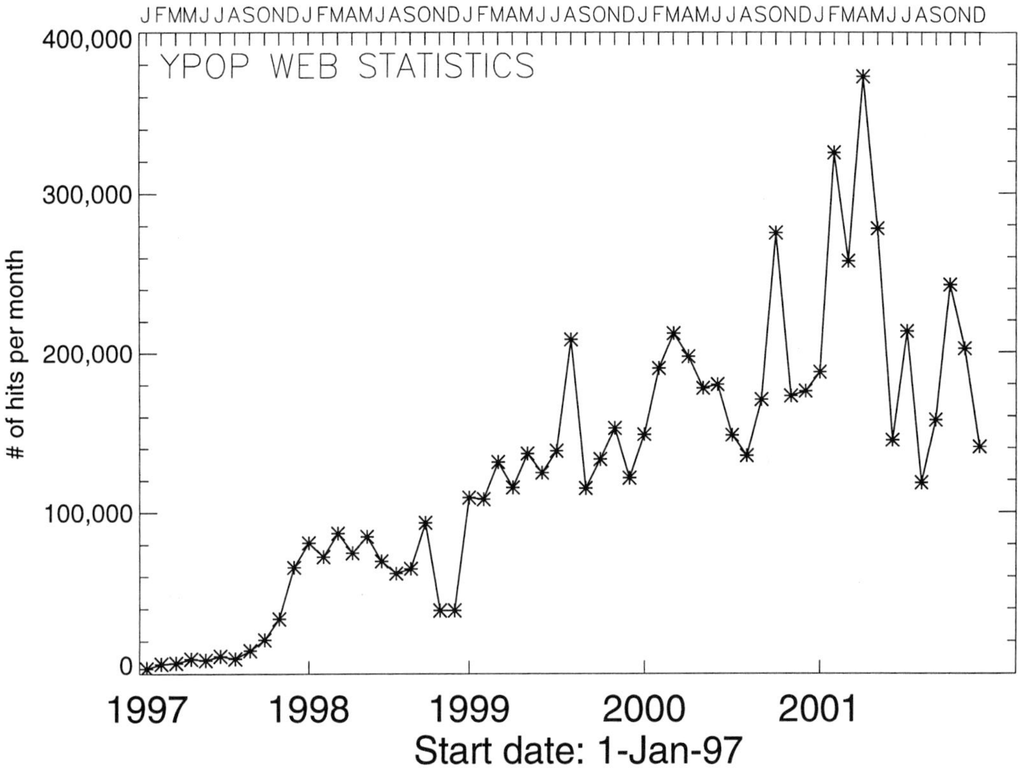

Fig. 1. YPOP usage has shown a consistent growth since its inception

Since the inception of the public web-site, YPOP has continued to grow in popularity as witnessed by the chart in Figure 1.

SOLAR WEEK

Another major E/PO activity resulting from the impact of *Yohkoh* is *Solar Week.* Solar Week was created with the primary aim of addressing the under-representation of women in the research and teaching of the physical sciences by specifically targeting middle-school girls. Solar Week is a daughter-site to YPOP, feeding off the YPOP Classroom activities. However, it also incorporates a number of novel features: a lower level of interaction via games in addition to activities, direct interaction with scientists via online Q&A sessions, and use of role models to encourage participation: all Solar Week scientists are female.

Solar Week is now in its 3rd year and has been run four times since October 2000. Solar Week is now a regular feature supporting the Sun-Earth Day activities with the total number of participants now surpassing 120 teachers, 12,000 students in 30 states and 11 different countries. The key focus of Solar Week is the direct participation of scientists to answer questions posed by the students taking part. All of the scientists are women and are introduced via an online biography. Solar Week would not have been as successful and effective without the contributions from: Mitzi Adams (Marshall Space Flight Center), Sallie Baliunas (Harvard-Smithsonian Astrophysical Observatory), Lyndsay Fletcher (University of Glasgow), Nicola Fox (Johns Hopkins University), Sarah Gibson (High Altitude Observatory), Karen Harvey (Solar Physics Research Corp.), Mandy Hagenaar (Lockheed Martin Solar and Astrophysics Lab.), Judy Karpen (Naval Research Lab.), Terry Kucera (Goddard Space Flight Center), Judith Lean (Naval Research Lab.), Dawn Myers (Goddard Space Flight Center), and Kris Sigsbee (Goddard Space Flight Center).

The structure of Solar Week is, as the name implies, based on a week long set of daily topics that introduce different aspects of solar physics. Currently, the topics are *The Sun as a Star*, emphasizing the Sun in the context of other stars, *Solar Close-ups*, introducing the collection of data and the concept of resolution, *The Active Sun*, which focuses on "solar storms", *Solar Eclipses*, and *Careers.* On each of the first four topics the scientific content is paramount and all activities, games and Q&A sessions center on the science topic of the day. Typically, we feature one or two "Star Scientists" who are currently performing research in the day's topic. The Careers day (usually Friday) is different in that it focuses on the role that women play in the sciences. All participating scientists are available to answer questions on the Careers day and the major activities are centered around the online biographies. This allows the students, particularly the female students, to learn more about the scientists, what it takes to become a scientist and what it is like to be a woman in a predominantly male field.

Each "day" within Solar Week has the same structural format to engender a familiarity with the site. The main area contains content (text, images and movies) on the day's topic, a game designed to be a low level and short duration interaction with the content, one or two activities that provide hands-on learning associated with the particular topic, information for teachers, interaction with the scientists, a brief link from the solar science to a life science topic, and hypertext links to other web-sites of particular value to the discussion. The daily topics are designed to be modular in order to change any particular topic with the minimum of impact to the rest of the site.

Two particular aspects of Solar Week bear further discussion.

Games as an Introduction to the Content

Typical science teachers have their class for about 40 minutes per day which is too short for the activities highlighted in the YPOP Classroom. We addressed this issue in Solar Week by developing a number of games that serve to highlight the topic being discussed. Some of the games amount to no more than a simple scavenger hunt or crossword puzzle but the *Solar Close-ups* and *Active Sun* games provide an interactive means to explore various aspects of observational resolution and the evolution of a solar storm, respectively.

The idea behind each of these games is to provide an entertaining forum in which to learn some aspect of space science. The *Solar Close-ups* game teaches the students about the importance of spatial resolution in the observation of astrophysical objects by having them interact with digital data in a "Guess the Celebrity" game. Pictures of ten celebrities, hopefully still high in the teen popularity rankings, are pixelated to different levels with the object of the game to identify the celebrity as early as possible for the most points. As the resolution improves the identification gets easier (see Figure 2). Through the content part of this topic the game is related directly to the role digital imaging plays in modern astronomical observation and is graphically illustrated by a comparison of *TRACE, SOHO/EIT* and *Yohkoh*/SXT images.

The *Active Sun* game makes the link between different solar phenomena and their participation in a solar storm. This online Yahtzee-like game randomly produces a set of four solar images on four die-faces (call them e-dice for convenience). You can choose to "hold" any number of the images for your second "roll". The purpose, like Yahtzee, is to score points by collecting pairs, three of a kind, CMEs, sunspots etc. However, the biggest scores are achieved by building up a sequence of either three or four images. In this game a

Fig. 2. Guess the Solar Celebrity

sequence is defined as a physically connected set of solar circumstances: e.g. filament – filament eruption – CME – aurora. To win the game the students have to develop some form of strategy to ensure a score in all of the relevant sections. To make a sequence happen, other than by pure luck, the student has to recognize a developing sequence and "hold" the relevant e-dice to achieve the completed sequence. This game is related to the "Story of a Storm" that describes the events surrounding the April 2 2001 flare. A static version of this game was developed for the most recent *Yohkoh* poster (*Yohkoh: A Decade of Discovery*) that was designed to celebrate the *Yohkoh* 10th anniversary (see Figure 3).

Scientists as Role Models

Aside from the fun aspects of Solar Week, the real impact is generated by the direct interaction of the students with the scientists. Not only are the scientists available online during Solar Week to answer questions about the Sun, dancing, music, scientific careers etc., but each participating scientist has a biography online that highlights their particular path to becoming a scientist. The purpose of these biographies is not to detail the scientist's current job and research but how they got there and to give a little insight into their personality and interests. By providing these positive role models, the students can see for themselves that women have an important role to play in scientific research and that it is not only one kind of woman who succeeds. The Solar Week scientists come from a wide range of backgrounds with a wide range of interests and attitudes. All of this provides a positive experience for the student, male or female, and gives them an appreciation of the diversity of people who ultimately become scientists.

One of the most pleasing aspects of the student/scientist interaction is the short-term bonds that sometimes occur. In the first run of Solar Week (October 2000) a number of students selected a particular scientist and began an online conversation about their common interests in dance, music, or cars, for example. These connections, more than the science content, are what makes Solar Week worth pursuing. The scientists who have participated in Solar Week over the last couple of years (see above) are to be commended for their hard work and dedication. The comments we have had from teachers clearly indicate that the biggest benefit they receive from their participation is the enthusiastic response to their students' questions by the scientists.

FEEDBACK, ASSESSMENT AND EVALUATION

One of the main metrics of the impact of any E/PO effort is the assessment of how the students' knowledge or understanding has improved after their participation. Meaningful assessment for any project, however, is extremely difficult to achieve, especially for those E/PO efforts relying on the part-time participation of scientists. The ultimate aim of most of our efforts is the hope that there is a budding scientist out there who needs a little encouragement and inspiration to choose science as an education and/or career goal and that our projects provide that inspiration or encouragement. The success of this aim is difficult to assess or evaluate.

In projects like YPOP and Solar Week we do not pretend to provide in-depth assessment but we do attempt to achieve some form of evaluation via an online guestbook, in the case of YPOP, and the completion of a questionnaire by both scientists and teachers, in the case of Solar Week. These provide some sense of how well we are doing but more importantly they allow us to continually improve our E/PO and our interaction with the students. For instance, it is clear from the feedback that YPOP has served a diverse population of users: teachers, students, home schoolers, general public and planetaria. This is a far wider audience than we had originally expected.

LESSONS LEARNED

A number of lessons learned can be highlighted from the YPOP and Solar Week E/PO efforts. Please also refer to the paper by Michelle Larson in this proceedings for additional discussion on what we learned from YPOP.

- Focused funding
  - YPOP was funded solely as an E/PO effort centered around the *Yohkoh* data but as a separate contract with its own goals and objectives. Adequate and dedicated funding is crucial to attain the necessary focus and perspective. It was clear in YPOP that having an appropriate support level allowed us to accomplish a significant set of goals and thereby increase the impact of our project.
- Diverse expertise in development team
  - From the very beginning YPOP understood the importance of bringing together a production team comprising scientists, education specialists and teachers. The diverse nature of the team

was crucial to producing a well-balanced, scientifically accurate, age-appropriate, classroom-tested product.

- Direct involvement of teachers
  - YPOP hosted two teacher writing workshops, where we spent time with ?master teachers? to design and implement our classroom activities. The project supported teachers? travel, accommodation and provided a stipend. This resulted in activities that were both scientifically accurate and written at an age-appropriate level.
- Feedback
  - Constructive feedback is difficult for a multi-use product like YPOP. However, it is important to let the end users compliment and criticize the content. YPOP does this via an online Guestbook, while Solar Week utilizes a questionnaire for teachers and scientists.

CONCLUSIONS

The principal conclusion of this paper is that if we want the general public to gain a deeper appreciation of our science they must have their interest stimulated by informed access to the data and its scientific interpretation. This can only be achieved by the direct participation by the scientists themselves at whatever level they have time for and deem appropriate. If nothing else, a scientist is required to ensure the scientific accuracy of the information being disseminated. Not only will this ultimately improve the scientific literacy of the general public but it may also help encourage and inspire future scientists.

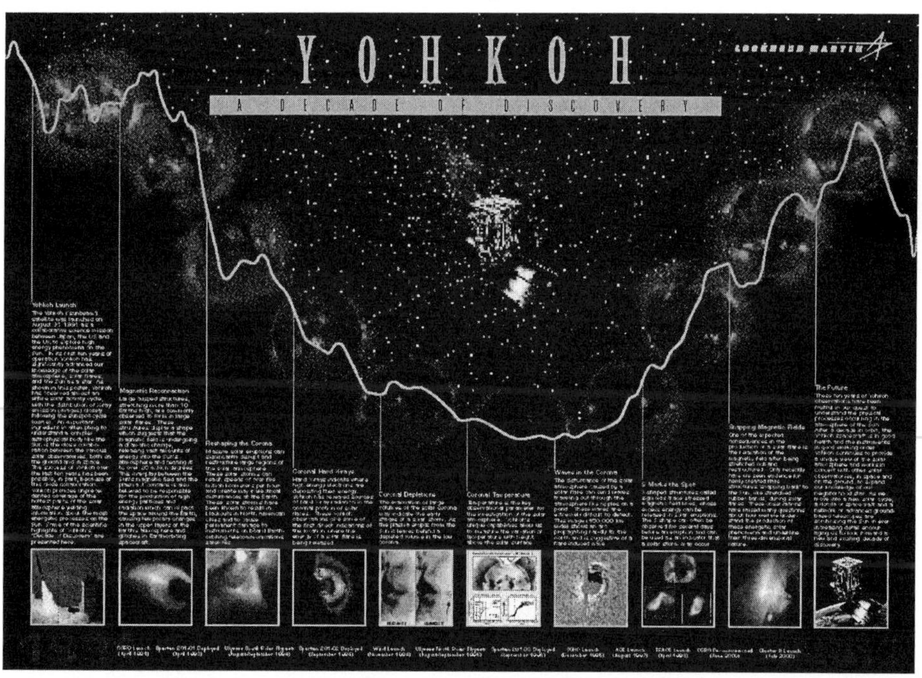

Fig. 3. *Yohkoh*: A Decade of Discovery

It is clear that the public wants access to solar, space science and astronomical data products but they have to be in a form that can be understood by the non-expert. These products can ultimately have a large impact on the scientific literacy of children around the world.

More and more scientists are participating in E/PO, either because it is often required by their grant or, more usually, because they see the value of it and, although often difficult to admit, they actually enjoy it. *Yohkoh* in its ten years of operation has been an inspiration to many children through easy access to the data via posters, videos, articles and web-sites. In some ways, *Yohkoh* was a pioneer for solar physics education and public outreach, facilitated by the intrinsic appeal of the SXT images and movies and coming at a time when the use of the internet was exploding. Other missions have learned from the *Yohkoh* experience and improved upon it over the years. We are now in a period of unprecedented interest in the Sun and its interaction with the Earth, thanks mainly to the efforts of *Yohkoh, SOHO, TRACE* and the ISTP. News stories and magazine articles appear almost every day. Now is the time to get involved and promote your science and science in general. No excuses!

ACKNOWLEDGEMENTS

YPOP is NASA funded and succeeded with the support of Bill Wagner and Joe Bredekamp at NASA HQ. This presentation could not have been completed without the discussions and arguments over the years with the YPOP team: James Lemen, Sam Freeland, Loren Acton, Michelle Larson, David McKenzie, and Tim Slater. We would also like to thank Isabel Hawkins for a critical reading of the manuscript.

SHARING THE SUN-EARTH CONNECTION

D. Kisich[1], I. Hawkins[1], and R. Vondrak[2]

[1]*UC Berkeley, Space Sciences Laboratory, MC 7450, Berkeley, CA 94720-7450, USA*
[2]*NASA/Goddard Space Flight Center, Code 690, Greenbelt, MD 20771, USA*

ABSTRACT

Sharing the Sun-Earth Connection - the effects of our active Sun on geospace and beyond - is the primary goal of NASA's Sun-Earth Connection Education Forum (SECEF). SECEF is one of twelve national centers of space science education and public outreach (EPO) funded by NASA's Office of Space Science. SECEF is a partnership between NASA/Goddard Space Flight Center and UC Berkeley Space Sciences Laboratory. We coordinate individual education and public outreach efforts from NASA Sun-Earth Connection missions and research programs, facilitating scientist involvement. SECEF develops products and sponsors programs in partnership with the pre-college and informal education communities that maximize the impact of individual efforts. We will show several examples of high-visibility products and programs for broad audiences that adapt and highlight Sun-Earth Connection data, research results, and scientist involvement.

THE GOALS OF THE SUN-EARTH CONNECTION EDUCATION FORUM

The Sun-Earth Connection Education Forum's goals are to engage the space science community and enable sustained participation in education and public outreach, provide national coordination and dissemination of EPO activities and give long-term access to EPO products and programs. Another SECEF goal is to tap Sun-Earth Connection science knowledge and mission discoveries by providing tools, resources, and assistance for the K-14 community, as well as contribute to the scientific literacy of the general public through high-visibility public events. We feel that high leverage is the key to having national impact. To assist in meeting these goals, SECEF has partnered with several groups including: NASA Goddard Space Flight Center's Education Office, Ideum, Lawrence Hall of Science, Exploratorium, San Francisco & San Jose State Universities, Pompea & Associates and the American Institutes for Research.

NEW SUN-EARTH CONNECTION EDUCATION MATERIALS

The Sun-Earth Connection Education Forum has developed several educational materials in coordination with their partners, Sun-Earth Connection missions, and teachers in order to reach thousands of educators. Some of the materials developed are: the SEC lithograph set with topics from the core of the sun to solar effects on the earth, "Making Sun-Earth Connections" CD for visual/audio presentations on SEC science at various K-12 levels, and the Space Weather folder for holding mission-related education materials serves as a teaching tool in and of itself. SECEF has also produced in collaboration with the Lawrence Hall of Science, the first Great Explorations in Math and Science (GEMS) guide with a CD-ROM. The guide is called "The Real Reason for the Seasons: Sun-Earth Connections" and the CD-ROM features NASA resources. Lawrence Hall of Science curricula are used in 25% of school districts nationally. This guide has also received recognition from the National Science Teachers Association (NSTA).

SECEF has also partnered with the *ACE, ISTP, SOHO, and IMAGE* missions, the SUNBEAMS program, Space Science Institute, and the National Science Foundation to create the Space Weather Center traveling museum exhibit. This exhibit covers the themes of *The Dynamic Sun, Earth Within Solar Atmosphere,* and *Earth*

Responds to Changing Sun. It has interactive, hands-on elements with near-real-time imagery, news, and data. Scientists around the country have participated in this exhibit by giving presentations and answering questions while on the exhibit floor. The development team expects this exhibit to impact about 700,000 museum visitors per year. Museums that hosted the Space Weather Center exhibit in 2000 and 2001 were the Denver Museum of Natural History, Denver, CO; Discovery Museum, Sacramento, CA; Maryland Science Center, Baltimore, MD; Lawrence Hall of Science, Berkeley, CA; and Adler Planetarium, Chicago, IL. For more information on the Space Weather Center you can visit the website http://www.spacescience.org/SWOP/1.html.

HIGH VISIBILITY EVENTS FOR THE GENERAL PUBLIC

One way that SECEF has involved scientists and targeted a large population of the general public is through high-visibility events, such as total solar eclipses and Sun-Earth Day events. The popularity of total solar eclipses provides a unique opportunity for engaging the public. Solar eclipses serve as a hook for showcasing NASA Sun-Earth Connection science, exciting discoveries, and the people behind them. The Exploratorium and NASA's Sun-Earth Connection Education Forum have partnered since 1998 to produce live eclipse broadcasts from the path of totality, using this extraordinary natural event as the catalyst for bringing together NASA scientists, educators, students, and the general public to share the Sun-Earth Connection. This high-visibility public event has evolved from an initial prototype that involved a live audience at only one museum to a highly-leveraged model in which the eclipse feed was made available to hundreds of museums and other institutions around the world. Museums, planetaria, schools, girl scout troops and other community groups serve as venues for public programs featuring the eclipse, Sun-Earth Connection science, and space scientists. It has helped to bring together some exemplary partnerships such as the endorsement of the Eclipse 2001 event by the National Society of Black Physicists (NSBP) in which 23 of their members gave presentations at various museums, schools and on television. For example, Dr. Beth Brown (Pegues & Brown, 2001) spoke about the eclipse on CNN and Dr. Jason Best talked with students in Indiana. Some of the topics highlighted in the eclipse events were solar cycle, auroras, magnetosphere, solar maximum, habitability of space, space weather, and *Living with a Star*.

In 2001 SECEF held the first celebration of Sun-Earth Day. Participants included formal and informal education communities in classrooms, museums, planetaria, NASA centers, as well as attendees at hundreds of celebrations throughout the country and in Europe. In 2001 over 200,000 participants learned about the Sun and Sun-Earth Connection missions. The next Sun-Earth Day event, on March 20, 2002, will celebrate the Equinox and involve partners from the Native American communities. A webcast on Sun-Earth Day will feature a Native American celebration of the Equinox, live from the Black Hills of South Dakota. Also, clips from the IMAX movie "Solar Max" will give an additional historical perspective. Student activities have been selected for the program, which will be featured on the website to enhance knowledge about the Sun and seasons. The GEMS guide, "The Real Reason for Seasons," will be featured through the use of an activity for middle school students. The activity will enable students to discover the meaning of the equinoxes, as well as help educators address common misconceptions students may have. Visit the Sun-Earth Day website:
http://sunearth.gsfc.nasa.gov/sunearthday or go to http://solarevents.org, then click "Sun-Earth Day."

HOW SOLAR SCIENTISTS CAN PARTICIPATE

There are several ways in which scientists can participate in Sun-Earth Connection Education and Public Outreach. One way is to contribute science expertise as part of the development process of educational materials. Scientists can also give presentations on Sun-Earth Connection science content as part of workshops at national conferences and at professional development venues. Another venue for scientist participation is involvement in "solarevents" such as Sun-Earth Day, March 20, 2002 or our planned Eclipse 2006 events. To keep track of upcoming "solarevents," visit: http://sunearth.ssl.berkeley.edu.

REFERENCES

Pegues, L., & Brown, B., *NSBP Shines on African Eclipse*, Nat. Soc. of Black Physicists Newsletter, Fall 2001.

SCIENTIST INVOLVEMENT IN HIGH VISIBILITY EDUCATION AND PUBLIC OUTREACH "SOLAREVENTS"

D. Kisich[1] and E. Lewis[2]

[1]*UC Berkeley, Space Sciences Laboratory, MC 7450, Berkeley, CA 94720-7450, USA*
[2]*NASA/Goddard Space Flight Center, Code 630, Greenbelt, MD 20771, USA*

ABSTRACT

Our dynamic Sun offers exciting opportunities to share research discoveries of NASA's Sun-Earth Connection with the pre-college education and public outreach communities. NASA's Sun-Earth Connection Education Forum (SECEF), a partnership between UC Berkeley Space Sciences Laboratory and NASA/Goddard Space Flight Center, coordinates national programs for broad audiences that highlight solar and geospace missions and research programs. One of SECEF's primary goals is to facilitate scientist involvement in education and public outreach. We will show results of two high-visibility "SolarEvents" - Sun-Earth Day and Eclipse 2001. These events involved more than 100 space scientists in schools, museums, and other venues across the nation, including members of minority professional societies, for example, the National Society of Black Physicists. We will discuss the lessons we learned and future opportunities for scientist participation.

HIGHLIGHTS FROM THE "SOLAREVENTS" OF 2001

Scientists' participation in the 2001 "SolarEvents" included: giving museum floor presentations and demonstrations, recruiting members of their research team to help, identifying support materials for education purposes, participating in an advisory capacity, presenting content in workshops, webchats, mentoring students, and interacting with media.

These high-leverage national events provide scientists with opportunities to share the Sun-Earth Connection science by participating in: a high visibility event, sharing research and data, giving interviews, sharing stunning graphics, workshops for educators, and webcasts. These two events had wide-reaching audiences, as displayed in Table 1. You can find out more about upcoming solar events at the following website: http://solarevents.org.

SUN-EARTH DAY

NASA's Sun-Earth Connection Education Forum (SECEF) is developing and supporting Sun-Earth Day, an education and public outreach event on March 20, 2002, to celebrate the Equinox. This is the second Sun-Earth Day event SECEF is coordinating. Last year's Sun-Earth Day was a great success. Refer to Table 1 for 2001 event statistics. The 2001 Sun-Earth Day event allowed us to reach out to classrooms and share the excitement of discovery and knowledge generated by Sun-Earth Connection science, missions and research programs. Visit the Sun-Earth Day website for the update on the upcoming Sun-Earth Day event. Go to: http://sunearth.gsfc.nasa.gov/sunearthday , or go to http://solarevents.org , then click "Sun-Earth Day."

Table 1. SolarEvents Statistics

Sun-Earth Day 2001	*Eclipse Event 2001*
3365 educators trained through NASA Education Network	72 scientists participated in the Eclipse 2001 Event
10,000 packets distributed	23 scientists are members of the National Society of Black Physicists
10,000 students involved during the webcast	164 organizations registered
87 scientists registered	21 of the organizations were international museums
67,500 European Space Agency *SOHO* Anniversary events	62 US museums, planetaria, schools and universities participated
20,000 participated with museums and planetaria	71 Girl Scout Troops participated
42 Challenger Centers	Approximately 42,000 people participated at museums and planetaria, all over the world
282 Amateur Astronomers – 58 events in 25 states within the US	198,004 visits to the Exploratorium website on June 21, 2001
1000 participated with Solar System Ambassadors	420,737 unique IP addresses visited the Exploratorium Eclipse Webcast in the month of June 2001
Discovery Science Channel-aired 6 times	

LIVE FROM AFRICA ECLIPSE 2001

Eclipse 2001 was a popular, high-visibility natural event which allowed us to highlight current NASA Sun-Earth Connection events, research, data, and scientists through informal science venues for the benefit of the public at large. Science museums, planetaria, clubs, and organizations engaged audiences by highlighting NASA's Sun-Earth Connection. The 2001 eclipse event was officially endorsed by the National Society of Black Physicists (NSBP), whose members, along with other scientists and museum educators, participated in public outreach as discussed in the NSBP newsletter (Pegues & Brown 2001). We also gathered feedback and the lessons we learned from scientists and museum visitor evaluations, which were processed by the American Institutes for Research (Markris 2001). You can download a summary of the eclipse event and the lessons we learned from `ftp://sunearth.ssl.berkeley.edu/Eclipse2001.ppt`.

ACKNOWLEDGEMENTS

We would like to acknowledge Jay Friedlander at Goddard Space Flight Center for designing the graphic layout of the poster. We would also like to acknowledge John Whitworth at the UC Berkeley Center for Science Education for his updates of the Eclipse 2001 statistics on the poster.

REFERENCES

Markris, F., Levine, R., & DuBois, P. *Evaluation of SECEF Eclipse Events 2001: Results from Eclipse 2001 Event Surveys*, American Institute for Research, California (2001).

Pegues, L., Brown, B. *NSBP Shines on African Eclipse*, Natl. Soc. of Black Physicists Newsletter, Fall, (2001).

THE YOHKOH PUBLIC OUTREACH PROJECT

M. B. Larson[1,2], T. Slater[1,3], D. McKenzie[1], L. Acton[1], D. Alexander[4], J. Lemen[4], S. Freeland[4], and T. Metcalf[4]

[1]*Montana State University, P.O. Box 173840, Bozeman, MT 59717*
[2]*currently at Caltech, W.K. Kellogg Radiation Lab, MC 106-38, Pasadena, CA 91125*
[3]*currently at University of Arizona, Department of Astronomy, Tucson, AZ 85721*
[4]*Lockheed-Martin Solar and Astrophysics Lab, Bldg. 252 Org. L9-41, 3251 Hanover St., Palo Alto, CA 94304*

ABSTRACT

The NASA funded *Yohkoh* Public Outreach Project (YPOP) began in 1995 with the goal of providing public access to high quality *Yohkoh* SXT data via the World Wide Web (WWW). The project utilizes the intrinsic excitement of the SXT data, and in particular the SXT movies, to develop science learning tools and classroom activities. The WWW site at URL: `http://www.lmsal.com/YPOP/` uses a movie theater theme to highlight available *Yohkoh* movies in a format that is entertaining and inviting to non-scientists and well received by scientists. We will discuss the wide range of people YPOP has reached over the past six years, as well as lessons learned during the development of the project.

A COLLABORATION OF CULTURES

Despite the challenge of combining very different professional cultures, the *Yohkoh* Public Outreach Project (YPOP) represents a dynamic and fruitful collaboration for both science and education. Five faculty and graduate students at Montana State University, six industrial scientists at Lockheed Martin Space and Astrophysics Laboratory in Palo Alto, CA, and six elementary and secondary school teachers from across the country partnered to create the entity known as YPOP. The development of YPOP took place when the World Wide Web was relatively new, and YPOP was one of the first solar outreach projects to bring solar data directly to the public via the Internet. The YPOP site is found at URL: `http://www.lmsal.com/YPOP/` and mirrored at URL: `http://solar.physics.montana.edu/YPOP/` .

LESSONS LEARNED ABOUT SUCCESSFUL WWW-DEVELOPMENT PARTNERSHIPS

The diversity of the YPOP development team, and the differences in the audiences with which we interacted on a daily basis, led to a more comprehensive and user-friendly WWW site than any of us working individually could have created. Elements of success that grew out of the YPOP effort include:

Face to Face Meetings, Major and Minor Milestones, and Action Items. The project had 10 major milestones to be met by a team separated by almost 2000 miles. We met quarterly to talk about site structure, refocus developing ideas and set minor milestones. Probably the most useful approach was to assign action items for each individual with highly publicized due dates. At each meeting, action item completion was reviewed among the entire group, effectively motivating individuals to get tasks done promptly.

Proximity to the Data. In this project, there were hundreds of Mb of image data available. A critical aspect was to have at the table the satellite computer programmer on the project team who dealt with the raw data.

Narrowly Defined Goal and Audience. The goal of the YPOP project was to make *Yohkoh* data accessible to the public. This meant that "the public" had to be defined. We defined the public as a parent "surfing" with their child, an interested amateur astronomer with little background, and a bright 8th grade student. In general, the lesson learned about audiences was that when there is a potential for multiple audiences, you need multiple items, or take the time to develop different parts for a single item – do not try to do two for one.

Well Defined Site Structure - A Theme. The project really never took off until we had a unified vision as expressed through a theme – the *YPOP Movie Theater.* We selected the movie theater because the real highlight of the project is a long term, scientific quality, on-line movie of the Sun. Each aspect of the site is created to be consistent with this theme.

LESSONS LEARNED ABOUT COLLABORATING WITH TEACHERS

A valuable addition to YPOP was the inclusion of three elementary and three secondary school teachers. Although YPOP was much more focused on the general public than on classrooms, these teachers' reviews and opinions had an enormous impact on the site as a whole. Once the teacher-participants overcame their initial reluctance to criticize the "scientists", they left virtually no portion of the project URL unaffected.

Unlimited Ideas. Early in the project, it was decided to include two or three classroom lessons on the YPOP site. After showing the teacher-participants the colorful and dynamic images available and letting them brainstorm together for several days, the project ended up with nine classroom lessons. However, they believed that most teachers would not use these exciting resources without sample lesson plans and scientific background. This countered our original notion that "if we build it - the teachers will use it." We learned that most classroom teachers are far too busy to create new lessons based on newly available Internet data. If we wanted our data to be used in hundreds of classrooms, then we had to create the resources and infrastructure to make adoption easy.

Specific Grade Levels are Elusive. It seems that student ability and interest vary substantially. We learned that, by developing appropriate sections, we could take a powerful lesson idea and aim it at a variety of levels: novice, intermediate, and advanced. This also allowed us to support parents surfing with their children and home-schooled students, who, to our surprise, often utilize YPOP the resources. This emphasized the need for creating hands-on activities that required easily available and inexpensive resources.

Multiple Formats On-Line. It was amazing to us to find out how diverse school computer and Internet resources are. We found that many teachers would print out everything, rarely reading on-line. Alternatively, we discovered that many home-surfers would read material on screen. Therefore, we put lessons on-line in multiple formats.

In summary, the most important aspect that made YPOP a success was that the project brought together dedicated, self-motivated, and flexible people. A diversity of ideas was key, but would have been moot if project participants had not been interested in working together. Today, YPOP receives 100,000 unique "hits" every month and the guest book displays constant heart-warming and positive feedback, which may be viewed at the YPOP website.

OUR SUN – THE STAR OF CLASSROOM ACTIVITIES AND PUBLIC OUTREACH EVENTS

N. Craig[1] and M. B. Larson[1,2]

[1]*Space Sciences Laboratory, University of California in Berkeley, MC 7450, Berkeley, CA 94720*
[2]*Currently at W. K. Kellogg Radiation Laboratory, California Institute of Technology, MC 106-38, Pasadena, CA, 91125*

ABSTRACT

We showcase *RHESSI (Reuven Ramaty High Energy Solar Spectroscopic Imager)* EPO (Education and Public Outreach) resources developed at the University of California in Berkeley (UCB) and Goddard Space Flight Center (GSFC), and *STEREO* IMPACT Instrument EPO. Both projects are supported by the SEGway program infrastructure at UCB. SEGway is a national consortium of science museums, research institutions, and educators who work together to present the latest space science research for students, educators, and the general public and is an on-line educational resource center, adapting space science research and information for the benefit of broad audiences using web-based learning technologies. The San Francisco Unified School District (SFUSD), a SEGway collaborator, has functioned as a test bed for newly developed on-line lessons, establishing valuable local relationships with teachers, students and administrators who provide input, feedback, and evaluation. Additional information on SEGway can be found at: `http://cse.ssl.berkeley.edu/segway/`.

ACTIVITIES

We provide innovative classroom activities for grades 8-12, as well as public outreach web-based resources featuring solar data, mathematics, and solar scientist interviews. These resources have been developed through partnerships between NASA scientists and educators. The classroom activities are well aligned with National Science Education Standards.

*``X-Ray Candles: Solar Flares on Your Birthday"* allows students to discover the solar cycle by analyzing x-ray flare data and graphing the percentage of high energy flares over time. ``X-Ray Candles" was developed as a part of the *RHESSI* EPO effort and is featured in the March 2002 episode of the Emmy-winning NASA Connect broadcast program.

*``Sunspots"* -an inquiry-based resource, emphasizes mathematical connections through measurement, graphing, and analysis of satellite and student-acquired data. ``Sunspots" has been successfully classroom tested in diverse school districts around the country. The resource incorporates background information, including the importance of the Sun in ancient cultures, a historical account of sunspots observations, and current NASA research. In addition, it includes guidance for safe sunspots viewing and a Java interactive research tool that allows students to analyze possible correlations between sunspots and X-Ray active regions from *Yohkoh* images of the Sun.

RHESSI EPO at UCB

UCB's *RHESSI* EPO effort focuses on middle and high school teachers, their students, and the public. The formal education part of the program includes the training of a core of lead teachers from urban and rural school districts in the larger Bay Area. The *RHESSI* EPO ``Sunspots'' resource was used in five SFUSD middle schools for a three week summer course. *RHESSI* EPO resources were also presented at NSTA conferences for professional development workshops. The informal education component of the EPO program consists of dissemination of *RHESSI* web and hardcopy resources within the SEGway program's national partnership of science museums, and the SEGway's collaboration with UCB's outreach programs for middle school science teachers. *RHESSI*'s participation in the widely observed Live@Exploratorium webcast of the 1999 Total Eclipse from a location in Turkey was a successful implementation of scientist and broad public interface. Collaboration with NASA's Sun Earth Connection Education Forum provides an effective multiplier effect for the local *RHESSI*-based education efforts at UCB and at GSFC. For additional information see the website at:
`http://cse.ssl.berkeley.edu/HESSI/`

STEREO/IMPACT EPO at UCB

NASA's STEREO mission, yet to be launched, contributes to EPO program through a *STEREO* website hosted by the Exploratorium. As a part of UCB's *STEREO*-IMPACT EPO effort, *STEREO* mission scientists participated in the Total Eclipse 2001 event with interviews and by supporting science museum eclipse programs as ``docents''. The *STEREO* program also contributed to this high visibility event by distributing 6,000 *STEREO* posters and 3D glasses to 45 Science Centers/Museums nationwide and to Girl Scout Troops that participated in the Live Eclipse Program. *STEREO*'s participation in the Eclipse is highlighted by NASA's Sun Earth Connection Education Forum as part of the recently released ``SolarEvents'' DVD. Further information is at:
`http://cse.ssl.berkeley.edu/impact`

RHESSI EPO at GSFC

The GSFC EPO effort includes teacher workshops modeled after an ongoing GSFC summer teacher program where teachers work with scientists in the Solar Physics Branch for eight weeks each summer. These workshops are in collaboration with the SUNBEAMS (Students United with NASA Becoming Enthusiastic About Math and Science) program. Teachers develop educational materials motivated by *RHESSI* science and technology. The summer teacher's workshop phase of the SUNBEAMS program will be followed by each teacher bringing his or herclass to GSFC for a full week of total immersion in hands-on, inquiry-based, cooperative-learning activities. GSFC scientists, engineers, and technicians stay involved with the teachers through visits to their classrooms, guest lectures, and by providing a continuing flow of *RHESSI* information. For additional information see:
`http://hesperia.gsfc.nasa.gov/HESSI/outreach.htm`

PRODUCTS

RHESSI EPO has produced a variety of educational products, including web-based resources, a poster, a lithograph, a paper model booklet, the Sunspot resource and CD, and a book cover. A Science@NASA article was prepared with the collaboration of *RHESSI* scientists and the EPO program prior to the launch date. The *RHESSI* satellite, mission science and a *RHESSI*-EPO-developed solar flare activity were featured in the 'Having a Solar Blast' episode of the NASA Connect broadcast on NASA TV and PBS stations.

ACKNOWLEDGEMENTS

The resources described above are supported by *RHESSI* and *STEREO* EPO and by The Science Education Gateway (SEGway) Project, a NASA Space Science Supporting Research and Technology Program.

SOLAR PUBLIC OBSERVATIONS IN JAPAN

K. Yaji

Kawabe Cosmic Park, 2107-1 Wasa Kawabe-cho Hidaka-gun Wakayama, Japan

ABSTRACT

In Japan, solar telescopes are now operated in more than fifty astronomical educational facilities, for example, public observatories and science museums. Since most of these have the capability of observing the sun in Hα, active chromospheric phenomena such as solar flares and prominences are often presented to the public there. Though these telescopes must be mainly used for education and public outreach, they have good enough performance to contribute to professional solar research. The staff in most of the facilities don't know well how best to observe the sun and how to understand solar phenomena. We started two efforts in order to support their solar observations. One is the administration of the "Solar Telescope Mailing List" (solnet ML). The other is the arrangement of the "Solar Telescope Workshops".

SOLAR TELESCOPES OF PUBLIC OBSERVATORIES

At present in Japan, solar telescopes are operated in more than fifty astronomical educational facilities, such as public observatories and science museums. This shows high interest in solar observations among the patrons of such facilities. Though solar observations generally meant sunspot observations in white light historically, most of these institutions have the capability of observing the sun in Hα, so active chromospheric phenomena such as solar flares and prominences are often presented to the public there. Now, most of the solar telescopes comprise four kinds of telescopes, which take total and partial solar images in white light and in Hα simultaneously. In addition, some observatories and museums acquire solar observations with unique instruments. For example, Rikubetsu Astronomical Observatory has an aurora monitor, Katsushika Museum has a high dispersion spectrometer, and Nishiharima Astronomical Observatory makes solar radio observations.

For this solar maximum, it is said that solar observations in Hα are very important. Although the main purposes of these solar telescopes must be education and public outreach, they have good enough performance to contribute Hα observations to professional solar research. In fact, some of these observatories co-observed with the XUV Doppler Telescope in 1998.

But the staff in the most of these facilities don't know well how to observe the sun and how to understand solar phenomena. We started two efforts in order to support their solar observations. One is the administration of the "Solar Telescope Mailing List (solnet ML)". The other is the arrangement of the "Solar Telescope Workshops", which were held in 1999 and 2000.

SOLAR TELESCOPE MAILING LIST

We started the "Solar Telescope Mailing List" (solnet ML) in January 1998. The purpose of this mailing list is to facilitate exchanges of information on solar phenomena and solar observations. The detailed contents of the mailing list are prompt reports of solar phenomena, instruments of solar telescopes, educational techniques and solar articles in newspaper and magazines. At least one hundred solar observers attend to this mailing list and the members are composed of public observatory/museum staff, amateur astronomers, professional researchers, and school teachers.

SOLAR TELESCOPE WORKSHOPS

Solar Telescope workshops were held in 1999 and 2000. They provide a chance for staff in public observational facilities to study (a) basics of observational methods in white light, in Hα, in spectrum and in radio, (b) educational techniques using solar observations (Classroom lessons, Displays, Planetarium), and (c) to show their observational results on solar active phenomena. The attendees are not only public observatory/museum staff, but also amateur astronomers, professional researchers, and school teachers, as with the members of "solnet ML". In the workshops, professional solar researchers presented reviews entitled: "Today's Solar Physics", "Space Weather", "The 23rd Maximum Report", and "Hardware of Solar Telescopes". These reviews were good chances to give the attendees of the workshops new knowledge on solar physics. In addition, the two workshops also played a role in linking public observers with professional solar researchers.

SUMMARY

Now, in Japan, solar telescopes are operated in more than fifty astronomical educational facilities. Although the main purposes of these solar telescopes must be education and public outreach, they have good enough performance to contribute to professional solar research. We started two efforts in order to support their solar observations. One is the administration of the "Solar Telescope Mailing List" (solnet ML). The other is the arrangement of the "Solar Telescope Workshops". These two efforts play a role to link public observers with professional solar researchers. We need to continue such efforts and support to encourage solar public observers in future.

ACKNOWLEDGEMENTS

I would like to acknowledge staff of all public observatories in Japan and the members of Solar Telescope Mailing List (solnet ML).

REFERENCES

Yaji, Kentaro, Solar Telescopes in Japan and the Solar Telescope Mailing List in *Astronomical Education with the Internet*, ed. M.Okyudo, T.Ebisuzaki, M.Nakayama, p. 135, Universal Academy Press,Tokyo, Japan (1998).

Section IV.

Sigmoidality and Helicity

SINUOUS CORONAL LOOPS AT THE SUN

A. A. Pevtsov

National Solar Observatory/Sacramento Peak, Sunspot, NM 88349, USA

ABSTRACT

Sinuous coronal loops are commonly observed in *Yohkoh* soft X-ray telescope data. The S-shape of these loops is a manifestation of the helical structure of the coronal magnetic fields and, hence, follows the hemispheric helicity rule established for quiescent filaments and photospheric magnetic fields. Sigmoids are often associated with the CMEs; they exist prior to eruption and disappear after. In addition, active regions that exhibit sinuous loops are more likely to be eruptive than non-sigmoidal regions. Once erupted, sigmoids tend to produce stronger geomagnetic storms, and often the orientation of magnetic field in interplanetary disturbance can be directly linked to the coronal field of the sigmoid. We review the observational properties of sigmoids, current theoretical models, and application of sinuous loops to space weather forecasting.

INTRODUCTION

Sinuous coronal features – later termed sigmoids (Rust & Kumar 1996) – were first described in the *Skylab* observations. Sheeley *et al.* (1975) observed a sinuous emission feature in Fe XV 284 Å spectroheliograms. After the underlying filament erupted, the feature changed its shape to what nowadays we recognize as a cusp. The significance of the sinuous shape of the X-ray structures was not recognized until much later, and so, for instance, Kahler (1977) referred to the same feature as Sheeley *et al.* (1975) simply as a bright linear feature. Kahler (1977) found that these features are commonly observed in association with X-ray long duration events (LDEs). Webb (1985) described several types of X-ray enhancements preceding solar flares, including sinuous coronal structures. The "golden age" of sigmoids began when the *Yohkoh* soft X-ray telescope (SXT) produced numerous examples of sinuous coronal loops (Acton *et al.* 1992). Due to conditions in the solar corona, the X-ray structures outline coronal magnetic field lines. Thus, the sinuous loops may be indicative of non-potential magnetic fields, and hence, play an important role in solar activity. In the following sections we review the properties of sinuous coronal loops, their origin and the role they play in solar eruptions. Other reviews on sigmoids can be found in Sterling (2000), and Canfield *et al.* (2000).

SIGMOIDS AS WE SEE THEM

As with any new phenomenon, there is no clear definition of sigmoids. Moreover, many morphological properties, e.g., size, lifetime etc, that are so appealing to the heart of the traditional astronomer are not well defined. Hence, we begin this section with a description of the statistical properties of sinuous loops.

Types of Sigmoids

The term "sigmoid" was originally applied to both coronal brightenings and Hα filaments (Rust & Kumar 1996). However, we follow the most recent practice and define a sigmoid as a coronal structure that exhibits sinuous shape. The structure may consist of a single or multiple loops. We classify sigmoids using their appearance as four different types: multiple loops, interlocking loops, single loop, and inter-region sigmoids.

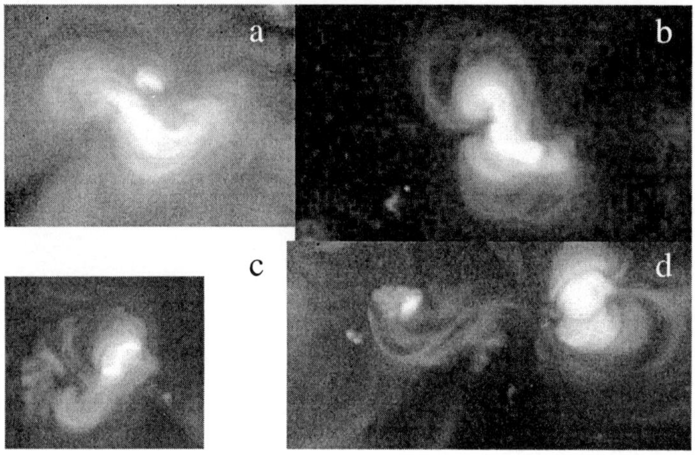

Fig. 1. Sigmoid classification. (a) – Multiple loops, 25-Jan-93, (b) – interlocking loops, 18-Oct-94, (c) – single loop, 24-Apr-93, (d) – inter-region sigmoid, 13-May-92.

"Multiple loops" sigmoids (Figure 1a) are comprised of many discrete coronal loops that collectively have a sinuous shape. The loops are not significantly different in their brightness. This type of sigmoid is typical for the quiet (non-eruptive) stage of active region evolution. "Interlocking loops" sigmoids (Figure 1b) consist of two (three) loops appearing as two fish-hooks laid side-by-side. These loops are significantly brighter than the surrounding coronal structures of the active region. Sigmoids of this type are usually present prior to eruption (e.g. Manoharan *et al.* 1996, Pevtsov *et al.* 1996). A "Single loop" sigmoid (Figure 1c) is a bright single distorted loop. Similar to "interlocking loops", sigmoids of this type are usually observed before the eruption (e.g. Rust & Kumar 1996). The "inter-region" sinuous loops connect different active regions (in the same or opposite hemispheres). In principle, their appearance fits our definition for "multiple/single loop" sigmoids. However, we put them in a separate category because of their properties. For instance, unlike the active region sigmoids, inter-region loops may exist even if the magnetic field is current-free (e.g. Pevtsov *et al.* 1997). With regard to their temporal visibility sigmoids can loosely be classified as "persistent" or "transient". The "persistent" sigmoids are present at any time during their lifetime. "Transient" sigmoids appear and disappear (but do not erupt) several times during their disk passage. The above classification is somewhat artificial. There are examples, when sigmoids evolve from one type to another (e.g. Figure 3).

Sigmoid Statistics

We employed several lists of sigmoids to study their statistical properties. The lists were composed using SXT full frame desaturated (SFD) images and cover the period from 1991–1998. In these surveys we identified all sinuous loops, independent of their association with numbered active regions, brightness and size. However, due to gaps in the observations, limited spatial resolution and subjectivity of sigmoid identification these lists may not include all sinuous loops that were present at the Sun.

Size and Shear Figure 2 shows the distribution of 185 sigmoids observed from 1991–1994 (Pevtsov *et al.* 1997). Only one observation (the best appearance) for a given region, was included in this list. For each of these sinuous loops we measured a separation L between footpoints, and shear angle γ in the middle part of the sigmoid. Using these two parameters we computed the coronal α of the linear force-free field, $\alpha = \frac{\pi}{L}\sin\gamma$ (Pevtsov *et al.* 1997). The footpoint separation (Figure 2a) varies from $\sim$ 11" to 640" with median value $\approx$ 150". The histogram shows a deficit of sigmoids with L < 50", which we contribute to the insufficient spatial resolution of *Yohkoh* observations. The coronal α varies between $1.62 \times 10^{-9}\ m^{-1}$ and

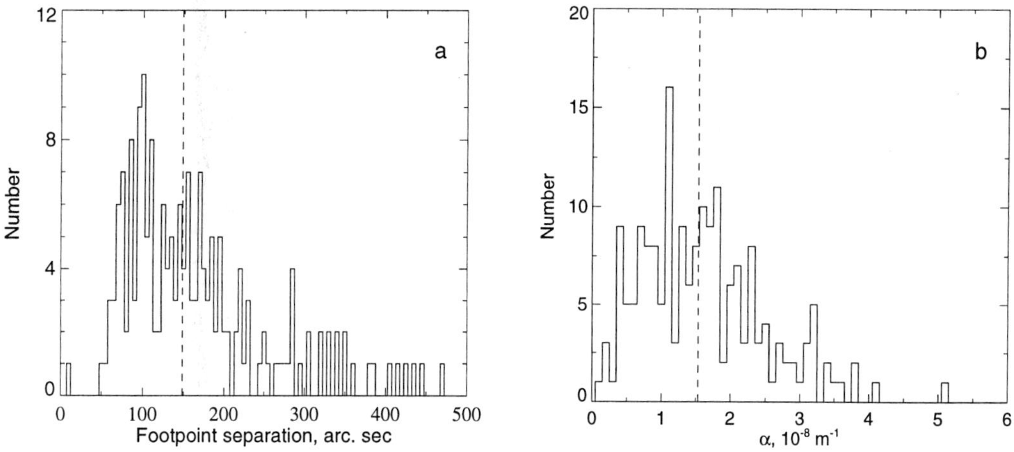

Fig. 2. Footpoint separation and force-free field α-coefficient for 185 sigmoids observed from 1991–1994. Vertical dashed line shows median values.

$2.43 \times 10^{-7}\ m^{-1}$, – typical values for active regions magnetic field (e.g. Pevtsov *et al.* 1997). The median footpoint separation L is also about the size of the typical active region. This suggests that sigmoids are mostly active region features. Indeed, 136 (74%) out of 185 sigmoids that we found in the 1991-1994 data were associated with numbered active regions. Moreover, the majority of active regions are non-sigmoidal. Another survey of 1991-1998 *Yohkoh* observations yields 362 sigmoidal active regions. In 221 of these cases the active regions exhibit sigmoidal shape in more than one *Yohkoh* image. The total number of active regions during the same period was about 1400. Thus, only a small fraction of numbered active regions (15%–26%) exhibits sigmoids.

Lifetime Table 1 shows the lifetime of 221 sigmoids. We define the "lifetime" as the difference between the first and last sigmoid appearance of the same active region. We disregard the fact that different images (of the same AR) may show different loops. Nearly half of the sigmoids persist < 1 day. Their median lifetime is about 1.5 d.

Table 1. Number of Sigmoids and Their Lifetime

	Days						Total
	1	2	3	4	5	6	
All data set	108(48.9%)	45(20.4%)	30(13.6%)	16(7.2%)	15(6.8%)	7(3.2%)	221
Restricted CMD	59(38.1%)	31(20.0%)	27(17.4%)	16(10%)	15(9.7%)	7(4.5%)	155

The fraction of short-lived sigmoids may be overestimated due to a contribution of sigmoids that emerge (disappear) behind the solar limb, but disappear (emerge) on the visible hemisphere. To make a rough estimate of the possible contribution of such sigmoids, we restricted our data set to those loops that appear in the eastern hemisphere and disappear west of -30° meridian. As expected, this subset of 155 sigmoids (Table 1, restricted CMD) shows a significant drop in short-lived sinuous loops. Since we did not register the fact of eruption as a cause of sigmoid disappearance, we cannot segregate between "true" (due to eruption) and "apparent" (due projection) sigmoid disappearance. Hence we believe that the maximum lifetime of sigmoids shown in Table 1 is an underestimate. Table 1 is based on observations of sigmoid regions during their disk passage. We did not attempt to identify returning active regions and their sigmoid appearance. Although some active regions may retain its sigmoid shape for several rotations (Figure 3, see also Glover *et al.* 2001), we believe that their presence will not change significantly our statistics in Table 1. For the majority of regions the sigmoid shape disappears when the region is still in the visible hemisphere.

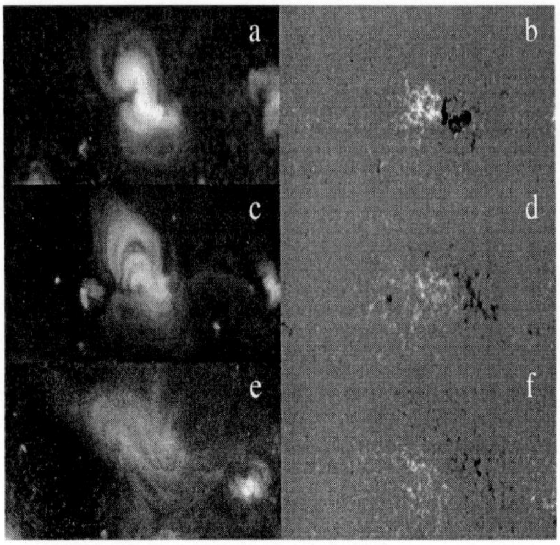

Fig. 3. Evolution of sigmoid region in three solar rotations, a, b – NOAA AR 7790, 18 October, 1994, c, d – 14 November 1994, e, f – 11 December 1994, SXT images (left column) and NSO/KP magnetograms (right) are not coaligned. (This example is courtesy of L.W. Acton).

Hemispheric Helicity Rule

As the majority of magnetic structures, sigmoids follow the hemispheric helicity (chirality) rule. The rule, stipulating a preferred sense of magnetic field twist for a given hemisphere, is observed in the quiet Sun (QS) network field, X-ray bright points (XBPs), quiescent filaments, strong magnetic fields of active regions (ARs), superpenumbral Hα filaments, X-ray arcades overlying quiescent filaments, large-scale magnetic field (LSMF), solar wind and magnetic clouds (MC) (Figure 4, for review, see individual chapters in Brown *et al.* 1999, Martin 1998). With respect to sinuous loops, Rust & Kumar (1996) found that the majority (69%) of those in the southern (northern) hemisphere were forward-S (inverse-S) shaped. This hemispheric preference is independent of solar cycle (Pevtsov *et al.* 2001).

Where Sigmoids Come From

Sigmoids (twisted magnetic field) may emerge from below the photosphere, or develop after the emergence as the result of surface motions. Subphotospheric origin of helicity has been proposed by several researchers (e.g. Seehafer 1990; Rust & Kumar 1996, Longcope *et al.* 1998). Leka *et al.* (1996) presented observational evidence that the magnetic field is already twisted, upon its emergence. However, we were unable to find an emerging sigmoid in SXT data that could support this idea. Lack of direct observations of emerging sigmoids can be explained by insufficient spatial resolution of SXT data. Thus, for instance, a sinuous loop with L=150" footpoint separation and force-free field coefficient $\alpha = 1.5 \times 10^{-8}\mathrm{m}^{-1}$ will exhibit a shear angle $\gamma \approx 30°$. The same loop with L=50" footpoint separation will have only 10° shear angle, which will be hard to detect on the *Yohkoh* half resolution images.

Wang (1996) studied evolution of magnetic helicity in NOAA AR 6233 and concluded that interaction of magnetic field and plasma flows in the photosphere is the major contributor to the helicity budget of this region. Solar differential rotation introduces systematic shear, which may transform a regular loop to a

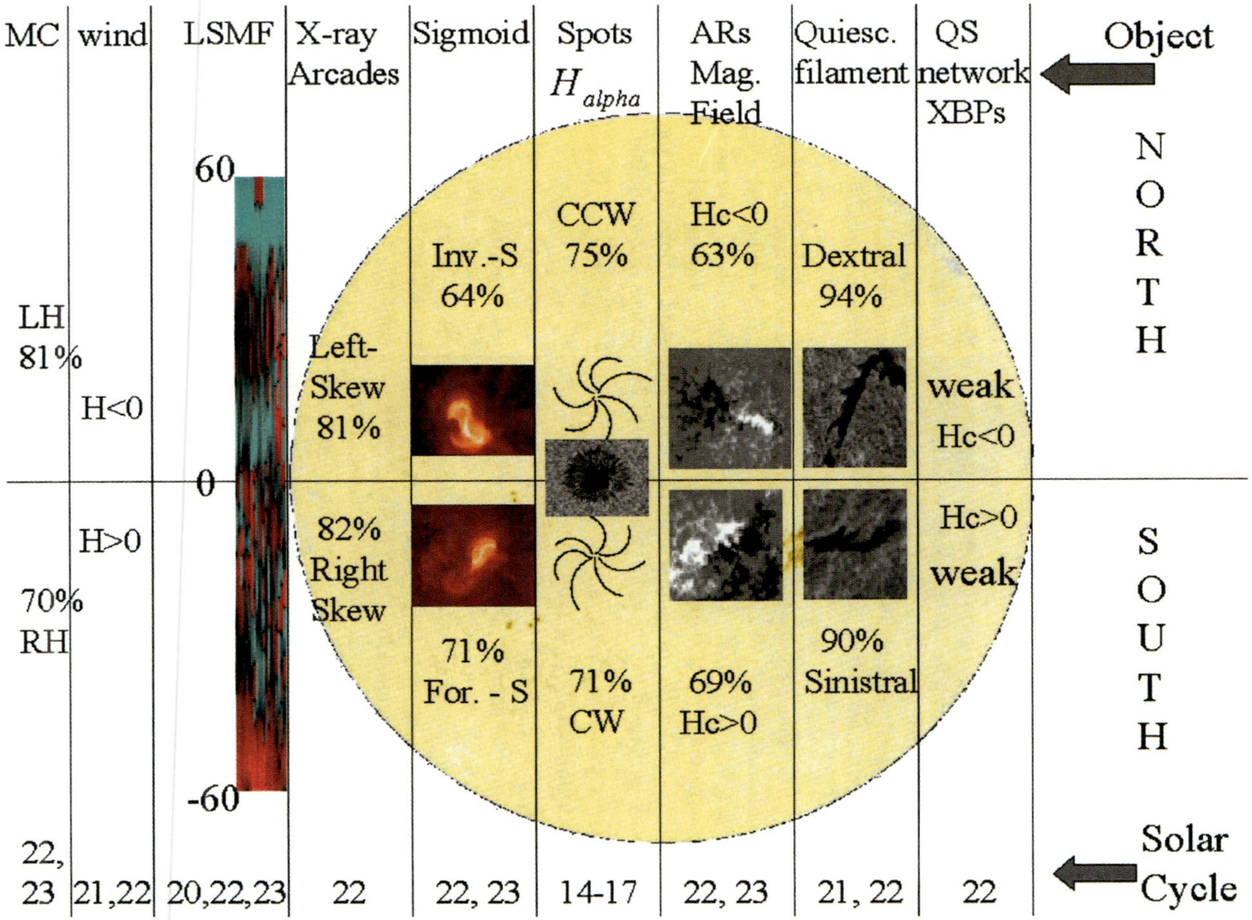

Fig. 4. Hemispheric helicity rule. H (H_c) is magnetic (current) helicity. CCW/CW is counter/clockwise. LH/RH is left-/right-hand twist. Other abbreviations explained in the text. The data for this plot are taken from published papers. LSMF data are from work in progress by the author. LSMF chart shows latitudinal ($\pm 60^\circ$) profiles of H_c for solar rotations 1910–1919. Red is positive H_c, blue is negative, and black is zero.

sigmoid. Van Ballegooijen (1999) has shown that this mechanism produces a correct hemispheric sign of helicity. DeVore (2000) and Berger & Ruzmaikin (2000) have found that the value of helicity created by differential rotation is in quantitative agreement with the observations. On the other hand, Figure 3 shows that the sigmoid is present at early stages of evolution of active region. There is no significant build-up of shear over three solar rotations, as one would expect from the differential rotation action. Glover *et al.* (2001) studied the evolution of magnetic and coronal structures in association with a large inter-region sigmoid. The sigmoid was formed during the second rotation of the studied area apparently as the result of both shear due to differential rotation and helicity transport from twisted emerging flux. Démoulin *et al.* (2002) have shown that the helicity budget of CME productive NOAA AR 7978 requires helicity injection far exceeding the contribution from solar differential rotation. Recently, Chae (2001) has shown that photospheric horizontal motions may generate a much larger helicity contribution than differential rotation. However, these motions are, in general, not hemisphere dependent. Moreover, some systematic motions (e.g. poleward drift of following polarity field) produce the wrong sense of shear for a given hemisphere. Raman *et al.* (2000, see also, Nightingale *et al.* this volume) have reported the development of a sigmoid brightening as the result of rotation of the umbra of a δ-spot. However, sunspot rotational motions are known to be hemisphere independent. Furthermore, the sunspots often change the direction of their rotation and exhibit so called sunspot torsional oscillations with periods of several days (e.g. Druzhinin *et al.* 1993). In addition to the above mechanisms, a "single loop" sigmoid may be formed by reconnection of two independent "interlocking" loops. The examples of such sigmoid formation with following eruption were reported in Manoharan *et al.* (1996) and Pevtsov *et al.* (1996).

Martens & Zwaan (2001) have shown that sinuous filaments may develop as the result of gradual "head-to-tail" reconnection between the leading polarity of an old flux system (e.g. decaying active region) and the following polarity of a new active region. In their model, however, the flux systems have no prior magnetic connection. In contrast, the majority of sigmoids are observed between the leading and following polarity of the same active region, not on the boundary between two independent regions. On the other hand, inter-region sigmoids may be formed via the mechanism proposed by Martens & Zwaan (2001).

FILAMENT-SIGMOID COMPLEX

The presence of chromospheric filaments in sigmoidal regions has been reported by many researchers (e.g. Sheeley *et al.* 1975, Aurass *et al.* 1999, Gibson & Low 2000, Sterling *et al.* 2000). Pevtsov (2002) found that 124 (91%) of 136 sigmoidal active regions also had a filament. However, the filament and sigmoid do not appear to represent different parts of the same structure. Figure 5 shows NOAA AR 8004 in different wavelengths. Both filament and sigmoid lay above the photospheric neutral line (see also Sterling & Hudson 1997, Zarro *et al.* 1999). However, the filament is significantly longer than the sigmoid. Furthermore, although the middle part of the sigmoid is parallel to the polarity inversion line, its footpoints are anchored in opposite polarity fields (cf. Figure 5c and d). In contrast, the entire filament body is situated on the polarity inversion line (cf. Figure 5a and d). There appears to be no continuous density structure linking filament and sigmoid (Figure 5a-c, e, f).

CMES AND MAGNETIC CLOUDS

Rust & Kumar (1996) studied the evolution of 103 sigmoids with short lifetimes ($<$ 12 hours). They found that the majority of these sigmoids evolve into an arcade of loops or diffuse cloud within $\approx$ an hour after the brightening onset. Canfield *et al.* (1999) studied the association between shapes of coronal structures (sigmoidal/non-sigmoidal), sunspot area, and the tendency toward eruption. They found that sigmoidal regions are more likely to erupt than non-sigmoidal ones. Glover *et al.* (2000) confirmed this finding in their study of 17 active regions. Hudson *et al.* (1998) found that a "sigmoid-arcade" transformation was commonly associated with a halo CME launch. Another CME precursor – coronal dimming, – was observed by several researchers in connection with sigmoid disappearance. Sterling & Hudson (1997) interpret coronal

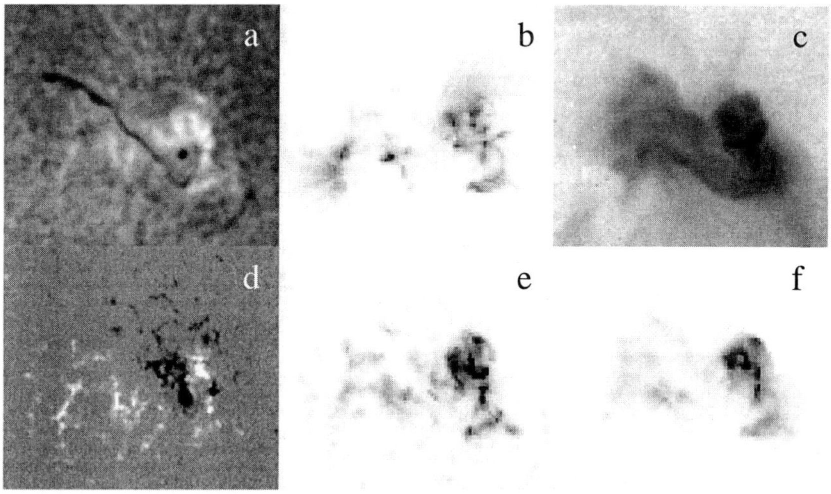

Fig. 5. NOAA AR 8004 in different wavelengths, 19 December 1996. (a) – NSO/SP H_α image, (b) – *SOHO*/EIT 171 Å, (c) – *Yohkoh* SXT, (d) – MDI magnetogram, (e) – *SOHO*/EIT 304 Å, (f) – *SOHO*/EIT 284 Å.

dimming as the result of the density depletion via expansion or ejection, which supports the flux-tube model of sigmoids. Pevtsov & Canfield (2001) studied properties of 18 geomagnetic storms associated with sigmoid eruptions. They determined the leading (north-/southward) component of the magnetic field in the source region, using the shape of sinuous loops and photospheric magnetograms. They found a significant correlation between the strength of the geomagnetic storm and field orientation in the source region. Cid *et al.* (2001) studied the properties of four magnetic clouds (MCs). Three MCs were associated with sigmoid eruptions and one originated from a non-sigmoidal region. Based on this Cid *et al.* (2001) concluded that the MC topology is not dependent on the solar source region. On the other hand, Leamon *et al.* (2002) found a strong correlation between the shape of sigmoids and helicity of MCs associated with sigmoid eruption. In 8 (73%) out of 11 cases the inverse-S sigmoids were associated with negative helicity MC. 17 (61%) of 28 S-shaped loops were associated with positive helicity MC. Leamon *et al.* (2002) also found that the erupting sigmoid will most likely produce a flux rope in interplanetary space and that the majority (86%) of sigmoid eruptions result in at least a moderate geomagnetic storm ($D_{st} > 50\ nT$).

SIGMOIDS AS WE MODEL THEM

In their appearance sinuous loops are reminiscent of magnetic field lines of linear force-free field (e.g. Canfield *et al.* 2000). A good correlation between α coefficients independently derived from photospheric and coronal data (e.g. Pevtsov *et al.* 1997) indicates a possible relationship between photospheric electric currents and sigmoid shape. Magara & Longcope (2001) have modeled the evolution of a twisted flux tube emerging through the photosphere. When the tube expands into the corona, its outer poloidal field lines form a potential-like arcade, and the inner toroidal field lines form a sigmoidal structure. Gibson & Low (2000) considered a flux rope model with a spheromak-type magnetic field. They showed that several key properties, e.g. coronal dimming, sigmoidal structure, sigmoid-filament association, and limb/disk CME geometry can be explained in the framework of this model. The sigmoid is formed by axial flux rope field lines supporting filament-sigmoid material. Thus, the filament and sigmoid should belong to the same density enhancement, which seems to be in disagreement with Figure 5.

Sinuous coronal loops may also be formed via interaction (reconnection) between emerging and pre-existing magnetic flux (e.g. Martens & Zwaan 2001, Gibson *et al.* 2002). Titov & Démoulin (1999) modeled the evolution of a twisted toroidal field emerging into a potential field. They found that the reconnection (and energy release) takes place at two specific boundaries – separatrices. The separatrices wrap around the twisted toroidal field forming a structure reminiscent of a sigmoid. To insure correct hemispheric handedness of the sigmoids toroidal flux tube twist must depend on hemisphere. The twist can be created by the differential rotation or have sub-photospheric origin. This model requires dynamical evolution (i.e. emergence) of twisted flux tubes relative to the background field.

Moore *et al.* (2001) invoked a tether-cutting model to explain sigmoid eruptions. In their model a sigmoid is formed by two "interlocking" independent loops with an overlying arcade. Due to external process (e.g. new flux emergence, photospheric motions) the interlocking loops reconnect and erupt. Several eruptive and confined events studied in detail were found in good agreement with the model.

HOW DOES IT ALL FIT TOGETHER?

SOHO/LASCO observations show intensity structures in erupting CMEs that can be interpreted as unwinding motions (e.g. Wood *et al.* 1999). The majority of CMEs associated with sigmoid eruptions result in a magnetic cloud with a simple flux rope-like magnetic structure. Is this flux rope ejected from the Sun's surface or is it formed during the eruption by the magnetic reconnection in the sheared arcade overlying the filament-sigmoid channel (Gosling 1999)? The high correlation between the helicity of MCs and the shape of sigmoids (Leamon *et al.* 2002) supports direct ejection of a flux tube from the solar corona. However, the amount of twist that one can infer from the LASCO limb observations is too high ($\gg 2\pi$). The photospheric/coronal flux tube should have become kink-unstable well before accumulating that much twist. Another potential difficulty of the flux tube model is partial filament eruptions (Tang 1986). If the filament sits at the bottom (concave-up) part of a horizontal flux tube, partial eruptions (sometimes associated with CMEs) mean that material can escape through the upper boundary of the tube without disturbing its lower boundary. Furthermore, multiple eruptions from the same active region imply that the flux tube does not completely erupt. Pevtsov (2002) described several CMEs associated sigmoid eruptions, which had little or no disturbance to the underlying filament. He explained these observations in the framework of the traditional two-ribbon flare model, which topologically is similar to the tether-cutting model (Moore *et al.* 2001). In this model the rising sigmoid pushes up to the overlying magnetic field. This brings magnetic field lines below the sigmoid closer to each other, and triggers their reconnection. Post-flare loops are formed below the sigmoid that continues to rise and eventually escapes. In this scenario the flux rope-like structure observed by LASCO is created during the eruption. The amount of stress remaining in the magnetic field will determine further eruptivity of the region. On the other hand, Gilbert *et al.* (2001) have considered the evolution of the flux rope prominence model under different reconnection scenarios. They demonstrated that reconnection between the flux rope magnetic field, the supporting filament, and the overlying arcade field may lead to a CME eruption without significant removal of filament material. Thus, the flux rope models may also explain partial (multiple) eruptions from the same region.

ACKNOWLEDGEMENTS

The National Solar Observatory (NSO) is operated by the Association of Universities for Research in Astronomy (AURA Inc.) under cooperative agreement with the National Science Foundation (NSF). *Yohkoh* is a mission of ISAS in Japan. *SOHO* is a project of international cooperation between ESA and NASA.

REFERENCES

Acton, L., S. Tsuneta, Y. Ogawara, R. Bentley, M. Bruner, *et al.*, The *Yohkoh* Mission for High-Energy Solar Physics, *Science,* **258**, 618 (1992).

Aurass, H., B. Vršnak, A. Hofmann, and V. Rudžjak, Flares in Sigmoidal Coronal Structures - a Case Study, *Solar Phys.,* **190**, 267 (1999).

Berger, M. A. and A. Ruzmaikin, Rate of Helicity Production by Solar Rotation, *JGR,* **105**, 10481 (2000).

Brown, M. R., R. C. Canfield, and A. A. Pevtsov, *Magnetic Helicity in Space and Laboratory Plasmas,* volume 111 of *Geophysical Monographs Series,* AGU, Washington, DC (1999).

Canfield, R. C., H. S. Hudson, and D. E. McKenzie, Sigmoidal Morphology and Eruptive Solar Activity, *GRL,* **26**, 627 (1999).

Canfield, R., H. S. Hudson, and A. A. Pevtsov, Sigmoids as Precursors of Solar Eruptions, *IEEE Trans. Plasma Science,* **28**, 1786 (2000).

Chae, J., Observational Determination of the Rate of Magnetic Helicity Transport through the Solar Surface via the Horizontal Motion of Field Line Footpoints, *ApJ Letters,* **560**, L95 (2001).

Cid, C., M. A. Hidalgo, J. Sequeiros, J. Rodríguez-Pacheco, and E. Bronchalo, Evidence of Magnetic Flux Ropes in the Solar Wind From Sigmoidal and non-Sigmoidal Active Regions, *Solar Phys.,* **198**, 169 (2001).

Démoulin P., C. H. Mandrini, L. van Driel-Gesztelyi, B. Thompson, S. Plunkett, *et al.*, What is the Source of the Magnetic Helicity Shed by CMEs? The Long-Term Helicity Budget of AR 7978, *A&A,* in press (2002).

DeVore, C. R., Magnetic Helicity Generated by Solar Differential Rotation, *ApJ,* **539**, 944 (2000).

Druzhinin, S. A., A. A. Pevtsov, V. L. Levkovsky, and M. V. Nikonova, Line-of-sight Velocity Measurements Using a Dissector-Tube. II. Time Variations of the Tangential Velocity Component in the Evershed Effect, *A&A,* **277**, 242 (1993).

Gibson, S. E. and B. C. Low, Three-dimensional and Twisted: an MHD Interpretation of on-disk Observational Characteristics of Coronal Mass Ejections, *JGR,* **105**, 18187 (2000).

Gibson, S. E., L. Fletcher, G. Del Zanna, C. D. Pike, H. E. Mason, *et al.*, The Structure and Evolution of a Sigmoidal Active Region, *ApJ,* in press (2002)

Gilbert, H. R., T. E. Holzer, B. C. Low, and J. T. Burkepile, Observational Interpretation of an Active Prominence on 1999 May 1, *ApJ,* **549**, 1221 (2001).

Glover, A., L. K. Harra, S. A. Matthews, K. Hori, and J. L. Culhane, Long Term Evolution of a Non-Active Region Sigmoid and its CME Activity, *A&A,* **378**, 239 (2001).

Glover, A., N. Ranns, L. K. Harra, and J. L. Culhane, The Onset and Association of CMEs with Sigmoidal Active Regions, *GRL,* **27**, 2161 (2000).

Gosling, J. T., The Role of Reconnection in the Formation of Flux Ropes in the Solar Wind, in *Magnetic Helicity in Space and Laboratory Plasmas,* ed. Brown, M. R., R. C. Canfield, and A. A. Pevtsov, p. 205, AGU, Washington, DC (1999).

Hudson, H. S., J. R. Lemen, O. C. St. Cyr, A. C. Sterling, and D. F. Webb, X-ray Coronal Changes During Halo CMEs, *GRL,* **25**, 2481 (1998).

Kahler, S., The Morphological and Statistical Properties of Solar Soft X-ray Events with Long Decay Times, *ApJ,* **214**, 891 (1977).

Leamon, R. J., R. C. Canfield, and A. A. Pevtsov, Properties of Magnetic Clouds Resulting from Eruption of Coronal Sigmoids, *Journal of Geophysical Research,* in press (2002).

Leka, K. D., R. C. Canfield, A. N. McClymont, and L. van Driel-Gesztelyi, Evidence for Current-Carrying Emerging Flux, *ApJ,* **462**, 547 (1996).

Longcope, D. W., G. H. Fisher, and A. A. Pevtsov, Flux Tube Twist Resulting from Helical Turbulence: the Σ-effect, *ApJ,* **507**, 417 (1998).

Magara, T. and D. W. Longcope, Sigmoid Structure of an Emerging Flux Tube, *ApJ Letters,* **559**, L55 (2001).

Manoharan, P. K., L. van Driel-Gesztelyi, M. Pick, and P. Demoulin, Evidence for Large-Scale Solar Magnetic Reconnection from Radio and X-Ray Measurements, *ApJ Letters,* **468**, L73 (1996).

Martens P.C. and Zwaan, C., Origin and Evolution of Filament-Prominence System, *ApJ,* **558**, 872 (2001).

Martin, S. F., Filament Chirality: A Link Between Fine-Scale and Global Patterns, in *New Perspectives on Solar Prominences,* ed. D. Webb, D. Rust and B. Schmieder, vol. 150 of *ASP Conf. Series,*, p. 419, San Francisco, ASP (1998).

Moore, R. L. A., C. Sterling, H. S. Hudson, and J. R. Lemen, Onset of the Magnetic Explosion in Solar Flares and Coronal Mass Ejections, *ApJ,* **552**, 833 (2001).

Pevtsov, A. A., Active Region Filaments and X-ray Sigmoids, *Solar Phys.,* in press (2002).

Pevtsov, A. A. and R. C. Canfield, Solar Magnetic Fields and Geomagnetic Events, *JGR,* **106**, 25191 (2001).

Pevtsov, A. A., R. C. Canfield, and S. M. Latushko, Hemispheric Helicity Trend for Solar Cycle 23, *ApJ Letters,* **549**, L261 (2001).

Pevtsov, A. A., R. C. Canfield, and A. N. McClymont, On the Subphotospheric Origin of Coronal Electric Currents, *ApJ,* **481**, 973 (1997).

Pevtsov, A. A., R. C. Canfield, and H. Zirin, Reconnection and Helicity in a Solar Flare, *ApJ,* **473**, 533 (1996).

Raman, S. K., K. B. Ramesh, R. Selvendran, P. S. Aleem, and K. M. Hiremath, Emergence of Twisted Magnetic Flux Related Sigmoidal Brightening, *J. Astrophys. Astr.,* **21**, 263 (2000).

Rust, D. M. and A. Kumar, Evidence for Helically Kinked Magnetic Flux Ropes in Solar Eruptions, *ApJ Letters,* **464**, L199 (1996).

Seehafer, N., Electric Current Helicity in the Solar Atmosphere, *Solar Phys.,* **125**, 219 (1990).

Sheeley, N. R., J. D. Bohlin, G. E. Brueckner, J. D. Purcell, V. E. Scherrer, *et al.*, Coronal Changes Associated With a Disappearing Filament, *Solar Phys.,* **45**, 377 (1975).

Sterling, A. C., Sigmoid CME Source Regions At the Sun: Some Recent Results, *Journal of Atmospheric and Solar-Terrestrial Physics,* **62**, 1427 (2000).

Sterling, A. C. and H. S. Hudson, *Yohkoh* SXT Observations of X-Ray "Dimming" Associated with a Halo Coronal Mass Ejection, *Apj Letters,* **491**, L55 (1997).

Sterling, A. C., H. S. Hudson, B. J. Thompson, and D. M. Zarro, *Yohkoh* SXT and *SOHO* EIT Observations of Sigmoid-to-Arcade Evolution of Structures Associated with Halo Coronal Mass Ejections, *ApJ,* **532**, 628 (2000).

Tang, F., The Two Types of Flare-Associated Filament Eruption, *Solar Phys.,* **105**, 399, (1986).

Titov, V. S. and P. Démoulin, Basic Topology of Twisted Magnetic Configurations in Solar Flares, *A&A,* **351**, 707 (1999).

van Ballegooijen, A. A., Photospheric Motions as a Source of Twist in Coronal Magnetic Fields, in *Magnetic Helicity in Space and Laboratory Plasmas,* ed. Brown, M. R., R. C. Canfield, and A. A. Pevtsov, p. 213, AGU, Washington, DC (1999).

Wang, J., A Note on the Evolution of Magnetic Helicity in Active Regions, *Solar Phys.,* **163**, 319 (1996).

Webb, D. F., Coronal X-ray Activity Preceding Solar Flares, *Solar Phys.,* **97**, 321 (1985).

Wood, B. E., M. Karovska, J. Chen, G. E. Brueckner, J. W. Cook, and R. A. Howard, Comparison of Two Coronal Mass Ejections Observed by EIT and LASCO with a Model of an Erupting Magnetic Flux Rope, *ApJ,* **512**, 484 (1999).

Zarro, D. M., A. C. Sterling, B. J. Thompson, H. S. Hudson, and N. Nitta, *SOHO*-EIT Observations of Extreme-Ultraviolet "Dimming" Associated with a Halo Coronal Mass Ejection, *ApJ Letters,* **520**, L139 (1999).

THE ORIGIN OF PROMINENCES AND THEIR HEMISPHERIC PREFERENCE FOR THE SKEW OF OVERLYING X-RAY LOOPS

P. C. H. Martens

Physics Department, Montana State University, P.O. Box 173840, Bozeman, MT 59717, USA

ABSTRACT

I review a "head-to-tail" linkage model for the formation, evolution, and eruption of solar filaments. The magnetic field structure of the model is based upon the observation that filaments form exclusively in filament channels with no apparent magnetic connections low above the polarity inversion line. The formation of a filament in this configuration is driven by flux convergence and cancellation, which produces loop-like filaments segments with a half-turn. Filament segments of like chirality may connect and form long quiescent filaments. I analyze the topology of the field in and around filaments formed through "head-to-tail" linkage and find consistency with the observed hemispheric preference for the skew of the X-ray emitting loops that vault the prominence, contrary to what one would expect from differential rotation.

INTRODUCTION

Systematic photographic and spectrographic observations of solar prominences stretch back all the way back to the 1860's with the work of father Secchi and de la Rue. For an excellent and concise review of the history of filament and prominence observations, see Tandberg-Hanssen (1998). Classical systematic studies can be found in the work of D'Azambuja & D'Azambuja (1948, 1949) and the review of Kiepenheuer (1953), which contains a succinct synopsis of the observations to that date. In the early 1960's it became clear that filaments are found in the zones of polarity inversion, exactly above strips of opposite dominant polarity. Fibrils and other elongated structures seen in Hα indicated that the dominant component of the magnetic field is along the polarity inversion line (PIL), not across it as one would expect from simple potential field extrapolations.

Models for filament formation usually take the above observations as starting conditions. However, filaments do not form along every polarity inversion line, and one observational aspect that is often ignored – although emphatically pointed out by observers since the 1960's – is that filaments only form in so-called filament channels ("plage couloirs"); areas that appear as nearly void in Hα filtergrams slightly redshifted from line-center. Gaizauskas *et al.* (1997), and Martin (1998) in a review, pointed out that in filament channels the magnetic fields on the opposite sides of the PIL are usually not part of the same original bipolar region. Indeed, the observed absence of fibrils crossing the PIL in filament channels (Foukal 1971a, b) indicates that there are no magnetic connections across the PIL at low levels[1]. Thus filament channels may be realizations of the "bald" spots (e.g. Priest & Forbes 2000) that theorists study.

[1]The presence of overlaying X-ray arcades, considered hereafter, shows that there are magnetic cross-connections at much larger scales.

Martens & Zwaan (2001) have proposed a scenario for filament formation, called "head-to-tail linkage", that for the first time explicitly takes into account the absence of local magnetic cross-connections in many filament channels. This model further relies on the well-known conditions of strong shear in the originating filament channel, as well as convergence and cancellation of magnetic field across its PIL. The former condition follows from the above mentioned observations of Foukal (1971a, b), and the latter has been discovered in the 1980's by Martin and co-workers (e.g. Martin, Livi, & Wang 1985), and has been confirmed independently many times over. Van Ballegooijen & Martens (1989) demonstrated in a numerical simulation that this process can indeed generate filament-like helical flux tubes that may erupt after sufficient photospheric flux has canceled.

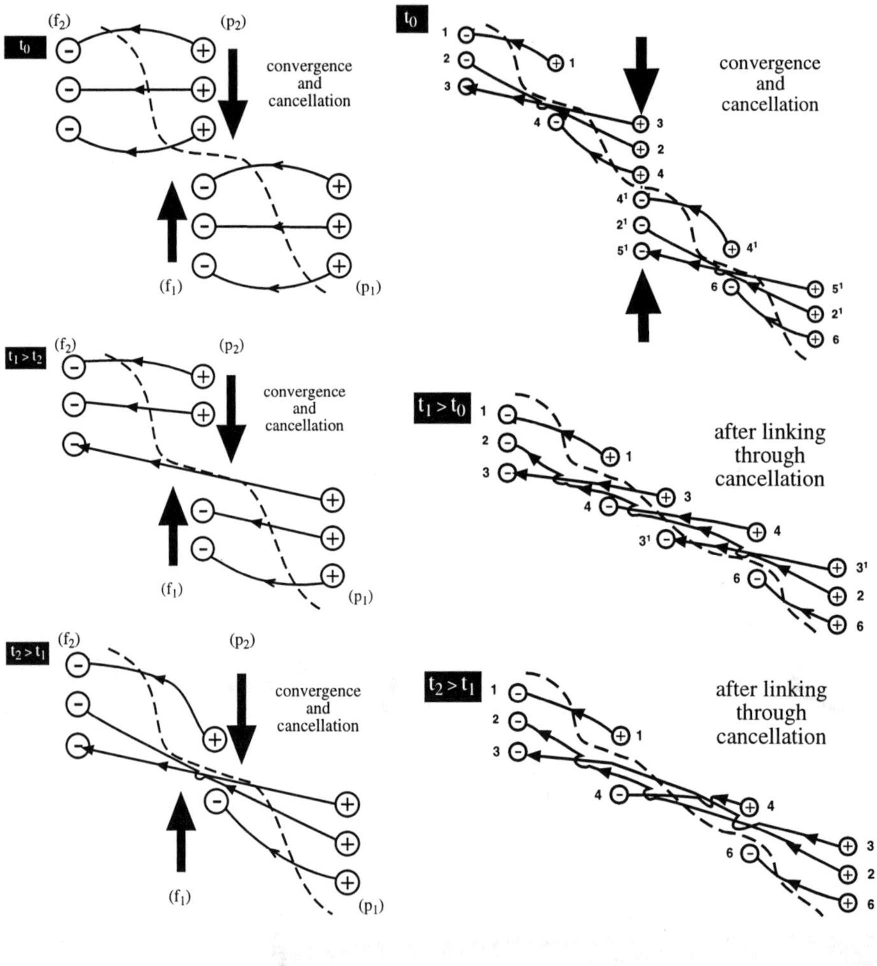

Fig. 1. Left: 'Head-to-tail" linkage of two sunspot pairs in the northern hemisphere, driven by convergence, cancellation, and reconnection, and leading to the formation of a filament with dextral chirality and an inverted S-shaped spine. Right: Growth of filaments through linkage of filament segments driven by the same process. The dashed line indicates the photospheric polarity inversion line. Field lines 3 and 5^1 represent straight filament axes above the "dipped" filament segment flux tubes 2 and 2^1. At time t_2, after convergence and reconnection, a helical flux tube with one-and-a-half turn has formed (from Martens & Zwaan 2001).

In this paper I will consider in detail what the "head-to-tail linkage" model predicts for the skew of SXT observed arcades of loops that are often observed vaulting prominences. The question is of considerable interest for the following reason: individual loops in the coronal arcades observed by SXT appear to cross the filament channel obliquely (Martin & McAllister 1996). The acute angle between the filament axis and the arcade loops is to the left of the filament axis in the northern hemisphere and to the right in the south. The authors call these orientations left-skewed and right-skewed. The observed hemispheric orientation is opposite to the skew that one would expect to be introduced by the solar differential rotation. Consider a filament stretched in the NE-SW direction, as is usually the case in the northern hemisphere (SE-NW in the southern hemisphere), and assume there is originally an arcade of loops crossing the PIL and filament axis at right angles, as one might expect in a potential field. Differential rotation will advance the equatorward footpoints of the arcade loops westward with respect to their poleward counterparts, and the acute angle defined above will be to the right in the northern hemisphere and vice-versa, in contrast to observations. According to McAllister *et al.* (1998) only the eastern sections of polar crown filaments appear to be have this differentially driven orientation in the second half of the cycle.

MODEL AND ANALYSIS

The left side of Figure 1, from Martens & Zwaan (2001), shows the basic element of the "head-to-tail linkage" model: two dipoles, originally unconnected, converge and cancel at their mutual PIL. The nearly horizontal section of the PIL satisfies the above mentioned conditions for a filament channel, and therefore is the only segment of the PIL where a filament can be formed according to observations. Through field cancellation in this filament channel a helical flux tube with only one-half turn is formed. Note that this simple configuration produces strongly sheared fields along the PIL without the need for large-scale shear flows, which are never observed except perhaps in rotating sunspots (Nightingale *et al.* 2002). The right side of the figure shows how the filaments thus formed can link up to form even longer filament segments with one whole turn added at each cancellation event. Linkage of filaments was first reported by McIntosh (1972), and a very clear occurrence in a sigmoid has been described by Pevtsov, Canfield, & Zirin (1996). From

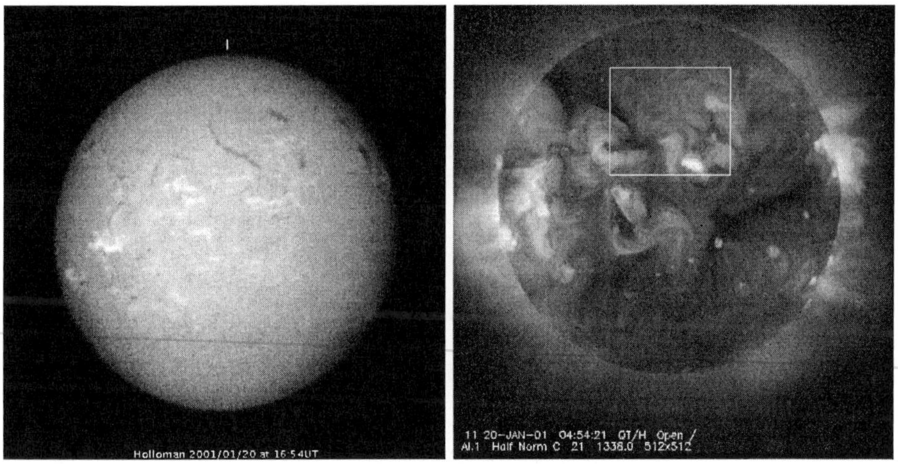

Fig. 2. Left: A full disk Hα image from Holloman solar observatory, showing a long filament near the meridian in the northern hemisphere in a typical NE-SW orientation. Right: A full disk SXT image taken about 12 hours earlier, showing the arcade of loops vaulting the filament in the left-skewed orientation that is typical for the northern hemisphere. These images were obtained from the February 2 2001 SXT Science Nugget by Aki Takeda.

Joy's law it follows that Figure 1 applies to the formation of a filament in the northern hemisphere. The overlaying loops are left skewed, in agreement with the observed hemispheric preference. Mirroring Figure 1 across the equator one obtains the scenario that applies to the southern hemisphere, with right-skewed arcades. Figure 2 shows a filament on the left and an overlaying soft X-ray arcade on the right that might have formed according to this scenario. As discussed above the NE-SW filament orientation and left-skewed arcade are typical for the northern hemisphere. I conclude that the Martens & Zwaan model correctly reproduces the observed hemispheric preferences for the skewedness of X-ray arcades overlaying filaments.

DISCUSSION

In our "head-to-tail" linkage model the direction of the axial (i.e. along the filament) magnetic field is the same inside the filament and in the overlying arcade. It is easily verified that when this is true and the hemispheric chirality rule is observed (see Martin 1998), the hemispheric skew preferences are automatically reproduced. Thus any model for filament formation that satisfies these conditions will yield the correct arcade skew. For example, the scenario advocated by Rust (2001) that filaments are flux tubes that rise in their entirety from below the photosphere will yield the same axial field component for filament and arcade – if one assumes these are part of the same flux tube – and the skew will be correct if the current helicity is negative in the northern hemisphere and vice-versa. Therefore the skew of arcades is not a good discriminator between models for prominence formation. However, the purpose of the present paper was merely to demonstrate the consistency of the Martens & Zwaan model with the *Yohkoh* observations of X-ray arcades. In future work I will consider an effective observational test of various prominence formation models, based upon the exceptional nature of the filament channels which are the breeding grounds for prominences.

REFERENCES

D'Azambuja, M. & D' Azambuja, L., *Ann. Obs. Paris-Meudon, Vol. 6, Fasc. VII* (1948).

D'Azambuja, M. & D' Azambuja, L., *Ann. Obs. Paris-Meudon, Vol. 7, Fasc. II* (1949).

Foukal, P., *Solar Phys.* **19**, 59 (1971a).

Foukal, P., *Solar Phys.* **20**, 298 (1971b).

Gaizauskas, V. Zirker, J.B., Sweetland, C. & Kovacs, A., *ApJ,* **479**, 448 (1997).

Kiepenheuer, K.O., *The Sun*, p. 322, University of Chicago Press (1953), Ed. G.P. Kuiper.

Martens, P.C.H., & Zwaan, C., *ApJ,* **558**, 872 (2001).

Martin, S.F., Livi, S.H.B., & Wang, J., *Austr. J. of Phys.*, **38**, 929 (1985).

Martin, S.F., & McAllister, A.H., *IAU Colloq. 153: Magnetodynamic Phenomena in the Solar Atmosphere - Prototypes of Stellar Magnetic Activity*, p. 497 (1996), Eds. Y. Uchida, T. Kosugi, & H.S. Hudson.

Martin, S.F., *Solar Phys.*, **182**, 107 (1998).

McAllister, A.H., Hundhausen, A.J., Mackay, D. & Priest, E.R., *ASP Conf. Seri50: IAU Colloq. 167: New Perspectives on Solar Prominences*, p. 430 (1998), Eds. D.F. Webb, D.M. Rust, & B. Schmieder.

McIntosh, P.S., *Rev. Geophys. Space Sci.*, **10**, 837 (1972).

Nightingale, R.W. *et al.,* these proceedings (2002).

Pevtsov, A.A., Canfield, R.C., & Zirin, H., *ApJ,* **473**, 533 (1996).

Priest, E.R., & Forbes, T.G. *Magnetic Reconnection: MHD Theory and Applications*, Cambridge University Press (2000).

Rust, D.M., *J. Geophys. Res.,* **106**, 25075.

Tandberg-Hanssen, E., *ASP Conf. Seri50: IAU Colloq. 167: New Perspectives on Solar Prominences*, p. 11 (1998), Eds. D.F. Webb, D.M. Rust, & B. Schmieder.

van Ballegooijen, A.A. & Martens, P.C.H., *ApJ,* **361**, 283 (1989).

TETHER-CUTTING ACTION IN TWO SIGMOIDAL FILAMENTS

K. Hori[1], A. Glover[2,†], M. Akioka[3], and S. Ueno[4]

[1]*National Space Science & Technology Center, 320 Sparkman Drive, Huntsville, AL 35805, USA*
[2]*ESA European Space & Technology Centre, Keplerlaan 1, Postbus 299, 2200AG Noordwijk, The Netherlands*
[3]*Hiraiso Solar Observatory, Communications Research Laboratory, 3601 Isozaki, Hitachinaka, Ibaraki 311-1202, Japan*
[4]*Hida Observatory, Kyoto University, Kamitakara, Gifu 506-1317, Japan*

ABSTRACT

The impressive S (or inverse-S) mark appearing in the lower corona called sigmoid, is understood as the manifestation of highly sheared magnetic structures (Rust & Kumar 1996). Recent studies using the *Yohkoh* Soft X-ray Telescope (SXT) have indicated that soft X-ray sigmoids, i.e., hot ($\geq$ 2MK) S-shaped features, are strongly linked with eruptive phenomena, such as filament eruptions and coronal mass ejections (CMEs) (Sterling & Hudson 1997, Hudson *et al.* 1998, Canfield *et al.* 1999, Glover *et al.* 2000). However, previous studies have focused on the magnetic topology of the sigmoid (e.g. helicity and shear buildup) and/or connection with the resultant CME (e.g. missing mass deduced from coronal dimming), and thus the physical process involved in sigmoid formation and eruption is still not well addressed (Pevtsov *et al.* 1996, van Driel-Gesztelyi *et al.* 2000).

We present observations of two sigmoidal filaments, in which the development of cool ($\sim 10^4$ K) sigmoids was well resolved in high-cadence Hα and microwave images ($\leq$30 s). One is an active region filament and the other is a non-active region filament. Both filaments delineated the magnetic neutral line and partially erupted in association with a *GOES* B class flare. Hα observations indicate that each filament consists of two curved segments that became active in a different manner. Here we focus on the role of the junction at which these two filament segments come together (i.e. the midpoint of the filament). Prior to or during the filament eruption we found a topological change at the junction that seems to suggest a linkage of the two filament segments. Throughout the events, magnetic flux cancellation was observed at the junction. Considering these features, we propose that there is a strong "tether" at the junction that is crucial for the bodily eruption of sigmoidal filaments. Possible magnetic reconnection that took place at the junction in the low atmosphere as the origin of the "tether-cutting action" is discussed.

OBSERVATIONS

Quiet Sun Filament

This is a huge ($\sim 5 \times 10^5$ km) S-shaped filament in the southern hemisphere formed in a quiet region between dispersed remnants of two earlier active regions. Long term evolution of this non-active region sigmoid was reported by Glover *et al.* (2001). The authors concluded that CMEs act to remove helicity from the sigmoid,

[†]Also at Rhea System SA, Avenue Einstein 2a, B-1348 Louvain-la-Neuve, Belgium

finally leaving an approximately potential filament channel. Here, we examine the filament evolution on May 8th, 2000, when the eastern half of the filament produced a *GOES* B6.8 two-ribbon flare.

Figure 1 top row shows the filament region observed in (a) soft X-ray, (b) Hα, and (c) 17 GHz by the Nobeyama Radioheliograph [NoRH] on May 7th, 2000, and the bottom row shows those images observed on the next day, prior to the filament eruption. The other filament lying south of the sigmoidal filament remained unchanged throughout the event. At this stage, soft X-ray features comprise many diffuse loops surrounding the S-shaped filament. These non-active region filaments usually appear as dark features at 17 GHz, if the brightness temperature (T_b) of the filaments is lower than that of the surrounding quiet region ($\sim 10^4$ K). As it heats up to $T_b \sim 10^4$ K, the filament gradually disappears in 17 GHz against the background emission, which often occurs prior to eruption. On May 8th the eastern end of the filament faded away in Hα, and the 17 GHz dark patch shrank toward the midpoint of the filament ((f) arrow). These features suggest that i) pre-flare heating already started along the filament from its eastern end toward the midpoint, and ii) the midpoint is the lowest temperature region in the filament during this early phase.

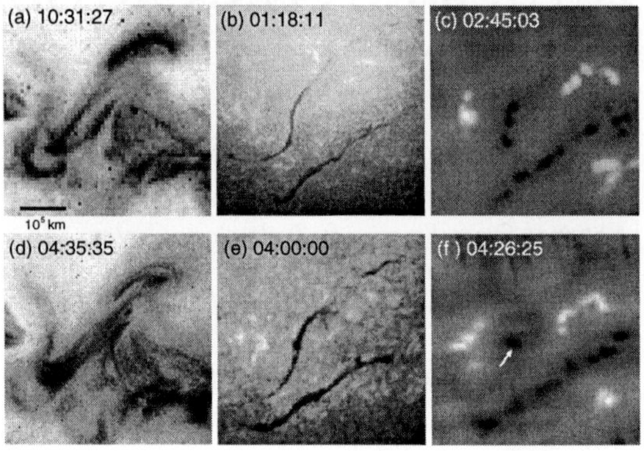

Fig. 1. (a),(d): *Yohkoh* SXT, (b): Hiraiso Hα, (c),(f): NoRH 17 GHz, (e): FMT Hα images on May 7th (top) and May 8th, 2000 (bottom). Time is given in UT. Solar north is up and east is to the left.

Figure 2 shows the Hα evolution of the sigmoidal filament from Hida flare monitor telescope (FMT). The filament became thicker and darker around the junction and then, suddenly bifurcated into two segments at 05:06 UT. The eastern segment immediately disappeared while the western segment remained visible, suggesting a partial filament eruption. Following this activity, the rest of the 17 GHz dark patch disappeared at 05:09 UT. Although no complete disappearance and/or eruption took place, extensive mass motion was observed in the western half of the filament. *SOHO* EIT 195 Å images show an EUV ejectum that started between 04:58 UT and 05:10 UT from the eastern end of the filament and moved westward along the filament. At 05:36 UT, a bright Hα flare kernel appeared in the location where the eastern half of the filament existed (dotted circle in Figure 2). Later, this brightening developed into a two ribbon flare. At the same time, a tip structure appeared at the junction (arrow), which was better observed in the Hα blue wing (center - 0.8Å). This tip gradually moved westward along the filament (~ 15 km s^{-1}) and disappeared around 06:00 UT.

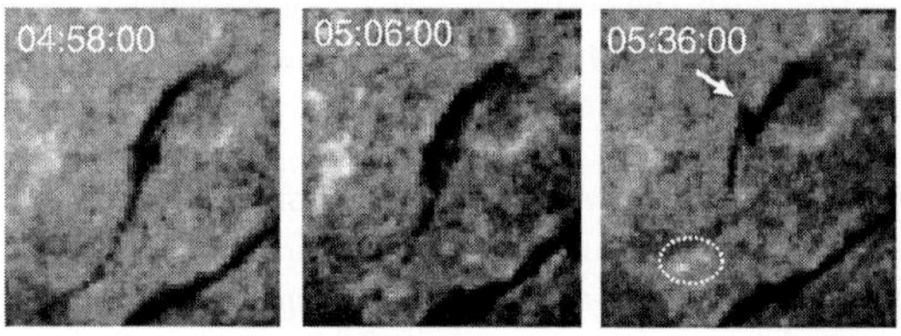

Fig. 2. Hida FMT Hα images on May 8th, 2000.

A soft X-ray arcade was formed over the Hα ribbons, which arranged itself into an S-shape along the sigmoidal filament. From 06:50 UT a faint partial halo CME was observed by the *SOHO* LASCO coronagraph, probably originating from the partial filament eruption (Glover *et al.* 2001). *SOHO* MDI observations of the photospheric magnetic fields show that the sigmoidal filament was formed along the magnetic neutral line, where positive and negative magnetic fragments were drifting towards the filament and canceling. Cancellation was especially obvious at the filament junction and the flare kernel in the eastern half of the filament prior to and during the event (from 05:00 to 08:00 UT). On the other hand, emergence of negative and positive magnetic flux was strong at the western and eastern ends of the filament, respectively.

Active Region Filament

This is an inverse-S filament in the northern hemisphere, surrounding a large leading spot in active region NOAA 8668. Chae *et al.* (2000, 2001) studied the evolution of photospheric motion and magnetic field in this AR over August 16th to 19th, 1999, when the filament was under formation (Figure 3). Gibson *et al.* (2002) presented a comprehensive study for this AR, in terms of a twisted magnetic flux rope that emerges into and equilibrates with overlying coronal structures. Starting August 14th this AR produced a series of *GOES* B and C class flares and CMEs while the northern and southern segments of the sigmoidal filament became active in a significantly different manner. For the *GOES* B9.5 flare that occurred nearby the northern filament segment on August 17th we made Hα Dopplergrams from Hiraiso Hα images. We found that the northern segment and the filament junction (see the dotted square in Figure 3) tend to show downward motion before and after the flare while the southern segment showed upward motion during the flare. In NoRH 17 GHz images, the big spot is the brightest throughout the event due to the gyro resonance emission ($T_b \sim 10^5$ K). Different from the previous event, no dark patch was observed along this active region filament. Instead, the southern segment was initially the brightest in the filament ($T_b \sim 3\times10^4$ K), which switched to the northern segment ($T_b \sim 5\times10^4$ K) when the flare occurred.

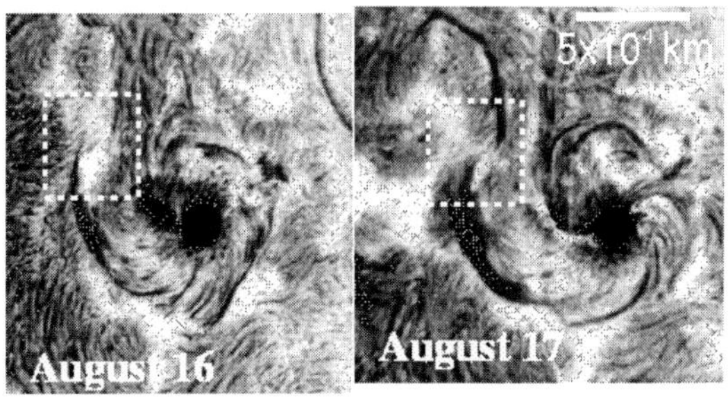

Fig. 3. BBSO Hα images of AR 8668 (From Chae *et al.* 2001)

The sequence of *TRACE* 195 Å images illustrates that each (inner) end of the two filament segments is anchored to the photosphere at the junction at this stage (Figures 4(a) and (b)). Following the next flare (C2.6; starting at 12:32 UT), however, these segments apparently joined at the junction and erupted together (c), resulting in a CME. A huge post flare arcade appeared, winding along the overall filament channel (d). Note that this was not a complete ejection of the whole filament; the sigmoidal filament was still visible in Hα (Gibson *et al.* 2002) while its shape has changed into smoother inverse-S, suggesting a complete link of the two segments. From Big Bear Solar Observatory (BBSO) and *SOHO* MDI observations, Chae *et al.* (2000, 2001) found a diverging mass motion (10-20 km s^{-1}), upflows, and weak magnetic cancellation in the southern segments while a converging mass motion and strong magnetic cancellation occurred at the junction. They concluded that these motions were caused by a large-scale rotation of the AR around the big spot.

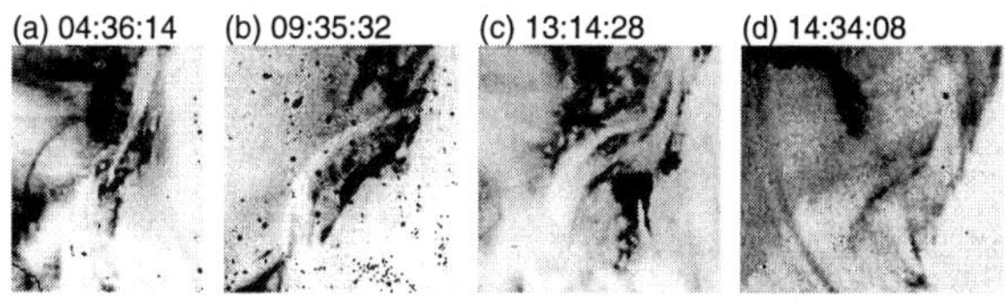

Fig. 4. Close-up of the junction from *TRACE* 195Å negative images on August 17th, 1999. Time is given in UT.

SUMMARY AND DISCUSSION

On the basis of the multi-wavelength observations of two sigmoidal filaments that produced *GOES*-B class flares, soft X-ray sigmoids, and CMEs, we examined the physical processes that were involved in the sigmoid formation and eruption. Each filament consisted of two curved structures that were not necessarily linked to each other as seen in their different behavior. We focused on the role of the junction, the midpoint of the filaments, where the two curved filament segments anchored to the photosphere. While the filament activity was basically triggered by a large-scale photospheric motion toward and/or along the magnetic neutral line, we confirmed remarkable magnetic flux cancellation at the junction for both of the events. We propose that there is a strong "tether" at the junction that holds the two filament segments to the photosphere and prevents a bodily ejection of the sigmoidal filament (see Figure 2 right and Figure 4(b)).

The idea of the "tether-cutting action" was originally proposed by Moore and Labonte (1980) and was further developed by Moore and Roumeliotis (1992) for flare/filament eruptions: while buoyant due to its magnetic pressure, the filament is line-tied to the photosphere by the magnetic tension force that holds the filament to the photosphere, just like a tether. Preflare slow reconnection, accompanying flux cancellation can cut the tether and produce the flare. In case of our tether at the junction the tether-cutting action corresponds to the coalescence of the two filament segments, which is presumably driven by magnetic reconnection in between the two filament segments in the low atmosphere as suggested by the observed magnetic cancellation at the junction (Chae *et al.* 2000, see also Wang & Shi 1993). The different behavior seen in the two filament segments may suggests that they have a different origin, rather than being formed as one continuous sheared filament below the photosphere (Martens & Zwaan 2001).

We would like to thank Ron Moore for helpful discussions. This work was partially supported by the Particle Physics and Astronomy Research Council, and was completed while KH held a National Research Council Research Associateship Award at NASA Marshall Space Flight Center in NSSTC.

REFERENCES

Canfield, R.C., Hudson, H.S., and McKenzie, D.E., *Geophys. Res. Lett.*, **26**, 627 (1999).

Chae, J., Denker, C., Spirock, T.J., Wang, H., and Goode, P.R., *Solar Phys.*, **195**, 333 (2000).

Chae, J., Wang, H., Qiu, J., Goode, O.R., Strous, L., and Yun, H.S., *ApJ*, **540**, 476 (2001).

Gibson, S.E., *et al.* submitted to *ApJ* (2002).

Glover, A., Ranns, N.D.R., Harra, L.K., and Culhane, J.L., *Geophys. Res. Lett.*, **27**, 2161 (2000).

Glover, A., Harra, L.K., Matthews, S.A., Hori, K., and Culhane, J.L., *A&A*, **378**, 239 (2001).

Hudson, H.S., Lemen, J.R., St. Cyr, O.C., Sterling, A.C., and Webb, D.F., *Geophys. Res. Lett.*, **14**, 2481 (1998).

Martens, P.C.H, and Zwaan, C., *ApJ*, **558**, 872 (2001).

Moore, R.L., and Labonte, B.J., in *Solar and Interplanetary Dynamics*, p. 207, eds. M. Dryer and E. Tandberg-Hanssen, Reidel, Dordrecht (1980).

Moore, R.L. and Roumeliotis, G., in *Lecture Notes in Physics*, **399**, 69, Eruptive Solar Flares, eds. Z. Svestka, B. Jackson, and M. Machado, Springer, Berlin (1992).

Pevtsov, A.A., Canfield, R.C., and Zirin, H., *ApJ*, **473**, 533 (1996).

Rust, D.M., and Kumar, A., *ApJ Letters*, **464**, L199 (1996).

Sterling, A.C., and Hudson, H.S., *ApJ Letters*, **491**, L55 (1997).

van Driel-Gesztelyi *et al.* in *3rd Advances in Solar Physics Conference*, ASP Conference Series, **184**, 302, eds. B. Schmieder, A. Hofmann, and J. Staude (2000).

Wang, J., and Shi Z., *Solar Phys.*, **143**, 199 (1993).

HELICITY LOADING AND DISSIPATION: THE HELICITY BUDGET OF AR 7978 FROM THE CRADLE TO THE GRAVE

L. van Driel-Gesztelyi[1,2,3,4], P. Démoulin[3], C. H. Mandrini [5], S. Plunkett[6], B. Thompson[7], Zs. Kővári[4], G. Aulanier[3], A. Young[7], M. López Fuentes[5], and S. Poedts[1]

[1]*Centre for Plasma Astrophysics, K. U. Leuven, 3001 Leuven, Belgium*
[2]*MSSL, University College London, Holmbury St. Mary, Dorking, Surrey, RH5 6NT, UK*
[3]*Observatoire de Paris, LESIA, F-92195 Meudon, France*
[4]*Konkoly Observatory, H-1525 Budapest, Hungary*
[5]*Instituto de Astronomía y Física del Espacio, Buenos Aires, Argentina*
[6]*Naval Research Laboratory, Washington, DC 20375, USA*
[7]*NASA Goddard Space Flight Center, Greenbelt, MD 20771, USA*

ABSTRACT

Through a multi-wavelength and multi-instrument analysis we evaluate the magnetic helicity budget of an isolated active region (NOAA 7978) from its emergence throughout its decay. Using *Yohkoh*/SXT images and linear force-free magnetic extrapolations carried out on *SOHO*/MDI magnetograms, we compute the relative magnetic helicity in the corona. Based on the observed magnetic field distribution we calculate the magnetic helicity injected by differential rotation. Then, using *SOHO*/LASCO & EIT and SXT images we identify all the 26 coronal mass ejections (CMEs) which originated from this active region during its lifetime and we estimate the amount of helicity which was shed via CMEs. Comparing these three values we find that the differential rotation can neither provide enough helicity to account for the diagnosed coronal helicity values, nor for the helicity carried away by CMEs. We suggest that the main source of the magnetic helicity must be the inherent twist of the magnetic flux tube forming the active region.

INTRODUCTION

An important role of CMEs is that they carry away magnetic helicity which would otherwise accumulate incessantly in the Sun (Rust 1994, Low 1996). Twisted flux tubes ejected in CMEs appear in interplanetary space as magnetic clouds, in most of which the twisted structure is still well observable. We attempt to trace magnetic helicity from the sub-photospheric layers to the Earth through an analysis of the long-term magnetic helicity budget of an isolated solar active region (NOAA 7978) between July-November 1996. For more details of this study see papers by Démoulin *et al.* (2002a, b).

RELATIVE MAGNETIC HELICITY

Helicity is one of the few global quantities which is conserved, even in resistive MHD on a timescale less than the global diffusion timescale (Berger 1984). Magnetic helicity is defined by a volume integral: $H_m = \int_V \vec{A}.\vec{B}dV$ where $\vec{A}$ is the magnetic vector potential, $\vec{B} = \vec{\nabla} \times \vec{A}$ is the magnetic field. It is physically meaningful only when $\vec{B}$ is fully contained inside the volume V. However, when this is not so ($B_n \neq 0$ along the boundary S), following Berger & Field (1984), a relative magnetic helicity can be computed by subtracting the helicity of a reference field B_0, which has the same B_n distribution on S as $\vec{B}$: $H_r = \int_V \vec{A}.\vec{B}dV - \int_V \vec{A}_0.\vec{B}_0 dV$, with $\vec{A}_0$ satisfying $\vec{B}_0 = \vec{\nabla} \times \vec{A}_0$. Since H_r is well conserved under solar conditions the only way helicity can be modified inside V is by helicity flux crossing the boundary S and/or generated along S (Berger & Field 1984): $\frac{dH_r}{dt} = -2 \int_S [(\vec{A}_0.\vec{v})\vec{B} - (\vec{A}_0.\vec{B})\vec{v}].d\vec{S}$, where $\vec{v}$ is the velocity of the plasma. The first term corresponds to helicity generation by plasma motion parallel to S, while the second term denotes inflow and outflow of helicity through the boundary S.

THE HELICITY BUDGET OF AR 7978

Computation of the Coronal Relative Magnetic Helicity

SOHO/MDI magnetograms taken close to the central meridian passages of the studied AR were used as boundary conditions for linear force-free field (lfff) magnetic extrapolations ($\vec{\nabla} \times \vec{B} = \alpha\vec{B}$; α = const). The extrapolated field lines were co-aligned with coronal loops observed with *Yohkoh*/SXT. Parameters of the best general fit between the models and observations were adopted for computation of the relative coronal helicity following Berger (1985). The magnitude of the latter depends on the photospheric flux distribution and on the value of α. The two values of H_{corona} given in Table 1 correspond to the minimum and maximum values of α in the best-fitting range. Even though α stays below its resonant value (which would give unrealistically high magnetic helicity) in all the extrapolations, we have used for the helicity computations a linearized expression in α (see Green *et al.* 2002). We have taken a computational box of the same extension centered on the AR in all the cases. Therefore we believe that though lfff models are imperfect representations of the coronal field, our computed helicity values are reasonably good estimates.

Computation of the Helicity Generated by Differential Rotation

For the computation of magnetic helicity generated by photospheric plasma motions, Berger (1984, 1988) derived an expression for dH_r/dt which depends only on observable photospheric quantities (B_n and $\vec{v}$). Berger (1986) showed that the helicity generation rate can be understood as the summation of the rotation rate of all the individual elementary flux pairs weighted by their magnetic flux. We applied the latter method to observations using *SOHO*/MDI data (Table 1, column 5 - the most accurate values in our helicity budget). Note that these helicity values are much smaller than the coronal ones (Table 1, columns 3 & 4). Démoulin *et al.* (2002b) noticed that photospheric plasma motions generate two different helicity terms: the rotation of each polarity introduces 'twist' helicity while the relative rotation of opposite polarity flux concentrations injects 'writhe' helicity. In the case of the differential rotation the generated twist and writhe helicities always have opposite signs, while their magnitudes are similar, thus they partially cancel.

Computation of the Helicity Ejected via CMEs

We identified all the CMEs which originated from AR 7978 during its entire evolution using *SOHO*/LASCO & EIT, and *Yohkoh*/SXT observations. Then we corrected these numbers for LASCO data gaps (26+5 CMEs). Due to a very low activity level in 1996, CMEs could be linked to AR 7978 even during its far-side locations. Assuming a one-to-one association between CMEs and magnetic clouds, i.e. interplanetary

Table 1. The magnetic helicity budget of AR 7978 listed per rotation. An interval of helicity is given for the corona (3rd column) and for the cloud estimations with the observed number of CMEs (considering the two lengths (0.5 and 2 AU) of the twisted flux tube in magnetic clouds. The budget is discussed in the text. All values are in units of 10^{42} Mx^2.

No. of rot.	Date 1996	$H_{\rm cor.}$	$\Delta H_{\rm cor.}$	$\Delta H_{\rm d.r.}$	$\Delta H_{\rm m.cl.}$ obs.	$\Delta H_{\rm m.cl.}$ cor.	$\Delta H_{\rm cor.} - \Delta H_{\rm d.r.}$
	07 Jul.	–					
1st				0.2	[16, 64]	[22, 88]	($\approx$ 7)
	03 Aug.	[5, 11]					
2nd			12	3.	[10, 40]	[10,40]	9
	30 Aug.	[17, 23]					
3rd			-9.5	3.	[4, 16]	[6, 24]	-13
	25 Sep.	[9, 12]					
4th			-5.5	1.	[10, 40]	[10, 40]	-7
	23 Oct.	[4, 6]					
5th			(-1)	0.8	[6, 24]	[8, 32]	-2
	19 Nov.	(4)					
6th			–	0.3	[6, 24]	[6, 24]	–
total		–		8.3	[52, 208]	[62, 248]	-6

twisted flux tubes (e.g. Webb *et al.* 2000) and taking a mean magnetic field B_0 (2×10^{-4} G) and radius R (2×10^{12} cm) of 18 magnetic clouds (Lepping *et al.* 1990), furthermore, using a numerically integrated form of Berger's equation (1999), we computed the relative helicity per unit length in the twisted interplanetary flux tube. For the length of the flux tube in the magnetic cloud two values were used: $L_1 = 0.5AU$ (DeVore 2000) which yielded $H_r \approx 2 \times 10^{42} Mx^2$ magnetic helicity, and $L_2 = 2AU$ (the cloud is still connected to the Sun; e.g. Richardson 1997) which gives four times as much helicity for one average-sized magnetic cloud, i.e. CME. These mean helicity values have to be multiplied by the number of the CMEs to obtain the total magnetic helicity ejected from this AR (Table 1, columns 6 & 7).

DISCUSSION AND CONCLUSIONS

AR 7978 was a classical bipolar AR oriented E-W, distorted only by the differential rotation. It had positive relative coronal helicity, and the differential rotation injected positive helicity as well throughout the studied six solar rotations (Table 1, 5th column). When looking at the changes in coronal helicity from one rotation to the next, and comparing the changes with the amount of helicity generated by the differential rotation (4th and 5th columns in Table 1), it is obvious that the differential rotation is an inefficient generator of helicity. This accentuates the importance of the helicity inflow and outflow through the boundaries of our coronal computational box. The total helicity budget of the active region may be written:

$$\Delta H_{\rm emergence} = \Delta H_{\rm corona} - \Delta H_{\rm diff.rot.} + N.H_{\rm CME},$$

where Δ denotes the variation of the helicity, N is the number of the CMEs, and $H_{\rm CME}$ is the mean helicity carried away per CME event. $\Delta H_{\rm emergence}$ can be computed adding the last column to either the 6th or 7th column of Table 1. In the helicity budget of AR 7978 the increase in coronal helicity during the first two rotations and the large amount of helicity carried away by CMEs during this period requires the largest input of helicity by the sub-photospheric layers. Indeed, major flux emergence episodes were observed during this period in the AR (Démoulin *et al.* 2002a). During later rotations the deficit in the helicity budget becomes smaller, but still not negligible. The total amount of helicity which we need to cover from twisted flux emergence (if we take the corrected CME numbers) can be estimated to be between $56 - 242 \times 10^{42} Mx^2$. For comparison, during the same period the differential rotation generated only $8.3 \times 10^{42} Mx^2$, so it clearly was a *minor contributor* to the magnetic helicity budget of AR 7978. Although all the helicity values, especially the amount carried away in CMEs, have large uncertainties (for a discussion see Démoulin *et al.* 2002a, and Green *et al.* 2002), the difference between the above mentioned total helicity numbers is significantly larger than any of the errors.

Thus we find that differential rotation is an inefficient generator of magnetic helicity; it can neither provide enough helicity to account for the diagnosed coronal helicity values, nor provide the helicity carried away by CMEs. This result is confirmed by a recent analysis of the helicity budget of another active region (NOAA 8100) by Green *et al.* (2002). We suggest that the main source of magnetic helicity must be the inherent twist of the magnetic flux tube forming the active region. This magnetic helicity is transferred from the sub-photospheric layers to the corona by magnetic flux emergence and replenished after relaxation (CME) events either by a slow continuous emergence of the flux tube or by torsional Alfvén waves (Longcope & Welsch 2000). This process can continue for several solar rotations until the helicity reserves of the flux tube are exhausted, or the flux tube is destroyed by convective motions.

ACKNOWLEDGEMENTS

The authors thank the *SOHO*/MDI, EIT and LASCO consortia for the *SOHO* data and the MSSL SURF for *Yohkoh*/SXT data. *SOHO* is a joint project by ESA and NASA. These results were obtained in the framework of the projects OT/98/14 (K.U.Leuven), G.0344.98 (FWO-Vlaanderen), and 14815/00/NL/SFe(IC) (ESA Prodex 6). LvDG is supported by Research Fellowship F/01/004 of the K.U.Leuven and by the Hungarian Government grants OTKA T032846 and T-038013. LvDG and PD acknowledge the Hungarian-French S&T cooperative program. PD and CHM acknowledge financial support from ECOS (France) and SETCIP (Argentina) through their cooperative science program (A01U04).

REFERENCES

Berger, M.A.,*Geophys. Astrophys. Fluid Dynamics*, **30**, 79 (1984).

Berger, M.A., *ApJ Supp.*, **59**, 33 (1985).

Berger, M.A., *Geophys. Astrophys. Fluid Dynamics*, **34**, 265 (1986).

Berger, M.A.,*A&A*, **201**, 355 (1988).

Berger, M.A., in *Magnetic Helicity in Space and Laboratory Plasmas*, Geophys. Monograph 111, AGU, p. 1 (1999).

Berger, M.A., and Field, G.B., *J. Fluid Mech.*, 147, 133 (1984).

Démoulin P., Mandrini C.H., van Driel-Gesztelyi L., Thompson B., Plunkett S., Kővári Zs., Aulanier G., and Young A., *A&A*, **382**, 650 (2002a).

Démoulin P., Mandrini C.H., van Driel-Gesztelyi L., López Fuentes, M.C., and Aulanier G., *Solar Phys.*, in press (2002b).

DeVore, C.R., *ApJ*, **539**, 944 (2000).

Green L.G., López-Fuentes, M.C., Mandrini, C.H., Démoulin, P., van Driel-Gesztelyi, L., and Culhane, J.L., *Solar Phys.*, in press (2002).

Lepping, R.P., Burlaga, L.F., Jones, J.A., *JGR*, **95**, 11957 (1990).

Longcope, D.W., and Welsch, B.T., *ApJ*, **545**, 1089 (2000).

Low, B.C.,*Solar Phys.*, **167**, 217 (1996).

Richardson, I.G., in *Coronal Mass Ejections*, Geophys. Monograph 99, AGU, p. 189 (1997).

Rust, D.M., *GRL*, **21**, 241 (1994).

Webb, D.F., Cliver, E W., Crooker, N.U., St. Cyr, O.C., and Thompson, B.J.,*GRL*, **105**, A4, 7491 (2000).

HEMISPHERIC HELICITY ASYMMETRY IN ACTIVE REGIONS FOR SOLAR CYCLE 21-23

M. Hagino[1] and T. Sakurai[2, 3]

[1]*Meisei University, 2-590 Nagabuchi, Ohme-shi, Tokyo 198-8655, Japan*
[2]*Department of Astronomical Science, The Graduate University for Advanced Studies, 2-21-1 Ohsawa, Mitaka-shi, Tokyo 181-8588, Japan*
[3]*National Astronomical Observatory, 2-21-1 Ohsawa, Mitaka-shi, Tokyo 181-8588, Japan*

ABSTRACT

We analyzed vector magnetograms obtained in the period of 1983–2000 with the Solar Flare Telescope (Mitaka: SFT) and the 65 cm Solar Telescope (Okayama: OAO) at the National Astronomical Observatory of Japan. We plotted current helicity against solar latitude and calculated a linear fit to the data. The slope is negative and confirms the hemispheric rule of previous studies. There is indication of time variability in helicity.

INTRODUCTION

Magnetic helicity observed at the surface carries information on invisible sub-surface processes such as internal rotation and the behavior of magnetic flux tubes in the convection zone. It has been recognized that current helicity shows a hemispheric rule (Pevtsov *et al.* 1995 [Mees Stokes Polarimeter], Bao & Zhang 1998 [Huairou SMFT]): the northern (southern) hemisphere tends to show negative (positive) helicity. It is also known that this rule does not change with solar cycle. Considering the importance of magnetic helicity, we present here the results based on our data set and compare them with results from other observatories.

DATA ANALYSIS

We analyzed vector magnetograms from the period of 1983–2000 obtained with the Solar Flare Telescope at Mitaka (1992–2000) and the 65 cm Solar Telescope at Okayama (1982–1995), respectively.

The current helicity was determined by two methods for 180 regions from the SFT data and 430 regions from the OAO data. The first method calculates the electric currents over active regions by a direct differentiation and then evaluates the average helicity, $\alpha = \sum(\nabla \times \vec{B})_z \cdot \mathrm{sign}(B_z)/\sum|B_z|$. The second method is the fitting of a linear force-free field $\vec{B}_{\mathrm{cal}}(\alpha)$ to the observed transverse field $\vec{B}_{\mathrm{obs}}$ which finds the best-fit α, $\alpha_{\mathrm{best}} \Leftrightarrow [\sum[\vec{B}_{\mathrm{cal}}(\alpha) - \vec{B}_{\mathrm{obs}}]^2/\sum B_{\mathrm{obs}}^2] = \mathrm{min}$.

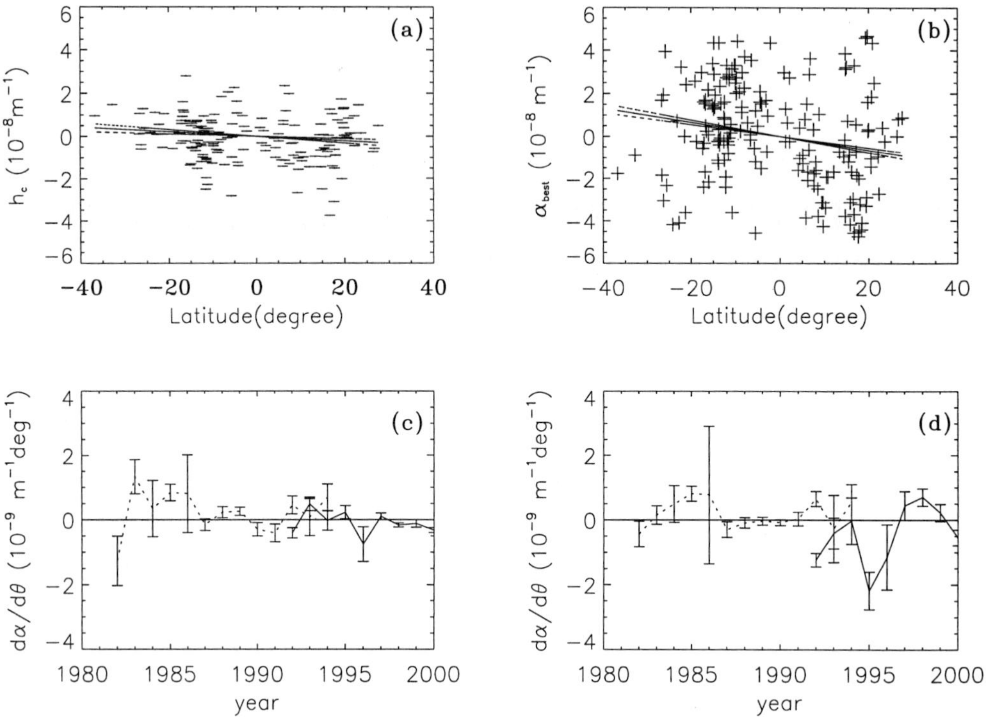

Fig. 1. Latitude distribution of current helicity obtained from SFT (top), and change of helicity gradient ($d\alpha/d\theta$) with time from 1983 to 2000 (bottom). In the lower panels dashed lines show OAO data and solid lines SFT data. Left-hand side panels show results from the direct differentiation method and right-hand side panels results from the fitting method.

RESULTS

Using the SFT data set, the slopes of the fit ($d\alpha/d\theta$) obtained from the first and the second methods are $(-1.08 \pm 0.51) \times 10^{-10}$ m^{-1} deg^{-1} and $(-3.30 \pm 1.14) \times 10^{-10}$ m^{-1} deg^{-1}, respectively. Using the OAO data set, the slopes are $(-1.08 \pm 0.51) \times 10^{-10}$ m^{-1} deg^{-1} and (-1.06 ± 1.04) m^{-1} deg^{-1}, respectively. Our results agree with previous studies and confirm the hemispheric rule. Although our data have large scatter, there is an indication of time variability in $d\alpha/d\theta$ (Fig. 1(c) and (d)). The hemispheric rule $\mathrm{d}\alpha/\mathrm{d}\theta < 0$ may not hold in some phases of the activity cycle.

REFERENCES

Bao, S.& H. Zhang, *Astrophys. J.*, **496**, L43 (1998).

Pevtsov, A.A., R.C. Canfield, & T.R. Metcalf, *Astrophys. J.*, **440**, L117 (1995).

CONCURRENT ROTATING SUNSPOTS, TWISTED CORONAL FANS, SIGMOID STRUCTURES, AND CORONAL MASS EJECTIONS

R. W. Nightingale[1], D. S. Brown[1,2] T. R. Metcalf[1], C. J. Schrijver[1], R. A. Shine[1], A. M. Title[1], and C. J. Wolfson[1]

[1]*Lockheed Martin Solar & Astrophysics Laboratory, Bldg. 252 Org. L9-41, 3251 Hanover Street, Palo Alto, CA 94304*
[2]*School of Mathematics, University of St. Andrews, St. Andrews, KY16 9SS, UK*

ABSTRACT

In an on-going study, several sunspots, rotating about their umbral centers, have been identified in *TRACE* photospheric white light images with accompanying twisting of coronal fans in the corresponding EUV (171, 195 Å) images. Three of these observations can be temporally and spatially associated with S or inverse-S shaped regions (sigmoid structures) appearing in *Yohkoh* SXT images and with concurrent coronal mass ejections (CMEs) and/or flares. We have determined the rotational speeds of these three events and present a table of the pertinent parameters. These observations provide information on the coupling of the solar magnetic field from the photosphere into the corona.

ANALYSIS

The leading sunspot in AR 9114 during August 8-10, 2000 is seen to be rotating in *TRACE*, *SOHO*-MDI, and Mees IVM data. In *Yohkoh*-SXT data an inverse-S shaped "sigmoid" structure is observed, appearing similar to the twisted coronal EUV loops in the *TRACE* images (Nightingale *et al.* 2000). A flash of possible reconnection was viewed late on August 9 in a *TRACE* EUV 171 Å movie. In a second event a CME was observed during August 15-18, 1999. This event also included an inverse-S shaped region in the SXT data, and a rotating sunspot and twisting coronal fans in the *TRACE* data. The large Bastille Day CME event of 14 July 2000 was accompanied by one or more apparently rotating sunspots as observed in *TRACE* white light (WL) data and by an inverse-S shaped region as seen in an SXT difference image. Movies and plots of these data have been shown (Nightingale *et al.* 2001a, 2001b), along with photospheric flow maps.

In order to determine the rotational speed of the above mentioned sunspots the data are corrected for solar rotation and clipped so that all the images are co-aligned. Despiking is performed on the data to remove cosmic ray and particle hits. The white light data about the rotating sunspot are "uncurled" by considering the sunspot in polar coordinates (r, θ), with an origin at the center of the sunspot, and then displaying the r, θ data in Cartesian coordinate plots. Time slices at different radii are utilized to better display the rotation, seen as inclined features, of the sunspot structure. The rotational speeds at the different times and positions are determined for each data set. In addition, 8 hour flow maps have been generated for the August 2000 data, using local correlation tracking techniques to show the flow within the penumbra.

Table 1. Rotational Rates and Other Parameters

date:	8-10 August 2000	15-18 August 1999	13-15 July 2000
active region:	AR 9114	AR 8667	AR 9077
latitude:	12°N	20°N	17°N
polarity, position in group:	positive, leading	leading	negative, following
direction of rotation:	anti-c/w	anti-c/w	anti-c/w
observed rotation:	120-150°	80-120°	70-120°
radius of max. rotation speed:	9″ or 6.7 Mm	12″ or 8.7 Mm	12″ or 8.7 Mm
peak of average rotation speed:	2.2°/hour or 67 m/s	1.3°/hour or 53 m/s	1.2°/hour or 49 m/s

The rotational rates and several pertinent parameters for the three events are presented in Table 1. All three sunspot groups were in the northern hemisphere and were rotating in an anti-clockwise direction. The maximum rotational speed appears to occur in the penumbra close to the umbral interface at about 9-12 arcsec or 6.5-8.7 Mm from the center of the sunspot with peak speeds of about 1.2-2.2°/hour or 49-67 m/s. The fastest rotation during the August 2000 event occurred in the middle of the second day, followed by a steep fall-off in speed over the last half of the day, and a more gradual falling off the following day. Near the end of the second day the SXT sigmoid structure dimmed and a flash occurred in the *TRACE* 171 Å data. The dimming and flash could be indicative of energy release.

SUMMARY

In this brief presentation we have described three rotating sunspot events in the photosphere, as observed by *TRACE* in WL, and MDI and Mees in magnetograms, associated with the foot points of twisted coronal fans of EUV loops seen in *TRACE* and below inverse-S shaped "sigmoid" structures imaged by SXT. Associated with these rotating sunspot events was the release of energy in the form of CMEs for two events and a weak release, such as a small flare, for the third. These concurrent observations provide information on the linkage of magnetic field between what is happening at the photosphere and in the corona above.

ACKNOWLEDGEMENTS

We thank the referee. Work on *TRACE* data was supported by NASA contract NAS5-38099.

REFERENCES

Nightingale, R.W., R.A. Shine, D.S. Brown, C.J. Wolfson, T.D. Tarbell, and A.M. Title, *TRACE* Observations of a Twisting Coronal Fan above a Rotating Sunspot, *Eos Trans. AGU,* **81**, Fall Meet. Suppl., Abstract SH11A-10 (2000).

Nightingale, R.W., D.S. Brown, R.A. Shine, C.J. Wolfson, Z.A. Frank, and A.M. Title, More Rotating Sunspot Observations by *TRACE* With Twisting EUV Coronal Fans, *Eos Trans. AGU,* **82**, Spring Meet. Suppl., Abstract SH41B-11 (2001).

Nightingale, R.W., D. Alexander, D.S. Brown, T.R. Metcalf, Energization of Rotating Sunspots, Twisted Coronal Fans, Sigmoid Structures, and Coronal Mass Ejections, *Eos Trans. AGU,* **82**, Fall Meet. Suppl., Abstract SH11C-0724 (2001).

HELICITY INJECTION INTO THE SOLAR CORONA

K. Kusano[1], T. Maeshiro[1], T. Yokoyama[2], and T. Sakurai[3]

[1] *Graduate School of Advanced Sciences of Matter, Hiroshima University, Higashi-Hiroshima, Hiroshima 739-8530, Japan*
[2] *Nobeyama Radio Observatory, NAOJ Minamimaki, Minamisaku, Nagano 384-1305, Japan*
[3] *National Astronomical Observatory, 2-21-1 Osawa, Mitaka, Tokyo 181-8588, Japan*

ABSTRACT

We developed a new methodology to measure magnetic helicity as well as magnetic energy injection into the solar corona from magnetograph observations, and we studied the relationship between helicity injection and X-ray activity in the solar corona. In order to calculate the gauge-invariant helicity flux and the Poynting flux across the photosphere, first the velocity tangential to the photospheric surface is constructed by applying the correlation tracking technique on *SOHO*/MDI observations, and second the normal velocity is calculated by solving the induction equation as an inverse-problem using data from *SOHO*/MDI and the Solar Flare Telescope at NAOJ in Tokyo. The helicity injection as well as the free energy build-up is analyzed for active region NOAA 8100 from November 1 to 4, 1997. The results indicate that the emerging flux and the shear flow inject magnetic helicity of opposite sign (positive and negative, respectively) into the active region prior to a series of flares. Furthermore it is found that during the helicity injection process magnetic free energy in the amount of 5×10^{32} erg was supplied, which is much more than the X-ray flux emitted by the flares.

INTRODUCTION

Although it has been theoretically pointed out that magnetic helicity is an important quantity for solar flares (Kusano 1999), the connection between magnetic helicity and flare activity is not yet clearly resolved. One reason for this is the fact that it is not easy to measure helicity. In this study, we not only propose a new methodology, which enables the measurement of the total injection rate of magnetic helicity into the solar corona across the photosphere, but also demonstrate the application of that method for active region NOAA 8100.

The gauge-invariant relative helicity flux across a plane S is given by $F_H = 2\int(\boldsymbol{A_P}\cdot\boldsymbol{B})\boldsymbol{V}\cdot d\boldsymbol{S} - 2\int(\boldsymbol{A_P}\cdot\boldsymbol{V})\boldsymbol{B}\cdot d\boldsymbol{S}$, where $\boldsymbol{V}$, $\boldsymbol{B}$ and $\boldsymbol{A_P}$ are the velocity, the magnetic field, and the vector potential of the potential field that has the same boundary condition as the vertical component of $\boldsymbol{B}$ (Berger & Field 1984). The first term of the above equation is caused by the normal velocity across S, which is relevant for the process of flux emergence, while the second term is generated by the horizontal shear motion. Therefore, in order to derive the total helicity flux across the photosphere, all three components of the velocity are necessary in addition to the vector magnetic field. On the other hand, the normal velocity can be uniquely determined from the condition that $\boldsymbol{V}$ and $\boldsymbol{B}$ satisfy the induction equation of the normal component on S, if the tangential velocity is specified. So, through the following procedure we can derive the photospheric velocity. First, by applying the correlation tracking technique on the sequential data observed by *SOHO*/MDI, the tangential velocity is defined. Second, the normal velocity is calculated by solving the induction equation

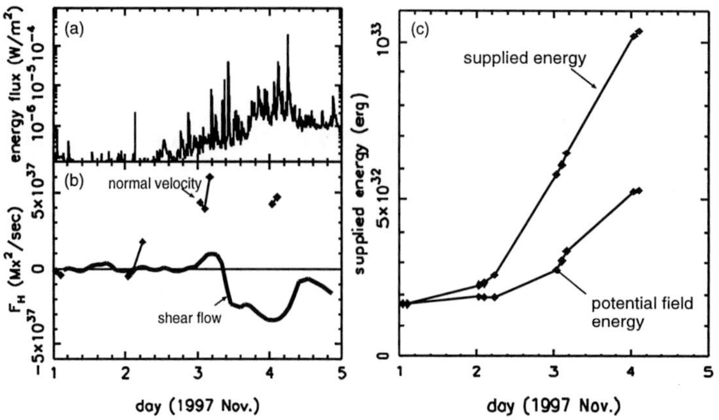

Fig. 1. (a) The variation of the *GOES* X-ray flux, and (b) the variation of the helicity injection rate for NOAA 8100. The helicity injection caused by normal velocity and by shear motion are respectively shown by thin and thick lines. (c) The integration of the Poynting flux across the photosphere is plotted together with the potential field energy calculated from *SOHO*/MDI data.

as an inverse problem, in which the data from *SOHO*/MDI and from the vector magnetograph of the Solar Flare Telescope (SFT) at NAOJ in Mitaka, Tokyo are used for the normal and the tangential components of the magnetic field, respectively.

RESULTS AND DISCUSSION

We analyzed helicity injection in active region NOAA 8100 from November 1 to 4, 1997. As shown in Figure 1(b), the helicity injection caused by flux emerging first starts on November 2 at 3:00 UT. This process continues at least until November 4, supplying a positive helicity of about 5×10^{42} Mx^2. On the other hand, the photospheric shear motion injects negative helicity from November 3 to 4. Note that the helicity injected by the two processes has opposite sign, but a similar absolute value. Furthermore we point out that X-ray activity increases and a series of flares begins with a delay of about a day after the start of the helicity injection. This suggests a relationship between helicity injection and the activation of the flares. We also integrated the energy supplied into the coronal field by interpolating the data gap in SFT observations. Figure 1(c) shows that the supplied energy has an excess of 5×10^{32} erg above the potential field energy. Since energy release is not taken into account here, the excess energy must give the maximum free energy build-up for this period. It is much bigger than the integrated *GOES* X-ray flux which is released in the series of flares.

REFERENCES

Berger, M.A., and Field, G.B., *J. Fluid Mech.*, **147**, 133 (1984).

Kusano, K., in Geophysical Monograph 111, *Magnetic Helicity in Space and Laboratory Plasmas*, p. 149 eds. M.R. Brown, R.C. Canfield, and A.A. Pevtsov, AGU, Washington DC (1999).

Section V.

Reconnection in Flares

SIGNATURES OF RECONNECTION IN ERUPTIVE FLARES

D. E. McKenzie

Montana State University, P.O. Box 173840, Bozeman, MT 59717, USA

ABSTRACT

Solar flares are defined by magnetism. The energy that is stored up, transferred, and released is done so in and by the magnetic fields of the Sun; the structures in which the flares occur are wholly dependent on the configuration of magnetic connections; and it is the rearrangement of these connections that we believe plays such a large and important role in many of the processes observed in flares. Many of our theoretical pictures of flare mechanisms rely in some part on magnetic reconnection, and one by one the observable signatures of these models are being uncovered in chromospheric and coronal data. These data, in turn, introduce observational constraints that help to drive the models. I will summarize some key observations that have helped to support the case for reconnection, and that have helped us to peer more deeply into the behavior of coronal plasmas and magnetic fields.

INTRODUCTION

This article is a review of observations, in particular observations that resemble our expectations of reconnection. By the term "magnetic reconnection", I am referring to the process by which magnetic flux is swept into a small area, where oppositely directed components annihilate each other, and the residual magnetic tension in the newly-reconnected field causes the field and plasma to be expelled from the reconnection region. The resulting conversion of magnetic energy into kinetic energy is responsible for the observables in a flare — the particle acceleration, plasma heating, motions, etc. The cartoon shown in Figure 1 is one representation of this fundamental idea; the arguments in this article will be built upon the geometry shown in the commonly invoked picture due to Carmichael (1964), Sturrock (1968), Hirayama (1974), and Kopp & Pneuman (1976) — hereinafter called CSHKP, a good summary is in Shibata (1999) — but it should be made clear that this is only one orientation. Many other orientations are possible, variations on a common theme, but this article will not attempt to consider them all.

Roughly speaking, reconnection is what happens after the creation of a current sheet. I will not attempt to discuss how the current sheet is arrived at, nor how energy is stored in the magnetic field. I will not discuss the "breakout model" or "tether cutting" as flare initiation scenarios — although both of these scenarios may be supported by observational evidence, they are beyond the scope of this article.

In what follows, I will briefly review the major components of the standard reconnection model; a lengthy explanation of each is not necessary, since the references offered in this article provide much more complete treatments than I can offer, as do many of the other articles in these proceedings. Afterwards, observations which seem to show examples of each of these signatures will be summarized. Some of the observed signatures are not precisely as were anticipated by theoretical predictions. I will try to make the case that these data provide constraints for the theoretical models. Finally, many of our cartoons of the reconnection model have

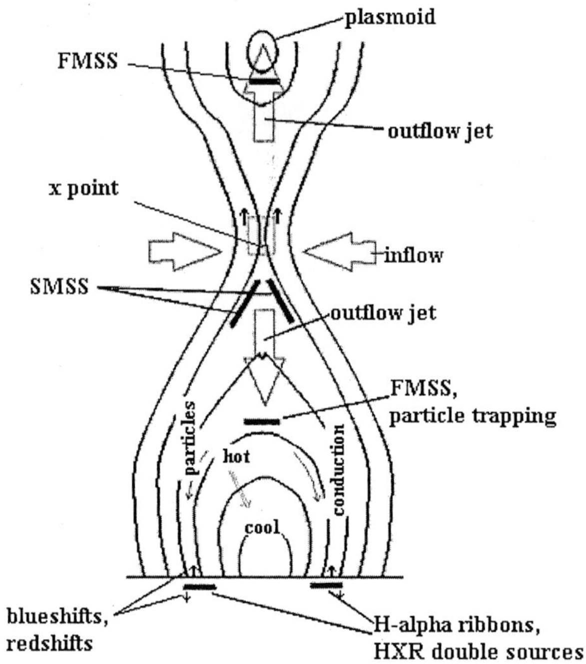

Fig. 1. Two-dimensional cartoon of the CSHKP model, wherein post-eruption coronal arcades result from magnetic reconnection. Some possibly observable features are annotated and discussed in the text.

perforce been drawn in two dimensions, and have nonetheless been extremely successful. Some aspects of the observations, though, require three-dimensional treatment. I will give some examples of this situation.

REVIEW OF THE CSHKP PICTURE

The fundamental parts of the model are annotated in Figure 1. Magnetic field is shown moving into the middle region with some inflow velocity. In this diffusion region, characterized as an X-point, magnetic fields of opposing directions cancel out, and the tension that exists in the external field causes the expulsion — upwards and downwards — of outflow jets. A detached plasmoid of closed magnetic field is shown moving upwards, away from the Sun, while the downward expelled field piles up to form the post-eruption arcade. There may be slow-mode standing shocks (SMSS) at the upper and lower edges of the diffusion region, but only the lower are shown; fast-mode standing shocks (FMSS) may exist where the outflow jets collide with previously closed field — a termination shock. As successive field is introduced from the sides, the location of the X-point moves steadily higher and higher. The released energy may go into accelerating particles, which stream down along the magnetic field into the chromosphere below, or thermal energy which is conducted downward along the magnetic field. The introduction of this energy to the chromosphere yields Hα ribbons, hard X-ray double sources, and chromospheric evaporation. The latter means that a pressure imbalance is generated which results in chromospheric material being heated and piped up into the coronal portions of the closed magnetic fields; it is that heated material which appears as the long-lasting soft X-ray arcade.

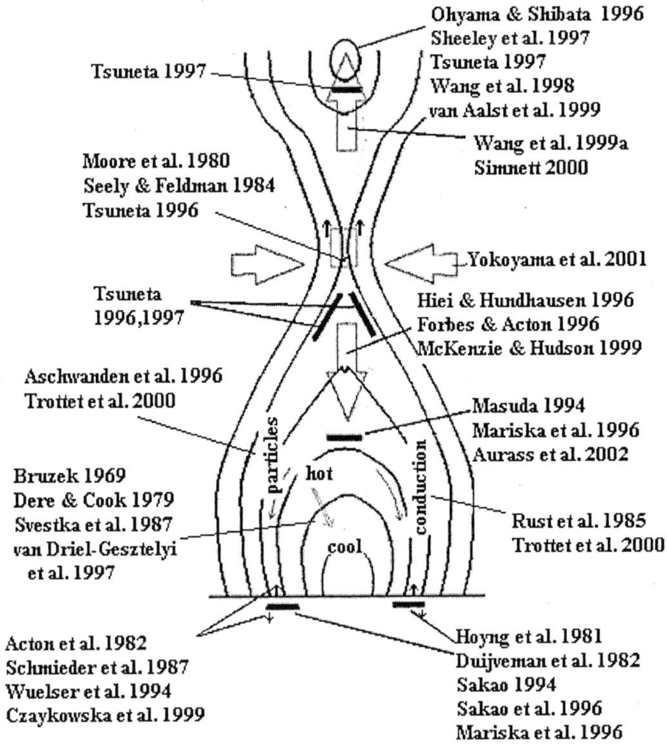

Fig. 2. Bibliographical rendition of Figure 1. The references indicate possible observations related to the annotated features of Figure 1.

Not all of these elements will necessarily be clearly observable in each and every flare, and in some instances the importance of each may be different, depending on the particulars of magnetic field strength, amount of helicity and interconnectedness, amount of stored energy, and plasma density. This makes the observational task more challenging, but piece by piece over the years many of the elements of the standard picture have revealed themselves. In Figure 2, a bibliographical representation of the same CSHKP picture is introduced, highlighting key observational papers which describe these signatures.

OBSERVED SIGNATURES

From Figure 2 one can judge that many of the elements of the standard picture appear to have been observed, though a few of the identifications are still tentative. Brief commentary about some of these is in order.

Yohkoh/SXT observations reported by Ohyama & Shibata (1996) describe dense collections of X-ray emitting material ejected from flares: these ejecta may well represent the closed-field plasmoid downstream of the reconnection (top of Figures 1 and 2). Further out, the plasmoid may have been observed by Sheeley *et al.* in 1997, though their first paper does not explicitly make that identification. The subsequent analyses by Wang *et al.* (1998) and van Aalst *et al.* (1999) interpret the "streamer blobs" in terms of reconnection artifacts. More directly, Tsuneta (1997) describes a set of *Yohkoh* observations that seem to have several of the elements of Figure 1. In addition to an apparent closed-field plasmoid, Tsuneta (1997) shows images which seem to include a region of hot plasma above the X-point and below the plasmoid. This feature

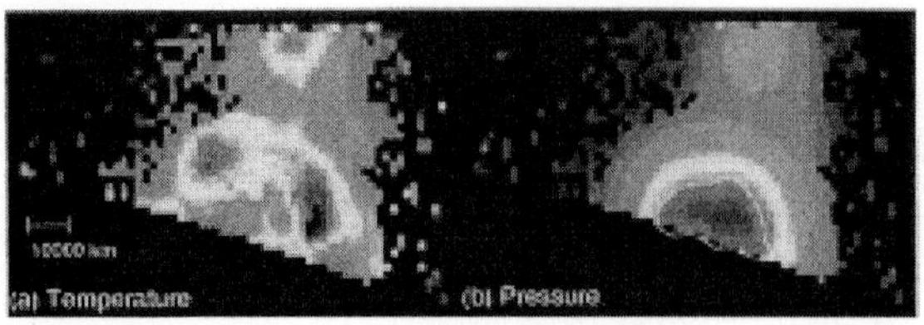

Fig. 3. Adapted from Tsuneta (1997), displaying a hot source above the X-point and below the closed-field plasmoid. The possible fast-mode shock feature is at the top of the frame; the plasmoid has left the field of view.

(Figure 3) moves slowly upwards in subsequent images, and may be evidence of a fast-mode shock in the region downstream of the reconnection.

All three of the papers referenced in regards to the rising of the X-point (Moore *et al.* 1980; Seely & Feldman 1984; Tsuneta 1996) report the well-known fact that X-ray loops get taller and taller during the flare. This is a proxy for the upward motion of the X-point, since the X-point itself is not directly observable. The growth rate is on the order of 20 – 50 km s^{-1}, but the rise speed of the X-point is probably greater than that (and in three-dimensional reconnection, difficult to define).

The velocity field surrounding the X-point has been a long-sought observational signature, since the speeds and plasma parameters in the inflow and outflow regions (upstream and downstream of the reconnection, respectively) can reveal information about conditions inside the diffusion region. The first tentative observations of outflows are those of Hiei & Hundhausen and Forbes & Acton, both in 1996, and both indicating rather slow outflow speeds — Hiei & Hundhausen on the order of less that 5 km s^{-1}. This flow, in retrospect, seems more like the thermal shrinkage described by Švestka *et al.* (1987, see below) than a shrinkage due to the relaxation of magnetic tension (i.e. "thermal shrinkage" rather than "magnetic shrinkage"). A report by Sterling *et al.* (1996) described *Yohkoh*/BCS data with simultaneous red- and blue-shifts in the upper reaches of a far-beyond-the-limb flare. Sterling *et al.* argued that the observed line shifts are probably not due to reconnection outflows, but the uncertainties attending their assumptions cause this observation to remain contentious.

The observations described by McKenzie & Hudson (1999) and McKenzie (2000) show many features moving downward into the tops of flare arcades at speeds between 40 and 500 km s^{-1}. These flow fields have been interpreted as signatures of reconnection outflows, but the speeds are slower than typically expected. There are some other characteristics of the "supra-arcade downflows" which will be discussed below. Wang *et al.* (1999a, 1999b) and Simnett (2000) have demonstrated many examples of features both up-flowing and down-flowing in LASCO image sequences, with speeds of a few hundred km s^{-1}. These also have been interpreted as signatures of reconnection outflows; some are found at sector boundaries in times of quiescence, but some seem to be part of the post-CME rearrangement.

Just as important, Yokoyama *et al.* (2001) have described an observation by *SOHO*/EIT that closely resembles the expected appearance of reconnection inflow. The inflow speed is 1 – 5 km s^{-1}, and the resemblance to the two-dimensional reconnection picture is remarkable.

The hot plasmas observed by Tsuneta (1996, 1997) (Figure 4) can be interpreted as due to heating in the slow-mode standing shocks. The long extension of the two hot lobes along the vertical direction, separated by a cool "tower", is consistent with conduction of heat energy along the magnetic field lines. The cool, dense "tower" has been interpreted as the result of catastrophic cooling — thermal instability leading to plasma condensation (e.g. Tsuneta 1996, Forbes & Acton 1996).

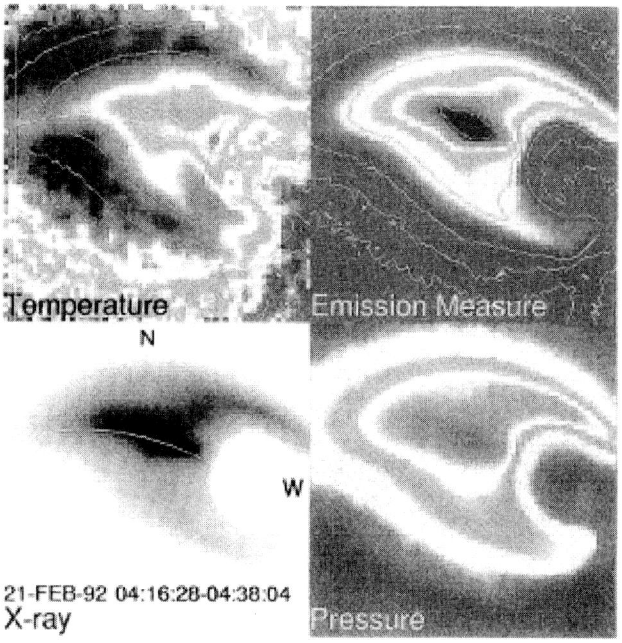

Fig. 4. Adapted from Tsuneta (1996); the results are consistent with plasma heating due to slow-mode standing shocks, coupled to heat conduction along (but not across) the magnetic field lines. "The contours in the temperature and emission measure maps indicate the locations of 83%, 20%, 5%, 2%, and 1% levels of the peak X-ray intensity of the map" (Tsuneta 1996).

The Masuda-type sources are very well known (Masuda 1994; Masuda *et al.* 1994; Mariska *et al.* 1996), and depicted in Figure 5 (reprinted from Masuda 1994). Though initially described in terms of thermal radiation from superhot plasma, the consensus now seems to favor a non-thermal population of particles, trapped in the looptop for some time before escaping to the footpoints. A very good treatment of the data, and review of analyses, is found in Alexander & Metcalf (1997).

The paper by Aurass *et al.* (2002) argues that if a Type II radio burst indicates a fast-mode shock, then a standing Type II suggests a fast-mode standing shock (FMSS). The authors report an observation which suggests such a standing shock, at a plasma density of a few times 10^9 cm^{-3}; this is tentatively identifiable as evidence for a termination shock.

Hot plasmas cool. As they do so, they may contract a little. Additionally, as new field gets swept into the reconnection region, the newly reconnected loops are generally higher than the previously reconnected loops. So one might expect to see hotter loops above the cooler loops, and loops shrinking (slightly) as they cool. Observations of this nature are not unknown (Bruzek 1969; Dere & Cook 1979; van Driel-Gesztelyi *et al.* 1997); see especially the paper by Švestka *et al.* (1987).

At the chromosphere, a variety of signatures are observed. Little discussion of these signatures is offered in this article, primarily because more complete descriptions appear elsewhere in these proceedings. Hard X-ray double sources are common, demonstrating the connectivity of the footpoints. An especially good example is shown in Masuda *et al.* (2001), whose analysis reveals two hard X-ray ribbons; most observations show

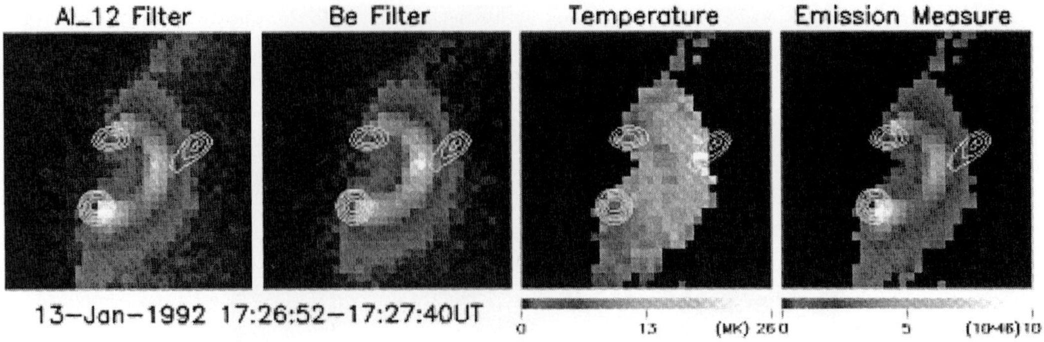

Fig. 5. Adapted from Masuda (1994), revealing a hard X-ray source at the top of the flaring loop, in addition to the more common footpoint sources. The grey scale images are from *Yohkoh*/SXT, and the contours from *Yohkoh*/HXT, energy passband 32.7 – 52.7 keV.

HXR sources of considerably smaller extent. The co-location of these sources with the Hα ribbons is further support for the standard picture, particularly in light of the observation by Wuelser *et al.* (1994) indicating blueshifted soft X-rays coincident with redshifted Hα. That observation, as well as the extended system of red- and blueshifts of Czaykowska *et al.* (1999), are strongly suggestive of chromospheric evaporation. The analysis of Acton *et al.* (1982) concluded that evaporation may be driven by either electrons and conduction.

OBSERVATIONS TO QUESTIONS

Despite the argument that these observations appear to support the standard reconnection picture, it must be understood that the match is in some cases not perfect. Perhaps these "imperfect matches" offer a chance to refine either some details of the standard picture, or our understanding of the physical parameters of the plasmas in and around the flares.

The downward flows observed by McKenzie & Hudson (1999) and Wang *et al.* (1999) are slower than nominally expected: 50 – 500 km s^{-1}, not thousands of km s^{-1}. If these flows are to be interpreted as reconnection outflows, then either the reconnection outflow is not necessarily Alfvénic, or the Alfvén speed is not what we thought it was. The former solution may be a result of plasma parameters in the downstream region — Priest & Forbes (2000) demonstrated that the outflow speed and the reconnection rate may be affected by the plasma conditions downstream of the reconnection — or possibly of interaction with a return current at the downstream edge of the diffusion region (Biskamp 1984). On the other hand, lower-than-expected Alfvén speeds may result from our incomplete knowledge of the plasma parameters and magnetic field in a region where the magnetic field is un-measurable and partly annihilated, the temperature and density of the plasma is unknown, and the velocity is poorly known, changing, and possibly turbulent. Additionally, the LASCO outflows which move upwards away from the Sun have speeds between 150 – 300 km s^{-1}; these characteristics must tell us something about the physical conditions at these high altitudes.

Similarly, what about the possible FMSS observed by Tsuneta (1997) and Aurass *et al.* (2002)? If the standard model is correct, then why is this signature not observed more often? The visibility must place some limits on the outflow speed, the density of the downstream plasma, and the velocity of the X-point. Aurass *et al.* suggest that the rarity of the observation indicates that "for a majority of flares the plasma parameter β is comparatively high." Alternatively, the FMSS is simply not a common feature in reconnections.

The sole, tentative observation of reconnection inflow, by Yokoyama *et al.* (2001) points to a reconnection rate on the order of $M_A \approx 0.001 - 0.03$. According to Yokoyama *et al.*, the implication is Petschek-type reconnection, rather than Sweet-Parker, though the precision of the measurements is pretty coarse. If more of these observations can be acquired, then perhaps they will provide a useful test of the reconnection rate, as well as constraints on the modeling of the atmosphere and conditions up- and downstream of the diffusion region.

DIMENSIONALITY

Finally, a word on dimensionality. Many of the diagrams and models in the literature have been strictly two-dimensional, and have been extremely successful. Surprisingly, several of the analyses which have made the greatest impact have dealt with flare structures that closely resemble the 2D cartoons (e.g. Tsuneta 1996, 1997; Forbes & Acton 1996; Aschwanden *et al.* 1996; Yokoyama *et al.* 2001). At least one of the observations mentioned herein bears a striking resemblance to a 2.5D representation: Czaykowska *et al.* (1999). But some flare observations can only be described in 3D (e.g. Švestka *et al.* 1998; McKenzie 2000; Fletcher *et al.* 2001), and so the models must reproduce the "strictly-3D" features.

For example, in the region above some post-eruption coronal arcades, there are "spikey" structures extending high up into the corona. These tall features, first analyzed by Švestka *et al.* (1998), are extremely hot, may persist for many hours, and evolve on fairly short time scales (see also McKenzie & Hudson 1999; McKenzie 2000). The mechanism which produces such a hot fan above an arcade is still to be determined.

A model that may be worth considering builds upon the results of Forbes & Malherbe (1986), as emphasized in Forbes & Acton (1996). In particular, these studies indicated that arcades with strong magnetic fields may be distinguishable from those with weaker fields by the presence of supermagnetosonic downflow jets. In events with weaker fields, Forbes & Acton (1996) suggest that most of the reconnection outflow may be directed upwards (i.e. away from the Sun and the arcade), and that thermal condensation at the location of the termination shock is less likely to occur. Figures 1 and 2 of Forbes & Acton (1996) demonstrate the possible distinction, including high-temperature signatures above the cusped loops in the high-B case but not in the low-B configuration. A possible 3D model might interleave these two essentially 2D constructions — by allowing the field strength to vary along the length of the arcade, such a model might recover the spikey appearance of the supra-arcade region.

Moreover, the supra-arcade downflows (McKenzie & Hudson 1999; McKenzie 2000) are traced by discrete X-ray features with a characteristic size. If the downflows are reconnection outflows, then these tracers strongly suggest that the reconnection takes place between discretized collections of magnetic flux, e.g., flux tubes. Such a conclusion would indicate a piecewise, or "patchy" reconnection, recalling Arons (1984): "rather rarely does one see the theorists' idealization of steady reconnection flow on a scale large compared to the local dissipation region."

A cartoon to depict the patchy reconnection and subsequent supra-arcade downflow is shown in Figure 6. This cartoon is created in the spirit of reconnection *à la* Klimchuk (1997, his Figure 5) leading into McKenzie & Hudson (1999, their Figure 3): reconnection occurs in small patches, creating discrete flux tubes, which then individually shrink to relax the magnetic tension. The shrinking flux tubes dipolarize to form the post-eruption arcade. If models can be developed which have features like the supra-arcade spikes and downflows, then comparison with the observations should tell us how the patchiness and characteristic size (and even the presence or absence of supra-arcade structures) depend on parameters like field strength, helicity, etc.

Apart from predictive modeling, observational analyses that attempt to recover the three-dimensional configurations of coronal structures are making considerable advances. In analysis like that by Mandrini *et al.* (1997), separators and/or quasi-separatrix layers indicate where flare brightenings should occur, with reasonable accuracy. In Fletcher *et al.* (2001), observed changes in magnetic topology (changes in separatrix position or configuration) seem consistent with the timing and morphology of the observed flare brighten-

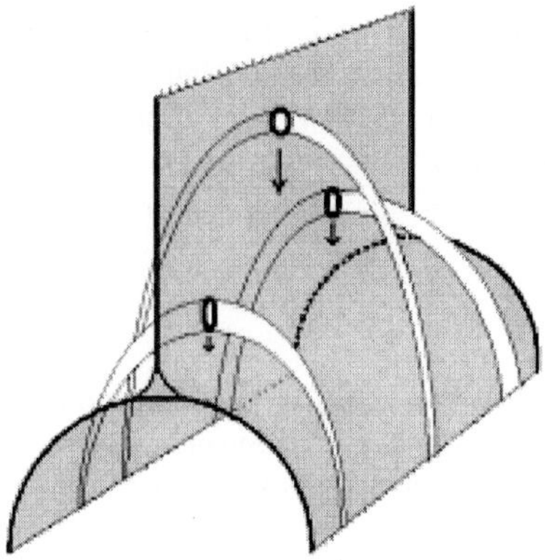

Fig. 6. Cartoon to depict supra-arcade downflows resulting from patchy reconnection (see text).

ings. These studies indicate significant advances in the quantitative treatment of flare morphologies since the initial diagrams of the CSHKP model.

CONCLUSION

Many observations seem to support the validity of the magnetic reconnection model of solar (and stellar) flares. They are, at the very least, consistent with the expectations of that model. But the appropriate use of observational data includes not only the confirmation (or refutation) of theoretical models, but also the obligation to push the models to ever-higher levels of sophistication. In the present case flare observations certainly exist which are difficult to describe in terms of the current "standard picture". It behooves us, then, to study these "nonconforming" observations and learn what they are telling us about the physics of magnetic plasmas.

REFERENCES

Acton, L.W., R.C. Canfield, T.A. Gunkler, H.S. Hudson, A.L. Kiplinger, & J.W. Leibacher, Chromospheric Evaporation in a Well-Observed Compact Flare, *ApJ,* **263**, 409 (1982).

Alexander, D. & T.R. Metcalf, A Spectral Analysis of the Masuda Flare Using HXT Pixon Reconstruction, *ApJ,* **489**, 442 (1997).

Arons, J., Astrophysical Implications of Reconnection, in *Magnetic Reconnection in Space and Laboratory Plasmas,* (Geophys. Monogr. 30), ed. E.W. Hones, Jr., p. 366, American Geophysical Union, Washington, D.C. (1984).

Aschwanden, M.J., T. Kosugi, H.S. Hudson, M.J. Wills, & R. Schwartz, The Scaling Law Between Electron Time-of-flight Distances and Loop Lengths in Solar Flares, *ApJ,* **470**, 1198 (1996).

Aurass, H., B. Vršnak, & G. Mann, Shock-excited Radio Burst from Reconnection Outflow Jet?, *A&A,* **384,** 273 (2002).

Biskamp, D., Validity of the Petschek Model, in *Magnetic Reconnection in Space and Laboratory Plasmas,* (Geophys. Monogr. 30), ed. E.W. Hones, Jr., p. 369, American Geophysical Union, Washington, D.C. (1984).

Bruzek, A., in *Solar Flares and Space Research,* eds. C. de Jager & Z. Švestka, p. 61, North-Holland, Amsterdam (1969).

Carmichael, H., in *AAS-NASA Symposium on the Physics of Solar Flares*, ed. W.N. Hess, p. 451, NASA SP-50 (1964).

Czaykowska, A., B. De Pontieu, D. Alexander, & G. Rank, Evidence for Chromospheric Evaporation in the late Gradual Flare Phase from SOHO/CDS Observations, *ApJ*, **521**, L75 (1999).

Dere, K.P., & J.W. Cook, The Decay of the 1973 August 9 Flare, *ApJ*, **229**, 772 (1979).

Duijveman, A., P. Hoyng, & M. Machado, X-ray Imaging of Three Flares During the Impulsive Phase, *Sol. Phys.*, **81**, 137 (1982).

Fletcher, L., T.R. Metcalf, D. Alexander, D.S. Brown, & L.A. Ryder, Evidence for the Flare Trigger Site and Three-Dimensional Reconnection in Multiwavelength Observations of a Solar Flare, *ApJ*, **554**, 451 (2001).

Forbes, T.G., & J.M. Malherbe, A Shock Condensation Mechanism for Loop Prominences, *ApJ*, **302**, L67 (1986).

Forbes, T.G., & L.W. Acton, Reconnection and Field Line Shrinkage in Solar Flares, *ApJ*, **459**, 330 (1996).

Hiei, E., & A.J. Hundhausen, Development of a Coronal Helmet Streamer of 24 January 1992, in *Magnetodynamic Phenomena in the Solar Atmosphere — Prototypes of Stellar Magnetic Activity*, eds. Y. Uchida, T. Kosugi, & H.S. Hudson, p. 125, Kluwer Academic Publishers, Dordrecht, The Netherlands (1996).

Hirayama, T., Theoretical Model of Flares and Prominences I: Evaporating Flare Model, *Sol. Phys.*, **34**, 323 (1974).

Hoyng, P., A. Duijveman, M.E. Machado, D.M. Rust, Z. Švestka, A. Boelee, C. de Jager, K.J. Frost, H. Lafleur, G.M. Simnett, H.F. van Beek, & B.E. Woodgate, Origin and Location of the Hard X-ray Emission in a Two-Ribbon Flare, *ApJ*, **246**, L155 (1981).

Klimchuk, J.A., Post-Eruption Arcades and 3-D Magnetic Reconnection, in *Magnetic Reconnection in the Solar Atmosphere*, ASP Conference Series, Vol. 111, eds. R.D. Bentley and J.T. Mariska, p. 319 (1996).

Kopp, R.A., & G.W. Pneuman, Magnetic Reconnection in the Corona and the Loop Prominence Phenomenon, *Sol. Phys.*, **50**, 85 (1976).

Mandrini, C.H., P. Démoulin, L.G. Bagalá, L. van Driel-Gesztelyi, J.C. Hénoux, B. Schmieder, & M.G. Rovira, Evidence of Magnetic Reconnection from Hα, Soft X-ray and Photospheric Magnetic Field Observations, *Sol. Phys.*, **174**, 229 (1997).

Mariska, J.T., T. Sakao, & R.D. Bentley, Hard and Soft X-ray Observations of Solar Limb Flares, *ApJ*, **459**, 815 (1996).

Masuda, S., Ph.D. Thesis, University of Tokyo (1994).

Masuda, S., T. Kosugi, H. Hara, S. Tsuneta, & Y. Ogawara, A Loop-Top Hard X-Ray Source in a Compact Solar Flare as Evidence for Magnetic Reconnection, *Nature*, **371**, 495 (1994).

Masuda, S., T. Kosugi, & H.S. Hudson, A Hard X-ray Two-Ribbon Flare Observed with Yohkoh/HXT, *Sol. Phys.*, **204**, 55 (2001).

McKenzie, D.E., & H.S. Hudson, X-ray Observations of Motions and Structure Above a Solar Flare Arcade, *ApJ*, **519**, L93 (1999).

McKenzie, D.E., Supra-Arcade Downflows in Long-Duration Solar Flare Events, *Sol. Phys.*, **195**, 381 (2000).

Moore, R., D.L. McKenzie, Z. Švestka, K.G. Widing, and 6 co-authors, The Thermal X-ray Flare Plasma, in *Solar Flares: A Monograph from Skylab Solar Workshop II*, ed. P.A. Sturrock, p. 341, Colorado Associated University Press (1980).

Ohyama, M, & K. Shibata, X-ray Plasma Ejection in an Eruptive Flare, in *Magnetodynamic Phenomena in the Solar Atmosphere — Prototypes of Stellar Magnetic Activity*, eds. Y. Uchida, T. Kosugi, & H.S. Hudson, p. 525, Kluwer Academic Publishers, Dordrecht, The Netherlands (1996).

Priest, E., & T. Forbes, in *Magnetic Reconnection: MHD Theory and Applications*, p. 125, Cambridge University Press (2000).

Rust, D.M., G.M. Simnett, & D.F. Smith, Observational Evidence for Thermal Wave Fronts in Solar Flares, *ApJ,* **288**, 401 (1985).

Sakao, T., Ph.D. Thesis, University of Tokyo (1994).

Sakao, T., & T. Kosugi, Non-thermal Processes and Superhot Plasma Creation in Solar Flares, in *Magnetodynamic Phenomena in the Solar Atmosphere — Prototypes of Stellar Magnetic Activity,* eds. Y. Uchida, T. Kosugi, & H.S. Hudson, p. 169, Kluwer Academic Publishers, Dordrecht, The Netherlands (1996).

Schmieder, B., T.G. Forbes, J.M. Malherbe, & M.E. Machado, Evidence for Gentle Chromospheric Evaporation During the Gradual Phase of Large Solar Flares, *ApJ,* **317**, 956 (1987).

Seely, J. F., & U. Feldman, Direct Measurement of the Increase in Altitude of the Soft X-ray Emission Region During a Solar Flare, *ApJ,* **280**, L59 (1984).

Sheeley, Jr., N.R., Y.-M. Wang, S.H. Hawley, G.E. Brueckner, and 15 co-authors, Measurements of Flow Speeds in the Corona Between 2 and 30 R_{SUN}, *ApJ,* **484**, 472 (1997).

Shibata, K., Evidence of Magnetic Reconnection in Solar Flares and a Unified Model of Flares, *Ap&SS,* **264**, 129 (1999).

Simnett, G.M., Studies of the Dynamic Corona from SOHO, in *High Energy Solar Physics: Anticipating HESSI,* eds. R. Ramaty & N. Mandzhavidze, ASP Conference Series, vol. 206, p. 43 (2000).

Sterling, A.C., H.S. Hudson, & J.R. Lemen, Yohkoh SXT and BCS Observations of the "Reconnection Region" of a Solar Flare, in *Magnetic Reconnection in the Solar Atmosphere,* ASP Conference Series, Vol. 111, eds. R.D. Bentley and J.T. Mariska, p. 177 (1996).

Sturrock, P.A., A Model of Solar Flares, in *Structure and Development of Solar Active Regions,* ed. K. Kiepenheuer, p. 471, Paris: IAU (1968).

Švestka, Z. F., J.M. Fontenla, M.E. Machado, S.F. Martin, D.F. Neidig, & G. Poletto, Multi-Thermal Observations of Newly formed Loops in a Dynamic Flare, *Sol. Phys.,* **108**, 237 (1987).

Švestka, Z., F. Fárník, H. Hudson, & P. Hick, Large-Scale Active Coronal Phenomena in Yohkoh SXT Images — IV. Solar Wind Streams from Flaring Active Regions, *Sol. Phys.,* **182**, 179 (1998).

Trottet, G., E. Rolli, A. Magun, C. Barat, A. Kuznetsov, R. Sunyaev, & O. Terekhov, The Chromospheric Response to Particle Beams During a Gamma-ray Flare, in *High Energy Solar Physics: Anticipating HESSI,* eds. R. Ramaty & N. Mandzhavidze, ASP Conference Series, vol. 206, p. 419 (2000).

Tsuneta, S., Structure and Dynamics of Magnetic Reconnection in a Solar Flare, *ApJ,* **456**, 840 (1996).

Tsuneta, S., Moving Plasmoid and Formation of the Neutral Sheet in a Solar Flare, *ApJ,* **483**, 507 (1997).

van Aalst, M.K., P.C.H. Martens, & A.J.C. Belien, Can Streamer Blobs Prevent the Buildup of the Interplanetary Magnetic Field?, *ApJ,* **511**, L125 (1999).

van Driel-Gesztelyi, L., J.E. Wiik, B. Schmieder, T. Tarbell, R. Kitai, Y. Funakoshi, & B. Anwar, Post-Flare Loops of 26 June 1992, *Sol. Phys.,* **174**, 151 (1997).

Wang, Y.-M., N.R. Sheeley, Jr., J.H. Walters, G.E. Brueckner, R.A. Howard, D.J. Michels, P.L. Lamy, R. Schwenn, & G.M. Simnett, Origin of Streamer Material in the Outer Corona, *ApJ,* **498**, L165 (1998).

Wang, Y.-M., N.R. Sheeley, Jr., R.A. Howard, N.B. Rich, & P.L. Lamy, Streamer Disconnection Events Observed with the LASCO Coronagraph, *Geo. Res. Letters,* **26**, 1349 (1999a).

Wang, Y.-M., N.R. Sheeley, Jr., R.A. Howard, O.C. St. Cyr, & G.M. Simnett, Coronagraph Observations of Inflows During High Solar Activity, *Geo. Res. Letters,* **26**, 1203 (1999b).

Wuelser, J.-P., R.C. Canfield, L.W. Acton, J.L. Culhane, A. Phillips, A. Fludra, T. Sakao, S. Masuda, T. Kosugi, & S. Tsuneta, Multispectral Observations of Chromospheric Evaporation in the 1991 November 15 X-class Solar Flare, *ApJ,* **424**, 459 (1994).

Yokoyama, T., K. Akita, T. Morimoto, K. Inoue, & J. Newmark, Clear Evidence of Reconnection Inflow of a Solar Flare, *ApJ,* **546**, L69 (2001).

SXT AND EIT OBSERVATIONS OF A QUIET REGION LARGE-SCALE ERUPTION: IMPLICATIONS FOR ERUPTION THEORIES

A. C. Sterling[1,2], R. L. Moore[2], and B. J. Thompson[3]

[1]*United Applied Technologies, Inc., Huntsville, AL;*
Current address: ISAS, 3-1-1 Yoshinodai, Sagamihara, Kanagawa, 229-8510 JAPAN
[2]*NASA/MSFC, SD50/Space Science Dept., Huntsville, AL 35812, USA*
[3]*NASA/GSFC, Code 682, Greenbelt, MD 20771*

ABSTRACT

In order to understand what drives solar eruptions, it is critical to test theoretical scenarios with data from observations. Here we investigate the idea that "tether cutting" reconnection in a core magnetic field region is responsible for the onset of solar eruptions. We do this by looking to see whether observational signatures of reconnection precede signatures of the eruption, as expected if tether cutting does indeed drive the eruption. Using data from the Extreme ultraviolet Imaging Telescope (EIT) instrument on *SOHO,* we examine the evolution of a slow filament eruption occurring in a quiet region. We find that brightenings associated with reconnection become obvious *after* the start of the rapid ejection of the erupting filament, suggesting that, at least for this case, the tether cutting is a byproduct of, rather than the cause of, the eruption.

INTRODUCTION

There are a variety of ideas for the cause of the onset of solar eruptions. Ultimately, we would like to use observations to confirm which scenarios best describe reality, and to refute those which are clearly at odds with the observations. As a first step, we hope at least to be able to place constraints on the various theories through observations. Here we consider the eruption scenario known as the "tether cutting model" (e.g. Sturrock 1989, Moore & LaBonte 1980, Moore *et al.* 2001), which is a fundamentally bipolar mechanism in which the eruption onset is due to reconnection of strongly-sheared magnetic field lines ("tethers") in the core of the erupting region.

If tether cutting is responsible for the onset of the eruption, then we would expect the reconnection in the core to begin *prior to* the expulsion of material associated with the eruption. This suggests a method for testing the model: compare the timing of the onset of the mass expulsion, e.g., as characterized by a filament eruption, with the timing of the onset of the tether-cutting reconnection.

Here we make such a timing comparison, using the ejection of a filament as a signature of the mass expulsion, and EUV intensity brightenings in the core region as a proxy for the reconnection. We are assuming that the reconnecting core loops produce a nearly immediate (within some tens of seconds) intensity response at their footpoints, as a result of particles or heat generated during the reconnection generating brightenings in the dense atmosphere at the base of the loops; this is a reasonable assumption, considering that the core loops are likely to be short (a few times 10^3 km), and the propagation speed would be at least the sound speed in the coronal loops (a few 100 km s^{-1}).

DATA AND RESULTS

We observed a filament eruption and associated weak flaring from a quiet region on 1999 April 18, near 7 UT. Here we present data from the *SOHO* EIT 195 Å filter; elsewhere we have also examined EIT 304 Å and *Yohkoh* Soft X–ray Telescope (SXT) images (Sterling *et al.* 2001). This eruption likely resulted in an Earthward-directed partial-halo CME detected shortly afterwards by the *SOHO* Large angle and Spectrometric Coronagraph (LASCO).

Compared to typical active region events, the evolution of this event was very slow, making it ideal for our timing analysis. Figures 1a and 1b show that the filament is already in motion several hours prior to eruption; cf. the filament's location relative to the brightpoint (marked 'bp' in Figure 1a) in the two panels. By the time of Figure 1c, the filament has erupted completely, leaving expanding postflare loops in its wake.

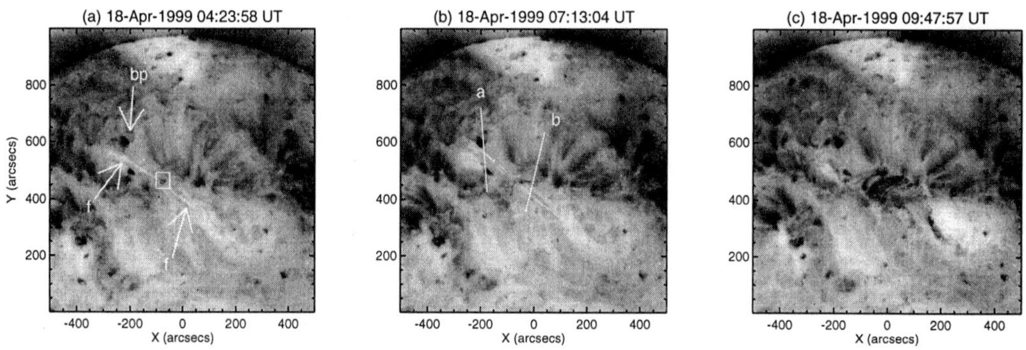

Fig. 1. Erupting filament in EIT 195 Å images. (a) Filament, marked by 'f' arrows, prior to eruption. Arrow 'bp' marks a bright point. (b) The filament has moved upward relative to the bright point indicated in (a) Fiducials 'a' and 'b' are used to plot the evolution of the filament in Figure 2. (c) After filament eruption, postflare loops develop and expand along the filament channel.

We can examine the filament's evolution in more detail using the fiducial lines labeled 'a' and 'b' in Figure 1b. Figure 2 plots the location of the filament along these two fiducials as functions of time. Relatively slow (≈ 1 km s^{-1}) upward motion along path a begins near 00:00 UT on 18 April, with a rapid (≈ 15 km s^{-1}) rise starting between 06:36 and 06:48 UT (indicated by vertical lines in Figure 2); we call these slower and faster periods of evolution Stage 1 and Stage 2, respectively.

The segment of the filament along path b shows very little motion prior to the time of Stage 2; this could be because its motion is along the line-of-sight, or it could be that the two halves of the filament behave independently (e.g. Martens & Zwaan 2001; Hori *et al.,* these proceedings). During Stage 2, this portion of the filament violently erupts outward along with the rest of the filament. This leg of the filament seems to erupt largely along the line-of-sight, which would explain why the path b trajectory in Figure 2 ends without showing an upturn; the material becomes too diffuse for us to measure a location at the point where that trajectory ends.

We can compare this filament motion with the brightenings in the central "core" region of the eruption, where the putative tether-cutting reconnection is expected to take place; this location is indicated by the box in Figure 1a. Indeed, the earliest eruption-associated brightenings in EUV do occur in this location, indicating that tether cutting may be taking place. The question is, what comes first: filament eruption or brightenings due to reconnection? Figure 2 shows the light curve from the box. The increase in intensity of the box begins after, or at best simultaneous with, the onset of Stage 2 of the filament's eruption. Moreover, the rapid increase in the intensity does not occur until well after the onset of Stage 2. (We selected the box in Figure 1a to cover the region which shows the earliest obvious brightening; we are implicitly assuming that the footpoints of at least some of the key reconnecting core loops reside within this box.)

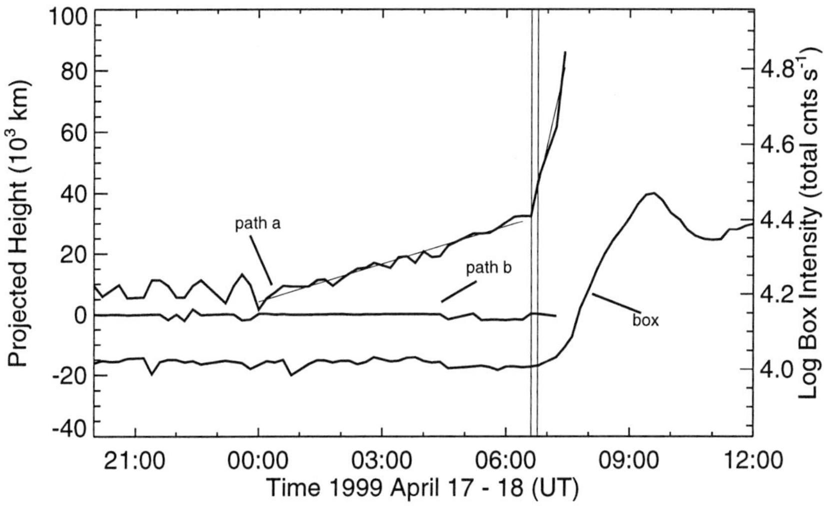

Fig. 2. Top two curves are the projected height of the filament as a function of time. Paths a and b correspond to motion along the respective fiducials in Figure 1b; thin lines overlaid on path a are best-fit lines during the two stages of the eruption. The lowest curve, labeled 'box,' is the integrated intensity from the region defined by the box in Figure 1a, and represents the region where tether-cutting reconnection is expected to take place. The two vertical lines bracket the time of the onset of Stage 2 of the eruption.

DISCUSSION AND CONCLUSIONS

It is hard to understand the relative timings of the curves in Figure 2 under the assumption that tether cutting initiates the eruption. Rather, it seems as if the tether cutting is *a consequence* of the eruption. Some other idea for eruption onset, such as the the Breakout model (Antiochos *et al.* 1999), or non-resistive models (e.g. Rust & Kumar 1996), may be more appropriate for explaining the eruption trigger mechanism.

Preflare filament activation, which likely corresponds to our filament's Stage 1 evolution, has been observed for a long time in chromospheric spectral lines. Also, observations from SXT show the ejection of plasmoids prior to the onset of hard X-rays in flares (e.g. Ohyama & Shibata 1997, Nitta & Akiyama 1999) which display two stages of evolution. In our terminology, these SXT events have Stage 1 and Stage 2 velocities of $\sim$ 10 km s^{-1} and several hundred km s^{-1}, respectively, which are higher than those we observe. This difference is likely due to the SXT events being associated with active regions, whereas here we observe an eruption in a quiet region. Generally speaking, if, as we are suggesting here, tether cutting does not trigger eruptions, then we expect that active regions would also show that Stage 2 evolution begins prior to the onset of the main flaring activity. In fact, Ohyama & Shibata (1997) found the onset of the rapid eruption of the plasmoids to occur just before or about simultaneous with the onset of the hard X-ray burst (where they had a time resolution of about 2.5 min); this is consistent with our findings. Time scales, however, are much more rapid in the active region events, and therefore the differences in onset times between the start of Stage 2 and the hard X-ray burst may be more difficult to determine in such events. It should be possible to address the timing question more effectively using data from *TRACE* combined with hard X-ray data from *RHESSI.*

A more complete discussion of the work presented here appears in Sterling *et al.* (2001).

ACKNOWLEDGEMENTS

ACS and RLM were supported by funding from NASA's Office of Space Science through the SR&T and Sun-Earth Connection GI Programs.

REFERENCES

Antiochos, S. K., DeVore, C. R., Klimchuk, J. A., A Model for Solar Coronal Mass Ejections, *ApJ,* **510,** 485 (1999).

Martens, P. C., & Zwaan, C., Origin and Evolution of Filament-Prominence Systems, *ApJ.,* **558,** 872 (2001).

Moore, R. L., & LaBonte, B. J., The filament eruption in the 3B flare of July 29, 1973: onset and magnetic field configuration, in *Solar and interplanetary dynamics, IAU Symp. 91,* ed. M. Dryer and E. Tandberg-Hanssen, p. 207, Reidel, Boston (1980).

Moore, R. L., Sterling, A. C., Hudson, H. S., & Lemen, J. R., Onset of the Magnetic Explosion in Solar Flares and Coronal Mass Ejections, *ApJ.,* **552,** 833 (2001).

Ohyama, M., & Shibata, K., Preflare Heating and Mass Motion in a Solar Flare Associated with Hot Plasma Ejection: 1993 November 11 C9.7 Flare, *Pub. Astro. Soc. Japan,* **49,** 249 (1997).

Nitta, N., & Akiyama, S., Relation between Flare-associated X-Ray Ejections and Coronal Mass Ejections, *ApJ,* **49,** 249 (1999).

Rust, D., M., Kumar, A., Evidence for helically kinked magnetic flux ropes in solar eruptions, *ApJ,* **464,** L199 (1996).

Sterling, A. C., Moore, R. L., & Thompson, B. J., EIT and SXT Observations of a Quiet Region Filament Ejection: First Eruption, Then Reconnection, *ApJL,* **561,** L219 (2001).

Sturrock, P. A., The Role of Eruption in Solar Flares, *Solar Phys.,* **121,** 387 (1989).

3 GHz FLUX VARIATIONS OF THE APRIL 7, 1997 FLARE AND CURRENT-LOOP COALESCENCE MODEL

F. Fárník and M. Karlický

Astronomical Institute of the Academy of Sciences of the Czech Republic, 25165 Ondřejov, Czech Republic

ABSTRACT

Regular time variations of the 3 GHz radio flux observed during the April 7, 1997 flare are presented. These observations are interpreted using the current loop coalescence model: the main period of the 3 GHz radio flux (about 100 s) corresponds to repetition of the current loop coalescence, and the radio double peaks are associated with maximum of the electric field component perpendicular to the interaction plane. The plasma β parameter in the current loop coalescence process is estimated as 0.63.

INTRODUCTION

Collisions between current-carrying loops are considered as a cause of some solar flares (Sakai & de Jager, 1996). Based on the loops' orientations and the size of interaction region three types of current loop interactions are distinguished: a) 1-D coalescence (I-type), b) 2-D coalescence (Y-type), and c) 3-D X-type coalescence. There are several papers showing observational indications of these processes (e.g., Shimizu *et al.*, 1994). A numerical and 1-D analytical model of the coalescence process of the current-carrying loops was presented by Tajima *et al.* (1987). In the present paper regular variations of the 3 GHz radio flux and accompanying flare effects observed during the April 7, 1997 flare are interpreted using this interaction loop model.

OBSERVATIONS

The April 7, 1997 flare (start at 13:50, maximum at 14:07, end at 14:19 UT) classified as C6.8/3N was observed in NOAA AR 8027. In Figure 1 its 3 GHz radio flux record is shown. In particular, the periodic double peaks in the 13:56:30-14:02:00 UT interval are interesting. The mean period of these 3 double peaks is 100 s, while the mean time interval between peaks in the double peak structures is 44 s, and in the last double peak somewhat longer (about 55 s). After these 3 double peaks there is some radio flux decrease, followed by the most intense radio burst at 14:07:30 UT, which also has the double peak structure. On lower frequencies (40-800 MHz) these 3 GHz radio flux variations were followed by several branches of the type II radio burst. The frequency drift of the low-frequency branches of the type II burst is -0.24 MHz s^{-1}. Thus, the estimated type II shock speed in the 4-fold Newkirk model of the solar atmosphere is 1200 km s^{-1}. Unfortunately, *Yohkoh*/SXT observed this flare only after the flare maximum; the first *Yohkoh* image was obtained at 14:10:58 UT . These images show an S-shape structure of the flare and a two-loop interaction (I-type) showing a brightness maximum at the interaction place.

INTERPRETATION

Observations presented above show that the current-loop coalescence model of solar flares of Tajima *et al.* (1987) can be used. In this model the time variations of magnetic field and plasma parameter changes are expressed through the scale factor $a(t)$. Assuming that the plasma is quasi-neutral, then the $a(t)$ is given by the second-order differential equation:

$$\ddot{a} = -\frac{v_A^2}{\lambda^2 a^2} + \frac{c_s^2}{\lambda^2 a^3} \tag{1}$$

where v_A and c_s are the Alfvén and sound velocities, and λ is the magnetic field scale length. This equation has a periodic solution with period $T = 2\pi(-2E)^{-3/2}t_A^{-2}$, where $t_A = \lambda/v_A$, and E is the initial "energy". Now, knowing $a(t)$ as the solution of Eq. 1, the electric field components E_x, and E_z (in the interaction plane, and in the perpendicular direction to the interaction plane, respectively) accelerating electrons during the periodic current-loop coalescence process are computed as:

$$E_x = \left(-\frac{m_i}{e\lambda}\frac{v_A^2}{a^3} + \frac{P_{0e}}{e\lambda a^4 n_0}\right)\frac{x}{\lambda}, \quad E_z = -\frac{B_{00}\dot{a}x^2}{ca^3\lambda} - \frac{B_{00}m_e c\dot{a}}{4\pi n_0 e^2 \lambda a^2},$$

where x is the space coordinate in the interaction plane, B_{00} and n_0 are the magnetic field and density constants, m_i and m_e are the proton and electron masses, P_{0e} is the initial electron pressure, e is the electron charge, and c is the speed of light.

It is obvious that the observed main period ($\sim$ 100 s) in the 3 GHz radio flux variations should correspond to the period T. But the question arises how to explain the double peaks of the 3 GHz flux variation. We assume that the individual peaks are connected with electric field variations. The electric field accelerates electrons which generate the radio emission. In principle, there are two components of the electric field (see above). Although the paper by Tajima *et al.* (1987) (their Figure 12c) offers an explanation of the double peaks through the time variation of E_x, Eq. 1 does not give results which fit the observed data for reasonable parameters. On the other hand, the time variation of E_z can fit the double peak structure easily. The best fit was obtained for the values: $v_A^2/\lambda^2 = 0.001$, $c_s^2/\lambda^2 = 0.00063$, $E = -0.79 \times 10^{-3}$ s^{-2}. Therefore in this case the plasma β parameter is 0.63, i.e. the plasma pressure effects are not negligible.

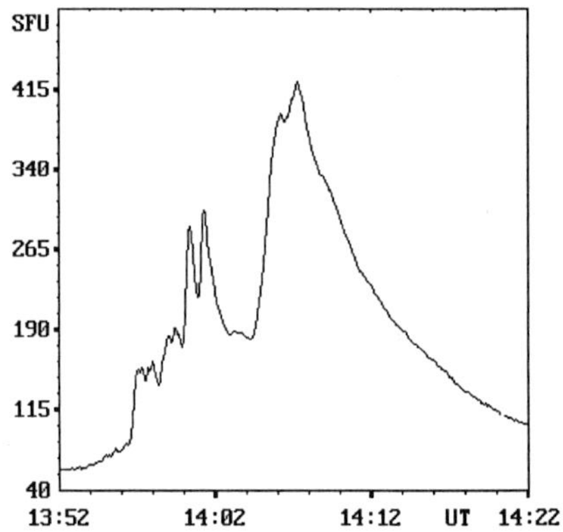

Fig. 1. 3 GHz radio flux observed by the Ondřejov radiometer.

ACKNOWLEDGEMENTS: This work was supported by the grants S1003006 and A3003202 of the Grant Agency of the Academy of Sciences of the Czech Republic under the key project K2043105.

REFERENCES

Sakai, J.I., C. de Jager, *Space Sci. Rev.* **77**, 1 (1996).

Shimizu, T., T. Tsuneta, L.W. Acton, J.R. Lemen, Y. Ogawara, Y. Uchida, *ApJ* **422**, 906 (1994).

Tajima, T., J.I. Sakai, H. Nakajima, T. Kosugi, F. Brunel, M.R. Kundu, *ApJ* **321**, 1031 (1987).

STATISTICAL STUDY OF THE RECONNECTION RATE IN SOLAR FLARES

H. Isobe, T. Morimoto, S. Eto, N. Narukage, and K. Shibata

Kwasan and Hida Observatories, Kyoto University, Yamashina, Kyoto, 607-8417, Japan

ABSTRACT

Reconnection rate defined by (inflow velocity)/(Alfvén velocity) is one of the key parameters for understanding the physics of magnetic reconnection. In this paper we utilize the method for determining the reconnection rate from observational data suggested by Isobe *et al.* (2002) for 7 solar flares and 2 giant arcades. We have found that the reconnection rate is 0.001–0.2, which is consistent with the fast reconnection model by Petschek.

INTRODUCTION

The Soft X-ray Telescope (SXT) aboard the *Yohkoh* spacecraft has established that the driving mechanism of solar flares is magnetic reconnection. However, the physics of reconnection has not been clarified yet. One of the current puzzles is: what determines the reconnection rate? The reconnection rate is defined as reconnected magnetic flux per unit time or, in dimensionless form, the ratio of inflow speed (into reconnection point) to Alfvén velocity. The reconnection rate is one of the most important physical quantities in reconnection physics. However, observations have not yet succeeded in statistically determining the reconnection rate because direct observation of reconnection inflow and coronal magnetic field is difficult. Actually, clear evidence of inflow in a limb flare found by Yokoyama *et al.* (2001) is the only direct observation made so far. So we need an indirect method to determine the reconnection rate in many flares from observational data. In this paper we utilize a method presented by Isobe *et al.* (2002) to determine the reconnection rate. We analyze 7 flares and 2 giant arcades, and examine the characteristics of reconnection rate in solar corona.

METHOD

The method for determining the reconnection rate is described in detail by Isobe *et al.* (2002). Here we briefly summarize the method. We utilize following two equations:

$$H = 2\frac{B_{corona}^2}{4\pi}v_{in}A_r \quad , \tag{1}$$

$$B_{corona}v_{in} = B_{foot}v_{foot} \quad , \tag{2}$$

where H is heating rate, $B_{corona}$and B_{foot} are coronal and photospheric magnetic field, v_{in} and v_{foot} are inflow velocity and separation velocity of footpoints of flare loops, and A_r is the area of the reconnection

Table 1. List of events and reconnection rate M_A. Reconnection rates are those in the impulsive phase unless otherwise indicated.

Event date	GOES class	M_A	Event date	GOES class	M_A
2000 Nov. 24	X2	0.002	1998 Oct. 15	Giant arcade	0.08
2000 Nov. 24	X1	0.07	1997 May 12	C2 (decay)	0.005
1999 Aug. 1	M1	0.05	1993 May 14	M4	0.1
1998 Dec. 14	C1	0.2	1993 May 14	M4 (decay)	0.001
1998 Nov. 8	M1	0.1	1993 Jan. 26	Giant arcade	0.02

region. Eq. 1 shows that the released magnetic energy comes from the Poynting flux into the reconnection region. Eq. 2 is conservation of magnetic flux, i.e., the left-hand side is the reconnected magnetic flux per unit time in the reconnection region, and the right-hand side is the corresponding magnetic flux in the photosphere.

Considering the balance between the heating and radiative and conductive losses, H is given by $H = dE_{th}/dt + n^2Q(T)V + \kappa_0 T^{7/2}V/L^2$, where E_{th} is the total thermal energy in the flare loops, $n^2Q(T)V$ is radiative loss, and $\kappa_0 T^{7/2}/L^2$ is conductive loss. Temperature T and loop length L are measured from SXT data. Density n and volume V are also measured from SXT data by assuming the line-of-sight length Hence H can be determined. v_{foot} is obtained by measuring the separation velocity of the footpoints, and B_{foot} is obtained from a photospheric magnetogram. Finally, A_r is assumed to be equal to the apparent area of the flare loops (arcade). Then, v_{in} and B_{corona} (and hence V_A) can be determined from Eq. 1 and 2.

RESULT

We have analyzed 7 flares and 2 giant arcades listed in Table 1. In most flares and arcades the reconnection rate has been determined in the impulsive phase in which the separation velocity is easily measured. For the May 12, 1997 and May 14, 1993 flares we have determined the reconnection rate in the decay phase. A more detailed analysis of the May 12, 1997 flare was made by Isobe *et al.* (2002).

We have found that the reconnection rate is 0.001-0.1, which is consistent with the fast reconnection model by Petschek (1964). This is the first statistical result of an observational derivation of reconnection rate, although it is quite a rough estimate. We need to increase the number of events to examine the characteristics of the reconnection rate, such as its dependence on magnetic Reynolds number.

REFERENCES

Isobe, H., Yokoyama, T., Shimojo, M., *et al.*, Reconnection rate in the decay phase of an LDE flare on 1997 May 12, *ApJ*, **566**, 528 (2002).

Petschek, H. E. in *AAS-NASA Symp. on Solar Flares*, ed. W. N. Hess (NASA SP-50), 425 (1964).

Yokoyama, T., Akita, K., Morimoto, T., *et al.*, Clear evidence of reconnection inflow of a solar flare, *ApJ*, **546**, L69 (2001).

DRIFTING PULSATIONS, 3 GHz OSCILLATIONS AND LOOP INTERACTIONS IN THE JUNE 6, 2000 FLARE

M. Karlický[1], H. S. Sawant[2], F. C. R. Fernandes[2], J. R. Cecatto[2], F. Fárník[1], and H. Mészárosová[1]

[1]*Astronomical Institute of the Academy of Sciences of the Czech Republic, CZ-25165 Ondřejov, Czech Republic*
[2]*Instituto Nacional de Pesquisas Espaciaias, INPE, C.P. 515, 12201-970 Sao Jose dos Campos, SP, Brazil*

ABSTRACT

The radio observations of the June 6, 2000 flare reveal two impulsive phases, 15:06-15:40 UT and 16:26-16:40 UT. At the beginning of both these phases, i.e. at 15:06:46-15:07:00 UT and 16:26:34-16:26:42 UT, two drifting pulsation structures (DPS) were observed in the 1.0-1.7 GHz range. Furthermore, during both these phases the quasi-periodic oscillations with periods of 160 and 11 s, respectively, were observed at 3 GHz. The first and main impulsive phase was also characterized by the 2-4.5 GHz broadband pulsations and the continuum, during the second impulsive phase the 2.5-3.5 GHz pulses, consisting of narrow band spikes, were observed. The images obtained with *SOHO*-EIT depict parallel flare loops whilst those from *Yohkoh*-SXT show a bright source located at their top which may originate between them. We have therefore interpreted the high-frequency 3 GHz oscillations as those in this loop system using the current-loop coalescence model. On the other hand, the DPSs are explained as in our previous papers, i.e. as the radio manifestation of the plasmoid ejection.

INTRODUCTION

There are long-duration flares that reveal more than one impulsive phase. Recently, for this type of flares it was found that drifting pulsation structures (DPSs) are observed at the onsets of the impulsive phases (Karlický *et al.* 2001). In the October 5, 1992 flare the DPS occurred during a plasmoid ejection (Kliem *et al.* 2000). On the other hand, there are several observations showing loop interactions (e.g. Šimberová *et al.* 1993). Both these features were observed during the June 6, 2000 flare.

OBSERVATIONS

The June 6, 2000 flare, classified as X2.3, was observed at 15:00-17:00 UT in the active region NOAA AR 9026. A full-halo coronal mass ejection and a type II burst were reported in association with this event. During the June 6, 2000 flare two impulsive phases were observed. Quasi-periodic oscillations with characteristic periods of about 160 s during the first phase and about 11 s during the second one were registered at 3 GHz. During the first phase (15:06-15:40 UT) the 2-4 GHz radio spectrum consists of broadband pulsations and continuum; during the second phase (16:26–16:40 UT) the pulses (2.5-3.3 GHz, duration $\sim$ 5s) consisting of many narrow band spikes were observed. These pulses correspond to quasi-periodic peaks observed at 3 GHz. In the 1.0-1.7 GHz frequency range the radio emission started with a group of type III-like radio bursts at 15:06:10-15:15:06:30 UT and with the DPS at 15:06:46-15:07:00 UT.

During the interval 15:15-15:40 UT, fibers and zebra patterns were observed. Just at the beginning of the second phase at 16:26:34-16:26:42 UT the second DPS was recorded. This flare was also observed by the *Yohkoh*-SXT and the *SOHO*-EIT. Their images, presented in Figure 1, are consistent with an interaction of parallel flare loops explaining the high-frequency 3 GHz oscillations.

INTERPRETATION

The DPSs were observed at the onsets of both impulsive phases as reported earlier for similar events by Karlický *et al.* (2001). The DPSs are therefore interpreted as the radio manifestations of the electrons accelerated in the current sheet during the reconnection process, connected with a plasmoid ejection. Two DPSs indicate two disruptions of the magnetic rope. After these processes, in the loop system (Figure 1) below the current sheet the parallel loops started to interact. At these times the oscillations at 3 GHz were recognized. These facts agree with the Tajima *et al.* (1987) model of interacting loops.

In such a model the observed 3 GHz oscillations are due to the periodic coalescence of current-carrying loops. Their period (Tajima *et al.* 1987) is: $T = 2\pi(-2E)^{-3/2}t_A^{-2}$, where $t_A = \lambda/v_A$, λ is the magnetic field scale length, E is the initial "energy" of the system, and v_A is the Alfvén speed. Assuming roughly that the zebra frequency corresponds to that of the plasma frequency, we estimate the electron density to be $\sim 1 \times 10^{11}$ cm^{-3} and the Alfvén speed $\sim$ 1000 km s^{-1} within the interaction region. Now, for λ = 10000 km, the initial "energy" E of the system is estimated to be – 2.68 × 10^{-3} s^{-2} in the main impulsive phase, and – 1.59 × 10^{-2} s^{-2} in the secondary one. The higher value of the initial "energy" E in the first phase means that at this phase a deviation of the loop system from the equilibrium state was greater than in the second phase.

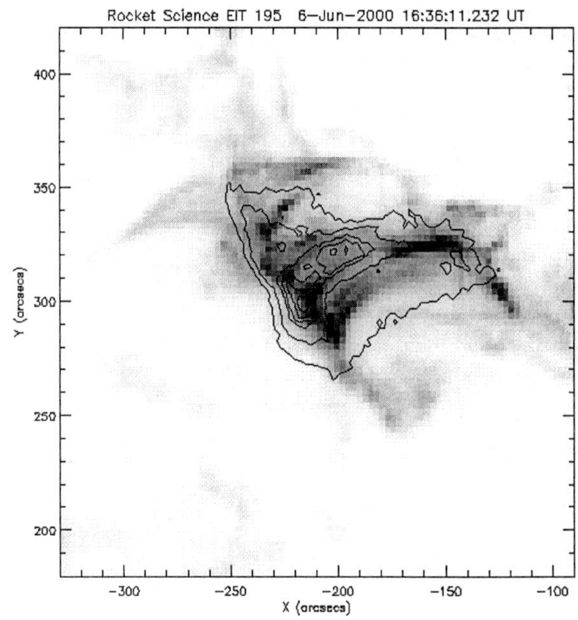

Fig. 1. The *Yohkoh*-SXT image (16:36:41 UT, contours) superimposed on the *SOHO*-EIT image at 16:36:11 UT.

ACKNOWLEDGEMENTS

M. K. thanks FAPESP authorities for supporting his visit to INPE (P.N. 01/00144-5). This work was also supported by the grant S1003006.

REFERENCES

Karlický, M., Y. Yan, Q. Fu, S. Wang, K. Jiřička *et al.*, *Astron. Astrophys.* **369**, 1104 (2001).

Kliem, B., M. Karlický, & A.O. Benz, *Astron. Astrophys.* **360**, 715 (2000).

Šimberová, S., M. Karlický, & Z. Švestka *Solar Phys.* **146**, 343 (1993).

Tajima, T., J. I. Sakai, H. Nakajima, T. Kosugi, F. Brunel *et al.*, *ApJ* **321**, 1031 (1987).

A STUDY OF MAGNETIC RECONNECTION USING SIMULTANEOUS SOHO/MDI AND TRACE DATA

J. L. R. Saba[1], T. Gaeng[2], and T. D. Tarbell[3]

[1]*Lockheed Martin Solar & Astrophysics Lab., NASA/GSFC, Code 682.3, Greenbelt, MD 20771, USA*
[2]*L-3 Communications Analytics Corp., NASA/GSFC, Code 682.4, Greenbelt, MD 20771, USA*
[3]*Lockheed Martin Solar & Astrophysics Lab., Bldg. 252, Org. L9-41, 3251 Hanover Street, Palo Alto, CA 94304, USA*

ABSTRACT

High resolution, high cadence images from the *Transition Region and Coronal Explorer* (*TRACE*) together with high quality magnetograms from the Michelson Doppler Imager (MDI) on the *Solar and Heliospheric Observatory* (*SOHO*) let us examine signatures of magnetic reconnection and attempt to infer associated physical parameters such as the electric field strength in the solar corona. We analyzed *TRACE* UV and MDI magnetogram data for a two-ribbon, *GOES* M1 class flare from NOAA active region 9236 at 2000 Nov 23 23:28 UT, with emphasis on dynamical development of the 1600 Å band flare ribbons in the *TRACE* images; we estimated the magnetic reconnection rate from the change in photospheric magnetic flux swept out by the evolving ribbons. For simple assumptions with standard coronal parameters, the reconnection appears to be fast, with the inferred inflow velocity a significant fraction of the Alfvén velocity. Some guidance from coronal imaging of the reconnection region or Doppler measurement of inflow is needed to sharpen the constraints on the length of the reconnecting current sheet and the coronal electric field.

DATA AND ANALYSIS

The primary data for this analysis are: (1) a pair of MDI high-resolution (0.6054 arcsec px) preflare photospheric images – a 5-min-average magnetogram and a continuum filtergram image at $\lambda \sim$ Ni I 6768 Å, (2) a pair of *TRACE* full-resolution (0.5 arcsec px) white light (WL) and 1600 Å band preflare images, and (3) three series of 200 high-cadence (1.45 s), 256 x 512 px, 1600 Å flare-response images, each centered on the brightest flare pixel at the start of the given series. The *TRACE* data contain no EUV/coronal lines; hence, *SOHO*/EUV Imaging Telescope (EIT) full disk coronal images (2.62 arcsec px) at $\lambda \sim$ 195 Å from before and during the flare are used for context and modeling; no data from the *Yohkoh* Soft X-ray Telescope were available. The data were reduced and analyzed and binary ribbon masks created using the Interactive Data Language (IDL) and SolarSoft libraries; movie cubes were created using ana/browser. The MDI continuum and *TRACE* WL images, and the *TRACE* WL and 1600 Å band images, were aligned in the flare vicinity by iterative blinking and shifting. The MDI continuum and magnetogram images are co-registered *ab initio* to much better than a pixel.

To estimate the magnetic reconnection rate, we chose two obvious flare ribbons which grew rapidly in the first 290 s sequence of high cadence 1600 Å flare images. These ribbons could be separated with a simple binary mask from ejecta and other emission. They were located on strong fields of opposite polarity and grew rapidly, then faded in place, suggesting emission low in the atmosphere and well aligned with the photospheric footpoints of field lines reconnecting in the corona. Thus we assume that the reconnection rate can be determined from the changing photospheric magnetic flux swept out by the evolving ribbon mask.

RESULTS AND DISCUSSION

The initial estimates of reconnection rates for the rise phase of the flare may be summarized as follows: (1) the reconnection rate is very noisy, with little correlation between time steps 1.45 s apart; (2) the peak instantaneous rate R_{peak} (summed over the whole flare ribbon) is $\sim 5 \times 10^{18}$ Mx/s; (3) the average rate over the whole ribbon for the 290 s interval is $\sim 5 \times 10^{17}$ Mx/s; and (4) the peak local rate is $\sim 2 \times 10^{18}$ Mx/s. A simple estimate of the uncertainties obtained by shifting the magnetograms by 2 px and redoing the mask-creation steps and statistical analysis yields results within $\sim 20\,\%$ of the original values.

For a coronal magnetic field strength B $\sim$ 100 G in the reconnecting region, R_{peak} implies that all flux through an area of 5×10^7 km^2 (e.g. $5000 \times 10\,000$ km^2) perpendicular to the magnetic field **B** must reconnect in 10 s. Suppose there is an inflow velocity **V** bringing flux into the reconnection region (current sheet) between the two polarities. Let **L** be the length scale along the current sheet, perpendicular to both **V** and **B** (i.e. **L** is essentially the segment of the perimeter of the area which reconnects, along the interface between opposite polarities). Then VL = $R_{peak}/B \sim 5 \times 10^6$ km^2/s. An estimate of L would give us an estimate of V, for comparison with the Alfvén speed V_A and for an estimate of the electric field strength, E $\sim$ VB/c. For B $\sim$ 100 G and a typical active region coronal density $n_e \sim 3 \times 10^9$ cm^{-3}, $V_A \simeq 2.2 \times 10^{11} B\,(G)/\sqrt{n_e} \sim 4.0 \times 10^3$ km/s. Since the flare ribbons contain roughly 2 % of the flux in the active region (i.e. $\sim$2 % of the flux which in principle could reconnect along the neutral line has in fact reconnected), we might suppose that L is of order the same fraction of the length of the neutral line. This gives L $\sim$ 2000 km, V $\sim$ 2500 km/s, and E $\sim$ 0.8 cgs units. Thus the inflow velocity would be a significant fraction ($\sim$0.6 for our chosen parameters) of V_A, implying fast reconnection. With this value of L, even the average reconnection rate implies V $\sim$ 250 km/s, a few percent of V_A. Or, based on the bursty time dependence of the reconnection rate, we might assume L $\sim V_A \tau_c$, with τ_c the correlation time of the reconnection rate (a few seconds or less). for τ_c and the values above for B and n_e. For the peak reconnection rate, V $\sim$ 400 km/s ($\sim$ 0.1 times V_A) and E $\sim$ 0.13 cgs units, again implying fast reconnection.

Unfortunately, the geometry (or even topology) of the reconnection region in the corona is not known. Real flares in *TRACE* movies do not look much like the 2.5-D cartoons frequently used to analyze two-ribbon flares. The region could be highly corrugated, or even multiply connected with domains of opposite polarity intermingled on very small scales, approaching the diffusion scale. If present, these effects could increase L (and reduce V) by orders of magnitude. Some guidance on the reconnection region can be obtained from EIT coronal imaging, which indicates that the field lines connecting the studied ribbons are *not* aligned with the line of sight. Simple modeling using the potential field code of T. Metcalf also yields angled field lines. To improve our estimate of the coronal electric field strength using this approach, we need higher resolution coronal images, as well as better theoretical models, and possibly Doppler measurements of the inflow.

ACKNOWLEDGEMENTS

We thank S. Freeland for help with IDL/SolarSoft, R. Shine for help with ana/browser and creating movie cubes, and T. Metcalf for help with his potential field code. We acknowledge use of K. Reeves' on-line *TRACE* flare catalog and the *TRACE* searchable database made by D. Myers, M. Despres, and J. Marbourg. This work was supported by NASA contracts NAG5-8878, NAG5-10483 (MDI), and NAS5-38099 (*TRACE*).

3D STRUCTURE OF A MAGNETIC RECONNECTION JET: APPLICATION TO LOOPTOP HARD X–RAY EMISSION

S. Tanuma[1], T. Yokoyama[2], T. Kudoh[3], and K. Shibata[1]

[1]*Hida and Kwasan Observatories, Kyoto University, Yamashina, Kyoto 607-8471, Japan*
[2]*Nobeyama Radio Observatory, Minamimaki, Minamisaku, Nagano 384-1305, Japan*
[3]*Department of Physics and Astronomy, University of Western Ontario, London, Ontario, Canada N6A 3K7*

ABSTRACT

To study the structure of a reconnection jet, we perform 3D resistive MHD simulations of magnetic reconnection triggered by the secondary tearing instability under simple assumptions. The results of 2D simulations have already revealed that fast reconnection occurs after current sheet thinning by the secondary tearing instability, and that the plasmoid ejection determines the reconnection rate. By performing 3D simulations, we find that the current sheet thinning occurs the same as with the 2D models. The Rayleigh-Taylor (RT) instability occurs in the reconnection jet, when the reconnection jet collides with the magnetic loop and high pressure gas created by the magnetic reconnection. The 3D structure, especially the helical magnetic field, appears because of RT instability. In the actual Sun, the fast reconnection will occur after the fractal tearing instability, although the magnetic Reynolds number is very large. The hot gas and high energy particles created by the reconnection may be confined by the helical and turbulent magnetic field, which will be created by the RT instability at the flare loop top.

NUMERICAL SIMULATIONS AND RESULTS

We assume a parallel uniform magnetic field, an anti-parallel uniform magnetic field, and the current sheet between them as the initial condition. In the 2D models, we assume a point explosion outside the current sheet (Odstrčil & Karlický 1997; Tanuma *et al.* 2001). It is revealed that the magnetic reconnection does not occur immediately after the passage of the fast MHD shock wave (i.e. the explosion is only a perturbation for the current sheet). The current sheet, however, evolves according to the following phases: (i) The tearing instability occurs, and the current sheet becomes thin. (ii) The secondary tearing instability occurs in the thin current sheet. (iii) The plasmoid, which is created by the secondary tearing instability, is ejected. Immediately after the ejection, the inflow velocity increases, so that the current sheet becomes thin. As a result, anomalous resistivity sets in, and Petschek reconnection occurs.

In the 3D models, we assume a point explosion and a cylindrical one outside the current sheet. The results of 3D models, such as how the current sheet evolves and how the fast shock is created, are similar to those of the 2D models. The Rayleigh-Taylor instability, however, occurs after the collision between the reconnection jet and high pressure gas, which has been created by the magnetic reconnection (Figure 1). The 3D structure, then, appears at the top of the magnetic loop.

DISCUSSION

In the actual flare, the fast reconnection will occur after the fractal tearing instability (Shibata & Tanuma 2001). The reconnection jet collides with the magnetic loop, and creates the fast shock region. A helical magnetic structure, as well as a random one (Jamiec *et al.* 1998), will be created at the flare loop top, if the Rayleigh-Taylor (RT) instability occurs due to this type of collision. Furthermore, the Richtmyer-Meshkov instability may also occur at the head of the reconnection jet before the collision (Nakamura *et al.* 2001).

Hard X-ray emissions are detected at the loop top in impulsive flares such as the Masuda (1994) flare. Hot gas and high energy particles may be confined by the helical magnetic field, created by the RT instability at the loop top (Figure 2).

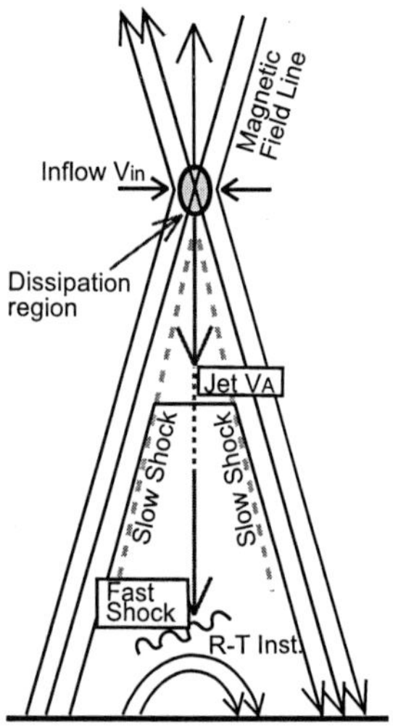

Fig. 1. Schematic illustration of the results of 3D simulations of magnetic reconnection. The reconnection jet collides with the magnetic loop created by the reconnection. The velocity of the reconnection jet decreases quickly, so that the Rayleigh-Taylor instability occurs at the flare loop top.

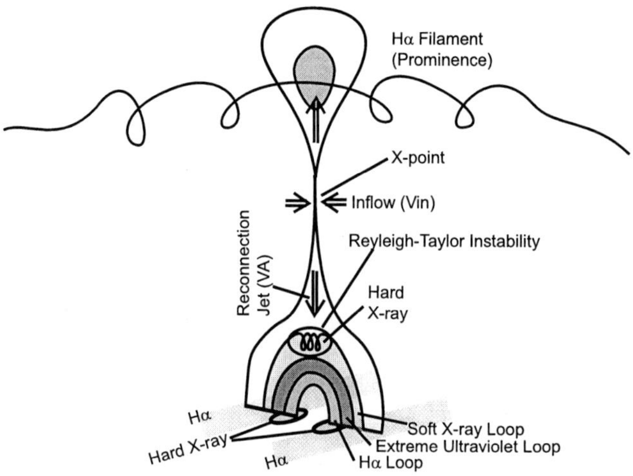

Fig. 2. Schematic illustration of the Rayleigh-Taylor(RT) instability of the reconnection jet in the solar flare. The RT instability will occur because of the collision between the reconnection downflow and magnetic loop, which has been created by the reconnection. The helical magnetic field will be created by the RT instability. High energy particles will be confined by the magnetic field at the loop top. It corresponds to the hard X-ray source of impulsive flares such as the Masuda (1994) flare.

REFERENCES

Jamiec, J. *et al.*, *A&A*, **334**, 1112 (1998).

Masuda, S. *et al.*, *Nature*, **371**, 495 (1994).

Nakamura, M. S. *et al.*, in *ISSS-6*, **303** (2001).

Odstrčil, D. & M. Karlický, *A&A*, 326, 1252 (1997).

Shibata, K. & S. Tanuma, *Earth, Planets, and Space*, **53**, 473 (2001).

Tanuma, S. *et al.*, *ApJ*, 551, **332** (2001).

Section VI.

MHD Simulations of Emergence and Eruptions

MODELS OF ARCADE FLARES IN VIEW OF OBSERVATIONS BY YOHKOH, SOHO/EIT, AND TRACE

S. Hirose and Y. Uchida

Department of Physics, Tokyo University of Science, 1-3 Kagurazaka, Shinjuku-ku, Tokyo 162-8601, Japan

ABSTRACT

High-quality observations by *Yohkoh*, *SOHO*/EIT, and *TRACE* have indicated that arcade flares are energy release phenomena through magnetic reconnection. Our next interest is to clarify the global situation of the magnetic field in which the magnetic energy is stored and released. Such a global model should also explain the dark filament support and its eruption associated with arcade flares. In this paper, we briefly review global models proposed thus-far and discuss them in relation to some essential observational features. One of the promising global models is "quadruple magnetic source model", in which a dark filament with longitudinal magnetic field is supported in the current sheet between two magnetic loop systems and prevents magnetic reconnection. The results of MHD simulation based on this model show that the squeezing out of the dark filament allows reconnection of the two magnetic loop systems, and the relation between the dark filament eruption and the following arcade flare can be given natural explanation.

INTRODUCTION

Arcade flares (including large-scale X-ray arcade formation events outside active regions) show characteristic features like cusp formation and the widening of the loci of the arcade footpoints. They are considered to be energy release process through magnetic reconnection in the "inverse Y-shape" configuration (e.g. Shibata 1995). Actually this *partial* view has been supported by the observations with *Yohkoh*, *SOHO*/EIT, and *TRACE*, but this is not the whole story.

Although we are aware that there remain unsolved problems with the reconnection process in a current sheet itself, we confine ourselves here to *global* models which describe how such a current sheet is formed in the global configuration of the magnetic field. Since a filament eruption is associated with arcade flares, the global model needs to explain in a consistent way the support of the dark filament before the eruption, and the relation between the dark filament eruption and the magnetic reconnection.

In the present paper we classify *global* models for eruptive arcade flares proposed thus far with their global magnetic configuration having the reconnection site within it, and briefly review them from the viewpoint of the storage and release of the magnetic energy. Then we proceed to discuss our "quadruple magnetic source model" with the results of MHD simulations based on it, which is one of the most promising global models favored by recent observations by *Yohkoh*, *SOHO*/EIT and *TRACE*.

GLOBAL MODELS PROVIDING THE SITE OF MAGNETIC RECONNECTION

CSHKP Model with a Single Bipolar Arcade

A model for arcade flares based on the early works by Carmichael (1964), Sturrock (1966), Hirayama (1974), and Kopp & Pneuman (1976) has been widely accepted by many people. In this so-called "CSHKP model", dark filament gas is suspended in a dip created by the weight of itself at the top of a bipolar magnetic arcade[1]. A current sheet is formed and the magnetic energy is stored when the closed magnetic arcade is opened up due to the rise of the dark filament. The stored energy is then released through magnetic reconnection, and heated arcade fields form through the reclosing process.

This idea is a straightforward explanation of the fact that the dark filament is supported above the polarity-inversion line and that the dark filament eruption is associated with the arcade flare, but it has some serious problems. Since the energy of the field opened up by the dark filament eruption is larger than that of the initial closed bipolar field (Aly 1991), the transition from the latter state to the former state cannot be spontaneous. How the dark filament can suddenly attain such a large amount of kinetic energy to rise, cutting through a strong part of the field that has supported it, is a difficult problem energetically. Another difficulty of the "CSHKP model" is that it cannot explain the observations of magnetic field in dark filaments seen from the side by Leroy *et al.* (1984) (i.e. the "inverse polarity" problem).

Bipolar Arcade Field Subject to Shearing or Twisting Footpoint Motions

One of the contexts for models with a bipolar arcade subject to shearing or twisting footpoint motions, is the explanation of prominence support. Priest *et al.* (1989) proposed a twisted flux tube model for prominences; when an isolated arcade flux tube is twisted sufficiently individual field lines acquire a convex curvature near their summits and provide support for prominence material. Amari *et al.* (1999) showed using three-dimensional MHD simulations that such twisted magnetic flux is generated in dipole loops subject to twisting footpoint motions. On the other hand, DeVore & Antiochos (2000), following Antiochos *et al.* (1994), numerically simulated a bipolar arcade subject to shear motions and showed the formation of a dipped, helical structure to support the prominence.

Another context is the energy buildup in the system and its release through magnetic reconnection for eruptive flares. Mikić *et al.* (1988) performed two-dimensional MHD simulations in which a laterally restricted dipole arcade is subjected to a shearing footpoint motion, and found plasmoid formation and upward ejection of it through magnetic reconnection. Based on the Woltjer-Taylor theorem, Kusano *et al.* (1995) showed that the minimum energy state in this type of restricted field geometry bifurcates into two different states, and that the energy release through magnetic reconnection is a transition process between the two states. On the other hand, Biskamp & Welter (1989) treated unrestricted systems and found that the eruptive development through reconnection is possible only when lateral restriction of adjacent arcades exists. Choe & Lee (1996a, 1996b) treated the evolution of similar single arcade undergoing shearing footpoint motions, but separated the ideal MHD evolution and the resistive evolution to see the effect of resistivity clearly.

From the observational point of view, however, there is no clear evidence for such *long-distance* shearing motion along the polarity inversion line. Morita *et al.* (2001) actually showed the absence of such a long-distance re-shearing between the successive events in a homologous flare series of 1992 February.

[1]This model is usually referred to as the Kippenhahn-Schluüter (1957) model, while CSHKP is normally reserved for a "flux rope system" (the editors).

Flux Rope System Losing Equilibrium

Many authors assume a magnetic configuration with a detached flux rope above the arcade for the energized state before the eruptive energy release.

Isenberg & Forbes (1993) proposed catastrophic evolution of a force-free flux rope for eruptive flares; they considered a force-free flux rope which is held-down at the photosphere by the overlying field. The gradual disappearance of the overlying field finally causes the flux rope to lose equilibrium and jump to a higher altitude of lower energy state (impulsive phase), which is followed by reconnection at the current sheet formed below the flux rope (gradual phase). Forbes & Priest (1995) considered a similar case, in which the loss of equilibrium occurs due to the buildup of magnetic pressure below the flux rope, not due to the disappearance of the holding-down fields.

Chen & Shibata (2000) proposed another trigger mechanism for the loss of equilibrium, reconnection-favored emerging flux, using two-dimensional MHD numerical simulations; when emerging flux appears within the filament channel, it cancels the magnetic field below the flux rope, leading to the rise of the flux rope owing to loss of equilibrium. Then magnetic reconnection at a current sheet formed below the flux rope induces the fast ejection of it. Amari *et al.* (2000) performed a three-dimensional MHD simulation and showed that a highly nonlinear force-free configuration consisting of a twisted magnetic flux rope cannot stay in equilibrium and is disrupted through a viscous relaxation process.

Although models of this type with flux rope have many interesting features and also are favorable for the inverse polarity problems, it has a difficulty in explaining the fact that dark filaments have "barbs", or legs connected to the solar surface (Martin *et al.* 1994). Also the mechanism of formation of such flux ropes themselves may still be an open question.

Models with a Quadruple Source Field Configuration

Another category of global model is a model with quadruple source configuration of the magnetic field.

This type of configuration was first mentioned by Sweet (1958), and was advocated by Uchida (1980) as the "quadruple magnetic source model" to avoid the energy difficulty of the "CSHKP model" discussed in the above. This was revisited by Uchida *et al.* (1999b) according to what they found based on careful examination of *Yohkoh* data (Uchida 1996, 2000, Uchida *et al.* 1999a). In this model a dark filament containing longitudinal magnetic fields is supported in a thin current sheet between two magnetic loop systems. An important point of this model is that the energy can be built up in the system within the vertical current sheet *before* the dark filament eruption due to squeezing. The dark filament acts to prevent the energy release by suppressing magnetic reconnection by its presence in the equilibrium state.[2] The strong part of the bipolar connections is already opened by the effect of the outer pair of the sources, which makes the dark filament eruption easier. The elongated magnetic structure along the dark filament is due to the dominance of the longitudinal magnetic field in the "neutral sheet" of the perpendicular component of the magnetic field, and thus a long-distance mechanical shearing, which is observationally absent, is not needed here. Based on this quadruple magnetic source model, Hirose *et al.* (2001) performed 2.5-dimensional MHD simulations and showed that (i) the energized state of the system is kept intact by the presence of the dark filament, and that (ii) the dynamical evolution of the system via magnetic reconnection occurs as the dark filament is squeezed out. These results will be presented in the section 4, after a summary in section 3 of the results of observations from *Yohkoh* and other satellites, supporting this model.

[2]In the "CSHKP model", the dark filament rise needs to open up the strong magnetic field, and when this is not possible due to energy paradox, no current sheet will be formed.

Recently the quadruple magnetic source characteristics have attracted greater attention. Vekstein *et al.* (1991) showed that shearing motions of a two-dimensional double arcade in general produce a current sheet along the separatrix. In Démoulin & Priest (1993) the prominence is modeled as such a current sheet with mass in equilibrium in a quadrupolar magnetic configuration.

Using MHD simulation, Cheng & Choe (1998) showed that an equilibrium field configuration containing a current sheet in a quadrupolar field geometry can be deformed into a configuration in which an inverse polarity prominence can stably reside. Galsgaard & Longbottom (1999) numerically treated the three-dimensional case and found that the reconnected field lines between two bipolar regions are able to lift plasma several pressure scale heights against gravity.

Other works treating the quadruple source situation are Karpen *et al.* (1995), who perform 2.5-dimensional MHD simulations of reconnection between two line dipoles in the context of chromospheric eruptions, and the "breakout" model of Antiochos *et al.* (1999) in which the dipole arcade subjected to shearing motion breaks the force balance of the quadruple source field, leading to a runaway eruption.

OBSERVATIONS SUGGESTING A QUADRUPLE SOURCE SITUATION

Observations by *Yohkoh* SXT and *SOHO* EIT of Loop Configurations Surrounding Dark Filaments

As discussed above, we claim, based on recent observation, that the quadruple magnetic source model is one of the most promising models for arcade flares. In this section, we discuss *Yohkoh* SXT observations which support the quadruple source situation (Uchida *et al.* 1996, 1999a). Morita *et al.* (2001) showed very clear evidence in which the rising dark filament is pulled out from the dip of the expanding loops in the shape of the *McDonald's* logo, rather than pulling up loops into a cusped shape as expected in the CSHKP model (also see Morita *et al.* in these proceedings).

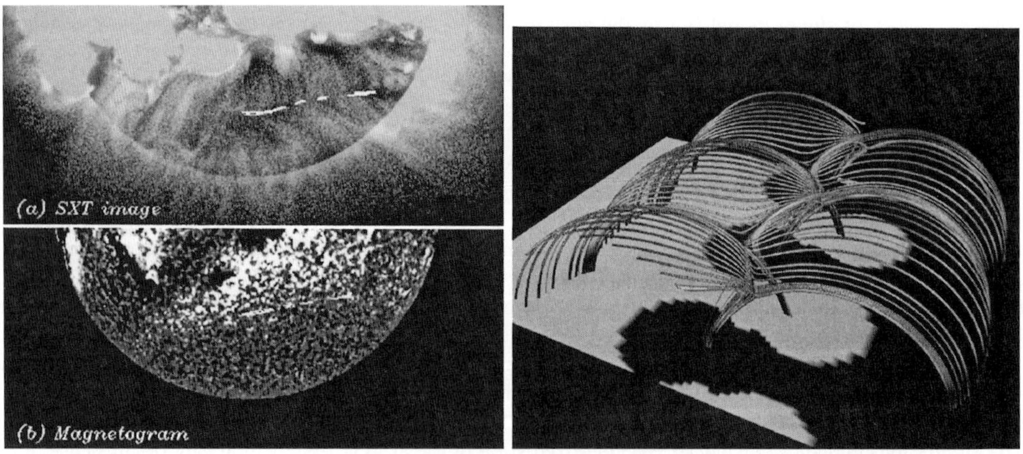

Fig. 1. *Left*: Coronal structure surrounding a dark filament in the pre-flare stage (February 18 1992). (a) *Yohkoh* SXT image. (b) Kitt Peak magnetogram image. The white curve is the locus of the top of the dark filament. *Right*: A 3D model with intermingled configuration for a case with exchanged patches of opposite polarity sources. (Uchida *et al.* 1999b).

Detailed information about the structure of the magnetic field surrounding a dark filament in the preflare stage is vital for clarifying the flare mechanism, and this was first obtained in a satisfactory form by *Yohkoh.* Figure 1 (left(a)) is a *Yohkoh* SXT image of such a coronal pre-event structure around a high-latitude dark filament. The loop configuration looks like a "dual arcade", one on each side of the polarity reversal line in the photosphere (Figure 1, left(b)). Furthermore the inside legs of each arcade land in the domain of the other arcade in an intermingled way, and the dark filament lies in the region in between where the legs of both arcades are crossing. These characteristic features are those of the case with exchanged patches of opposite polarity sources (Figure 1 (right)), which corresponds effectively to *quadruple* array of sources. Note that we can actually see this kind of exchanged patches in the magnetogram (Figure 1 (left(b)), where the "polarity inversion line" is not a clearly definable line, but there exists a belt of mixed polarity with many exchanged opposite polarity patches. This has been more definitively shown by Kichiraku *et al.* (in preparation) by examining the data from *SOHO*/EIT that the magnetic loop structure surrounding the high latitude dark filament is very clearly of a quadruple shape, corresponding to the configuration seen in Figures 3 and 4 (our numerical results).

Examples of Flares Observed by *TRACE*

Uchida *et al.* (in preparation) found that the detailed analysis of the flare of July 19 1999 gave evidence for our view that the arcade flares are basically quadruple in character. Interested readers are referred to our paper, and here we show two figures, Figure 2(a), and Figure 2(b). Figure 2(a) shows the magnetic field configuration in the photosphere around the region, showing the four magnetic regions involved in this event. Figure 2(b) shows the *TRACE* image of the flare in the 195 Å band at about 30 minutes after the dark filament eruption. A flare arcade develops only between sources B and C, and the overall structure connecting the sources A to D has gone at the time of the dark filament eruption, but the loops connecting the sources A to B, and C to D, remain. The high-lying loops in those closed regions on both sides *are moving towards the central line*, to each other. This is exactly the behavior predicted and actually seen in our quadruple source model (Figure 4 (model simulations)) as shown in the following section.

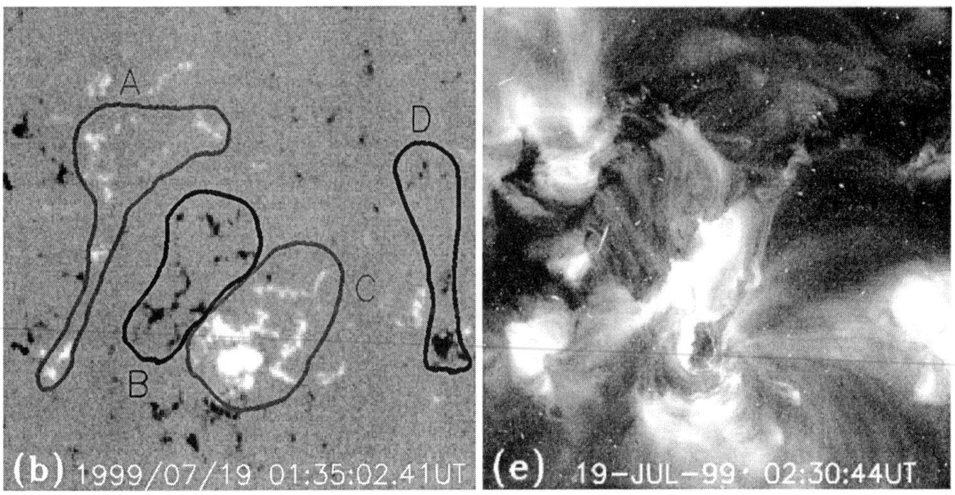

Fig. 2. *Left*: MDI magnetogram of 013502 UT showing four magnetic regions involved in the event. *Right*: *TRACE* 195 Å image of 023044UT showing that the higher parts of the flux tubes connecting A to B, and D to C are approaching each other.

MHD SIMULATION BASED ON THE QUADRUPLE MAGNETIC SOURCE MODEL

In this section, we briefly introduce our 2.5D MHD simulations (Hirose *et al.* 2001) based on the quadruple magnetic source model (Uchida 1980, Uchida *et al.* 1999a).

Energized State Before the Filament Eruption

We start the simulation with a current-free field due to the quadruple magnetic source belts plus the uniform field parallel to the polarity inversion line. Although in the initial model the strength of the field component parallel to the polarity inversion line was only a few percent of that of the perpendicular component, the parallel component dominates near the neutral sheet of the perpendicular field. The background plasma is isothermal and hydrostatic against constant gravity. To store magnetic energy in the system, we apply slow (about 1% of local Alfvén velocity) converging footpoint motions toward the central polarity inversion line. We also put a plasma blob between the two magnetic loop systems as the dark filament to be formed.

We arrive then at an energized quasi-equilibrium state of the system (Figure 3); The input plasma is distributed in the current sheet formed between the two magnetic loop systems, and is supported as a thin vertical sheet, preventing direct contact between the two magnetic loop systems. This "dark filament" in our numerical result has some observational features: it contains longitudinal fields, the origin of which are magnetic flux trapped between the two magnetic loop systems while these two systems are pressed together. The plasma reaches the photosphere along the arcade loops below (see also Figure 5(a), t=23.0). This leg part of the dark filament is actually observed as a "barb" in H_α (Martin *et al.* 1994). We note that the energy has already been stored in the system at this stage before the filament eruption. The presence of the dark filament plays an important role here in *preventing* energy release through magnetic reconnection throughout the energy build up phase.

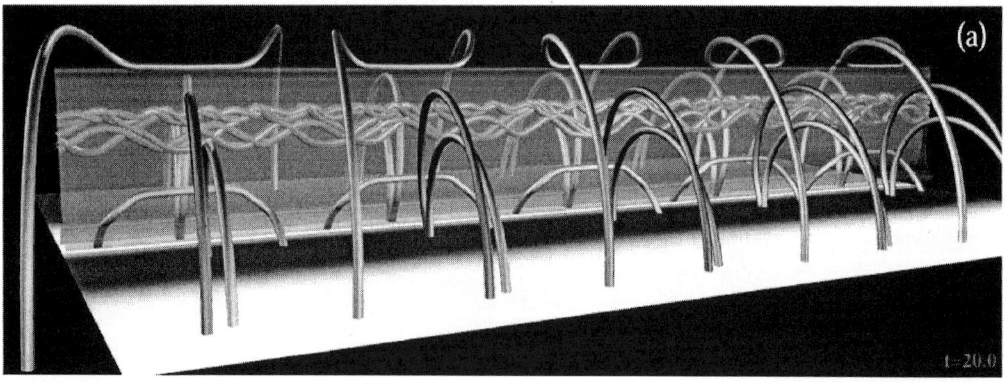

Fig. 3. 3D representation of the quasi-static energized state, where the dark filament in the current sheet between two magnetic loop systems is preventing magnetic reconnection. Colored tubes and transparent surface represent magnetic field lines and an iso-density surface, respectively. The grayscale on the bottom plane represents the magnetic polarity on the plane. Only selected field lines are drawn for clarity. (Hirose *et al.* 2001)

Energy Release After the Filament Eruption

We continue the converging footpoint motions until the dark filament is squeezed out from the current sheet, and we see the following dynamical evolution of the system with energy release through magnetic reconnection (Figure 4).

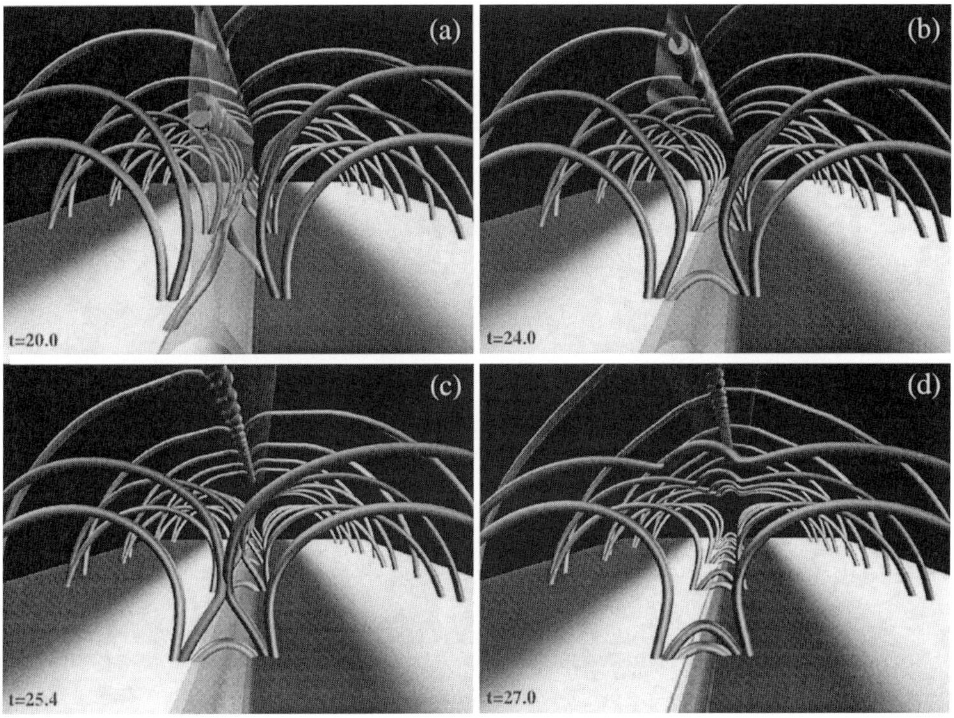

Fig. 4. 3D representation of the dynamical evolution via magnetic reconnection; the dark filament eruption and the associated arcade formation. Notations are same as in Figure 3, but the number of field lines drawn is reduced for the sake of clarity. (Hirose *et al.* 2001).

It is seen that the dark filament plasma is split into a higher and lower part, and they are expelled from the current sheet due to the magnetic pressure from both sides. Then the anti-paralleled field lines come into direct contact, and anomalous resistivity sets in there, which leads to magnetic reconnection between the two magnetic loop systems (we assume that anomalous resistivity is induced where j over ρ exceeds some threshold).

Note that the low central part of the system is *exactly* the same as the inverse Y-shaped magnetic configuration in the observed arcade flares discussed in section 1 (Figure 5). In the region below the reconnection point, the reconnected field lines shrink to form the magnetic arcade, onto which newly reconnected magnetic field lines having cusps continue to shrink. On the other hand, in the upper region plasma acceleration occurs at a V-shaped slow-mode shock wave (Figure 5b), where the magnetic field lines confined below the separatrix surfaces with high magnetic stress are allowed to expand upward owing to the magnetic reconnection. The ejected dark filament plasma is accelerated upwards by the expansion of the magnetic flux

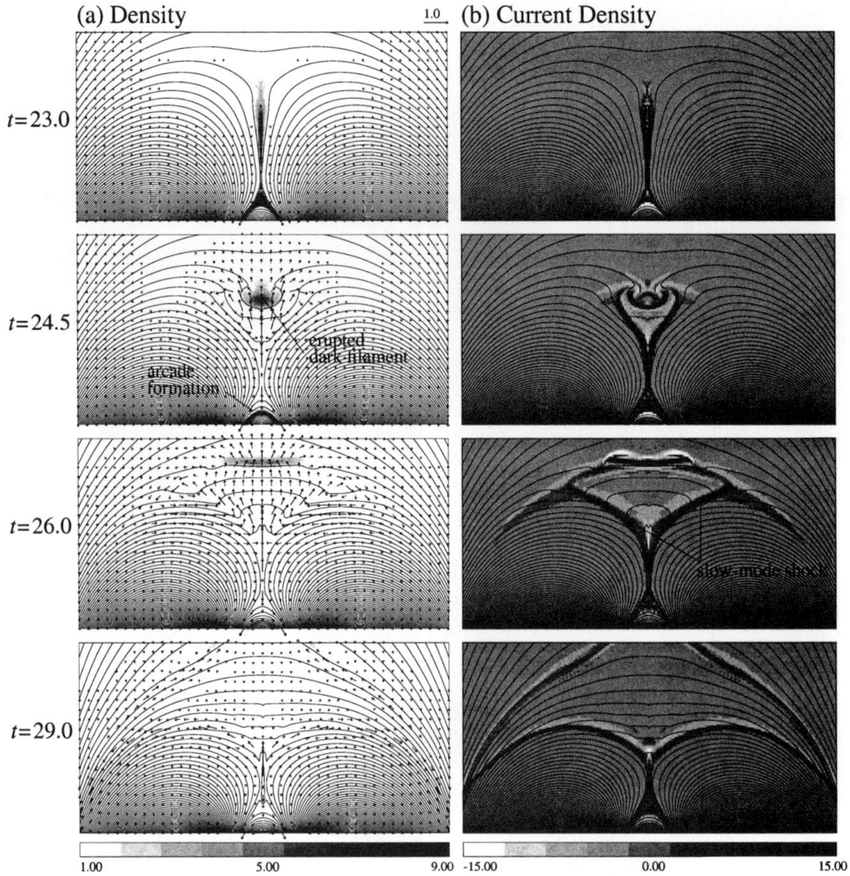

Fig. 5. Cross section of the system during the dynamical evolution. (a) Density (grayscale), velocity field (arrows), and magnetic field lines (solid lines). (b) Current density (grayscale) and magnetic field lines (solid lines). (Hirose *et al.* 2001).

released from the compressed region. In this case, most of the stored energy in the form of magnetic stress is converted to kinetic energy of the upward motion at the slow shock, and the rest is converted to arcade formation below. Through magnetic reconnection the energized state with the current sheet may approach a current free field, but not necessarily completely (Figure 5(b)).

SUMMARY AND DISCUSSION

We have reviewed current models for eruptive filaments from the point of view of their global configurations whether they can explain the energy storage and release aspects of the flare problem. That the "CSHKP model" has some problem in this point was noted by one of the present authors, and his alternative proposal was a quadruple magnetic source model (Uchida 1980). Careful examinations of the observational results from *Yohkoh*, and more recently from *SOHO* and *TRACE*, has shown that the events analyzed actually support this model (Uchida 1996, Uchida *et al.* 2000, Morita *et al.* 2001, Uchida *et al.* in preparation). As for the candidates for the sources forming a quadruple source configuration, there are many candidates in active regions (actually only those special ones with evolutionary relevance). We, however, worried first about the applicability of the quadruple magnetic source model to high latitude dark filament eruptions for which all the previous observers reported bipolar sources. We have spent some time examining these, and found that the magnetic structure surrounding high latitude dark filaments is *not at all of simple bipole* (see also Martens & Zwaan 2001). Magnetograms show that there exist many exchanged opposite polarity patches, and the loop structure above cannot be interpreted by simple bipolar regions but is explained well by the quadruple source model (Figure 1). Also recent examination of EIT images very clearly support quadruple magnetic source model for high latitude dark filament (Kichiraku *et al.*, in preparation).

We now discuss some of the remaining problems.

(i) Hirose *et al.* (2001) did a 2.5D simulation by continuing the slow squeezing of the footpoints, and the dark filament is squeezed out, leaving behind it opposite polarity loops in direct contact on both sides, and magnetic reconnection is allowed. We are now doing 3D simulations for realistic situations (Hirose *et al.* in preparation). In the actual process of destabilization, it is known that flux emergence near an otherwise very stable dark filament disturbs it, and the dark filament is slowly squeezed out, and starts rising rapidly when the arcade flare (magnetic reconnection) begins below. We are simulating this situation.

(ii) In the quadruple magnetic source model described in section 2.4 and 4, the rising dark filament does not have to cut the strong part of the magnetic field open as in the "CSHKP model", but still there is closed magnetic field above it. There can be a question whether those fields, although an order of magnitude weaker, might accumulate and eventually prevent the ejection of the dark filament. The case of a loop-type CME with a dark filament ejection, however, suggests that such overlying weak field is also already open because of the influence of the outlying sources, which makes the dark filament eruption easier (Uchida *et al.*, these proceedings).

(iii) There is a proposal for a unified flare model by Shibata (1996), but we believe that the mechanisms for small loop flares and active region transient loop brightenings are different from that for the large arcade flares described here. The former are likely to be due to dynamic helicity injections as seen from *Yohkoh* (Uchida *et al.* 2001, Miyagoshi *et al.* 2001). Also from the observations of active regions by *TRACE*, it is clear that there no previous deformation of the brightening loops occurs, and the brightening is due to the injection of the blob of emitting plasma coming up from the footpoint intermittently and moving along the fixed loop with a velocity of 500 km/s or so. These will supplement the energy-event frequency diagram obtained by Shimizu *et al.* (1996) from *Yohkoh* data.

(iv) The last and the most important point to make is that the physics of the local process of reconnection at the neutral sheet is still very far from being solved, more than 40 years after Sweet (1957). Many workers have addressed this problem, but none of them, including those using numerical simulations, has been completely successful thus far. Our intention in mentioning this here is that this should be clearly separated from advances in the *global* modeling problem which deals with the configuration that can store and release magnetic energy. The quadruple magnetic source model advocated in the above seems to be a most promising of the global models, but the solution of the process occurring in the local neutral sheet, which is the most important issue in the flare problem, has not yet been attained. The anomalous diffusivity has to be huge (or the diffusion regions impracticably small as in Petchek (1964)) so that the diffusive time scale is as short as the Alfvén time scale. This suggests that the traditional approaches to try to speed up the process by enhancing diffusion is wrong, and the initial rise of flares is likely to be *dynamic*, and not diffusive, in character. We will deal with this problem in the future on the basis of our global model.

REFERENCES

Aly, J.J., *ApJ Letters*, **375**, L61 (1991).

Amari, T., J.F. Luciani, Z. Mikic, and J. Linker, *ApJ Letters*, **518**, L57 (1999), *ibid* **529**, L49 (2000).

Antiochos, S.K., R.B. Dahlburg, and J.A. Klimchuk, *ApJ Letters*, **420**, L41 (1994).

Antiochos, S.K., C.R. DeVore, and J.A. Klimchuk, *ApJ*, **510**, 485 (1999).

Biskamp, D., and H. Welter, *Solar Phys.*, **120**, 49 (1989).

Carmichael, H., in *Proc. NASA Symp. on The Physics of Solar Flares*, ed. W.N. Hess, p. 451, NASA, Washington, DC (1964).

Chen, P.F., and K. Shibata, *ApJ*, **545**, 524 (2000).

Cheng, C.Z., and G.S. Choe, *ApJ*, **505**, 376 (1998).

Choe, G.S., and L.C. Lee, *ApJ*, **472**, 360 (1996a), *ibid* **472**, 372 (1996b).

Démoulin, P., and E.R. Priest, *Solar Phys.*, **144**, 283 (1993).

DeVore, C.R., and S.K. Antiochos, *ApJ*, **539**, 954 (2000).

Forbes, T.G., and E.R. Priest, *ApJ*, **446**, 377 (1995).

Galsgaard, K., and A.W. Longbottom, *ApJ*, **510**, 444 (1999).

Hirayama, T., *Solar Phys.*, **34**, 323 (1974).

Hirose, S., Y. Uchida, S. Uemura, T. Yamaguchi, and S.B. Cable, *ApJ*, **551**, 586 (2001).

Isenberg, P.A., and T.G. Forbes, *ApJ*, **417**, 368 (1993).

Karpen, J.T., S.K. Antiochos, and C.R. DeVore, *ApJ*, **450**, 422 (1995).

Kippenhahn, R., and A. Schlüter, *Zeitschrift für Astrophysik*, **43**, 36 (1957).

Kopp, R.A., and G.W. Pneuman, *Solar Phys.*, **50**, 85 (1974).

Kusano, K., Y. Suzuki, and K. Nishikawa, *ApJ*, **441**, 942 (1995).

Leroy, J.L., V. Bommier, and S. Sahal-Bréchot, *A&A*, **131**, 33 (1984).

Martens, P.C.H., and C. Zwaan, *ApJ*, **558**, 872 (2001).

Martin, S.F., R. Bilimoria, and P.W. Tracadas, in *Solar Surface Magnetism*, eds. R.J. Rutten and C.J. Schrijver, p. 303, Springer, New York, NY (1994)

Mikić, Z., D.C. Barnes, and D.D. Schnack, *ApJ*, **328**, 830 (1988).

Morita, S., Y. Uchida, S. Hirose, S. Uemura, and T. Yamaguchi, *Solar Phys.*, **200**, 137 (2001).

Priest, E.R., W. Hood, and U. Anzer, *ApJ*, **344**, 1010 (1989).

Shibata, K., S. Masuda, M. Shimojo, H. Hara, T. Yokoyama, S. Tsuneta, T. Kosugi, and Y. Ogawara, *ApJ Letters*, **451**, L83 (1995).

Shimizu, T., S. Tsuneta, A. Title, T. Tarbell, R. Shine, and Z. Frank, in *Magnetodynamic Phenomena in the Solar Atmosphere – Prototypes of Stellar Magnetic Activity*, eds. Y. Uchida, T. Kosugi, and H.S. Hudson, p. 37, Kluwer Academic Publisher, Dordrecht (1996).

Sturock, P.A., *Nature*, **221**, 695 (1966).

Sweet, P.A., in *Electromagnetic Phenomena in Cosmic Physics*, ed. B. Lehnert, p. 123, Cambridge University Press (1958).

Uchida, Y., in *Skylab Workshop, Solar Flares*, ed. P.A. Sturrock, p. 67 and p. 110, Colorado Associated University Press (1980).

Uchida, Y., *Adv. Space Res.*, **17**, 19 (1996).

Uchida, Y., K. Fujisaki, S. Morita, M. Torii, S. Hirose, and S. Cable, *PASP*, **51**, 53 (1999a).

Uchida, Y., S. Hirose, S.B. Cable, S. Morita, M. Torii, S. Uemura, and T. Yamaguchi, *PASP*, **51**, 553 (1999b).

Uchida, Y., in *Advances in Solar Research of Eclipses from the Ground and from Space*, eds. J.P. Zahn and M. Stavinschi, p. 105, Kluwer Academic Publisher (2000).

Vekstein, G., E.R. Priest, and Amari, T., *A&A*, **243**, 492 (1991).

NUMERICAL SIMULATION OF A FLARE

T. Yokoyama

National Astronomical Observatory, Minamimaki, Minamisaku, Nagano, 384-1305, Japan

ABSTRACT

Magnetohydrodynamic simulation of a flare including the effect of anisotropic heat conduction, chromospheric evaporation, and radiative cooling based on the magnetic reconnection model is performed. In the simulation model the coronal magnetic energy is converted into thermal energy of the plasma by magnetic reconnection. This energy is transported to the chromosphere by heat conduction along magnetic field lines and causes an increase in temperature and pressure of the chromospheric plasma. The pressure gradient force drives upward motion of the plasma toward the corona. i.e. chromospheric evaporation. This enhances the density of the coronal reconnected flare loops, and such evaporated plasma is considered to be the source of the observed soft X-ray emission of a flare. The flare loops filled with evaporated dense plasma cool down due to the radiative cooling effect. EUV emitting post-flare loops thus are reproduced in the simulation.

INTRODUCTION

Yohkoh/SXT observed a class of flares with cusp-shaped loops in soft X-rays. This observation is consistent with the magnetic reconnection model (e.g. Shibata 1999). In this model the energy of the coronal field is released around a magnetic X-point. The released energy is converted into thermal and bulk flow energy of coronal plasma. We have been performed two-dimensional MHD simulations of a solar flare based on this magnetic reconnection model (Yokoyama & Shibata 1998, 2001). We show here recent results of our simulations in which the effects of heat conduction, evaporation, and radiative cooling are taken into account. In this simulation we try to cover the period from the peak of the flare to the end of the decay phase. We pay special attention to the growth and cooling of post-flare loops. A light curve and profiles of differential emission measure as a function of time are also derived.

RESULTS

We solved the two-and-a-half dimensional MHD equations. Here 'two and a half dimensional' means that there is translation symmetry in one direction, but all the three components of magnetic and velocity fields are taken into account. The effects of non-linear non-isotropic Spitzer-type heat conduction (see Yokoyama & Shibata (2001) for the functional form) and cooling by optically thin radiation (e.g. Priest 1982) are taken into account. We ignore gravity for simplicity at this time. (To reproduce processes such as coronal rain, in the very late decay phase when the scale height of the cooled plasma becomes small, gravity should be included. This will be done in future work.) The initial plasma is in magnetohydrostatic equilibrium with anti-parallel magnetic fields, between which there is a current sheet. The magnitude of the magnetic field is prescribed by the plasma β, which is 0.2 for the typical case. We position a dense and cool plasma at the base, which is a model of the chromosphere. We imposed a localized resistivity to force magnetic reconnection (see Yokoyama & Shibata (2001) for the functional form of the resistivity). It is known that

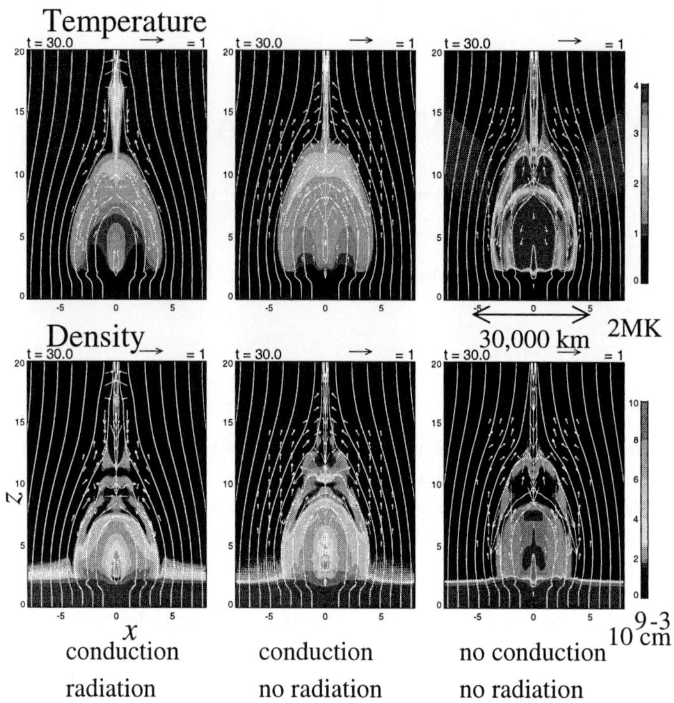

Fig. 1. Simulation results: Color maps in the top panels are for temperature, and those in the bottom panels are for density. Lines are magnetic field lines. Arrows indicate velocity. *(Left)* Results of the case with radiation and conduction. *(Middle)* Results for the case without radiation but including conduction. *(Right)* Results for the case without conduction and radiation.

reconnection becomes fast, i.e. proceeds on an Alfvén time scale when some localization mechanism for the resistivity is included (e.g. Ugai 1992). The Alfvén wave transit time is 100 s, the heat conduction time is 600 s, and the radiation cooling time is 16,000 s in the initial state. Note that the conduction and radiation time scales become shorter as the flare proceeds in the simulation.

Because of the locally enhanced resistivity, magnetic reconnection starts there. By the magnetic reconnection process magnetic energy is converted to thermal energy at the slow-mode MHD shocks extending from the magnetic neutral point in a Petschek-type configuration. Thermal energy is transferred along the field lines to the chromosphere by non-isotropic heat conduction. The chromospheric plasma is heated suddenly by this transferred energy, and flows up due to the pressure gradient.

In order to demonstrate the effects of heat conduction and radiative cooling we performed simulation runs without these effects (Figure 1). The right panel in Figure 1 is the case without conduction or radiation (the 'only MHD' case). The middle panel is the case without radiation but with conduction. The temperature of the flare loops in this case is smaller than in the 'only MHD' case. This is due to the cooling effect of heat conduction. Note also that the density of the flare loops is much higher, which is caused by the dense gas supplied by chromospheric evaporation. The left panel is the case with both radiation and conduction. The main difference from the previous case is the cool part in the center of the flare loops. This cool part is located exactly at the dense part of the flare loops. This indicates effective radiative cooling there.

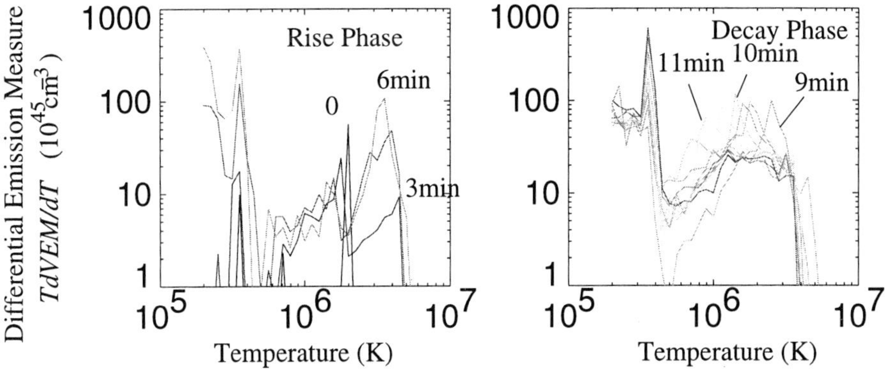

Fig. 2. Differential emission measure derived from the simulation results.

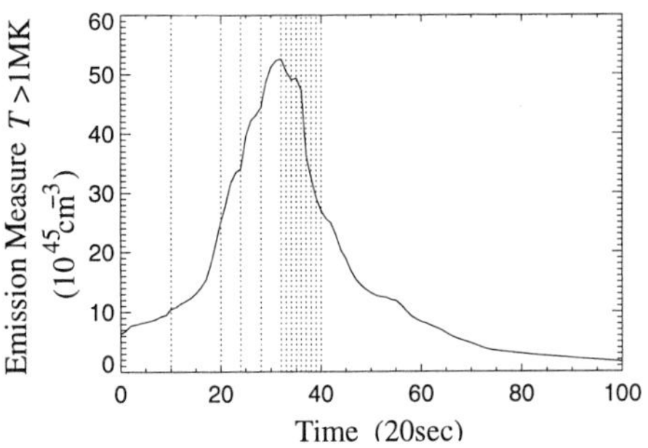

Fig. 3. Total emission measure (integrated only for gas elements whose temperature is over 1 MK) as a function of time. Vertical dotted lines indicate the times for the differential emission measure shown in Figure 2.

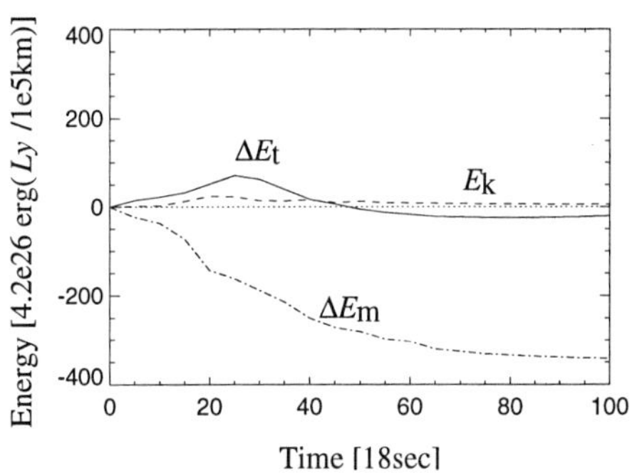

Fig. 4. Energy budget of the simulation box. ΔE_{t}, ΔE_{m}, and E_{k} are the thermal, magnetic, and kinetic energy increase/decrease from the initial state. The loss and gain of energy through the boundaries of the box is taken into account. L_y is the assumed length in the direction perpendicular to the simulation box.

We derived the differential emission measure (DEM) from the simulation results (Figure 2). The left panel is for the rise phase of the flare and the right is for the decay phase. Each line corresponds to a time step of the simulation. In the rise phase a rapid increase of the DEM of the hot plasma was seen, at constant temperature. On the other hand, in the decay phase the temperature of the DEM maximum decreases, while the form of the DEM remained constant. We can compare this result with the observations of Dere & Cook (1979). This observation is limited to the initial part of the decay phase, but the simulations qualitatively reproduce the observations.

Figure 3 is the derived light curve, namely the total emission measure as a function of time. Figure 4 shows the energy as a function of time. By comparing these figures, we found that the energy release continues even in the decay phase. We also found that the total amount of released magnetic energy is several times the thermal energy derived from the snapshot at the peak of the flare.

ACKNOWLEDGEMENTS

This study is a collaboration with Prof. Shibata from Kyoto University.

REFERENCES

Dere, K.P., and J. W. Cook, The Decay of the 1973 August 9 Flare, *ApJ*, **229**, 772 (1979).

Priest, E.R., *Solar Magnetohydrodynamics*, D. Reidel, Dordrecht, p. 88 (1982).

Shibata, K., Evidence of Magnetic Reconnection in Solar Flares and a Unified Model of Flares, *Astrophys. and Space Science*, **264**, 129 (1999).

Shibata, K., and T. Yokoyama, Origin of Universal Correlation between Flare Temperature and Emission Measure for Solar and Stellar Flares, *ApJ Letters*, **526**, L49 (1999).

Shibata, K., and T. Yokoyama, An HR-like Diagram for Solar/Stellar Flares and Coronae - Emission Measure vs. Temperature Diagram, *ApJ*, submitted (2002).

Ugai, M., Computer Studies on Development of the Fast Reconnection Mechanism for Different Resistivity Models, *PhysFluids B*, **4**, 2953 (1992).

Yokoyama, T. & K. Shibata, Two-Dimensional Magnetohydrodynamic Simulation of Chromospheric Evaporation Driven by Magnetic Reconnection in Solar Flares, *Astrophys. J.*, **494**, L113 (1998).

Yokoyama, T. & K. Shibata, Magnetohydrodynamic Simulation of a Solar Flare with Chromospheric Evaporation Effect Based on Magnetic Reconnection Model, *Astrophys. J.*, **549**, 1160 (2001).

THREE-DIMENSIONAL MHD SIMULATION OF AN EMERGING FLUX TUBE IN THE SUN

T. Magara and D. W. Longcope

Physics Department, Montana State University, P.O. Box 173840, Bozeman, MT 59717, USA

ABSTRACT

We have done 3-dimensional MHD simulations of a magnetic flux tube emerging through the solar atmosphere and studied not only the dynamics but also the geometric structure of emerging magnetic field. A flux tube twisted in the left-handed sense (Gold-Hoyle flux tube) is initially placed below the photosphere, and thereafter starts to rise with an upward motion induced to the middle part of the flux tube. When the outer edge of the flux tube crosses the photosphere, it is found that emerging field lines are almost perpendicular to the neutral line and the gas flow induced by the emergence diverges from the emerging region. As emergence proceeds, the direction of emerging field lines rotates to align with the neutral line and the flow direction changes from a diverging pattern (perpendicular to the neutral line) into a shearing pattern (parallel to the neutral line).

As for the dynamical behavior of emerging field lines, it is found that there are two kinds of evolutionary phases: an expansion phase and a gradual phase. Outer field lines of the flux tube, which emerge earlier than inner field lines, simply expand when they enter the solar atmosphere. On the other hand, inner field lines show a gradual phase in which the field lines rise slowly with a waving motion.

We have also investigated the structure of emerging field lines. Simulations show that outer field lines form an arcade over the neutral line that resembles a potential field. On the other hand, those inner field lines that are initially located just beneath the tube axis show an inverse-S structure. Observationally this structure is known as a sigmoid, an assembly of bright soft X-ray coronal loops, with an S or an inverse-S shape. By showing how current density is distributed around the footpoints of individual field lines, we provide a possible explanation for why a certain group of emerging field lines are illuminated to form a sigmoid.

INTRODUCTION

Emergence of magnetic field is one of the most important phenomena in the solar physics, because it is the fundamental process providing the magnetic field that causes various activities in the solar atmosphere. Observational studies have revealed that those activities are closely related to the magnetic field emerging into the corona. For example, Rust & Kumar (1996) and Canfield, Hudson, & Pevtsov (2000) investigated Yohkoh Soft X-ray images and found that a certain group of field lines become bright in the corona just before energetic events occur. Since these field lines have a particular structure (S or inverse-S shape), they are called 'sigmoids' and recognized as a precursor of energetic events.

From a theoretical viewpoint, emergence of magnetic field is generally a 3-dimensional dynamical process, making it difficult to clarify the details of the actual evolution. Fortunately, the power of recent super-computers has become so high that we can directly reproduce this process using 3-dimensional numerical simulation (Matsumoto *et al.* 1998, Fan 2001, Magara & Longcope 2001). In this paper we discuss the following two results obtained from our recent 3-dimensional MHD simulations: the dynamical behavior and the geometric structure of emerging magnetic field.

DESCRIPTION OF SIMULATION

For this simulation we have provided a Cartesian coordinate system, where the z-axis is directed vertically upward and other axes, x and y, form a horizontal plane. The total domain is (- 100, -100, -10) ¡ (x, y, z) ¡ (100, 100, 100) and the grid points are distributed nonuniformly in such a way as $(\Delta x, \Delta y, \Delta z) = (0.4, 0.4, 0.2)$ for $(-8, -12, -10) < (x, y, z) < (8, 12, 10)$ and each spacing gradually increases toward (4, 4, 4) along the axis.

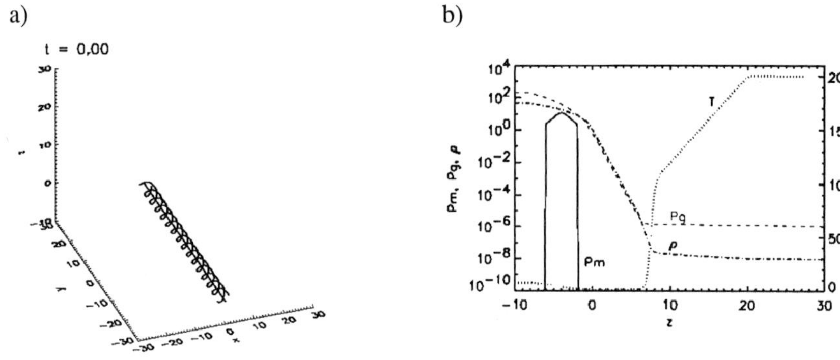

Fig. 1. -(a) Initial state of the simulation. (b) Initial distributions of magnetic pressure (Pm), gas pressure (Pg), gas density (*/rho*), and temperature (T). The units of gas pressure, density, and temperature are given by their photospheric values. The units of length and time are 300 km and 30 seconds, respectively.

Figure 1a is a snapshot of the initial state of the simulation where a straight magnetic flux tube is surrounded by unmagnetized background atmosphere. The model of background atmosphere is identical to the one used in Magara & Longcope (2001). The flux tube is given by a Gold-Hoyle flux tube twisted in the left-handed sense (radius = 2 (600 km) and twist parameter = 1). The center of the flux tube has a maximum field strength, about 8000 G, and a local value of plasma beta at the center is 3.3 (see Figure 1b). The flux tube is initially in mechanical equilibrium and forced to rise by inducing a vertical motion along the tube axis which drives the middle of the tube $(-15 < y < 15)$ upward and both ends downward. The subsequent adiabatic evolution is pursued by solving the time-dependent ideal MHD equations which include a uniform gravity but do not include any diffusive effects (resistivity and viscosity).

RESULTS

Dynamical Behavior of Emerging Magnetic Field

Figures 2a and 2b are snapshots that show the velocity field along an emerging field line. The curve and arrows represent a magnetic field line and velocity field on that line, respectively. Subpanels located at the top right show a side view of the snapshots. The field line of Figure 2a is initially on the outer shell of flux tube, overlying the field line of Figure 2b which originally corresponds to the axis of the flux tube.

These figures clearly exhibit the different dynamical behavior among emerging field lines that belong to the same flux tube. The overlying field line shows a simple expansion (Figure 2a), while the lower field line emerging later has a concave structure in the middle part of the field line and a plasma converges toward this concave part rather than expands outward (Figure 2b).

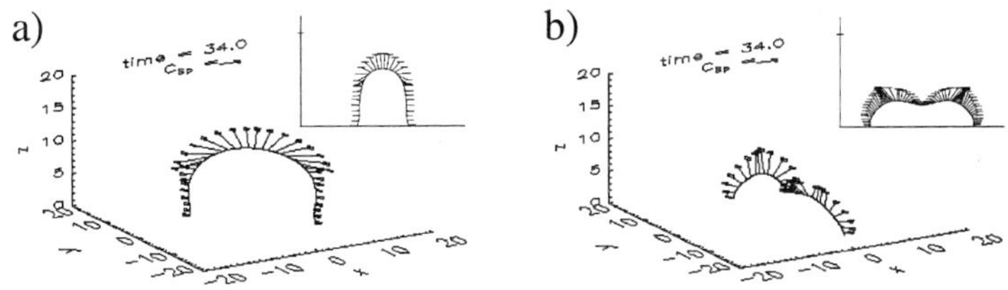

Fig. 2.-(a) Velocity field along the overlying field line at $t=34$. The subpanel located at the top right position shows the side view. The velocity unit is 10 km/s (photospheric sound speed). (b) Same as (a), except for the lower field line.

Structure of Emerging Magnetic Field

Figures 3a-3d are top-view maps showing how magnetic field and fluid velocity are distributed in the photosphere at two different times (t = 14 for the early phase and t = 40 for the late phase of emergence). Contour plots and arrows show the vertical magnetic field and horizontal velocity field, respectively. Gray-scaled maps display the vertical magnetic field for the top panels (Figure 3a and 3c) and the vertical velocity field for the bottom panels (Figure 3b and 3d). Curves in the bottom panels indicate magnetic field lines (Figure 3b and d). At the early stage of emergence, field lines are almost perpendicular to the neutral line (Figure 3b) and the photospheric flow shows a simple divergence from the emerging region (Figure 3a). As emergence proceeds, the outer part of emerging field lines expand to form an arcade structure while the inner part shows a more sheared structure (Figure 3d). This distribution of emerging magnetic field at the

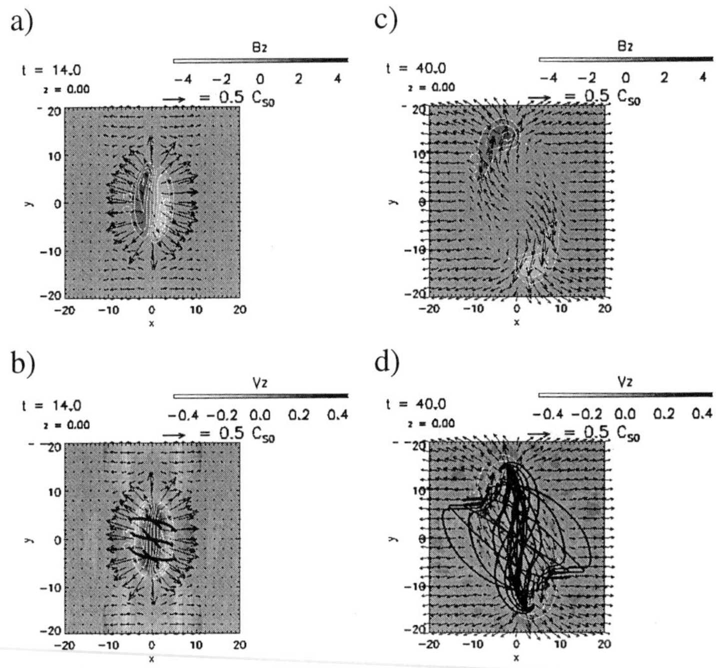

Fig. 3.-(a) A snap shot ($t=14$) of vertical magnetic field (contours and gray-scaled map) and horizontal fluid velocity (arrows) in the photosphere. (b) A snapshot (time = 14) of vertical magnetic field (contours), vertical fluid velocity (gray-scaled map), horizontal fluid velocity (arrows), and emerging field lines (black curves) on the photosphere. (c) Same for (a), except for $t=40$. (d) Same for (b), except $t=40$. The unit of magnetic field strength is 450 G.

late phase simply reflects the initial distribution of internal magnetic field of the flux tube, that is, the poloidal field which is dominant at the outer part of the flux tube form an arcade structure while the toroidal field (B_y) which is dominant around the center of the flux tube form a sheared structure. It is also found that the photospheric flow shows a shearing pattern at the late phase (Figure 3c). These results are consistent with the results reported in Fan (2001).

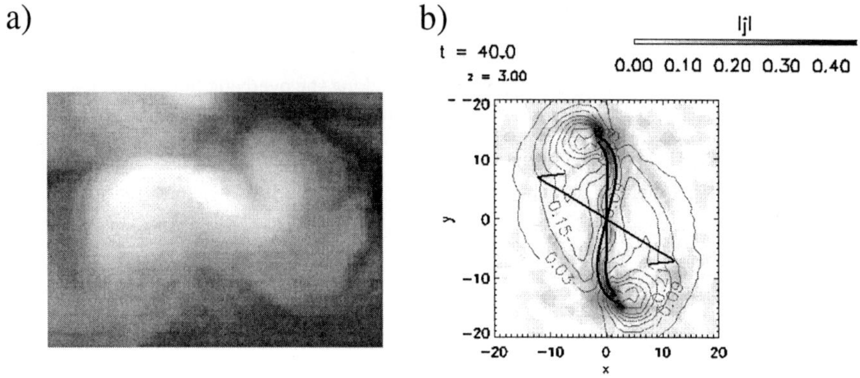

Fig. 4. (a) An example of a sigmoid. *Yohkoh*/SXT image, courtesy of the *Yohkoh* team. (b) A snapshot ($t = 0$) of the absolute value of the current density

Recently particular attention has been paid to a coronal structure called a 'sigmoid', an assembly of bright coronal loops with an S or an inverse-S shape (Figure 4a), because that structure is found to be a precursor of energetic events (Rust & Kumar 1996, Canfield, Hudson, & Pevtsov 2000) . Looking at Figure 3d, it is found that inner field lines have an inverse-S shape. We then seek the reason why a certain group of emerging field lines are illuminated to be observed as a sigmoid. What we have done is to investigate the distribution of current density around the footpoints of individual emerging field lines. The result is shown in Figure 4b, where we can clearly see that strong current density is distributed around the footpoints of inner sheared field line while there is little current around the footpoints of outer field line. On the assumption that the brightness of coronal loops reflects the concentration of currents around their footpoints, then Figure 4b suggests that only inner sheared field lines could become bright as a sigmoid in the corona, although the detailed mechanism to illuminate coronal loops is still unclear.

ACKNOWLEDGEMENTS

This work was supported by AFOSR grant F49620-00-1-0128. The numerical computations have been carried out with the help of the National Institute of Fusion Science in Japan and the National Center for Atmospheric Research in the US.

REFERENCES

Canfield, R. C., Hudson, H. S., and Pevtsov, A. A., *IEEE Transact. on Plasma Science*, **28**, 1786 (2000).

Fan, Y. *ApJ Letters*, **554**, L111 (2001).

Magara, T., and Longcope, D.W., *ApJ Letters*, **559**, 55L (2001).

Matsumoto, R., Tajima, T., Chou, W., Okubo, A., and Shibata, K., *ApJ Letters*, **493**, L43 (1998).

Rust, D.M., and Kumar, A., *ApJ Letters*, **464**, L199 (1996).

LOOP-TYPE CME PRODUCED BY MAGNETIC RECONNECTION OF TWO LARGE LOOPS AT THE ASSOCIATED ARCADE FLARE OCCURRING BETWEEN THE FOOTPOINTS OF A CME

Y. Uchida, J. Kuwabara, R. Cameron, I. Suzuki, T. Tanaka, and K. Kouduma

Physics Department, Science University of Tokyo, 1-3 Kagurazaka, Shinjuku-ku, Tokyo 162-8601

ABSTRACT

We concentrate in the present paper on loop-type CMEs, one of the two physically different types of CMEs we discussed elsewhere (Uchida *et al.* 2001). Loop-type CMEs are interpreted here as being due to a rising large magnetic loop that results from magnetic reconnection of two pre-existing loops which have connected the two outer magnetic regions to a region where an associated arcade flare will occur. We have found supporting evidence for this view in the faint structure-enhanced *Yohkoh* images before the CME event. Namely, the acceleration and the characteristic distortion of some of the rising loop-type CMEs can be explained in our model, and these may be driven in part by large amplitude torsional Alfven wave packets (TAWP's) coming up from the footpoints. Simultaneous release of TAWP's from both footpoints separated by a huge distance is not very natural to suppose, and therefore, we think that the TAWP's are released from the flare site. These TAWP's, bouncing back from the outer footpoints, propagate up along the large loop newly created in the reconnection process in the flare, and may cause helical instability, etc. Some intermediate results of our 3D MHD simulations of this process (Kuwabara *et al.* 2002) will be shown and discussed.

INTRODUCTION

The current general view of coronal mass ejections (CMEs) is that they are basically bubble-like in shape. We, however, claim that there are two physically different types of CMEs, namely, (1) loop-type CMEs which have the basic structure of an expanding loop with a preceding front, and (2) flare blast-associated bubble-type CMEs whose structure is basically an expanding dome. The former have two fixed footpoints, with acceleration of expansion, and in some cases, show a distortion of the loop shape very characteristic ofthe MHD helical instability in a relatively early stage of the expansion. The preceding front seems to be the compressed magnetic field ahead of the accelerating magnetic loop, and is likely to have an elongated paraglider-wing shape. This type of CME tends to have a wider range of fractional enhancement of density over the background corona, up to 20%. The latter, whose typical case is a halo-type CME when directed toward the Earth, does not have any fixed footpoints (their sweeping skirts are identified as Moreton waves, Uchida *et al.* 2001), and expands nearly isotropically in all directions. The expansion velocity is high ($\sim 10^3$ km/s), and more or less constant in time. The fractional enhancement in density is a few % in this type.

We based our arguments for this claim on our examination of LASCO data. Our attention has been focused on the simple fact that an isotropically expanding bubble should not have two *fixed* footpoints even viewed from any direction, whereas a structure with two fixed footpoints can not be associated with an almost isotropically expanding bubble.

We therefore specifically examined whether there are fixed footpoints in CMEs in the *SOHO*/LASCO data. The results of this examination showed that there actually exist two clearly definable types, (1) and (2) above, distinguishable not only by the "footpoint behavior" but also by other characteristics like the acceleration in their expansion velocity, or the occurrence of distortion in their shape, etc., in a way consistent with the above classification. Examining the *SOHO*/LASCO data of a given period during the maximum of solar activity, Jan-Mar 2001, we found that a large fraction of CMEs ($\leq$ 75%) have fixed footpoints at the solar surface (of course, not at the occulter, but when dually extrapolated to the solar limb), with fewer not having fixed footpoints, namely, the "footpoints" moved along the solar surface and finally lifted into the corona as the CME expands, even when observed near the limb (see Suzuki *et al.* 2002).

Due to the limited space here, we refer the readers to our papers (Uchida *et al.* 2001; Cameron & Uchida 2002a, b) for our view on the flare blast-associated disturbances, namely, bubble-type CMEs, Moreton waves, and EIT waves, while we confine ourselves here to the loop-type CMEs.

SOME OBSERVATIONAL PROPERTIES OF LOOP-TYPE CME'S

Since LASCO observes a number of so-called halo-type CMEs (first found by Howard (1979) with the P78-1 satellite), there must also be a corresponding number of that type seen near the limb. We consider this type as a flare blast-associated bubble-type front, and halo-type CMEs are observed when those are directed toward the Earth. Some confusion is generated, since the surface of such bubble-like structure seen from the sides will appear as a "loop". This is because the shell of the bubble seen from the sides has longer lines-of-sight penetrating the edge part than other parts.

Loop-type CMEs, selected according to our criteria given above often, have a frontal loop, a dark cavity between this, and a second loop which is likely to be related to the rising dark filament suspended in the middle, just as described in the results from previous satellites.

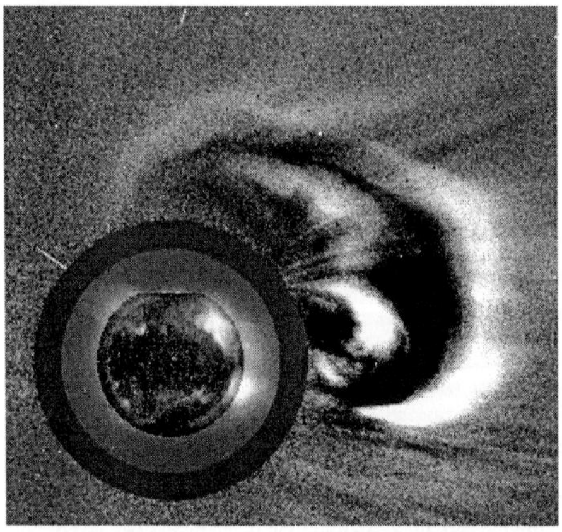

Fig. 1. An example (from Uchida 2000) of a loop-type CME of December 6, 1997, showing the clear structures of the first loop, cavity, and the second loop. The footpoints of the loops do not move, and the second loop with a dip in the middle follows the first front. The dip exactly corresponds to the location where the associated arcade flare occurred in the *Yohkoh* insert, suggesting that the dip is the anchored point released in the reconnection process in the flare.

In some examples with favorable viewing conditions (mainly the viewing direction) one notices that the second structure clearly has a dip in the middle, and that the location of the dip is right above the associated arcade flare occurring between the footpoints (Figure 1, with *Yohkoh* insert). A model of arcade flare and

X-ray arcade formation considering quadruple magnetic sources (Uchida *et al.* 1999b, Hirose *et al.* 2001), which has increasing support from observations (Uchida *et al.* 1999c, Morita *et al.* 2001), is very naturally compatible with this by forming a hexapole magnetic source (Uchida *et al.* 2001).

We also performed a deep survey of faint structures in the pre-event phase, using *Yohkoh* soft X-ray images a few days before the event for loop-type CMEs at the limb. We found that there exist faint connections between the locations of the footpoints-to-be of the CME and the region where the associated arcade flare will occur. Footpoints of CMEs of this type are therefore the *prescribed* points that are connected in the pre-event phase to the region where the associated arcade flare is to occur! It is natural then that the CME footpoints do not move freely and stay at the fixed points during the expansion of the loop. Many other very clear examples of this loop-type exist. Among them there are events whose footpoint distance is small, and remains small. These are the easy-to-tell examples of the loop-type, because they cannot be explained as an expanding bubble at all.

OUR PROPOSED MODEL FOR LOOP-TYPE CME'S

Our interpretation (Uchida *et al.* 2001, Kuwabara *et al.* 2002) is that the first-expanding structure is the compressed magnetic structure ahead of the driving loop. This second loop behind the dark region separating from the first is a magnetic structure whose central anchoring point has been released by the flare occurring somewhere between the footpoints, and behaves as the driver. The first front may have a shape of a *paraglider wing*. Ahead of this, also a very faint bubble-like front is seen to propagate, but this is different from the more or less complete dome-shaped front of the flare blast-associated bubble-type CMEs discussed in our previous papers.

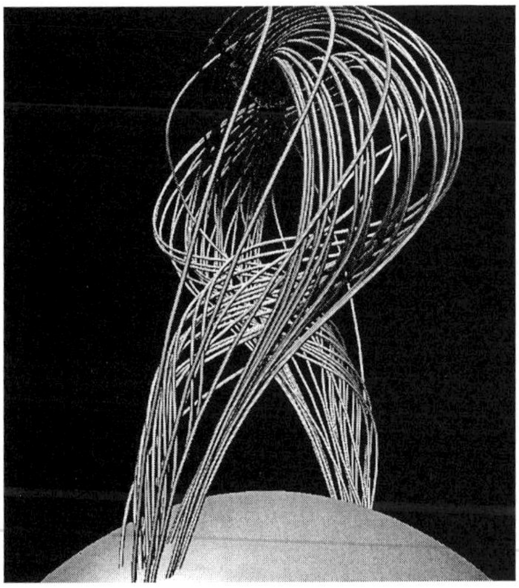

Fig. 2. Our MHD model for loop-type CMEs. We assume that the outer regions with + and - polarities are connected to the region where an arcade flare occurs later somewhere between the footpoints. Magnetic twists are released from the flare, and reflected back up along the large loop after being bounced from the outer ends of the initial connections. The propagating twists explain the acceleration and distortion of the loop (Kuwabara *et al.* 2002). Here, we show a preliminary case in which magnetic twists of the same sense are injected from the footpoints, and increase the helical instability of the loop when these collide at the looptop. Also, for the case with opposite sense, we find observational counterparts of the different type of distortions. Only relevant magnetic field lines are shown for clarity.

The two pre-existing loops, connecting the footpoints of the CME to the flare site, may become joined through reconnection in the flare. This reconnection process forms one long loop that is rising with a dark filament in the dipped part, and gradually becomes smoother and loses its dipped shape as it expands.

With an assumed release of magnetic twists from the flare site into each of the two loops connecting the flare and the footpoints, the rising large loop is accelerated by the thrust exerted by the propagating twists (torsional Alfven wave packets = TAWP), that were bounced from both footpoints back into the reconnected large loop. A helical instability of this large loop may occur when the amount of the toroidal field exceeds the criterion for helical instability of the loop (which corresponds to the Kruskal-Shafranov criterion in the linear case in the simpler straight geometry of a twisted flux tube. When the base flux tube is bent, this pushes the loop outward and drives the loop expansion). This may explain very nicely the characteristic distortion of the loop often seen in this type. The distortion will disappear when the rising magnetic twists (TAWP's) cancel each other on colliding at the looptop. A secondary wavefront expanding spherically may also be created in the collision of TAWP's at the looptop. In Figure 2 we show a case with TAWP's strengthening each other's effects, and a helical instability of the loop as a whole takes place. Acceleration of the rising loop will also be aided by magnetic buoyancy due to the outward gradient of the magnetic field intensity in the low corona, as well as due to this thrust given by the propagating torsional Alfven wave packets.

SUMMARY AND DISCUSSION

We have given our view on loop-type CMEs in which magnetic reconnection takes place in the arcade flare somewhere between the footpoints of CMEs. Harrison & Sime (1989) doubted the active role of the associated flare in the production of the CME because of their reversed order in time. This does not have to imply that the CME is the first thing and the flare is caused by that, if we note that there is a considerable delay in the flare brightening after the first dynamical, or magnetic, changes, especially in the case of a dark filament disappearance at high latitude. The time-lag sometimes amounts to tens of minutes.

Finally, Gosling *et al.* (1993) raised a point that flares have smaller energy than the energy of the related CMEs. This indicates that the energy deposited into thermal and the gas motion in the locality of the flare is only a fraction of the magnetic energy released. We think that a large part of the CME energy comes in the form of magnetic energy carried by TAWP's, assisted by the "melon seed" drive in the release of the large loop from the region of high magnetic stress in the low corona to the region of low magnetic stress in the outer corona and onwards.

REFERENCES

Cameron, R., and Uchida, Y., to be submitted to *PASJ* (2002a,b).

Gosling, J., *JGR*, **98**, 18937 (1993).

Harrison, R.A., and Sime, D.G., *JGR*, **94**, 2333-2344 (1989).

Hirose, S., Uchida, Y., Uemura, S., and Yamaguchi, T., *ApJ*, **551**, 586 (2001).

Howard, R., *et al.*, *ApJ Letters*, **228**, L45 (1979).

Hundhausen, A., *JGR*, **98**, 13177 (1993).

Hundhausen, A., in *The Many Faces of the Sun*, eds. K. Strong *et al.* (Springer), pp. 143-200 (1999).

Kuwabara, J., Cameron, R., and Uchida, Y., in preparation (2002).

Morita, S., Uchida, Y., Hirose, S., Uemura, S., and Yamaguchi, T., *Solar Phys.*, **200**, 137 (2001).

Uchida, Y., Hirose, S., Morita, S., Fujisaki, K., Torii, M., Tanaka, T., Yabiku, Y., Miyagoshi, T., Uemura, S., and Yamaguchi, T., *Astrophys. Space Sci.*, **264**, 145 (1999a).

Uchida, Y., Hirose, S., Cable, S., Morita, S., Torii, M., Uemura, S., Yamaguchi, T., *PASP*, **51**, 553 (1999b).

Uchida, Y., in *Advances in Solar Research at Eclipses from Ground and from Space*, eds. J.-P. Zahn, and M. Stavinschi (Kluwer), p. 105 (2000).

Uchida, Y., Tanaka, T., Hata, M., and Cameron, R., *Publ. Astron. Soc. Australia*, **18**, 345 (2001).

THREE DIMENSIONAL MHD SIMULATIONS FOR AN EMERGING TWISTED MAGNETIC FLUX TUBE

T. M. Miyagoshi[1] and T. Y. Yokoyama[2]

[1]*National Astronomical Observatory, 2-21-1 Osawa, Mitaka, Tokyo 181-8588, Japan*
[2]*National Astronomical Observatory, Nobeyama, Minamimaki, Minamisaku, Nagano 384-1305, Japan*

We have studied the behaviour of a twisted emerging flux tube from the upper convection zone to the corona by means of three-dimensional ideal MHD numerical simulations (see also Matsumoto *et al.* 1998, Magara & Longcope 2001, Fan 2001). The purpose of our study is to reproduce the evolution of flux tubes in the upper solar atmosphere, and to find the final structure of magnetic fields in the corona. The parameters of this simulation are the magnitude of the twists, the radius of the flux tube, and the wavelength of the perturbation. Here we present results for different magnitudes of the twist. An S-shaped structure is formed when the magnitude of the twists is large ($B_\theta = qrB_0/(1+(qr)^2), q \sim 1.0$), and a helical structure is formed when the magnitude of the twists is small ($q \sim 0.2$). Here B_θ is the toroidal field, B_0 is the poloidal field, r is the distance from the axis of flux tube, and q is a parameter measuring the magnitude of the twist.

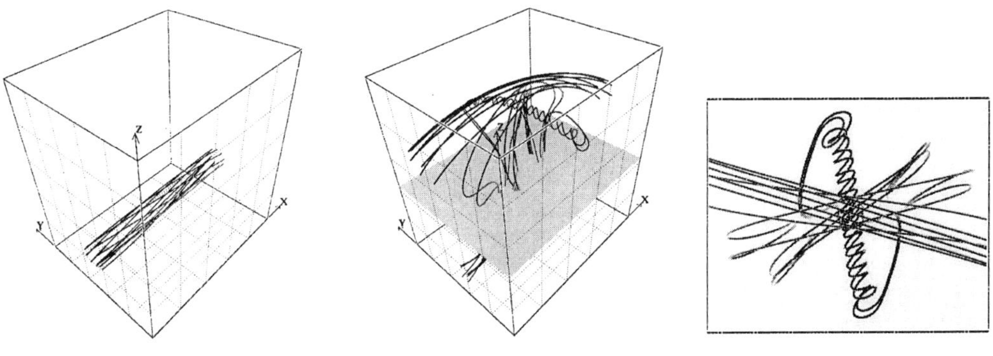

Fig. 1. Simulation results for an initially weakly twisted flux tube ($q = 0.2$). The left panel shows the initial condition, the center panel shows the structure after flux tube emerges, and the right panel shows a top view of the center panel. Solid lines show magnetic fields, and the gray surface (center panel) represents the surface of the sun. The scale of the boundary box is $25H \times 25H \times 25H$ (H is the scale height of the photosphere).

Initially a Gold-Hoyle force-free flux tube (in mechanical equilibrium) is embedded in the upper convection zone. It starts emerging by the Parker instability after an imposed small amplitude perturbation. The wavelength of the perturbation and the radius of the flux tube are 12.5 and 5.0 times the photospheric scale height respectively. The plasma β is 20 in the center of the flux tube. Figure 1 shows the numerical simulation results for weak twists ($q = 0.2$). In this case the magnetic fields are nearly straight. As the tube emerges, convection flow occurs near the footpoints of the flux tube. This convective motion pushes magnetic fields towards the center (Figure 2). As a result of this magnetic reconnection occurs below the

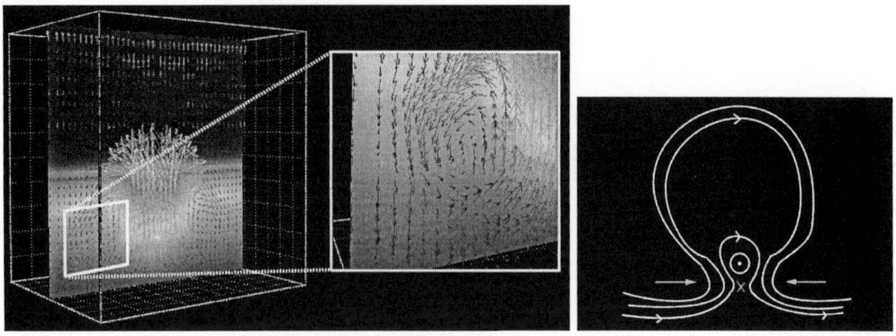

Fig. 2. (Left) Velocity fields in the simulation indicated by arrows. (Right) Schematic picture to explain the formation of the helical structure by magnetic reconnection. Arrows show the convective gas motion and solid lines show magnetic fields.

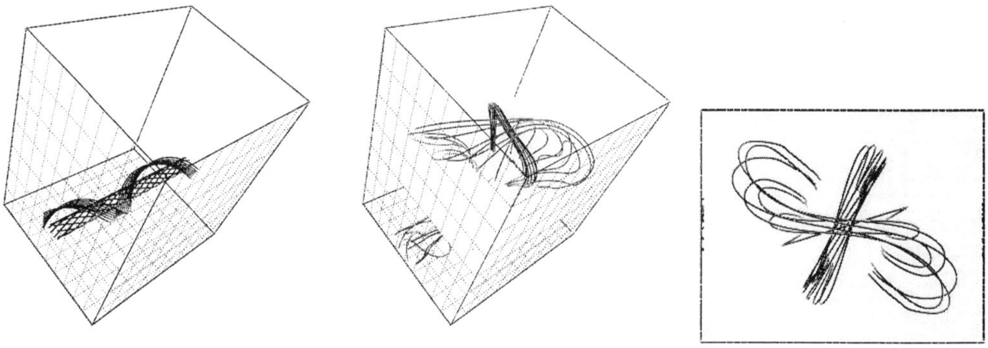

Fig. 3. Simulation results for the strongly twisted case ($q = 1.0$). An S-shaped structure is formed by the strongly twisted field.

corona (convective motion pushes the fields effectively into the high β region) and the helical structure of the magnetic field is formed (Figure 1, center panel). The right panel of Figure 1 shows the magnetic structure viewed from the top. Figure 3 shows the strongly twisted case ($q = 1.0$). In this case reconnection does not occur because no neutral point is formed by the converging flow, but an S-shaped structure is formed in the corona. From these results it is found that the magnitude of the twist has a strong effect on the coronal magnetic structure of emerging flux.

REFERENCES

Magara, T., and Longcope, D.W., *ApJ Letters*, **559**, L55 (2001).

Matsumoto, R., Tajima, T., Chou, W., Okubo, A., and Shibata, K., *ApJ Letters*, **493**, L43 (1998).

Fan, Y., *ApJ Letters*, **554**, L111 (2001).

PROPERTIES OF MAGNETIC RECONNECTION IN A STRATIFIED ATMOSPHERE

A. Takeuchi[1] and K. Shibata[2]

[1] *Yonago National College of Technology, Hikona 4448, Yonago, Tottori, 683-8502, Japan*
[2] *Kwasan Observatory, Kyoto University, Yamashina, Kyoto, 607-8471, Japan*

ABSTRACT

We investigate properties of magnetic reconnection in a stratified atmosphere such as the solar photosphere, performing 2-dimensional magnetohydrodynamic (MHD) numerical simulations. The reconnection is induced by an encounter of oppositely directed vertical magnetic flux sheets. It is found that the velocity of the upward reconnection jet is faster than the Alfvén velocity, while the velocity of the downward jet is slower than the Alfvén velocity. Moreover, the temperature of the upward jet is cooler than that of the downward jet.

RESULTS AND DISCUSSION

It is widely believed that magnetic cancellation is caused by photospheric magnetic reconnection. Since solar photosphere is a highly stratified atmosphere, the reconnection in the photosphere must be affected by the stratification. Thus, we investigate the nature of the reconnection in a stratified atmosphere.

We solve the resistive MHD equations numerically. An illustration of the initial state is shown in Figure 1, where the unit of length is the photospheric pressure scale height $H_p(= 160 \text{ km})$. We calculate the initial field configuration adopting the thin flux tube approximation (Roberts & Webb 1978). In the thin flux tube approximation, when the width of the flux sheet W is much smaller than H_p the initial field configuration becomes a magnetostatic equilibrium solution. We adopt $W = 0.3$ at the temperature–minimum region ($z = 0$) to maintain an initial force balance. Note that in the initial flux sheet the Alfvén velocity and $\beta = p_{gas}/p_{mag}$ are not functions of height but are constants, respectively, where we assume $\beta = 0.9$. Actual solar resistivity possesses a maximum at the temperature–minimum region (Kovitya & Cram 1983). We adopt a resistivity model in which the resistivity is described as a function of height with a maximum (where the magnetic Reynolds number = 1000) at $z \sim 0$. Due to this height variation of resistivity, magnetic reconnection is most preferred around $z \sim 0$. In Figure 2, we show the temporal evolution of velocity profiles of the reconnection jets, where the unit of velocity is the photospheric sound velocity C_s ($= 8$ km/s).

Although the velocity of a reconnection jet is the Alfvén velocity in an uns tratified medium, it is apparent that the velocity of the upward jet is faster than the Alfvén velocity, while the velocity of the downward jet is slower than the Alfvén velocity. In Figure 3, we show temporal evolution of temperature profiles of the jets, where the unit is the photospheric temperature ($T_0 = 6300$ K). The temperature of the upward jet is cooler than that of the downward jet. The reason why the upward jet is cooler than the downward jet is that the upward jet goes toward a low-pressure region and gets cooled through adiabatic expansion, while the downward jet goes toward a high-pressure region and gets heated by adiabatic compression.

Recently Galsgaard & Roussev (2002) investigated magnetic reconnection in stratified atmospheres and showed similar results to those shown here. In their results, however, the velocity of the upward jet does not exceed the Alfvén velocity. The differences of initial conditions between their models and our model are probably the cause.

The inflow speed of our simulation $v_i \sim 0.24$ km/s is nearly equal to the speed implied by observations of cancelling magnetic features on the photosphere(Garcia de la Rosa *et al.* 1989).

The authors thank Dr. T. Kudoh, Dr. T. Yokoyama and Dr. S. Nitta for useful discussions. Numerical computations were carried out on VPP5000 at National Astronomical Observatory, Japan.

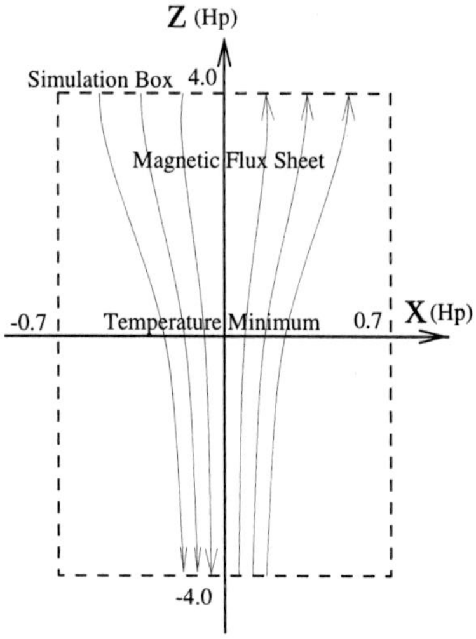

Fig. 1. Illustration of initial state.

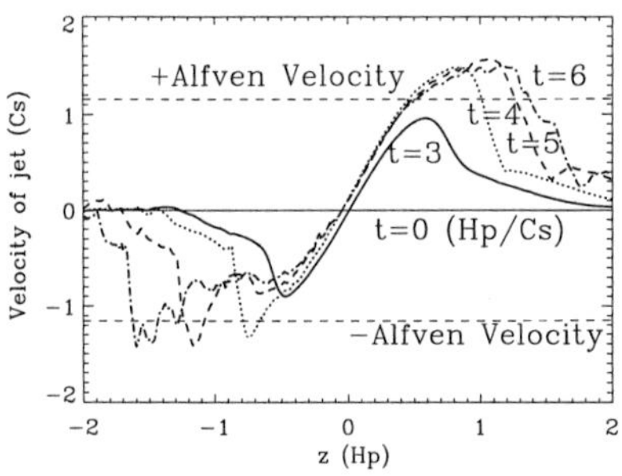

Fig. 2. Temporal evolution of velocity of reconnection jets. The velocity of the upward jet is faster than the Alfvén velocity.

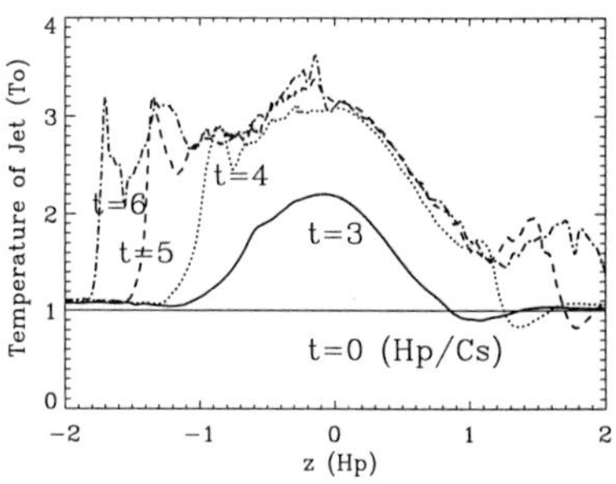

Fig. 3. Temporal evolution of temperature of the jets. The temperature of the upward jet is cooler than that of the downward jet.

REFERENCES

Galsgaard, K., & Roussev, I., *A&A*, **383**, 685 (2002).

Garcia de la Rosa, J.I., Aballe, M.A., & Collados, M., *Sol. Phys.*, **124**, 219 (1989).

Kovitya, P., & Cram, L., *Sol. Phys.*, **84**, 45 (1983).

Roberts, B., & Webb, A.R., *Sol. Phys.*, **56**, 5 (1978).

Section VII.

Fine Structure in Flares

HIGH RESOLUTION OBSERVATIONS OF SOLAR FLARES

B. Sylwester

Space Research Centre, Polish Academy of Sciences, Kopernika 11, 51-622 Wroclaw, Poland

ABSTRACT

Over the past few years, our team in the Space Research Center (SRC) of the Polish Academy of Sciences in Wroclaw has been interested in developing numerical image enhancement techniques with the aim of increasing the resolution on SXT flare images. An algorithm (ANDRIL) has been developed which allows for the image deblurring with oversampling. The application of this algorithm allows an increase in the resolution of the SXT images to the level of ~ 1 arcsec. We performed the deconvolution of a large number of flare sequences using ANDRIL algorithm. The analysis of these data allowed us to study the morphology of flaring plasma with a resolution comparable to *TRACE*. In order to infer the thermodynamic parameters of the plasma from the analysis of deconvolved images their precise co-alignment is required. Techniques which have been developed at SRC to achieve high accuracy of image co-alignment will be discussed. The application of the filter ratio technique to the deconvolved and co-aligned images allows the study of temperature and emission measure distributions of the flaring plasma in great detail. These maps better resolve flare kernels located both at the summits and footpoints of flaring loop structures. The main results obtained will be presented in this review.

OVERVIEW

From the early days of X-ray coronal observations it has been clear that there is an association between the location of the structures observed in the corona and the pattern of features seen on the solar disc. It has been understood that this connection relates with the magnetic activity. The most energetic phenomena associated with magnetic activity are solar flares usually observed within active regions. Even 30 years ago the spatial resolving power of ground based observations routinely achieved arcsec resolution. At that time the techniques of X-ray observations were only beginning to be developed and the resolution of the corona was only a few arcmins at best.

The era of satellite borne experiments with the X-ray detectors on-board begun in 1962 when the first (from a series of eight) *Orbiting Solar Observatories* (*OSO 1*) was launched. Among the significant achievements of this program were the first full-disc photographs of the solar X-ray corona and the first X-ray observations of a solar flare. The launch of the first manned observatory *Skylab* in 1973, marked the beginning of a new era in the investigation of solar short wavelength radiation. The *Skylab* X-ray and EUV images provided convincing evidence and confirmation of results from sounding rockets (obtained in the late 1960s and early 1970s), that the corona is comprised mostly of magnetic loops and arches (Sheeley *et al.* 1975, Foukal 1976). From that time coronal loops are often referred to as the fundamental building blocks of the magnetically dominated quiet and active corona. The disadvantages of the *Skylab* mission were that it took place during a period of low flare activity and that the image recording medium used was photographic plates. In February 1980 NASA's *Solar Maximum Mission* (*SMM*) was launched carrying eight instruments developed by international teams. The HXIS and FCS instruments on board *SMM* allowed for imaging flares in hard and soft X-rays. The BCS provided high-resolution spectra around Fe XXV-XXVI and Ca XIX resonance lines. The HXRBS allowed for observing the hard X-ray spectra of solar flares. The HXIS, FCS and BCS spectrometers on *SMM* were among the highest resolution instruments making systematic solar X-ray observations at that time.

The Japanese rotationally modulated telescope on board the *Hinotori* spacecraft, launched in 1981, and devoted to imaging solar flares in hard X-rays (5 - 40 keV) had similar resolution. The launch of *Yohkoh* constituted another milestone in the studies of high-temperature flare plasma. Important support to *Yohkoh* came from the EIT, CDS and SUMER on *SOHO*, and more recently from *TRACE*. In the meantime a number of sounding rocket flights carrying high resolution experiments have been launched. Although allowing for a few minutes of observing time only they made it possible to test new observing techniques in the X-ray and EUV ranges and provide the images with the best ever spatial resolution. Among these sounding rocket experiments we should mention the innovative HRTS (High Resolution Telescope and Spectrograph) instrument, the NIXT X-ray imager (Normal Incidence X-ray Telescope), the MSSTA (Multi-Spectral Solar Telescope Array), and SERTS (the Solar Extreme-ultraviolet Rocket Telescope and Spectrograph). In particular the disc X-ray flare recorded by NIXT on 11 September 1989 with 0.75 arcsec resolution has been extensively studied in many papers (Golub *et al.* 1990, Herant *et al.* 1991, Golub 1992, 1997, and Sams *et al.* 1992).

On the Eastern hemisphere experimental investigations of solar X-ray radiation have been undertaken by teams working in Russia (FIAN Lebedev Physical Institute in Moscow), the Czech Republic (Astronomical Institute in Ondrejov), and Poland (Space Research Center in Wroclaw). From the beginning of *Yohkoh* era the Wroclaw team took part in the analysis of data from the BCS, SXT, and HXT instruments. During the ten years of *Yohkoh* observations we have processed and analyzed a wealth of data on flares using special techniques of image reconstruction and handling, worked out in Wroclaw. In this review I will emphasize the major results obtained by our group working with *Yohkoh* solar flare data .

METHODS OF SXT IMAGE PROCESSING AND ANALYSIS

From the earlier analysis of the best high resolution flare X-ray images (Herant *et al.* 1991) and results of theoretical considerations (Gomez, Martens, & Golub 1993) it follows that the structuring and most of physical processes operating in solar flares act on small sub-arcsecond scales, i.e. below the nominal resolution of SXT images on *Yohkoh*. As the SXT pixel dimension and the FWHM of the point spread function (PSF) of the grazing incidence mirror are of similar sizes it is then possible to increase the effective resolution of the images by using image reconstruction techniques. A number of methods have been worked out and presented in the literature which allow the instrumental blurring to be removed. However, there was no well documented method (except the work of Roumeliotis 1995) which allows deblurring and over sampling to be performed simultaneously. Oversampling means that the recovered (deconvolved) picture contains more resolution elements than the original one. In the case of SXT images we have found that a fortunate compromise between FWHM and pixel size allows for subdivision of each SXT pixel into 5$\times$5 subpixels. Such an oversampling scheme results in the formal numerical increase of the nominal resolution of the SXT image down to $\sim$ 1 arcsec (the over-sampled size of subpixel is $\sim$ 0.5 arcsec). We developed an algorithm called ANDRIL (**A**ccelerated **N**oise **D**amped **Ri**chardson-**L**ucy) for SXT image processing. In constructing the algorithm we followed the ideas used for the reconstruction of images from the Hubble Space Telescope prior to its refurbishment. The details of ANDRIL can be found in the papers by Sylwester *et al.* (1996a) and Sylwester & Sylwester (1998a, 1999a) where extensive tests of the algorithm for various synthetic input models and various shapes of the PSF have been performed. An example of the application the ANDRIL algorithm to real SXT data is shown in Figure 1, where the comparison of original and deconvolved images is presented for two flares (17 April 1992, during the decay phase, and 24 March 1993, during the rise phase). The advantage of using ANDRIL image processing is obvious. Many fine structures of sub-arcsecond sizes

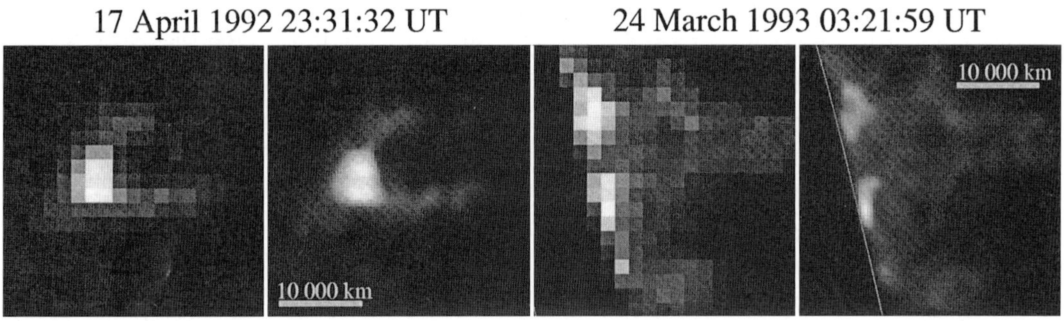

Fig. 1. Comparison of original and deconvolved Al12 images for two flares. The original and deconvolved images have been compressed in the same way in order to reveal weaker sources. The sizes of the original images are 18×18 SXT pixels which correspond to 90×90 subpixels after the deconvolution.

are revealed in the deconvolved images. Their physical reliability can be established when comparing the evolution on sequences of consecutive deconvolved images taken with different filters. Near the limb the looptop kernel of the relatively simple M1.1 flare on 17 April 1992 reveals a double structure during the decay phase. The structure of the limb C6.6 flare observed on 24 March 1993 looks more complicated. Deconvolution reveals a cusp-like structure with the main emission concentrated in one (south) foot during the rise phase.

A common interpretation of flare intensities as measured in a pair of images taken in slightly different energy bands provides information on the average temperature distribution of emitting plasma when the images are taken at the same time and the same plasma regions contribute to both bands. It is known that when image ratios are interpreted for regions of strong intensity gradients, derived values of temperatures are very sensitive to the exact co-alignment of the images in a pair. In order to better understand the dependence of the derived temperature distribution on possible small misalignments of Be119 and Al12 images due to fixed accuracy of spacecraft attitude information, Sylwester (1995) investigated to what extent derived T-maps depend on the possible inaccuracy of co-alignment. Siarkowski *et al.* (1996) have shown that the alignment procedures used in the standard SXT image co-alignment and processing routines may be inadequate when the temperature maps are in question. In order to improve the determination of the temperature pattern in the investigated structures, three numerical methods of Additional Pointing Corrections (APC) settlement have been proposed and investigated.

The uncertainty of temperature determination is much larger when deconvolved SXT images are to be used as the angular resolution is increased five-fold. In the paper by Sylwester & Sylwester (2001a) a method of co-alignment of deconvolved images has been proposed for flares partially occulted by the solar disc. For these flares the limb of the solar atmosphere occults the X-ray source at the very same location for images taken using individual filters. The description of modelling of X-ray occultation process can be found in Sylwester & Sylwester (2002). Taking these results into account it is possible to co-align the images to better than 0.1 arcsec accuracy which allows for temperature determinations on subpixel scales. We will refer to this method of co-alignment as the Limb Position Adjustment (LPA), as it is based on the limb as a reference. We have checked that the agreement between Al12 and Be119 images improves by using the LPA method so we have decided to adopt it routinely wherever possible.

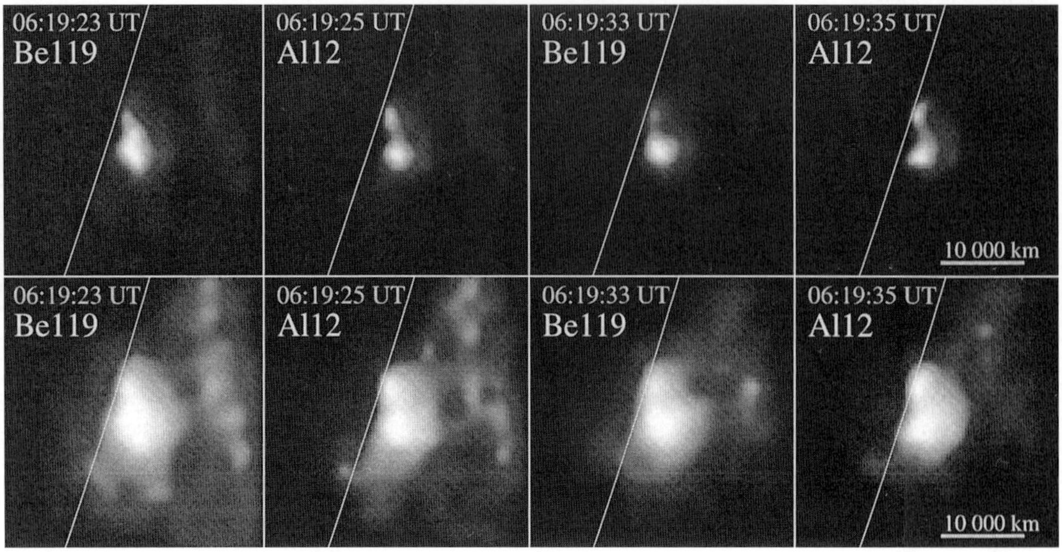

Fig. 2. Two pairs of consecutive deconvolved and compressed Be119 and Al12 images for the 5 November 1992 limb flare. The compression scheme for the lower row images allows one to see the weaker structures. The size of each image is 90×90 subpixels.

MAIN RESULTS

After testing the ANDRIL algorithm we realised that it provides a powerful tool for processing SXT images, and we began a massive analysis of flare observations. Detailed inspection of the sequences of deconvolved images revealed usually a number of fine structure details. In the most cases the flare (as seen in all SXT bands) consists of kernels which constitute well localized, small emitting sources. The radii of these fine X-ray structures are of the order of ~ 1 - 1.5 arcsec. Kernels are located at the summits as well as at the footpoints of flaring loops. They were identified on the original SXT images as well (Acton *et al.* 1992, Feldman *et al.* 1994a, b, Khan *et al.* 1995, Harra-Murnion *et al.* 1997, Jakimiec *et al.* 1998, Nitta *et al.* 2001). However, the deconvolved images revealed often their fine structure, which was barely seen in the original ones (Figure 1). The top bright regions appear shortly after the flare onset and persist well into the decay phase of the flaring compact and long duration events. The soft X-ray footpoint regions (kernels) are not as well observed as the summit ones. Footpoint kernels dominate only during a short time at the flare onset and, except a few cases (Masuda *et al.* 1994, McTiernan *et al.* 1993, Hudson *et al.* 1994, Sylwester & Sylwester 1998c) are not well covered by observations because of the flare trigger logging scheme used by SXT.

The deconvolved images reveal that the morphology of hot flaring plasma is usually more complicated that the theoretical flare models showing a simple loop or loop arcade connecting footpoints in the photosphere present. The flares presented in Figure 1 constitute good example. The Al12 images for the flare on 17 April 1992 during the decay and for the flare on 24 March 1993 during the rise phase reveal more complicated and nonuniform structures than those presented on theoretical sketches. The top region of the 17 April 1992 flare reveals a double structure of the kernel. The diameter of each component is ~ 1300 km and their centers are separated by ~ 2500 km. We have observed also flares composed of multiple coronal kernels (Sylwester *et al.* 1996b). The limb flare presented in the right panel of Figure 1 is relatively more complicated. The cusp-like structure, the dominating footpoint kernel, and many related small emission

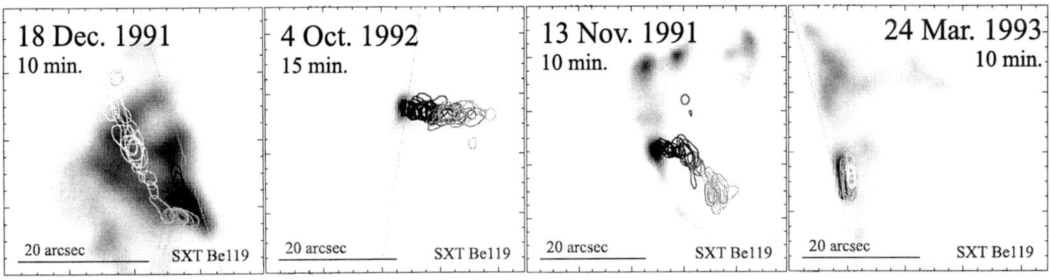

Fig. 3. The evolution of maximum brightness location presented as shaded contours drawn at 0.9 of maximum Be119 intensity for four flares. The contours are overlaid atop the emission pattern (in grey) of the first unsaturated Be119 image within the sequence. The duration of presented evolution pattern is indicated.

patches located in the corona are present. This flare has been analyzed by Sylwester & Sylwester (2001b). Such images convincingly suggest that a simple loop structure with the emission enhanced at the top and extending towards the double footpoints at the chromosphere is not the typical pattern observed for flares.

An interesting flare on 5 November 1992 (M2.0) has been analyzed by Sylwester & Sylwester (1999b). The two pairs of consecutive (Be119 and Al12) deconvolved images for this event are presented in Figure 2. In the top row only the dominant kernels are seen. The lower row shows the same images but compressed in a way that allows us to observe much weaker structures. It is seen that the main flaring region forms a small, low-lying nonuniform emitting structure with individual knots clearly distinguished. Besides this dominant region, the enhanced emission located above (higher in the corona) and forming the ridge can be seen. Individual structures composing the ridge are seen to be connected to the main loop top kernels throughout the weaker emission fringes. The ridge structure can be identified in all images obtained throughout the events evolution. However, it is most pronounced at the beginning of flare cadence. A large number of observations similar to these presented gave birth to the rather controversial idea that a possible loop hierarchical structure of magnetic coronal fields exists (Sylwester & Sylwester 1998b). According to this concept the magnetic field supporting the observed structures is organized into a hierarchical cascade where the larger structures are rooted at the summits (kernels) related to the smaller, lower lying structures. It is easy to interpret most of the high resolution observations of solar atmosphere within such a concept.

Inspection of sequences of SXT deconvolved images of many flares has revealed that soft X-ray flare emission patterns are not only spatially complicated but also fast evolving. This is especially apparent during the flare impulsive phase. The brightest fine structures observed at the summits and at footpoints change their position during flare evolution and the solar corona appears much more dynamic than was thought before. The time variations of soft X-ray emissivity pattern based on deconvolved SXT images have been investigated by Sylwester & Sylwester (1999c, d, 2001a). We have found it convenient to follow the changes of the flare brightness pattern by looking at the position of selected brightness iso-contours. The inspection of the history of iso-contour locations reveals "motions" of top and footpoint kernels during the flare. In Figure 3 the changes of maximum brightness location are presented for four events. Three of them occurred at the limb which makes it easy to distinguish between top and footpoint kernels while the flare of 13 November 1991 is a disc flare. It is seen that during 10 - 15 min of evolution presented in the figure the maximum emission changed its location suggesting a real motion of the kernels. For the 18 December 1991 event the observed pattern history of the 10 min evolution indicates a rather complicated arrangement of maximum brightness displacement. To clearly see the dynamics for this particular flare the reader is referred to the color animation at `www.cbk.pan.wroc.pl/publications/2002/high\_res/18dec91prod.avi`. For the flare of 4 October 1992 the movement of the maximum brightness location is observed in the direction away from the Sun. The quantitative measurement of the displacement observed using Be119 filter gives a distance of about 6000 km with a corresponding velocity of not less than about 6.6 km/sec. For the disc flare of 13 November 1991 the relatively large displacement of the most intense patch in the SW direction is pronounced. On the other hand, for the brightest kernel of the flare of 24 March 1993 only a small apparent displacement of maximum brightness is observed. In this case the radial motions are probably substantial. A similar pattern of evolution is observed for both the Al12 and Al01 images. At this point we should point out that it is not only the soft X-ray plasma that does change the location of its maximum emission during the flare evolution. When we analyze the MEM reconstructed HXT images and trace the maximum brightness location for different HXT energy channels a similar pattern of displacement as in the case of the SXT images is observed. This has been illustrated for the flare of 5 October 1992 in the paper by Sylwester & Sylwester (2001a). Feldman *et al.* (1994b) noticed that for non-impulsive flares the brightest region does not seem to move from its location at the top by more than one pixel. However, from our analysis of deconvolved images, both footpoint and summit kernels as a rule are displaced as the event progress. Occasionally we observe the maximum brightness location to be stable for relatively long time with a following rapid change of this location (as seen through all available SXT filters). We believe these displacements are related to the real motions of plasma blobs.

The physical conditions of plasma confined in flare kernels have been investigated by many authors. The analysis of SXT images (Acton *et al.* 1992, McTiernan *et al.* 1993, Feldman *et al.* 1994a, b, Doschek, Strong, & Tsuneta, 1995, Tomczak, 1997, Jakimiec *et al.* 1998) and of spectroscopic observations (Fludra *et al.* 1994, Khan *et al.* 1995, Doschek 1999) have been carried out. However, only the investigation of deconvolved images allows one to study the plasma parameters on small scales. We have studied (Sylwester & Sylwester 1998c, 1999b) the physical conditions within fine flaring structures observed at and near the limb. Several bright areas representing footpoint and summit kernels have been investigated. We concentrated on images in which the dominant kernel has reached its maximum intensity. In order to determine the kernel volume we assumed a "quasi-symmetric" geometry. The integration of Al12 and Be119 fluxes for all subpixels located within the selected kernel and using the filter ratio technique allowed us to determine the average electron temperature, emission measure, and density (assuming unit filling factor). Next we investigated the physical characteristics of plasma at this location and followed their time variations.

We have analyzed several limb, partly occulted flares in order to search for similarities and/or discrepancies between the temperature and emission measure patterns for the plasma confined in the footpoint and the summit volumes of flaring loops. It turned out that the kernels located at the summits are on average hotter by 2 - 3 MK in comparison with those located at the footpoint regions. Their average temperature is usually 10 - 15 MK during the rise phase and falls below 10 MK during the maximum and decay phase of the event. Their emission measures are commonly larger than those estimated for the footpoint kernels but the difference in the density is not large. However, the characteristic sizes of summit kernels are larger than those of footpoints, having typically a 1.5 - 2.5 arcsec diameter. The summit kernels are generally observed during the entire event evolution. The footpoint kernels as seen in soft X-rays are observed mostly during the rise and maximum phase and they usually fade-away during the decay phase. They constitute well localized, small sources with a characteristic size of usually less than 1.5 arcsec. The smallest diameter of footpoint kernels as obtained from the FWHM of the cross sectional brightness profile was ~ 0.8 arcsec. During the rise phase the footpoint kernels are composed of dense (maximum density $\sim 1 - 5 \times 10^{11}$ cm^{-3}) but not especially hot (6 - 9 MK) plasma. Usually the temperature of footpoint kernels is more or less constant during the phase of intense soft X-ray emission, in contrary to what is observed for summit kernels for which temperature tends to increase from the very beginning of the event. Later on (during the maximum and decay phase) the footpoint kernel temperature increases, becoming comparable with that of the top kernels. The time variations of the characteristic parameters (soft and hard X-ray fluxes, temperature, and density) for two footpoints and summit kernels are presented in Figure 4 for the "canonical" Masuda flare on 13 January 1992. It is worth noting that the time profile of the hard X-ray emission represented by HXT L channel flux resembles the soft X-ray emission (SXT Al12 and Be119 fluxes) for the Northern footpoint (left column) the best. The double maximum can be recognized in both the hard and soft X-ray lightcurves at the same time. The observed differences in physical conditions at the footpoint and summit (coronal)

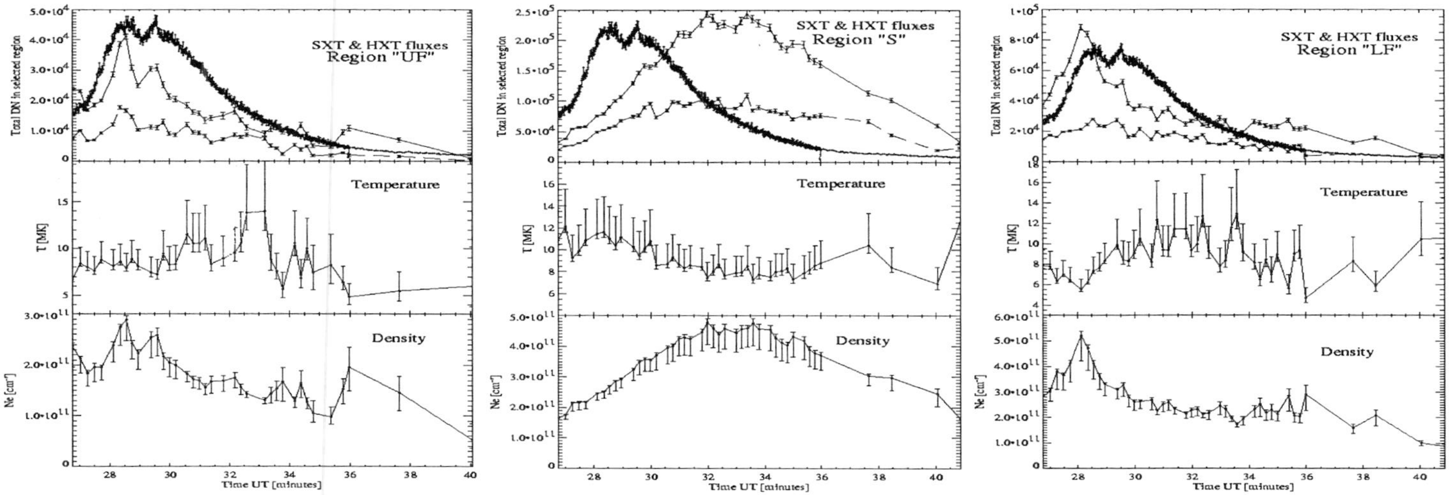

Fig. 4. Time variations of some important characteristics for individual structures of the Masuda flare on 13 January 1992. UF (left column), S (central column) and LF (right column) denote the North footpoint (upper), summit and South footpoint (lower) kernels respectively. In the upper row the following lightcurves are displayed: HXT L channel flux multiplied by the appropriate factor in order to bring it to scale (thick line), Al12 (middle) and Be119 (bottom) SXT fluxes. The middle row shows the course of kernels' plasma temperature and the bottom row the corresponding variations of plasma density.

areas can be useful in order to distinguish them for disc flares for which there height may not be obvious because of projection effects. Identification of kernels for disc flares can lead to better reconstruction of the magnetic field line "connectivity" pattern in the corona.

Using the deconvolved images and incorporating the fine image alignment technique we were able to determine maps of the distribution of important physical parameters in flares. An example is shown in Figure 5 (right panel). In the left panel the deconvolved soft (SXT Be119) and MEM reconstructed hard (HXT LO channel) X-ray images for three selected times during the evolution of the flare on 18 December 1991 are shown. At the bottom the lightcurves (*GOES* and HXT LO and M1 channels) are presented. The two upper rows on the right panel present maps of plasma temperature and emission measure at three times. The areas embraced by the solid curve are those for which the count rate is above 30 DN/subpixel (i.e. 750 DN/pixel) in both the Al12 and Be119 images. In the lowest row the "quasi" differential emission measure distributions (QDEM) together with the average temperature of the region are presented. The QDEM histogram has been constructed based on the temperature and emission measure maps. First the elements (subpixels) have been identified where the temperature falls into $(T, T+1)$ MK range. Next the emission measures for those elements have been summed and plotted on the vertical axis of the histogram for appropriate temperature ranges. We find that the bulk of the plasma in the flaring region is characterized by a relatively high temperature around 10 MK. (One must remember that in using the Al12 and Be119 filter ratio technique only plasma with a temperature higher than 5 MK can be probed.) The average temperatures for the entire areas selected are indicated in the figure. As might be expected the average temperature is the lowest (8.7 MK) during the flare decay phase. The corresponding average densities are: 7.6 x 10^{10}, 1.1 x 10^{11} and 1.4 x 10^{11} cm^{-3} respectively. The presence of a small amount of plasma with temperatures above 20 MK during the rise phase (the left and middle histogram) is common. This high temperature tail in the QDEM distribution is not seen during the decay phase. The high temperature component fades away after the main hard X-ray phase. It may be responsible for the radiation observed in the Fe XXV emission line in the spectra recorded by BCS.

The other important problem which can be investigated with deconvolved SXT images is the detailed examination of the spatial relationship between the soft and hard X-ray emission patches. As is presently known they do not exactly overlap. The hard X-ray sources are usually placed above the corresponding maxima of softer emission both for the summit and footpoint regions. This can be seen in the two upper rows of the left panel in Figure 5. In Figure 6 the relative location of the soft X-ray (seen in deconvolved SXT images) and hard X-ray emission (seen in the MEM reconstructed HXT images) is presented, which confirms these earlier findings unambiguously. In the figure the three frames obtained during the rise phase of the flare of 24 November 1992 at $\sim 10:02$ UT are shown. It is seen that the soft and hard X-ray emission patches are not co-spatial and the distance of their centers increases with time during the ~ 70 sec of evolution presented.

As a last point we would like to stress that the presence of persistent bright regions seen in the most flares observed by *Yohkoh* is hard to accommodate within proposed simple flare scenarios, as has been pointed out by many authors. The team from the Astronomical Institute of Wroclaw University (headed by Prof. J. Jakimiec) has worked out a flare model based on the hypothesis that turbulent energy release in solar flares takes place inside the entire volume of each kernel. This concept can help in resolving the problem of incompatibility between existing flare observations and theoretical scenarios. In a series of contributions (Jakimiec 1998, 1999, 2001a, b) it has been argued that flare kernels are strongly turbulent regions with tangled magnetic lines undergoing volumetric reconnections in many places at the same time. Khan *et al.* (1995) have added important supporting evidence to this hypothesis. They found significant non-thermal (turbulent) motions for loop-top flare kernels when investigating limb flares with only top kernel visible from behind the limb. In the Jakimiec model the turbulent flare kernels are supposed to be filled up with a number of tiny turbulent reconnection/acceleration regions, and thus the energy is expected to be released simultaneously in many closely packed transient current sheets within the kernels volume. Inside, the kernels' energy is transported by turbulent plasma motions. It has been shown that it is easy to produce a

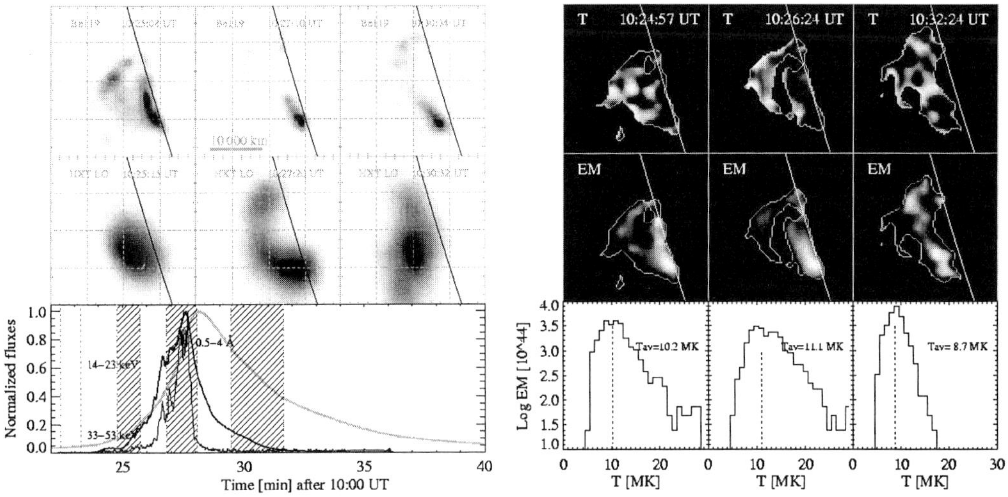

Fig. 5. *Left*: Composite of Be119 SXT deconvolved images (top row) and corresponding in time MEM reconstructed LO HXT images (middle row) for three moments during the 18 December 1991 flare evolution. The soft (*GOES*) and hard (LO and M1 HXT) X-ray normalized lightcurves are displayed in the bottom row. The hatched areas denote time intervals when the HXT fluxes have been integrated in order to reconstruct the images displayed in the middle row. *Right*: The distribution of main thermodynamic parameters (temperature - top row, emission measure - middle row and "quasi-differential" emission measure histogram - bottom row) for three moments (respective columns) during the flare evolution. The size of each frame is (40×40) arcsec.

turbulent kernel in a typical active region magnetic topology. Mixed magnetic polarities and their dynamics in young active regions cause the probability of interaction between three or even more flux tubes to be high there. If this happens a cascade of reconnections may develop with many transient reconnection regions where the magnetic energy is dissipated efficiently. According to this hypothesis MHD turbulence is the basic mechanism of flare energy release. It also gives a chance to explain the huge flux of the accelerated electrons which is responsible for the generation of the observed hard X-ray flare radiation. It is postulated by Jakimiec that the regions responsible for the most efficient particle acceleration are kernel boundaries, where the high-speed turbulence interacts with the surrounding magnetic fields. The plasma turbulence at the looptop has been regarded as a possible acceleration mechanism for electrons produced during solar flares also in a paper by Fletcher (1999).

CONCLUDING REMARKS

Results of inspection of imaging data from recent missions like *Yohkoh*, *SOHO* and *TRACE* emphasize the well-known fact that the solar corona is highly structured. In addition, interpretation of *Yohkoh* data has resulted in a consensus that the kernels are the dominant emitting structures in flare events. One may conclude that there are no simple flares. The simplicity of *GOES* soft X-ray lightcurves for flares is misleading. The complexity of the soft X-ray sources is evident from SXT, *SOHO*, and *TRACE* observations. High resolution ground based observations and hard X-ray flux variability provide additional arguments in favor of elementary flare complexity. The persistent presence of looptop kernels is hard to understand within the widely accepted chromospheric evaporation flare scenario. The studies of morphology and thermodynamics of flaring regions based on deconvolved SXT images led us to the conclusion that the non-uniformities are present in both density and temperature within these kernels. They must be a direct result of spatial and

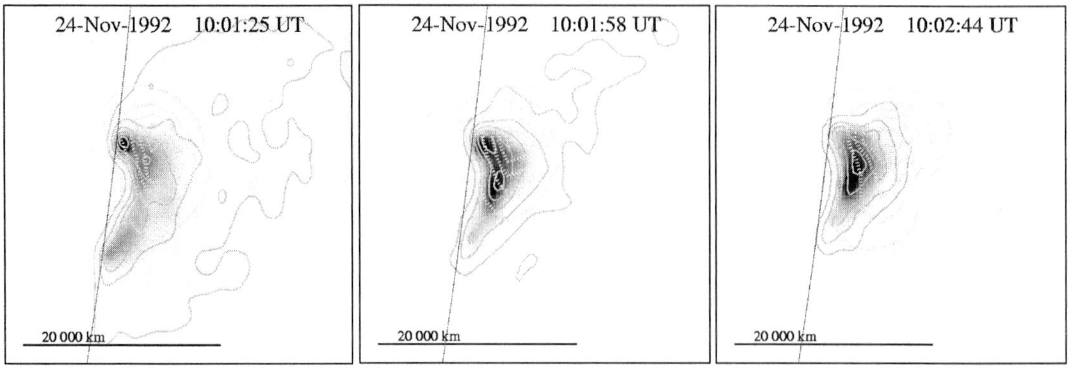

Fig. 6. Comparison of the soft and hard brightest area locations for three moments during the rise phase of the flare on 24 November 1992 at $\sim$ 10 UT. At the base the Al12 deconvolved images are presented for the times indicated. The solid line contours correspond to Be119 SXT images while the dashed contours for the M1 channel of HXT images.

temporal variations in the rate of volumetric flare heating. Despite the fact that *major* progress has been made recently in the field of solar flares we are still far from a complete understanding of the detailed mechanism of energy build up and release. It is evident that current observations of flares still lack the spatial resolution to verify present theoretical models. It is said again and again that fast imaging with better than 0.1 arcsec resolution over a small field of view, together with a wide energy coverage, and with support of spectrographic observations seems to be needed in order to investigate the distribution of temperature and density together with magnetic and velocity fields. This will allow energy release, transport, and dissipation processes to be studied in more detail. A major effort of the international solar physics community for the next decade will be the Solar-B space mission. However the presence of a turbulent regime within the flaring kernels can not be observationally confirmed even by Solar-B, since the range of scales associated with the dissipation of energy and the cascade are expected to operate on spatial scales beyond the present-day spatial resolution limit. Only missions currently in the planning stage aimed for an angular resolution of the order of a tenth of an arcsec would possibly permit for such observations.

ACKNOWLEDGEMENTS

This work has been supported by the Grant 2.P03D.024.17 of Polish Committee for Scientific Research.

REFERENCES

Acton, L.W., Feldman, U., Bruner, M.E., Doschek, G.A., Hirayama, T. *et al.*, The Morphology of 20×10^6 K Plasma in Large Non-Impulsive Solar Flares, *PASP,* **44**, L71, (1992).

Doschek, G.A., The Electron Temperature and Fine Structure of Soft X-ray Solar Flares, *ApJ,* **527**, 426, (1999).

Doschek, G.A., Strong, K.T., and Tsuneta, S., The Bright Knots at the Tops of Soft X-ray Flare Loops: Quantitative Results, *ApJ,* **440**, 370, (1995).

Feldman, U., Hiei, E., Phillips, K.J.H., Brown, C.M., and Lang, J., Very Impulsive Solar Flares Observed with the Yohkoh Spacecraft, *ApJ,* **421**, 843, (1994a).

Feldman, U., Seely, J.F., Doschek, G.A., Strong, K.T., Acton, L.W. *et al.*, The Morphology of the 10 MK Plasma in Solar Flares: Nonimpulsive Flares, *ApJ,* **424**, 444, (1994b).

Fletcher, L., Looptop Hard X-ray Sources, *ESA SP-448*, 693, (1999).

Fludra, A., Jakimiec, J., Tomczak, M., Culhane, L.J., and Acton, L.W., Long Duration Events in Magnetic Arcades and Large Loops, *NRO, Report No. 360,* 393, (1994).

Foukal, P.V., The Pressure and Energy Balance of the Cool Corona over Sunspots, *ApJ,* **210**, 575, (1976).

Golub, L., Structure of the Solar X-ray Corona, *ASP Conference Series,* **26**, 193, (1992).

Golub, L., Difficulties in Observing Coronal Structures, *Solar Phys.,* **174**, 99, (1997).

Golub, L., Herant, M., Kalata, K., Lovas, I., Nystrom, G. *et al.*, Sub-arcsecond Observations of the Solar X-ray Corona, *Nature,* **344**, 842, (1990).

Gomez, D.O., Martens, P.C.H., and Golub, L., Normal Incidence X-ray Telescope Power Spectra of X-ray Emission from Solar Active Regions. II. Theory, *ApJ,* **405**, 773, (1993).

Harra-Murnion, L.K., Culhane, J.L., Hudson, H.S., Fujiwara, T., Kato, T., and Sterling, A.C., Isolating the Footpoint Characteristics of a Solar Flare Loop, *Solar Phys.,* **171**, 103, (1997).

Herant, M., Pardo, F., Spiller, E., and Golub, L., Flares Observed by the Normal Incidence X-ray Telescope on 1989 September 11, *ApJ,* **376**, 797, (1991).

Hudson, H.S., Strong, K.T., Dennis, B.R., Zarro, D., Inda, M. *et al.*, Impulsive Behavior in Solar X-radiation, *ApJ Letters,* **422**, L25, (1994).

Jakimiec, J., What we Can learn from X-ray Observations of Stellar Flares, *Physica Scripta,* **T77**, 137, (1998).

Jakimiec, J., Tomczak, M., Falewicz, R., Phillips, K.J.H., and Fludra, A., The Bright Loop-Top Kernels in Yohkoh X-ray Flares, *A&A,* **334**, 1112, (1998).

Jakimiec, J., Turbulent Energy Release in Solar Flares, *ESA SP-448,* 729, (1999).

Jakimiec, J., Small-Scale Reconnection in Solar Flares, *Adv. Space Res.,* in press, (2001a).

Jakimiec, J., Energy Release in Solar Flares, *Adv. Space Res.,* in press, (2001b).

Khan, J.I., Harra-Murnion, L.K., Hudson, H.S., Lemen, J.R., and Sterling A., Yohkoh Soft X-ray Spectroscopic Observations of the Bright Loop-Top Kernels, *ApJ Letters,* **452**, L153, (1995).

Masuda, S., Kosugi, T., Hara, H., Tsuneta, S., and Ogawara,Y., A Loop-Top Hard X-Ray Source in a Compact Solar Flare as Evidence for Magnetic Reconnection, *Nature,* **371**, 495, (1994).

McTiernan, J.M., Kane, S.R., Loran, J.M., Lemen, J.R., Acton, L.W., *et al.*, Temperature and Density Structure of 1991 November 2 Flare Observed by the Yohkoh Soft X-ray Telescope and Hard X-ray Telescope, *ApJ Letters,* **416**, L91, (1993).

Nitta, N.V., Sato, J., and Hudson, H.S., The Physical Nature of the Loop-Top X-ray Sources in the Gradual Phase of Solar Flares, *ApJ,* **552**, 821, (2001).

Roumeliotis, G., A Novel Maximum-Likelihood Method of Image Reconstruction with Increased Sampling, *ApJ,* **452**, 944, (1995).

Sams III, B.J., Golub, L., and Weiss, N.O., X-ray Observations of Sunspot Penumbral Structure, *ApJ,* **399**, 313, (1992).

Sheeley, Jr. N.R., Bohlin, J.D., Brueckner, G.E., Purcell, J.D., Scherrer, V., and Tousey, R., XUV Observations of Coronal Magnetic Fields, *Solar Phys.,* **40**, 103, (1975).

Siarkowski, M., Sylwester, J., Jakimiec, J., and Tomczak, M., Improvement of SXT Image Alignment in Order to Obtain High-Resolution Temperature Maps, *Acta Astron.,* **46**, 15, (1996).

Sylwester, J., High-Resolution Temperature Maps of Flares from Deconvolved SXT Images, *JOSO Annual Report '95,* 131, (1995).

Sylwester, J., Sylwester, B., and Siarkowski, M., ANDRIL Algorithm for Deconvolution of SXT Images, *ASP Conference Series,* **111**, eds. R.D. Bentley and J.T. Mariska, 244, (1996a).

Sylwester, B., Sylwester, J., and Siarkowski, M., Fine Structures Observed on Deconvolved SXT Images, *ASP Conference Series,* **111**, eds. R.D. Bentley and J.T. Mariska, 249, (1996b).

Sylwester, J., and Sylwester, B., ANDRIL - Maximum Likelihood Algorithm for Deconvolution of SXT Images, *Acta Astron.,* **48**, 519, (1998a).

Sylwester, J., and Sylwester, B., Evolution of Flaring Structures, *ASP Conference Series,* **155**, eds. C.E. Alissandrakis and B. Schmieder, 381, (1998b).

Sylwester, B., and Sylwester, J., Physical Conditions in Flaring Loops, *JOSO Annual Report '97,* 97, (1998c).

Sylwester, J., and Sylwester, B., Reconstruction of Images with Poisson Noise, *Acta Astron.,* **49**, 189, (1999a).

Sylwester, B., and Sylwester, J., Reconstruction of Coronal Magnetic Fields from Deconvolved SXT Images, *JOSO Annual Report '98,* 105, (1999b).

Sylwester, B., and Sylwester, J., Flaring Structures Observed in Deconvolved SXT Images, *Acta Astron.,* **49**, 85, (1999c).

Sylwester, B., and Sylwester, J., The Properties of Flares Produced within AR 6919, *ESA SP-448,* 895, (1999d).

Sylwester, B., and Sylwester, J., Physical Conditions within Flare Kernels, *Adv. in Space Res.,* in press, (2001a).

Sylwester, B., and Sylwester, J., Thermodynamics of Partly Occulted Limb Flares, *ESA SP-477,* in press, (2001b).

Sylwester, J., and Sylwester, B., Behind the Limb X-ray Sources as Seen by Yohkoh, *these proceedings,* (2002).

Tomczak, M., The Impulsive Phase of the Arcade Flare of 28 June 1992, 14:24 UT, *A&A,* **317**, 223, (1997).

FINE STRUCTURE INSIDE FLARE RIBBONS AND ITS TEMPORAL EVOLUTION

A. Asai[1], S. Masuda[2], T. Yokoyama[3], M. Shimojo[3], H. Kurokawa[1], K. Shibata[1], T. T. Ishii[1], R. Kitai[1], H. Isobe[1], and K. Yaji[4]

[1]*Kwasan and Hida Observatories, Kyoto University, Yamashina-ku, Kyoto 607-8471, Japan*
[2]*Solar-Terrestrial Environment Laboratory, Nagoya University, Toyokawa, Aichi 442-8507, Japan*
[3]*Nobeyama Radio Observatory, Minamisaku, Nagano, 384-1305, Japan*
[4]*Kawabe Cosmic Park, Kawabe, Wakayama, 649-1443, Japan*

ABSTRACT

We observed an X2.3 flare, which occurred on 10 April 2001, in Hα with the Sartorius Telescope at Kwasan Observatory, Kyoto University. Thanks to the short exposure time used for the flare, the Hα images showed the fine structure in the flare ribbons. First, we examined the temporal and spatial evolution of the Hα kernels. We identified the conjugate footpoints in each flare ribbon by calculating cross-correlation functions of the light curves of the Hα kernels. We found that these footpoints are really connected by the flare loops seen in extreme-ultraviolet images obtained with the *Transition Region and Coronal Explorer*. We also followed the evolution of the energy release site during the flare. Then, we compared the spatial distribution of the hard X-ray (HXR) sources with those of the Hα kernels. While many Hα kernels are found to brighten successively in the development of the flare ribbons, the HXR sources are locally confined to some special Hα kernels where the photospheric magnetic field is sufficiently strong. We estimated the energy release rates at each radiation source, and found that they are high enough at the HXR sources to explain the difference in appearance between the Hα and HXR images.

INTRODUCTION

Non-thermal particles are thought to be accelerated near the flare energy release site, and to flow with very high speed along the flare loops. The speed is as high as about one-third of the speed of light ($\sim 10^5$ km s^{-1}). Due to this high speed they bombard the chromospheric plasma at both the footpoints of the flare loops almost simultaneously. The temporal evolutions of the intensities at both footpoints are very similar (Kurokawa *et al.* 1988, Sakao 1994). Although the mechanism responsible for the particle acceleration remains open to debate, the location and time of the acceleration and/or the energy release can be determined by identifying the highly-correlated pairs of footpoints and the precipitation times of the non-thermal particles.

In this paper we first determine the precise location and precipitation times using Hα data of the April 10 2001 flare, and examine the evolution of the energy release sites. The locations and light curves of the HXR sources show high correlations with those of the Hα kernels, because the bombardments stimulate the excitation and ionization of hydrogen atoms which in turn causes enhanced Hα emission in a short time (Kurokawa *et al.* 1988, Trottet *et al.* 2000. Wang *et al.* 2000. Qiu *et al.* 2001). Furthermore, flare observations with higher spatial resolution are achieved by using Hα images rather than by using HXR

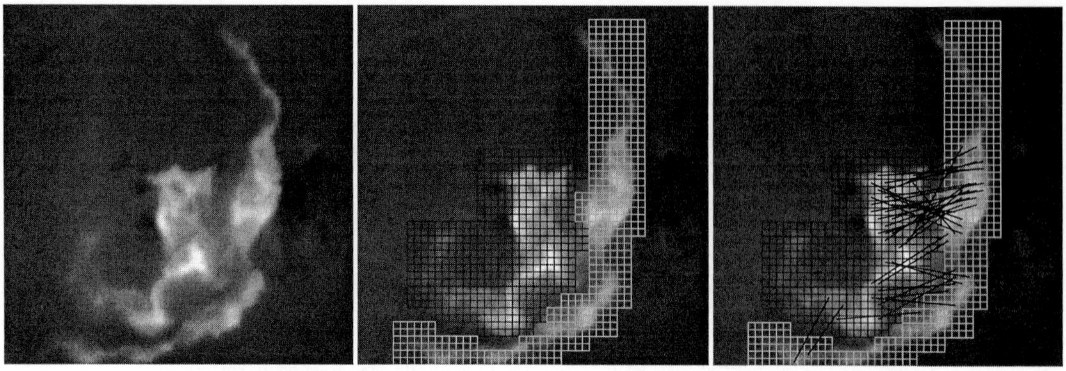

Fig. 1. Method for analyzing the Hα data. Each image is $164'' \times 164''$ in area. (The color images are available on the URL site "http://isass1.solar.isas.ac.jp/nuggets/2002/020125/020125.html")

and/or microwave images. Therefore, we can investigate the precipitation sites of the particles with higher spatial resolution using Hα images (Kitahara & Kurokawa 1990).

Second, we compare the locations of the HXR sources with those of the Hα kernels and discuss what is essential for the energy release at the HXR sources. HXR images obtained with the hard X-ray telescope (HXT; Kosugi *et al.* 1991) aboard *Yohkoh* (Ogawara *et al.* 1991) often show only a few HXR sources, except in the case of the HXR flare ribbons in the Bastille Day event on 14 July 2000 (Masuda 2001, private communication). HXR sources are accompanied by Hα kernels in many cases, but many Hα kernels are not accompanied by HXR sources. This difference of spatial distributions between Hα and HXR sources can be explained by the difference of the amount of released energy at each source and the low dynamic range of the HXT. The energy release rate in solar flares is considered to depend on magnetic field strength. To check that idea, we measured the photospheric magnetic field strength at each source and investigated its relation with the amount of released energy.

CONJUGATE FOOTPOINTS

The X2.3 flare occurred in the active region NOAA 9415 at 05:10 UT, 10 April 2001. The flare showed a typical two-ribbon structure. The Hα images of the flare were obtained with the Sartorius Telescope (*Sartorius*) at Kwasan Observatory, Kyoto University. The left panel in Figure 1 shows an Hα image of the flare. During the observation the exposure time was properly regulated so that the fine structure inside the flare ribbons, like Hα kernels, can be clearly seen without saturating.

To find the highly-correlated pairs of Hα footpoints we devised a new method for analyzing Hα data. Firstly, we divided the Hα flare ribbons into fine meshes (see the middle panel in Figure 1). The dark gray mesh in Figure 1 shows the positive polarity and the light gray mesh shows the negative polarity of the ribbons. Secondly, using cross-correlation functions of the light curves, we identified the conjugate points in each mesh. The black lines in the right panel in Figure 1 connect these highly-correlated pairs. Figure 2 shows two examples of the light curves of the highly-correlated pairs. The solid lines and dotted lines in Figure 2 show the light curves from the positive side and from the negative side, respectively. Thirdly, we confirmed whether the highly-correlated pairs were really connected by the flare loops seen in the extreme-ultraviolet (EUV) images obtained with the *Transition Region and Coronal Explorer* (*TRACE*; Handy *et al.* 1999, Schrijver *et al.* 1999). The *TRACE* 171 Å images clearly show the post-flare loops which confine 1 MK-plasma. We found that 171 Å flare loops really connected almost all the pairs of Hα kernels.

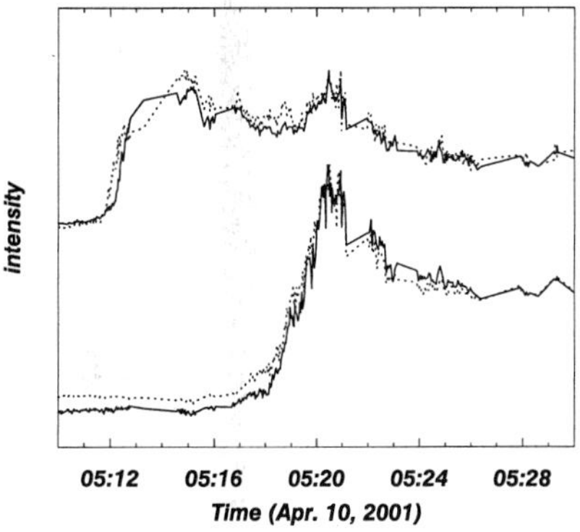

Fig. 2. Examples of the light curves of the highly-correlated pairs (scaled arbitrarily). Solid lines show the light curves with positive polarity and dotted lines show those with negative polarity.

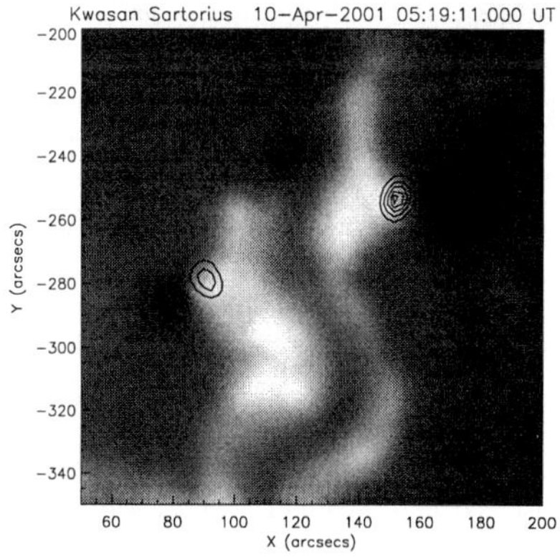

Fig. 3. Comparison of the spatial distribution between Hα kernels and HXR sources. A contour image of HXT (H band; 53 - 93 keV) is overlaid on an Hα image obtained with the *Sartorius*. Contour levels are 20%, 40%, 60%, 80%, and 95% of the peak intensity.

From the times and the locations of the brightenings of each Hα pair we know where and when the flare loops were connected, that is where and when the energy release occurred. This gives us information about the site and the time of energy release. Thus we can follow the whole history of the energy release.

Hα KERNELS AND HXR SOURCES

Figure 3 shows images of the flare in Hα and HXR. In the Hα images we can see many Hα kernels all over the flare ribbons. On the other hand, the HXT images show only one or two sources (see the contour images in Figure 3). The HXR sources correspond to one of the Hα pairs. This difference of appearance is due to the low dynamic range of the HXT images. Only the strongest sources are seen. The weaker sources are buried in the noise of the HXT images. The dynamic range of the HXT images is about 10. If the released energy at the HXR sources is at least 10 times larger than that at the Hα kernels without accompanying HXR emission then the difference of appearance can be explained.

To examine the difference in the amount of the released magnetic energy, we measured the photospheric magnetic field strengths of each Hα kernel with the Michelson Doppler Imager (MDI; Scherrer *et al.* 1995) aboard the *Solar and Heliospheric Observatory* (*SOHO*; Domingo *et al.* 1995). The magnetic field strengths in the Hα kernels are higher than in the other regions, 400 G on average, and those at the HXR sources are especially high ($\sim$ 1200 G). The photospheric magnetic field strength in the HXR sources is about 3 times larger than the field strength in the Hα kernels without accompanying HXR emission.

Here we assume that the HXR intensity observed with the HXT is proportional to the energy release rate $\frac{dE}{dt}$ due to magnetic reconnection. This rate is proportional to the product of Poynting flux into the reconnection region and the area of the region (Isobe *et al.* 2002). The area of the reconnection region is not thought to change so much and is thought to be independent of the magnetic field strength. Therefore, $\frac{dE}{dt} \propto B^2 v_i$, where B is the magnetic field density in the photosphere and v_i is the inflow velocity into the reconnection region. v_i has some dependence on B. The Sweet-Parker type reconnection suggests $v_i \propto B^{0.5}$, which is the smallest dependence. On the other hand, the largest dependence is attained if the reconnection rate is constant implying $v_i \propto B$. Therefore, $\frac{dE}{dt} \propto B^\gamma$ where $2.5 \leq \gamma \leq 3.0$. Since B at the HXR sources is 3 times larger, the energy release is 16 - 27 times stronger for the Hα sources accompanied by HXR sources than for the other Hα kernels. This is sufficiently larger than the dynamic range of the HXT data, and can explain the difference in appearance.

SUMMARY

We have developed a new method for analyzing Hα data. Using the Hα images we have investigated the precipitation of non-thermal particles into the chromosphere with higher spatial resolution than actually possible with hard X-ray and microwave data. This method enables us to follow the evolution of the energy release sites. We have also examined the difference between the magnetic field strength of the HXR sources and of the ordinary Hα kernels. The magnetic field strengths in the Hα kernels accompanied by HXR sources are sufficiently higher than those at the other Hα kernels, and the energy release rates are also sufficiently strong to explain the difference between the appearance of the Hα and HXR images.

ACKNOWLEDGEMENTS

We first would like to acknowledge the anonymous referee for his/her careful reading and useful comments that have improved this paper. We wish to thank all the members of Kwasan and Hida observatories for their support during our observations. We also would like to thank all the members who attended the *Yohkoh* 10th Anniversary Meeting for fruitful discussions. We made extensive use of the *TRACE* Data Center, *Yohkoh* Data Center and *SOHO* MDI Data Service.

REFERENCES

Domingo, V., Fleck, B., & Poland, A.I., *Solar Phys.*, **162**, 1 (1995).

Handy, B.N., Acton, L.W., Kankelborg, C.C., Wolfson, C.J., Akin, D.J., *et al.*, *Solar Phys.*, **187**, 229 (1999).

Isobe, H., Yokoyama, T., Shimojo, M., Morimoto, T., Kozu, H., *et al.*, *ApJ*, **566**, 528 (2002).

Kitahara, T., & Kurokawa, H., *Solar Phys.*, **125**, 321 (1990).

Kosugi, T., Masuda, S., Makishima, K., Inda, M., Murakami, T., *et al.*, *Solar Phys.*, **136**, 17 (1991).

Kurokawa, H., Takakura, T., & Ohki, K., *PASJ*, **40**, 357 (1988).

Ogawara, Y., Takano, T., Kato, T., Kosugi, T., Tsuneta, S., *et al.*, *Solar Phys.*, **136**, 1 (1991).

Qiu, J., Ding, M.D., Wang, H., Gallagher, P.T., Sato, J., *et al.*, *ApJ*, **554**, 445 (2001).

Sakao, T., Ph.D. thesis, University of Tokyo (1994).

Scherrer, P.H., Bogart, R.S., Bush, R.I., Hoeksema, J.T., Kosovichev, A.G., *et al.*, *Solar Phys.*, **162**, 129 (1995).

Schrijver, C.J., Title, A.M., Berger, T.E., Fletcher, L., Hurlburt, N.E., *et al.*, *Solar Phys.*, **187**, 261 (1999).

Trottet, G., Rolli, E., Magun, A., Barat, C., Kuznetsov, A., *et al.*, *A&A*, **356**, 1067 (2000).

Wang, H., Qiu, J., Denker, C., Spirock, T., Chen, H., *et al.*, *ApJ*, **542**, 1080 (2000).

3-D STRUCTURE OF ARCADE TYPE FLARES DEDUCED FROM SOFT X-RAY OBSERVATIONS OF A HOMOLOGOUS FLARE SERIES

S. Morita, Y. Uchida, and S. Hirose

Physics Department, Science University of Tokyo, 1-3 Kagurazaka, Shinjuku, Tokyo 162-8601, Japan

ABSTRACT

There occurred three flares in active region NOAA 7070 over seven days in February 1992. One of them was seen at the east limb of the Sun, and the others were seen on disk. After confirming the homology among these flares, we used them to derive the 3-D coronal structure by making use of the images at some common phases, but seen along three different lines of sight by *Yohkoh*/SXT. We found some new characteristic features: The so-called "cusped arcade" turned out to be different from the "flare arcade" as widely conceived. Rather it is a thin "elongated arch" seen with a shallow oblique angle. Four magnetic sources participate in these flares in an essential way. An X-ray "blob" with a heated surface (rising dark filament) was *pulled out* by the expansion of high loops, which connect the top of the preflare cusp back to the photosphere on both sides, in the preflare phase of February 21. This opposite to the X-ray "blob" pulling the surrounding loops open as assumed in the "classical" flare models.

INTRODUCTION

In the solar flare problem no satisfactory model that really explains the observed characteristics has been established. One of the reasons for this is the restrictions in the observational data: they lack information about the third dimension. There have some effort to determine 3-D structures of flares, but it was difficult.

In the present paper we note the homology of three flares which occurred in active region NOAA 7070 late in February 1992, namely, February 21 (limb flare), 24, and 27 (disk events). Since we found that these three flares constituted a homologous series, we could derive a 3-D coronal structure by comparing the images obtained at some common phases from these three different lines of sight by *Yohkoh*/SXT. In the limb event of February 21, we could use the distribution of photospheric magnetic sources to identify the positions of the footpoints of the coronal loops quite reasonably.

EXAMINATION OF THE HOMOLOGY

Detailed comparison of these three events (Figure 1) tells us that the three flares are extended entities. The structure extended toward the north-east and south-west coherently brightened and faded, showing that these are also important parts of these flares. It is seen in Figure 2 that four magnetic regions are involved in these events. We made transformed maps of the photospheric magnetic field distribution, using the data from Kitt Peak, for the 8 days covering the period of these flares. By transforming the observed maps to maps viewed from above (Morita *et al.* 2001) it was found that the distribution of the magnetic field had

(i) four major sources (**a**, **b**, **c**, and **d**, **b=b1+b2**) that were involved in these flares, and (ii) no remarkable changes occurred in the large scale distribution **a**, **b**, **c**, and **d** during the period. There was, for example, no large scale re-shearing before the next event. There are, instead, many small scale changes near the main magnetic sources in the central region including changes in **b2**. Since the coronal loop structures are rooted in the photospheric magnetic field below, we could also see the homology in the X-ray images of the February 24 and 27 events. We found that the shapes of heated coronal loops and the development are fairly similar in these two (Figure 1). With this in mind, we compare these with the frames of the corresponding phases of February 21 event. We found that there exist corresponding coronal features in the February 21 event, and this suggests that these features are the corresponding structures viewed from different directions (Figure 2).

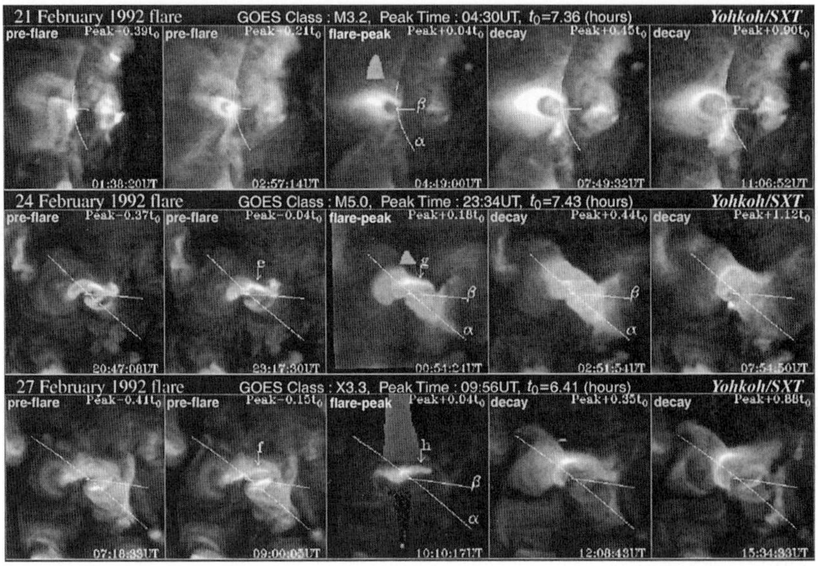

Fig. 1. Time development of the three flares in a homologous flare series of February 1992, observed by *Yohkoh*/SXT.

3-D STRUCTURE OF ARCADE TYPE FLARES COMPOSED FROM THE VIEWS FROM THREE DIFFERENT SIGHT LINES

Based on the confirmation of the homology among these three flares of 1992 February 21, 24, and 27, we examined the 3-D coronal structures of these arcade type flares. We consider the images of these flares as the projections of the "same" structure on planes perpendicular to these three different lines of sight. We arranged the images of the three flares according to their time development in Figure 1. We defined the corresponding stages using the normalized times. The top, the middle and the bottom rows show the time development of the 1992 February 21, 24, and 27 flares respectively. The columns correspond from left to right, to two preflare stages, the peak stage, and two decay phase stages. From the detailed analyses of these we obtained several new findings that follow:

(1) The so-called "cusped arcade" at the maximum phase in the well-known 1992 February 21 flare is, contrary to the general view, a "thin elongated arch" seen with a shallow oblique angle. It is *not* the "flare arcade" seen on axis as widely conceived! The line-of-sight to the flare cusp of February 21 (see the lines named β in Figure 1) made an angle of about 30 degrees with the direction of the axis of the true "flare arcade" that comes up only in later phases (see the lines named α in Figure 1). Moreover, this elongated arch coincided roughly with a diagonal of the main body of the "flare arcade" that came up later.

(2) The magnetic structure causing the flare as a whole turned out to be a structure with quadrupolar magnetic sources (**a**, **b**, **c**, and **d** in Figure 2) confirming Uchida *et al.*'s (1999) model, initially proposed in Uchida (1980). The relative locations of these four characteristic sources remained unchanged throughout the period of this homologous flare series, determining the fundamental behavior of this homologous series.

(3) An X-ray "blob" was *pulled out by the expansion of high loops* which connect the top of the preflare cusp back to the photosphere on both sides in the preflare phase of February 21 flare. It is *not* that the rising blob pulls open the loops as supposed in the "classical" models, but rather that the blob is pulled by the rising loops out of the valley near the top of the preflare cusp (Figure 3).

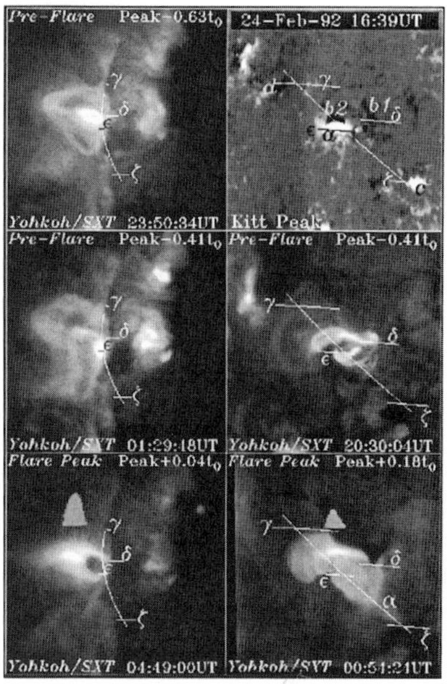

Fig. 2. Defining the location of footpoints of the coronal structures.

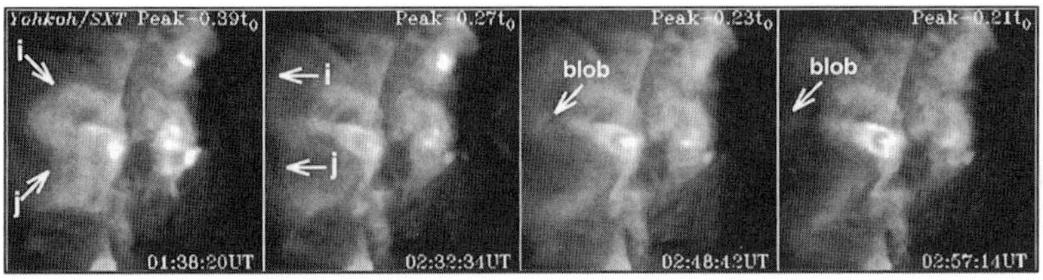

Fig. 3. *Yohkoh*/SXT images of expansion of high altitude faint connections with a rising blob.

After this analysis we tried to identify the detailed positions of the footpoints of some characteristic "loops" by examining the 3-D structures of those loops (see Figure 2: In the left column, γ, δ, ϵ, and ζ are the footpoints of some characteristic "loops" seen at the limb. The lines of sight passing through these points are rotated properly in the other images and properly projected.). Considering the above findings, and by looking into the 3-D structure of these arcade-type flares, we arrived at, amongst others, the following understanding: this flare series has a "cusp", and we take this to mean that there is a magnetic X-point located above the "cusp". In addition, from the above examinations of detailed positions of the footpoints of characteristic "loops", the positions of the footpoints of the "cusp" (which is not the flare arcade as generally thought as shown above!) correspond to the positions of the both ends of the preflare reversed-S-shape structure. This means that the magnetic X-point is located above the reversed-S-shape structure in the preflare stage (in the top view of this configuration). From this, we could deduce the 3-D structures of the bright loops seen in the preflare stage of these homologous arcade-type flares by simple geometrical

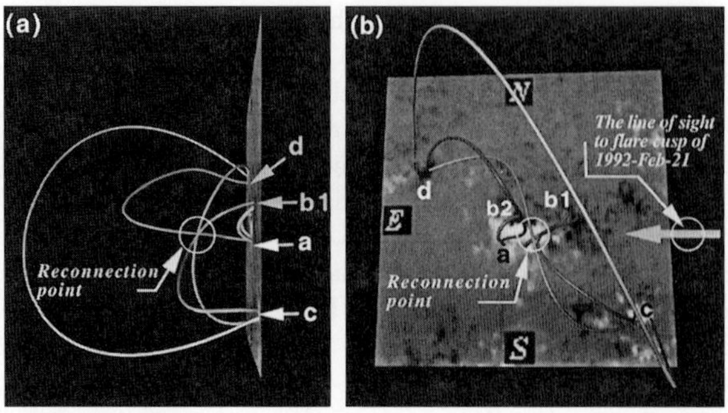

Fig. 4. 3-D schematic pictures of an empirical model at a preflare stage. This structure can explain several findings, if reconnection takes place at the point indicated in (a) and (b).

considerations as the structure connecting **a** to **b1**. Figure 4 gives the schematic pictures of the 3-D structures empirically derived for the preflare stage from our "stereoscopic" observation. These two pictures are for two different lines-of-sight, (a) for the line-of-sight of February 21, and (b) for line-of-sight of February 24. Needless to say, these loops are only the empirically derived counterparts of bright loops, and other magnetic field lines that are less visible fill the rest of the space between these flux tubes in Figure 4 in a consistent way.

DISCUSSION AND CONCLUSION

We deduced a 3-D empirical model for the bright loops in the preflare stage by using three flares in a homologous series in February 1992. In this empirical model there is a magnetic X-point between the two flux tubes, which are the connected to the photospheric sources **a** to **d** and **b1** to **c**, respectively. If we postulate that magnetic reconnection occurred at the X-point, the lower parts of the two flux tubes are reconnected to form a new loop **a** to **b1** with a shallow angle to the line-of-sight of February 21, and may well explain our elongated arch which was previously taken in error to be the flare arcade seen axis-on (see Figure 4 (a)). The upper counterpart of the product of the reconnection, a connection from **d** to the reconnection point, and from there to **c**, will be seen to rise as an expanding loop *pulling a "blob" out of the dip* upward at the flare-onset (see marks **i** and **j** in Figure 3).

We also examined another similar event in the July 19 1999 flare by using *TRACE* data, and we can see some characteristic features that are suggestive of the existence of a similar quadrupolar magnetic field configuration causing flares (see Uchida *et al.* 2002).

REFERENCES

Morita, S., Uchida, Y., Hirose, S., Uemura, S., and Yamaguchi, T., *Solar Phys.*, **200**, 137 (2001).

Uchida, Y., in *Solar Flares, Skylab Workshop*, ed. P.A. Sturrock, p. 67, University of Colorado Press (1980).

Uchida, Y., Hirose, S., Cable, S., Morita, S., Torii, M., Uemura, S., and Yamaguchi, T., *PASJ*, **51**, 553 (1999).

Uchida, Y., Title, A., Kubo, M., Morita, S., and Hirose, S., in preparation (2002).

DYNAMICS OF CORONAL MAGNETIC FIELDS INFERRED FROM MULTI-FREQUENCY RADIO OBSERVATIONS OF A SOLAR FLARE

E. Correia[1], J.-P. Raulin[2], G. Trottet[3], and P. Kaufmann[2]

[1]*INPE-CRAAM, Universidade Presbiteriana Mackenzie, Rua da Consolação 896, 01302-907 São Paulo, SP, Brazil*
[2]*CRAAM, Universidade Presbiteriana Mackenzie Part time researcher at CCS/Unicamp, Campinas, Brazil*
[3]*Observatoire de Paris, Section de Meudon, DASOP, CNRS-UMR 8645, F-92195 Meudon France*

ABSTRACT

We analysed multi-frequency radio and hard X-Ray observations of an M1.5 solar flare that occurred on November 5, 1998. The event was observed at various discrete radio frequencies from decimetric to microwave wavelengths, covering a wide range of coronal altitudes. Using combined imaging radio and hard X-Ray emissions it was possible to follow the dynamics of solar magnetic fields from the low corona up to about 0.5 solar radius above the photosphere. The radio spectra show different characteristics during event evolution, suggesting the appearance of new sources. The imaging analysis at hard X-Rays and decimetric/metric waves confirmed the presence of various and distinct emitting sources during the event, suggesting a dynamic and complex magnetic topology, probably associated with different injections of non-thermal electrons.

OBSERVATIONAL RESULTS

The event was observed at 7 GHz with the Itapetinga Radio Polarimeter, at various fixed radio frequencies with the Radio Solar Telescope Network (RSTN), and from 435 to 164 MHz at decimetric range with the Nançay Radio Heliograph (NRH/France). It was also observed at hard X-Rays with *CGRO*/BATSE, and SXT and HXT experiments on board *Yohkoh* (Figure 1). Based on the radio spectra in Figure 2a, the event was divided into four time intervals as marked in time profiles. The radio spectra in interval I show only a plasma emission component, without significant emission at microwaves; while the subsequent intervals show also a gyrosynchrotron component with turnover frequency at 5, 7 and 9 GHz, respectively, and also an enlargement. The time profiles at dm/m range show various fast drift bursts during the impulsive phase. Most of them, are bidirectional type III bursts, which permitted estimating the plasma density in the acceleration site. The results suggest that the bursts become more complex and the start frequency increases, as the event evolves. The hard X-Ray (HXR) spectra, from BATSE data, were obtained near the maximum of the main flux peaks. The spectra were fitted by a double power law which show an energy break from 80 to 120 keV, for the consecutive peaks, and a hardening of the spectral indices. The radio images, which have significant differences, are shown in Figure 2b. The source spatial distribution suggests a very distinct magnetic topology in each time shown. In interval II, it suggests that the energetic electrons are propagating in an open large scale loop, extending in the westward direction. Afterwards there is an appearance of a new 435 MHz source, in the northwest, suggesting an interaction between two large scale loops anchored in the two 435 MHz sources. The last image suggests that only one large scale loop is present, which is anchored in the newest 435 MHz source. Figure 2c shows the HXR images, for the main peaks, obtained from *Yohkoh*. They show two sources at L energy channel, which are well separated in the last two peaks, while in the higher energy channels only the left source is present. These images suggest that the two sources at L channel are the tops of small loops, where only the left one has emission from the footpoints, which are not resolved.

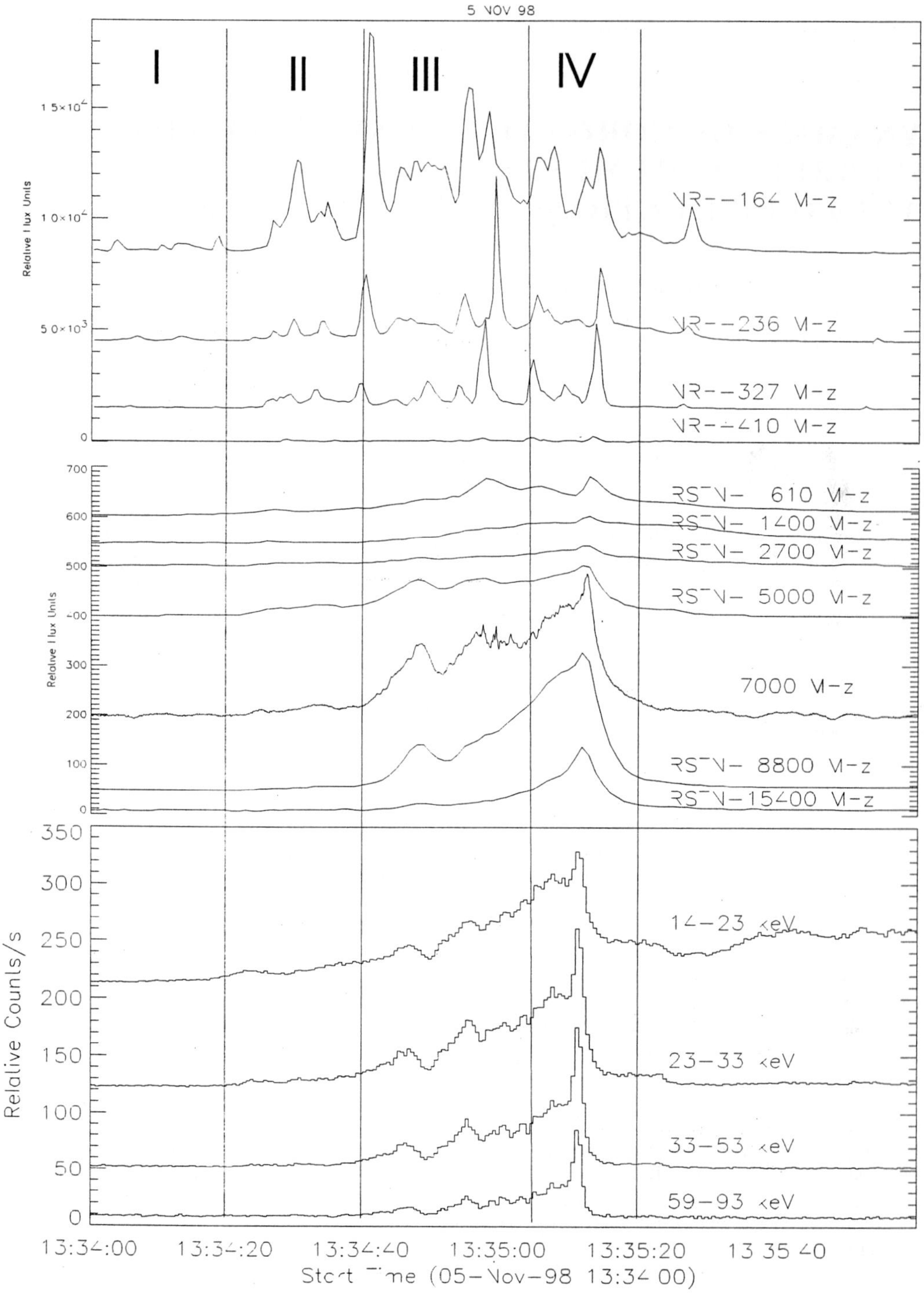

Fig. 1. Time profiles at various radio frequencies and at X-Rays from *Yohkoh*.

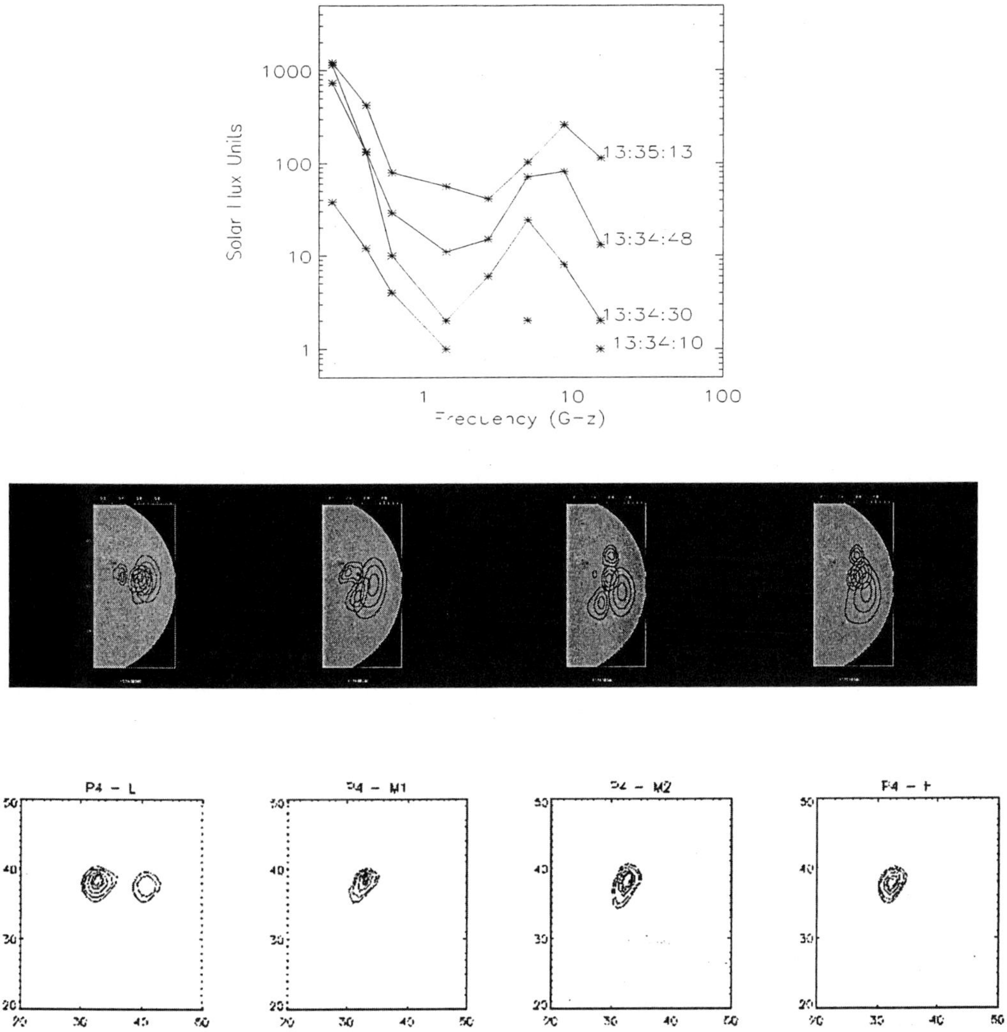

Fig. 2. (a: top) The radio spectra for the different time intervals. (b: center) NRH Radio images at 13:34:32, 13:34:48, 13:34:58 and 13:35:13 UT, and (c: bottom) HXR *Yohkoh* images at 13:35:13 UT for L, M1, M2 and H energy channels.

DISCUSSION

The results suggest that various episodes of energy release have occurred, associated with different sources belonging to distinct magnetic topologies, as evidenced by the radio spectra and images during the different time intervals. The start frequency and the frequency drift rates of type III bursts, and the energy of the non-thermal electrons increased during the event evolution, suggesting that the acceleration site is going down in the solar atmosphere. The radio images suggested that the energy releases were produced by the interaction of small and large scale loops, in a very dynamic process.

ACKNOWLEDGEMENTS

This research is partially supported by the Brazilian agency FAPESP, contracts No. 98/12108-9 and 01/07485-2.

ACKNOWLEDGMENTS

MULTIPLE-LOOP STRUCTURE OF A SOLAR FLARE FROM MICROWAVE, EUV, AND X-RAY IMAGING DATA

V. I. Garaimov and M. R. Kundu

Astronomy Department, University of Maryland, College Park, MD 20742

ABSTRACT

We present the results of the analysis of a flare event of importance M2.8 that occurred at 00:56 UT Aug 28, 1999. The analysis is based upon observations made with the Nobeyama Radio Heliograph (NoRH) and Polarimeters(NoRP), *TRACE*, *SOHO*/MDI, EIT and *Yohkoh*/SXT and shows a very complex morphology of the flaring region.

OBSERVATIONS AND RESULTS

Using multiwavelength data, we have studied the morphology of the multiple loop configuration of a complex flare event which occurred at 00:56 UT August 28 1999. At least four flaring loops participated in emitting radio radiation in different phases of the flare event.

The impulsive phase microwave emission which is clearly nonthermal in nature originated primarily from two loops – one overlying the main spot (main spot loop) and the other from the loop designated as the main flare loop. The latter has two foot points ending in opposite polarity magnetic regions (as seen in MDI magnetograms), and therefore it is a bipolar loop. The main spot loop is an unresolved compact region; it is polarized, and may be a small bipolar loop as delineated by the small EIT loop (see Figure 1c).

We believe that it is the interaction between these two loops that is responsible for the flare onset. Note also that there was new magnetic field emergence near these flaring regions. Clearly there must be reconnection of magnetic fields near this region, which ultimately results in the impulsive flare onset and the release and acceleration of energetic electrons (of 100's KeV energy) responsible for the nonthermal microwave emission in the impulsive phase.

Flaring region 3, designated as the second flare loop, seems to radiate mostly thermal microwave radiation. This is evident from the close spatial similarity between the microwave loop and the EIT flare loop and the gradual nature of the emission. We conclude that the flare energy is released in a region near the interface of two loops (1 and 2) – the main spot loop and the main flare loop.

The hard X-ray emission time profiles in the 50-100 KeV and 100-300 KeV channels have close similarity with the microwave time profiles at 17 and 34 GHz from the sunspot flare loop, which suggests that the hard X-ray emission originates in this flare loop. There is a sharp hard X-ray spike at 00:56:42 UT, which corresponds to the overall strongest microwave peak, but the peak, at this time from the main spot region, is not spiky. This may suggest that there may be contributions to the hard X-ray emission from energetic

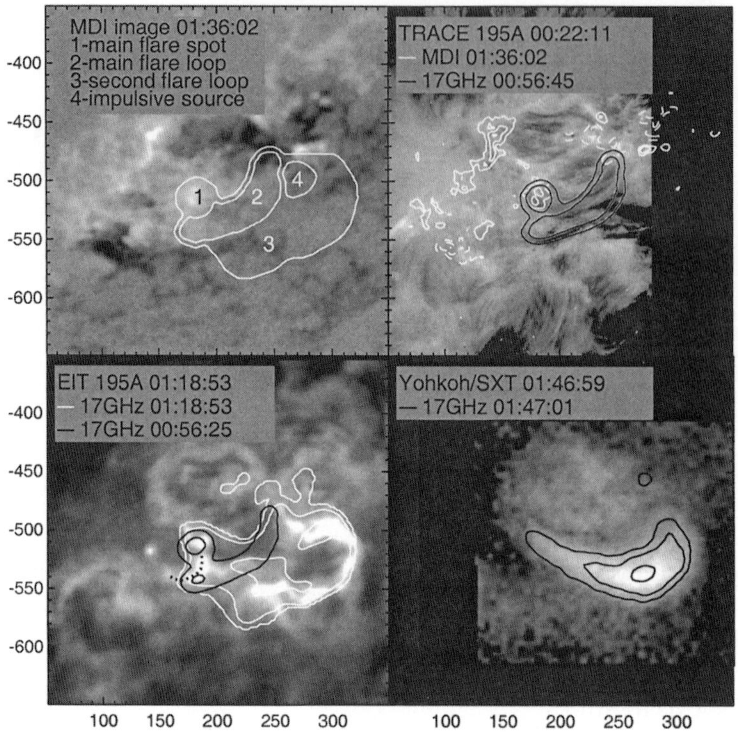

Fig. 1. Superpositions of images obtained from different instruments with radio contours at 17 GHz during the event. Upper left panel: MDI image at 01:36:02 UT and a schematic of the locations of selected regions in the 17 GHz radio maps. Upper right panel: *TRACE* 195 Å image obtained at 00:22:11 UT superposed with *SOHO*/MDI contour (white) obtained at 01:36:02 UT and 17 GHz contour (black) at 00:56:45 UT. Lower left panel: *SOHO*/EIT 195 Å image at 01:18:53 UT superposed with 17 GHz contours obtained at 00:56:45 UT (black) and at 01:18:53 UT (white). The dotted line shows the location of the small flare loop. Lower right panel: *Yohkoh*/SXT image at 01:46:59 UT superposed with 17 GHz contour at 01:47:01 UT.

electrons in the main flare loop.

ACKNOWLEDGEMENTS

This research at the University of Maryland was supported NSF grants ATM99-09809 and INT 98-19917, and NASA grant NAG5-8192.

FLARING IN MULTIPOLAR REGIONS ON THE SUN: THE 1999 JULY 19 FLARE

M. Sersen

Astronomical Institute, Comenius University, Faculty of Mathematics, Physics, and Informatics, 842 15 Bratislava, The Slovak Republic

ABSTRACT

This presentation focuses on a possibility of creation of a twisted magnetic flux system in the multipolar active region NOAA 8636 on 1999 July 19 as a subsequence of interaction of low-lying coronal loops. It is speculated that the flare that occurred in the active region destabilized the overlying magnetic arcade and triggered a coronal mass ejection, which was observed with the Large Angle and Spectrometric Coronagraph (LASCO) instrument on board the *SOHO* satellite. The flare was also observed in H-α at Modra Astronomical Observatory in Slovakia.

THE FLARE AND ACTIVE REGION CHARACTERISTICS

The flare, of importance M5.8, started at 08:16 UT and peaked at 08:46 UT. The topology of magnetic field in the corona in the flaring region may be inferred from the plasma-filled loops imaged by *Yohkoh*/SXT and *SOHO*/EIT that trace the field configuration. Inspection of the images indicates that at least two, more probably several, compact low-lying loops interacted under the overlying arcade. The interaction led to the reconnection and the H-alpha flare which was also observed at Modra Astronomical Observatory from 08:30 UT to 10:30 UT (H-alpha filter 0.7A and 1.5A FWHM). The H-alpha observations (Figure 1c) show numerous knots that are footpoints and intersection points of the flare loops. The data also show the low-lying flare loops. The H-alpha observations will be depicted in more details in a subsequent paper that is to be published in a refereed journal, shortly.

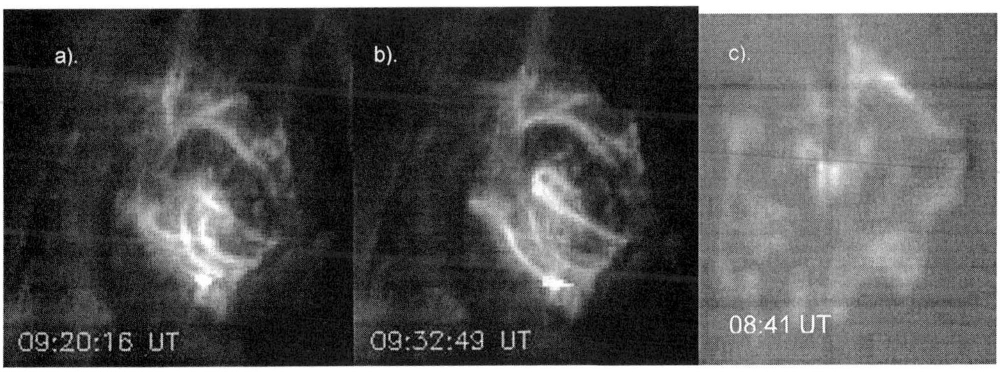

Fig. 1. (a,b): *SOHO*/EIT 195A, and (c): Modra Astronomical Observatory H-alpha images, of the active region NOAA 8636 on 1999 July 19.

The cross-loop structures (Figure 1a and 1b) that were observed with *SOHO*/EIT are consistent with the model of Amari *et al.* (1999). The model proposes that it is possible to produce a configuration that consists of a twisted magnetic flux system embedded in an overlying, almost potential, arcade such that high electric currents are confined to the inner twisted magnetic flux rope (Figure 2). Using three-dimensional MHD simulations the authors showed that such twisted magnetic flux may be generated from dipole arcade loops subjected to a twisting footpoint motion.

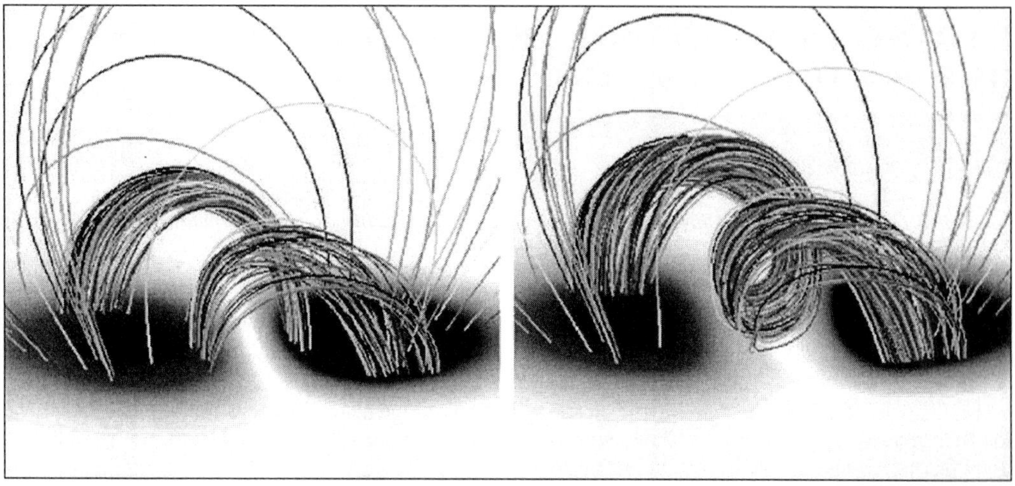

Fig. 2. Evolution of the configuration from an arcade-like topology to a twisted flux rope-like topology (Amari *et al.*, 1999)

SUMMARY

It is suggested that the formation of a twisted magnetic structure from a dipole arcade loop system in the active region NOAA 8636 created a magnetic instability that led to the flare on 1999 July 19. It is also suggested that the flare subsequently destabilized the overall magnetic arcade in the high corona and caused a coronal mass ejection that was observed with the LASCO instrument onboard *SOHO*. The CME itself would have been the erupted filament originally involved with the interaction and reconnection of magnetic flux. The plasma morphology of this dynamic event, we suggest, follows the scenario of Amari *et al.* (1999) in which a twisted flux rope-like structure is created from an arcade-like dipole.

REFERENCES

Amari, T., Luciani, J. F., Mikic, Z., and Linker, J., Three-dimensional solutions of magnetohydrodynamic equations for prominence magnetic support: twisted magnetic flux rope, *ApJL,* **518**, 57 (1999).

Section VIII.

Preflare Phenomena

OBSERVATIONS OF PRE-FLARE ACTIVITY WITH TRACE AND YOHKOH

H. P. Warren

Harvard-Smithsonian Center for Astrophysics, MS 58, Cambridge, MA 02138, USA

ABSTRACT

Many theories and observations appear to indicate that a simple bipolar structure with strong shearing along the neutral line can produce a coronal mass ejection (i.e. "tether cutting"). The "breakout reconnection" model, in contrast, states that multi-polar magnetic fields are necessary for a CME. Here I briefly review *TRACE* and *Yohkoh* observations related to these two competing views on the amount of magnetic complexity needed for a coronal mass ejection.

INTRODUCTION

Despite several decades of observational and theoretical effort, a complete understanding of how magnetic reconnection works in the solar atmosphere remains elusive. It has been especially difficult to understand how the evolution of the Sun's magnetic fields can trigger the sudden release of energy. Perhaps the best source of detailed information on this process is the observation of large solar flares. Of particular importance to understanding the build-up and release of magnetic energy is the observation of the "pre-flare" corona.

Since there have been so many studies of pre-flare activity this brief review is focused mainly on the theory and observational evidence for two competing ideas on the origin of coronal mass ejections: Tether cutting models (e.g. Moore & Labonte 1980, Sturrock 1989) and the magnetic breakout model (e.g. Antiochos *et al.* 1999). For the purposes of this talk tether cutting refers to all models based primarily on bipolar fields. In these models almost all of the important dynamics occur near the neutral line where shearing footpoint motions can drive the build-up of magnetic energy. The breakout model, in contrast, insists that for a coronal mass ejection to occur the fields must be multipolar in nature. Thus, in addition to the reconnection associated with shearing footpoint motions, there must also magnetic reconnection of some of the external magnetic fields. The grouping of models and observations into tether cutting and breakout is not meant to be a formal classification scheme for coronal mass ejections, such have been proposed by (Forbes 2000) and (Klimchuk 2001). It is simply a way of organizing the existing literature around the question: How much magnetic complexity is needed for a coronal mass ejection? Achieving both a theoretical understanding of this process as well as a predictive capability would not only be a great science achievement but would also have important practical implications in space weather forecasting.

Almost all models related to the origin of flares and coronal mass ejections rely on magnetic reconnection. This, unfortunately, presents us with a difficult problem. Since there is no clear observational signature of reconnection we are forced to assume that the pre-flare activity we see is the result of magnetic reconnection. For example, in some of the studies I am going to discuss here coronal loop brightenings and motions, filament activations, as well as chromospheric ribbon brightenings are simply assumed to be the result of reconnection. More work needs to be done to establish the observational signature of magnetic reconnection.

DIPOLAR MODELS

For some time it has been recognized that footpoint shearing motions near the neutral line of a magnetic flux system can lead to the formation of a suspended flux rope (e.g. van Ballegooijen & Martens 1989, Antiochos *et al.* 1994). The numerical simulations of Mikić & Linker (1994) illustrate a bipolar CME model driven by this mechanism. Mikić & Linker (1994) solve the MHD equations for a time-dependent, three dimensional system with translational symmetry. Footpoint shearing motions are applied along the neutral line. Initially their model is ideal and the field simply expands outward as the magnetic energy of the system increases. After some time they introduce a finite resistivity to the simulation. This leads to the formation of a current sheet and a disconnected plasmoid which rises through the computational domain. The change in the resistivity might mimic the development of a tearing mode or a microinstability which lead to anomalous resistivity (Forbes 2000).

Forbes & Priest (1995) have shown that an ideal system can also evolve to form a current sheet and disconnected plasmoid. They consider a magnetic flux system comprised of a flux rope and two photospheric sources. They find that when the photospheric magnetic sources are driven together by converging flows a current sheet forms once a critical separation is reached. In the absence of resistivity the system settles into a new equilibrium with the flux rope at a larger height and a current sheet extending from the flux rope to the photosphere. In the presence of a finite resistivity Lin & Forbes (2000) calculate that the plasmoid will escape to infinity provided the flows into the current sheet are sufficiently fast.

In these models, and models like them, translational symmetry along the axis is assumed, and the systems considered are two dimensional. Since flux ropes on the Sun are connected to the photosphere and not suspended above the surface, it is not clear that these models are applicable to real coronal mass ejections. In particular, it is not clear if a three-dimensional system will have the same stability properties as a two-dimensional system. A fully three dimensional, analytic model of a twisted flux rope embedded in a magnetic field has been developed by Titov & Démoulin (1999). They find that the flux rope becomes unstable when it becomes too long. A time-dependent, three-dimensional numerical simulation of flux rope formation due to shearing footpoint motions has been performed by Amari *et al.* (2000). As shown in Figure 1 the system evolves to include a twisted flux rope enclosed by an arcade of nearly bipolar loops. Furthermore, Amari *et al.* (2000) find that the emergence of a small bipole near the neutral line leads to a disruption of the system. At present, however, it is not clear if the strong shearing motions assumed in the simulation are observed on the Sun (Forbes 2000).

An important aspect of the three-dimensional models of Titov & Démoulin (1999) and Amari *et al.* (2000) is their ability to produce flux ropes that have an S-shape. These S-shaped or "sigmoid" brightenings, such as the one shown in Figure 2, are common features of eruptive active region (e.g. Rust & Kumar 1996). Comprehensive studies with *Yohkoh* have shown that sigmoidal active regions are much more likely than non-sigmoidal active regions to erupt. For example, Canfield *et al.* (1999) found that 51 of the 61 (83%) sigmoidal active regions they studied ultimately erupted. In contrast, only 28 of the 56 (50%) of the non-sigmoidal active regions in their study produced an eruption.

The evolution of several sigmoid eruptions observed with SXT has recently been studied in detail by Moore *et al.* (2001). All 6 of the events in their study initially show core fields that are twisted into a sigmoidal shape and are enclosed within a single, overlying bipole. Four of the events are ejective while two of the events are ultimately confined within the overlying fields. The early phases of all of the events, however, are similar: brightenings occur first near the highly sheared fields then spread out as an arcade gradually forms oriented orthogonally to the neutral line. There is no evidence for any reconnection with overlying magnetic fields in these flares. Moore *et al.* (2001) conclude that these events are driven by runaway tether-cutting of the core fields. The "sigmoid-to-arcade" evolution of several active regions has also been been investigated by Sterling *et al.* (2000) using *Yohkoh* SXT and *SOHO* EIT images. The events they studied, which were all associated with halo coronal mass ejections, also show initial brightenings near the neutral line.

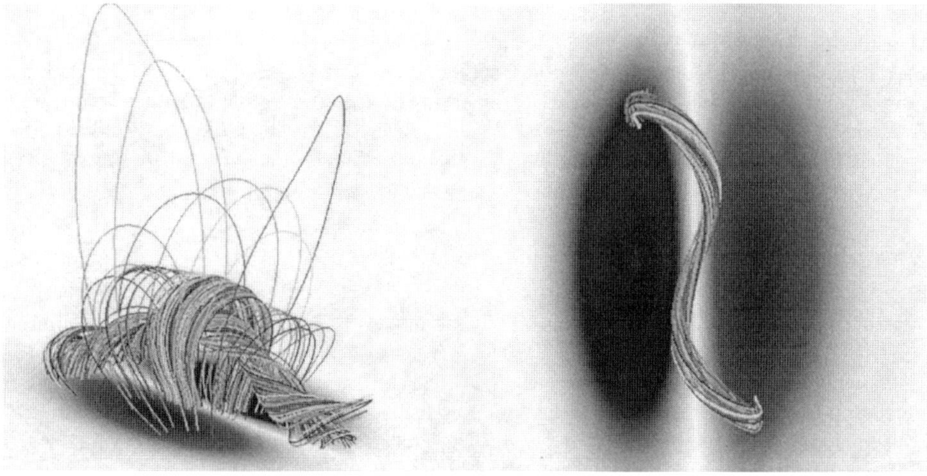

Fig. 1. (*left panel*) Field lines showing the formation of a twisted flux rope along the neutral line from footpoint shearing motions. (*right panel*) A top view of the twisted flux rope showing the inverted S-shape. From Amari *et al.* (2000)

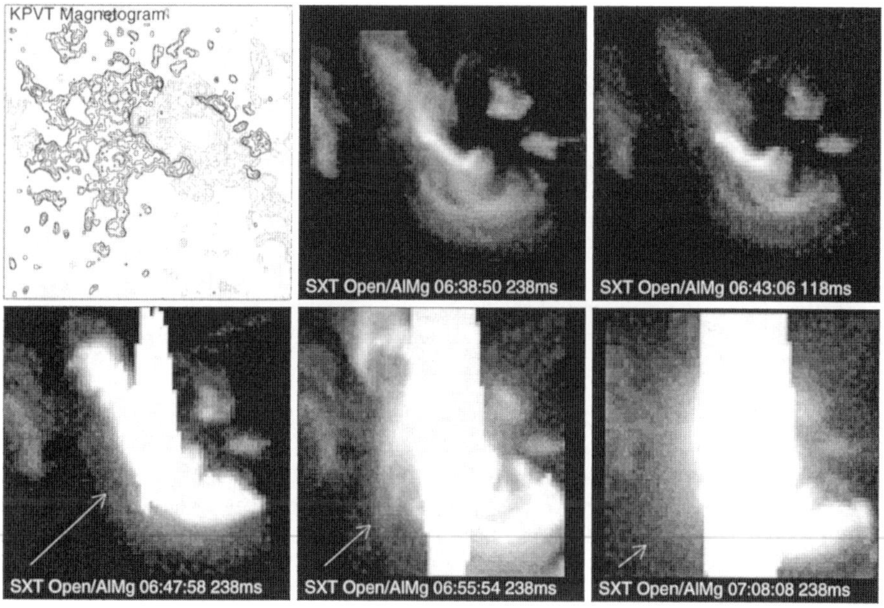

Fig. 2. Data from the September 23, 1998 M7 flare, which is an example of an erupting sigmoidal active region observed with SXT and *TRACE*. The *GOES* fluxes peaked at about 07:13 UT. (*top left panel*) A Kitt Peak Vacuum Telescope magnetogram. Contours are drawn at $\pm$ 50, 100, 250, 400, 600 G. (*top and bottom panels*) SXT half-resolution AlMg images from the event. The final three images show evidence for an ejection.

These theoretical and observational studies paint a fairly complete picture of the coronal mass ejection. Footpoint shearing motions can lead to the formation of a highly stressed magnetic field which is composed of a twisted flux rope embedded in nearly bipolar, overlying fields. The introduction of a small perturbation, such as the development of anomalous resistivity, emerging flux, or converging flows, leads to the formation of a current sheet. Reconnection along the current sheet allows the flux rope to escape and the flux system to relax into a lower energy state. In this view, the magnetic flux external to the core fields plays only a minor role. These fields may, for example, act to confine some sigmoid eruptions (Moore *et al.* 2001).

THE BREAKOUT MODEL

Antiochos *et al.* (1999) have argued that these bipolar models of CMEs are "doomed to failure." They believe that these models will never produce a true CME because of the Aly-Sturrock conjecture (Aly 1984, Sturrock 1989). The Aly-Sturrock conjecture suggests that the maximum energy state of any force-free field is fully open field. This means that the mechanism powering a CME must not only heat and accelerate the plasma but must also raise the magnetic energy in the system. Since magnetic energy is supposedly driving the CME we must conclude that CMEs are impossible.

To circumvent this problem Antiochos *et al.* (1999) proposed that CMEs must be associated with multipolar flux systems. In their "breakout" model some of the flux associated with the sheared arcade is transferred to an adjacent flux system instead of opening. Also, the overlying fields act as a "lid" allowing the footpoint shearing motions to build-up magnetic energy in the stressed fields near the neutral line. The overlying fields must be removed before the eruption can occur.

Aulanier *et al.* (2000) have further refined the concept of breakout reconnection to be "the opening of initially low-lying sheared field, triggered by reconnection at a null point that is located high in the corona and defines a separatrix enclosing the sheared fields." They go on to present a detailed analysis of the July 14, 1998 Bastille Day flare. This event peaked at about 12:59 UT and was classified as a *GOES* M4.6 flare. *TRACE* observations show dimming, which suggests that there was a CME associated with this event. Unfortunately there are no *SOHO* observations for this flare so the CME cannot be confirmed.

As shown in Figure 3, a potential field extrapolation of the Kitt Peak Vacuum Telescope (KPVT) magnetogram taken several hours after the event reveals the presence of a magnetic null point and its associated "spine" and "fan" field lines. *TRACE* 171 Å images taken before the event show evidence for loop motions near the spine. Just before the main part of the flare begins a brightening propagates along one of the field lines close to the spine. This ordering of events, "breakout" reconnection along the spine before the onset of the flare, is consistent with their model.

Further evidence in support of the breakout model has been given by Sterling *et al.* (2000). They have found several events where a "crinkle-like" pattern forms in EIT 195 Å images prior to the onset of a flare. These "EIT crinkles" are formed well away from the core fields and suggest the pre-impulsive phase reconnection of the overlying fields. It is not clear if these crinkles are loops with temperatures near 1.5 MK or the footpoints of hot loops (i.e. "moss," Berger *et al.* 1999). Because of the relatively small field of view in the available SXT images, Sterling *et al.* (2000) were not able to follow the evolution of the crinkles at higher temperatures, nor were they able to follow the evolution of the EIT crinkles and the core region simultaneously. Several of the EIT crinkle events were also observed at high spatial and temporal resolution in Hα (Sterling *et al.* 2001). These observations show that EIT crinkles have a clear chromospheric signature, as do the core brightenings. Comparison of the Hα light curves from the external and core regions show evidence for the reconnection of the external fields before the onset of the flare.

Aulanier *et al.* (2000) did not consider chromospheric brightenings a source of information on the relative timing of core and external reconnection as Sterling *et al.* (2001) did. As shown in Figure 4, there are, however, *TRACE* 1600 Å images for this event. These data suggest that the core brightenings in this

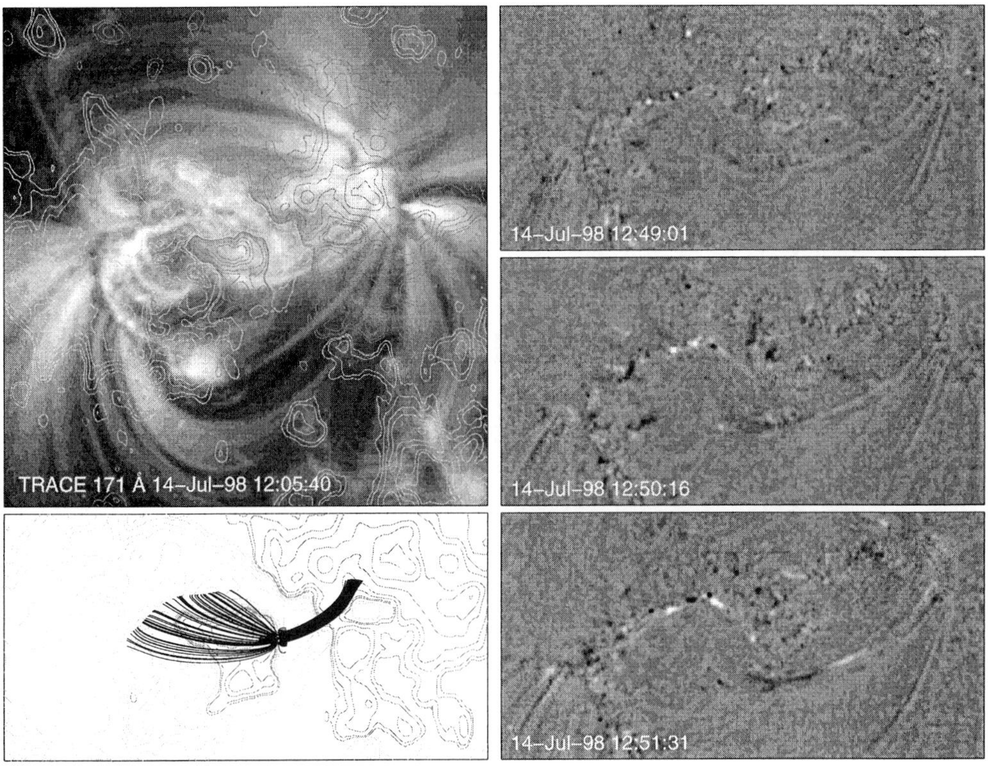

Fig. 3. The magnetic field topology of the 1998 Bastille Day flare. (*top left panel*) *TRACE* 171 Å image prior to the onset of the event. Contours are from a Kitt Peak magnetogram. Levels are drawn at ± 50, 100, 250, 400, 600 G. (*bottom left panel*) Spine and fan field lines that pass close to the null in the magnetic field as determined from a potential field extrapolation. (*right panels*) *TRACE* 171 Å running difference images showing the propagation of a disturbance along a field line close to the null. The disturbance begins at approximately 12:51:31 UT.

flare actually precede the chromospheric brightening associated with activity near the spine field lines. Furthermore, the core brightenings are evident in the 1600 Å images as early as 12:46 UT and appear to precede the loop motions seen in the *TRACE* 171 Å images as reported by Aulanier *et al.* (2000).

Inspection of the hard X-ray data from the *Compton Gamma Ray Observatory* Burst and Transient Source Experiment for the 1998 Bastille Day flare shows no evidence for hard X-ray emission prior to 12:55 UT. The onset of the hard X-ray emission is coincident with a dramatic brightenings in the 1600 Å images. It is clear from the *TRACE* images for this flare, however, that there is considerable activity in the core region before the onset of the hard X-rays. In a study of 9 other events, Warren & Warshall (2001) found a similar trend of the UV ribbon brightenings to precede the onset of the hard X-ray emission. One of events they studied was observed by *TRACE* at very high cadence (~ 2 s). This allowed the footpoint light curves to be correlated with the hard X-ray light curves. This analysis suggested that the energetic particle precipitation occurred only on field lines that showed no pre-flare activity. The relationship between the pre-flare and flare chromospheric brightenings and the magnetic field topology was not considered in this paper. Almost all of the pre-flare brightenings in their study, however, occurred near the core fields and did not appear to be associated with the pre-flare reconnection of external fields.

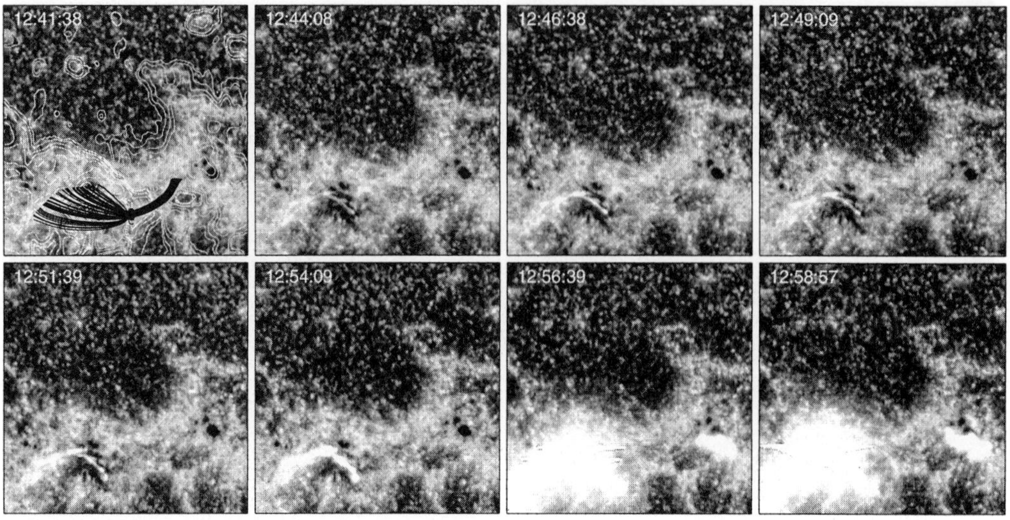

Fig. 4. *TRACE* 1600 Å images from the 1998 Bastille Day flare. On the first panel contours from the KPVT magnetogram are shown. Also shown are field lines that pass close to the null point determined from the potential field extrapolation.

The September 23, 1998 flare presents another potential problem for the breakout model. As illustrated by the *TRACE* 1550 Å images displayed in Figure 5, this flare shows the activation of the filament and development core brightenings as well the development of external brightenings to the north and south of the core. The pre-flare 195 Å images, such as the one shown in Figure 5, suggest that the south ribbon may connect to the region near the core. It is also possible that these external brightenings connect to each other. In either case, it is clear from the *TRACE* 1550 Å images that the core brightenings begin before any brightenings associated with the external fields. The core dynamics also precede the onset of the hard X-ray emission measured with HXT (Warren & Warshall 2001).

Furthermore, it is not clear that there is a null point present in the coronal magnetic field that has any relevance to this event. A search of the potential field extrapolation does not yield such a point. This field is not purely bipolar so there are other topological features, such as separatrix surfaces, which may play a role in pre-flare reconnection. A magnetic null point, which was stressed by Aulanier *et al.* (2000), does not appear to be involved, however.

NONTHERMAL BROADENING

Another topic of potential relevance to understanding what triggers flares and coronal mass ejects is the pre-flare nonthermal broadening measurements taken with the Bragg Crystal Spectrometer (BCS) instrument on *Yohkoh*. In has long been recognized that the widths of high-temperature emission lines observed during flares are much larger than can be accounted for by thermal broadening alone (e.g. Doschek *et al.* 1979). The largest nonthermal velocities are typically observed very early in the flare and the line widths tend to decreases as the emission measure increases.

The origin of this nonthermal broadening remains unclear. Recent studies with BCS have shown that in some flares the largest nonthermal velocities actually occur before the onset of significant hard X-rays (Alexander *et al.* 1998, Mariska & McTiernan 1999, Ranns *et al.* 2001, Harra *et al.* 2001). This suggests that

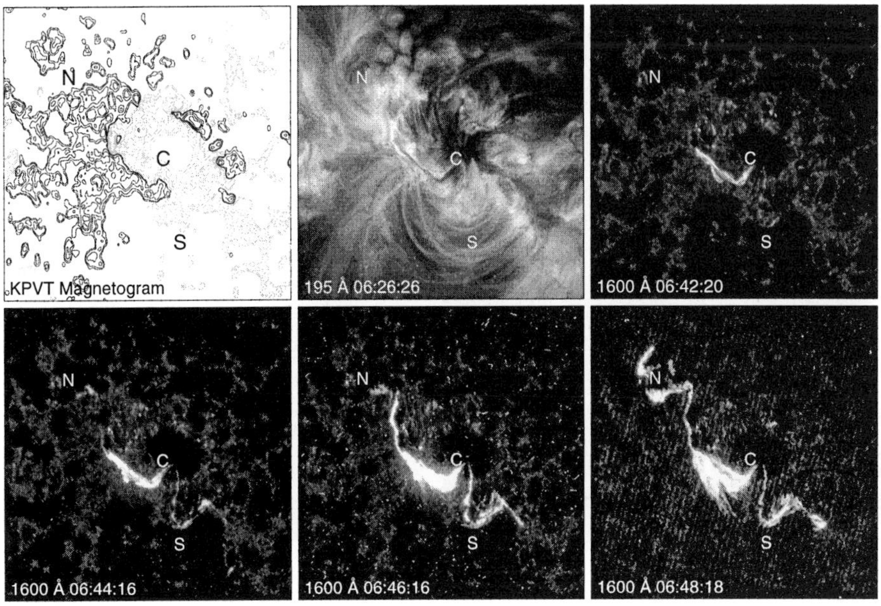

Fig. 5. Images from the September 23, 1998 M7 flare. (*top left panel*) KPVT magnetogram. Contours are drawn at ± 50, 100, 250, 400, 600 G. (*top middle panel*) A *TRACE* 195 Å image taken before the onset of the flare. (*remaining panels*) *TRACE* 1550 Å images. SXT images from this event are shown in Figure 2. The letters indicate the approximate position of the core ribbon and north and south ribbons in each image.

the nonthermal broadening may be a signature of the energy release process, such as plasma turbulence, rather than an indirect effect, such as evaporation (Alexander *et al.* 1998).

The BCS S XV data shown in Figure 6 illustrates the evolution of the nonthermal velocity as seen in the September 23, 1998 M7 flare. The earliest S XV profiles that can be measured for this event are from about 06:40 UT. After this time the nonthermal velocities rise rapidly from about 50 km s^{-1} to about 150 km s^{-1}. As shown in Figure 6, the SXT AlMg images from this period show evidence for a brightening in the sigmoid core. The apparent correlation between the brightening of the sigmoid and the increase in nonthermal velocity suggests that line broadening may be signature of the reconnection process.

SUMMARY AND DISCUSSION

In this brief paper I have summarized some of the *TRACE* and *Yohkoh* observations of pre-flare activity related to the onset of flares and coronal mass ejections. Observations such as these have lead to significant advances in our undestanding of the magnetic topologies that are responsible for eruptions. The SXT sigmoid studies, for example, have clearly demonstrated the importance of magnetic stresses near the neutral line in creating potentially instable magnetic field configurations. Furthermore, the recent observational and three-dimensional modeling suggests that the core dynamics are sufficient to produce eruptions. More work is needed, however, before detailed comparisons can be made between theory and experiment.

The evidence for necessity of multipolar flux systems in these *TRACE* and *Yohkoh* studies, as required by the breakout model, is somewhat ambiguous. The 1998 Bastille Day flare has a null point associated with it and there is clear evidence for pre-flare dynamics in the vicinity of the spine field lines. The *TRACE*

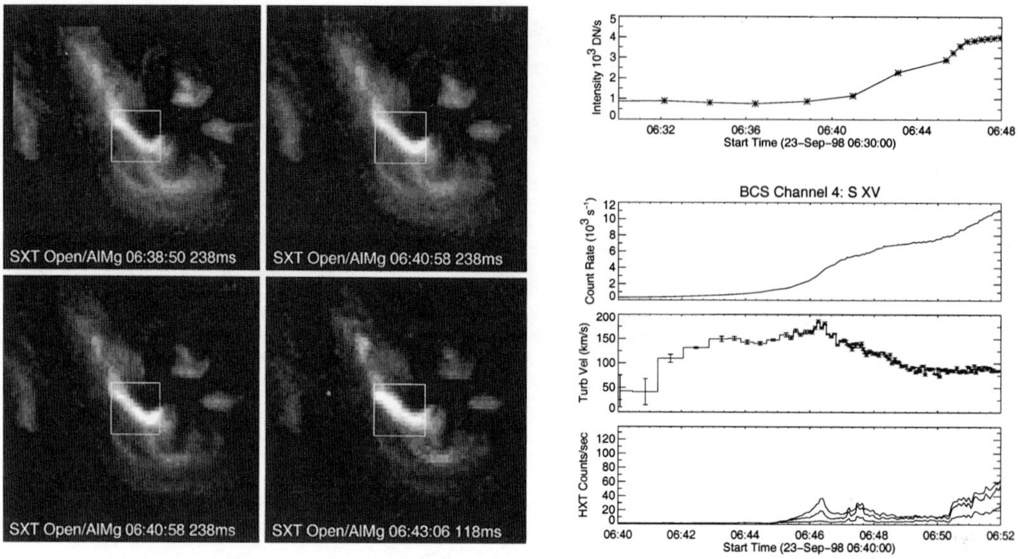

Fig. 6. SXT, BCS, and HXT observations from the rise phase of the September 23, 1998 M7 flare. (*left panels*) Half-resolution SXT AlMg images. The white box indicates the region used to compute the light curve for the sigmoid. (*top right panel*) Average intensity of the core sigmoid region as a function of time. (*middle right panels*) Total counts and nonthermal velocities measured from the BCS S XV channel. (*bottom right panels*) Integrated counts in the HXT L, M1, and M2 channels.

1600 Å images, however, suggest that reconnection in the core actually precedes this activity. Similarly, core reconnection appears to occur first in the September 23 1998 M7 flare. The events presented by Sterling *et al.* (2000) and Sterling *et al.* (2001), in contrast, show brightenings associated with the external fields that precede activity in the core region. A larger sample of flares is needed to address this question definitively. It is particularly important that future studies address the magnetic topology of the event along with the relative timing of the pre-flare brightenings.

One weakness in all of the studies presented here is the uncertain relationship between pre-flare brightenings and the reconnection of the magnetic field. Coronal loop motions and chromospheric brightenings are simply assumed to be proxies for reconnection, there is no clear evidence that the observed pre-flare activity actually is the result of magnetic reconnection. Better observational signatures of magnetic reconnection are clearly needed. Spectroscopic measurements, such as the nonthermal velocity measurements with BCS, hold particular promise for providing important additional information on the reconnection process.

ACKNOWLEDGEMENTS

TRACE is supported by Contract NAS5-38099 from NASA to LMATC. *Yohkoh* is a mission of the Institute of Space and Astronautical Sciences (Japan), with participation from the U.S. and U.K.

REFERENCES

Alexander, D., L.K. Harra-Murnion, J.I. Khan, and S.A. Matthews, Relative Timing of Soft X-Ray Non-thermal Line Broadening and Hard X-Ray Emission in Solar Flares, *ApJ,* **494**, L235 (1998).

Aly, J.J, On Some Properties of Force-Free Magnetic Fields in Infinite Regions of Space, *ApJ,* **283**, 349 (1984).

Amari, T., J.F. Luciani, Z. Mikic, J. Linker, A Twisted Flux Rope Model for Coronal Mass Ejections and Two-Ribbon Flares, *ApJ,* **529**, L49 (2000).

Antiochos, S.K., R.B. Dahlburg, J.A. Klimchuk, The Magnetic Field of Solar Prominences, *ApJ,* **420**, L41 (1994).

Antiochos, S.K., C.R. Devore, and J.A. Klimchuk, A Model for Solar Coronal Mass Ejections, *ApJ,* **510**, 485 (1999).

Aulanier, G., E.E. DeLuca, S.K. Antiochos, R.A. McMullen, and L. Golub, A Model for Solar Coronal Mass Ejections, *ApJ,* **510**, 485 (2000).

Berger, T.E., B. de Pontieu, C.J. Schrijver, and A.M. Title, High-resolution Imaging of the Solar Chromosphere/Corona Transition Region, *ApJ,* **519**, L97 (1999).

Canfield, R.C., H.S. Hudson, and D.E. McKenzie, Sigmoidal Morphology and Eruptive Solar Activity, *GRL,* **26**, 627 (1999).

Doschek, G.A., R.W. Kreplin, and U. Feldman, High-Resolution Solar Flare X-Ray Spectra, *ApJ,* **233**, L157 (1979).

Forbes, T.G., A Review on the Genesis of Coronal Mass Ejections, *JGR,* **105**, 23153 (2000).

Forbes, T.G., and E.R. Priest, Photospheric Magnetic Field Evolution and Eruptive Flares, *ApJ,* **446**, 377 (1995).

Harra, L.K., S.A. Matthews, and J.L. Culhane, Nonthermal Velocity Evolution in The Precursor Phase of a Solar Flare, *ApJ,* **549**, L245 (2001).

Klimchuk, J.A., Theory of Coronal Mass Ejections. In: Space Weather. Vol. 125 of Geophysical Monograph. 143 (2001).

Lin, J., and T.G. Forbes, Effects Of Reconnection on The Coronal Mass Ejection Process, *JGR,* **105**, 2375 (2000).

Mariska, J.T., and J.M. McTiernan, Hard And Soft X-Ray Observations of Occulted And Nonocculted Solar Limb Flares, *ApJ,* **514**, 484 (1999).

Mikić, Z., and J.A. Linker, Disruption of Coronal Magnetic Field Arcades, *ApJ,* **430**, 898 (1994).

Moore, R.L., Labonte, B.J., The Filament Eruption in the 3B Flare Of July 29, 1973 - Onset and Magnetic Field Configuration. In: IAU Symp. 91: Solar and Interplanetary Dynamics. Vol. 91. 207 (1980).

Moore, R.L., A.C. Sterling, H.S. Hudson, and J.R. Lemen, Onset of the Magnetic Explosion in Solar Flares and Coronal Mass Ejections, *ApJ,* **552**, 833 (2001).

Ranns, N.D.R., L.K. Harra, S.A. Matthews, and J.L. Culhane, The Timing of Non-Thermal Soft X-Ray Emission Line Broadenings in Solar Flares, *A&A,* **379**, 616 (2001).

Rust, D.M., and A. Kumar, Evidence for Helically Kinked Magnetic Flux Ropes in Solar Eruptions, *ApJ,* **464**, L199 (1996).

Sterling, A.C., H.S. Hudson, B.J. Thompson, and D.M. Zarro, *Yohkoh* SXT and *SOHO* EIT Observations of Sigmoid-to-Arcade Evolution of Structures Associated with Halo Coronal Mass Ejections, *ApJ,* **532**, 628 (2000).

Sterling, A.C., and R.L. Moore, EIT Crinkles as Evidence for the Breakout Model of Solar Eruptions, *ApJ,* **560**, 1045 (2001).

Sterling, A.C., R.L. Moore, J. Qiu, and H. Wang, Hα Proxies for EIT Crinkles: Further Evidence for Preflare "Breakout"-Type Activity in an Ejective Solar Eruption, *ApJ,* **561**, 1116 (2001).

Sturrock, P.A., The Role of Eruption in Solar Flares, *Solar Phys.,* **121**, 387 (1989).

Titov, V.S., and P. Démoulin, Basic Topology of Twisted Magnetic Configurations in Solar Flares, *A&A,* **351**, 707 (1999).

van Ballegooijen, A.A., and P.C.H. Martens, Formation and Eruption of Solar Prominences, *ApJ,* **343**, 971 (1989).

Warren, H.P., and A.D. Warshall, Ultraviolet Flare Ribbon Brightenings and the Onset of Hard X-Ray Emission, *ApJ,* **560**, L87 (2001).

THE MAGNETIC FREE ENERGY AND A CME IN ACTIVE REGION 8299

T. R. Metcalf[1], D. L. Mickey[2], B. J. LaBonte[3], and L. A. Ryder[1]

[1]*Lockheed Martin Solar & Astrophysics Laboratory, Dept. L9-41, Bldg. 252, 3251 Hanover St., Palo Alto, CA 94304*
[2]*Institute for Astronomy, University of Hawaii, 2680 Woodlawn Dr., Honolulu, HI 96822*
[3]*Johns Hopkins University Applied Physics Laboratory, 11100 Johns Hopkins Road, Laurel, MD 20723-6099*

ABSTRACT

We calculate the magnetic free energy as a function of time for NOAA active region 8299 on 1998 August 11 using vector magnetic field measurements in the Na I 5896Å spectral line observed with the Imaging Vector Magnetograph at Mees Solar Observatory. The free energy dipped to a value consistent with zero for one hour during the observation. *Yohkoh*/SXT images reveal that around the time of this dip in the free energy the coronal structure of AR 8299 and the nearby AR 8297 changed significantly. SXT observed the brightening of a cusp connecting AR 8299 and AR 8297 and observed coronal dimming in both active regions, suggesting that a gradual CME was launched as the magnetic energy dipped. Further, the magnetic connection between AR 8299 and AR 8297 was temporarily severed during the energy dip and the magnetic field took on a more open configuration afterwards. Unfortunately, LASCO data were not available to confirm the existence of a halo CME. However, the circumstantial evidence points to the magnetic free energy as the energy source for the postulated event.

INTRODUCTION

The magnetic field permeating the solar atmosphere is generally thought to provide the energy for much of the activity seen in the solar corona, such as flares, CME's, etc. Some models also invoke the magnetic field to energize the hot coronal plasma seen in EUV or X-rays. To directly understand the role that the magnetic field plays in energizing the solar corona, it is necessary to measure the amount of free magnetic energy that is available in the active region as a function of time. A method for observing the magnetic free energy in active regions was pioneered by Metcalf *et al.* (1995) using observations of the chromospheric Na I D-line at 5896 Å. This magnetically sensitive spectral line is formed in the chromosphere, high enough so that the magnetic field is force-free but low enough so that the field is sufficiently strong for a reliable measurement. In this paper we apply the same technique to a series of observations of an active region spanning several hours with the goal of studying the time variability of the magnetic free energy.

OBSERVATIONS AND ANALYSIS

We observed AR 8299 on August 11 1998 using the Imaging Vector Magnetograph at Mees Solar Observatory (Mickey *et al.* 1996). The IVM uses a Na I prefilter and a Fabry-Perot to observe 68 mÅ redward from the center of the Na I 5896Å spectral line. The full-width spectral bandpass is 60 mÅ, so the selected wavelength is far from line center. This is the same wavelength used by Metcalf *et al.* (1995) in the original study of the Na I line and it was shown in that paper that the magnetic field measured at this height in the atmosphere is force-free. This is confirmed in the observations presented below.

The vector magnetic field is computed using the derivative method described by Jefferies, Lites, & Skumanich (1989) after applying the standard corrections to the IVM data such as corrections for atmospheric blurring, scattered light, polarimetric calibration, etc. (LaBonte *et al.* 1999). The derivative method is a weak field approximation; however, since the Na I line is very broad, the approximation is extremely good for this spectral line with the magnetic fields typically found on the Sun. The 180 degree ambiguity in the transverse field was resolved using the "minimum energy" method described by Metcalf (1994).

The magnetic virial theorem (e.g. Molodenskii 1969, Aly 1984, Low 1984, Aly 1989) gives the total (force-free) magnetic energy as

$$E_{ff} = \frac{1}{4\pi} \int_{z=z_0} (xB_x + yB_y)B_z \, dxdy \ , \tag{1}$$

where B_x, B_y, and B_z are the horizontal and vertical components of the magnetic field, x and y are the horizontal coordinates on the Sun (e.g. Gary 1990), and z_0 is the lower, chromospheric boundary.

Since Eq. (1) contains the coordinates x and y, it appears at first that the magnetic energy will depend on the coordinate system chosen. However, if the magnetic field is force-free on the boundary, the horizontal Lorentz forces are zero. Hence, Eq. (1) is independent of the coordinate system and valid when the field is force-free. As shown by Metcalf (1995), we can use the x and y dependence in Eq. (1) to simultaneously estimate both the random and the systematic error on the measurement of the total magnetic energy by recalculating the total energy for many different origins of the coordinate system. This error estimate is indicative of both the statistical errors and the applicability of the virial theorem. If the field is not force-free or flux is leaving the field-of-view, this will be reflected in the error bars. Hence, a statistically significant energy measurement implies the validity of the virial theorem.

Figure 1 shows the total magnetic energy as a function of time for AR 8299. The energies of the equivalent potential field and open field (Eq. (14) of Aly (1984)) were computed using the observed vertical field as a boundary condition. To be physically meaningful, the total magnetic energy must lie between the potential energy and the open field energy.

To use the virial equation, the field should be force-free and all the magnetic flux at the chromospheric boundary must be accounted for, i.e. none of the flux should connect outside our field-of-view. To the extent that the field is forced or flux is leaving the IVM field-of-view, a systematic error is introduced in the calculation of the magnetic energy. We have estimated the error incurred due to this loss of flux and find that the 8% flux loss in this observation is acceptable. Further, any flux loss will appear in the virial theorem as a pseudo- Lorentz force and, as with any real force, the associated systematic error will be included in the errors described above. This explains why the error bars are so large, though we reiterate that a statistically significant energy measurement implies the validity of the virial theorem.

The most interesting feature in Figure 1 is the dip in the magnetic energy late in the observation. This remarkable drop in the magnetic free energy observed between 20:30 and 21:30 UT is not accompanied by any significant flare activity on the Sun as evidenced by the *GOES* X-ray emission. Late in the observation, the total magnetic energy approached the open field limit, suggesting that the field was in a substantially more open configuration compared to the field before the dip in the magnetic energy.

The dramatic drop in free magnetic energy observed in AR 8299 amounts to approximately 10^{33} ergs. This in an enormous amount of energy given that there was no indication of any flares at the time of the energy dip. Where did the energy go?

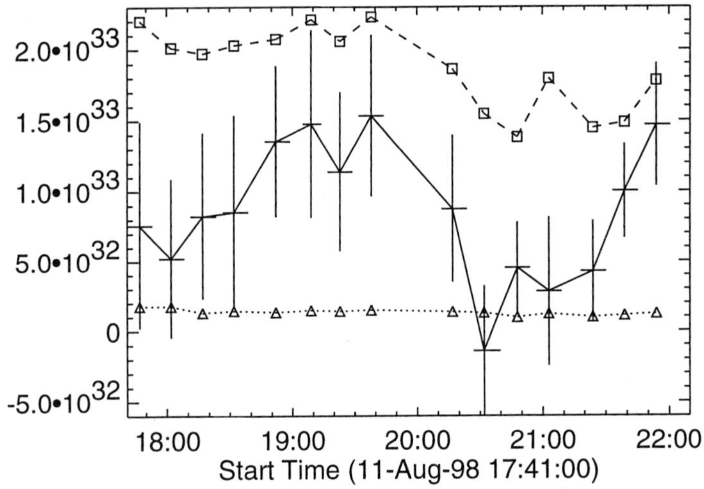

Fig. 1. The total magnetic energy (ergs) above the chromosphere in AR 8299 (solid line). The vertical error bars indicate both statistical and systematic errors. The horizontal error bars indicate the time interval over which each observation was averaged. The dotted line shows the energy of the equivalent potential field and the dashed line shows the equivalent open field energy.

While the evidence is only circumstantial, the change in the SXT topology observed during the energy dip (Figure 2) is reminiscent of the changes that are typically observed at the onset of gradual CME's, particularly the cusp and the dimming that occur in AR 8299 and AR 8297 (e.g. Sterling *et al.* 2000, Hudson *et al.* 1998). While it is unlikely that a slow CME could account for the full 10^{33} ergs (e.g. Antiochos, Devore, & Klimchuk 1999), the topological change associated with a CME might redistribute the free energy out of the active region we observed. Further evidence for this is the rapid restoration of the free energy starting at 21:00 UT. The increase in free energy from 21:00 to 22:00 is about 10^{33} ergs and implies a power input of 3×10^{29} ergs s^{-1}. While possible, it is unlikely that this increase in the free energy could come from photospheric motions over such a short timescale since photospheric motions are typically only 1 km s^{-1} (e.g. Eq. (27) in Aly (1989)). There is no observational evidence for emerging flux which might stress the coronal field more quickly. Hence, the redistribution of free energy in the AR 8299/8297 system is a more likely explanation.

If a CME did occur in association with the drop in magnetic free energy, this would constitute the first direct evidence that CME's are powered by free magnetic energy. Since the SXT data show that the energy dip was associated with a magnetic interaction between AR 8299 and AR 8297, and since this connection was temporarily severed during the dip in the magnetic energy before recovering to a more open state, our observations suggest that the magnetic breakout model could explain the postulated CME.

Ultimately, we hope to make advance predictions of solar activity by observing variations in the available magnetic energy. For a CME with little or no signature in X-ray light curves, such as the event postulated here or the January 6 1997 event studied by Webb *et al.* (1998), a dramatic change in the free energy may be the only surface indication of a CME.

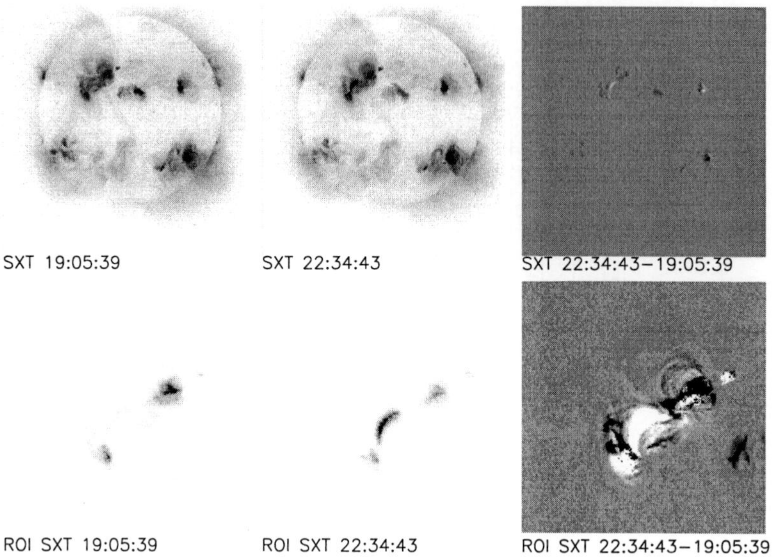

Fig. 2. Composite, negative Yohkoh/SXT X-ray images of the full disk (logarithmically scaled) and AR 8299/8297 (linearly scaled) using the AlMg analysis filter. Left: before the energy dip. Middle: after the energy dip; Right: difference image.

REFERENCES

Aly, J.J., *ApJ*, **283**, 349 (1984).

Aly, J.J., *Solar Phys.*, **120**, 19 (1989).

Antiochos, S K., DeVore, C.R., and Klimchuk, J.A. 1999, *ApJ*, **510**, 485 (1999).

Gary, G.A., and Hagyard, M.J., **Solar Phys.**, **126**, 21 (1990).

Hudson, H.S., Lemen, J.R., St. Cyr, O.C., Sterling, A.C, and Webb, D.F., *GRL*, **25**, 2481 (1998).

Jefferies, J.T., Lites, B.W., and Skumanich, A., *apJ*, **343**, 920 (1989).

LaBonte, B.L., Mickey, D.L., and Leka, K.D., *Solar Phys.*, *189*, 1 (1999).

Low, B.C. 1984, in *Measurements of Solar Vector Magnetic Fields*, ed. M.J. Hagyard, NASA CP 2374, 49 (1984).

Metcalf, T.R., Jiao, L., McClymont, A.N., and Canfield, R.C., 1995, *ApJ*, **439**, 474 (1995).

Metcalf, T.R. 1994, *Solar Phys.*, **155**, 235 (1994).

Molodenskii, M.M., *Soviet Astron.-AJ*, **12**, 585 (1969).

Mickey, D.L., Canfield, R.C., LaBonte, B.J., Leka, K.D., Waterson, M.F., and Weber, H.M., *Solar Phys.*, **168**, 229 (1996).

Sterling, A.C., Hudson, H S., Thompson, B.J., and Zarro, D.M., *ApJ*, **532**, 628 (2000).

Webb, D.F., Cliver, E.W., Gopalswamy, N., Hudson, H.S., and St. Cyr, O.C., *GRL*, **25**, 2469 (1998).

ANATOMY OF A FLARE AND CORONAL MASS EJECTION

C. R. Foley, L. K. Harra, J. L. Culhane, K. O. Mason, K. Hori, S. A. Matthews, and R. H. A. Iles

Mullard Space Science Laboratory, University College London, Holmbury St Mary, Dorking, Surrey RH5 6NT, UK

ABSTRACT

We present observations of an X2.3 flare which was observed on the April 10, 2001. This was subject of a letter by Foley *et al.* (2001). In this short paper we present further observations obtained with the *TRACE* spacecraft of the evolution of the EUV flare ribbons and of the flux rope's intensity variation. These are discussed in the content of the Standard Flare model.

INTRODUCTION

CMEs which originate within active regions tend to be brighter, faster and larger than CMEs with origin's outside active regions (see Andrews & Howard 2001). However the link between CMEs and solar flares is one which has stimulated much discussion and lively debate, but remains as yet unresolved. Understanding the initiation and onset of CMEs, remains one of the primary challenges in studying the Sun and its corona.

The discoveries which have been made possible by the *Yohkoh* spacecraft's excellent compliment of instruments have refined the 'Standard model' used to describe our understanding of solar flares (see Shibata *et al.* 1995). In this standard model flares and CME are linked through the presence of a plasmoid or flux rope. In the model the plasmoid which goes on to form the core of a CME may be driven in a similar way to the famous loop top sources discovered by Masuda *et al.* (1996). However, direct observations of this process are relatively rare and the acceleration unresolvable in most cases due to the relatively short duration of the acceleration during the impulsive phase of solar flares (e.g. Ohyama & Shibata 1998). The first spectroscopic observations of the ejection and acceleration of a flux rope to form the core of part of a halo CME were recorded with the Coronal Diagnostic Spectrometer (CDS) onboard *SOHO*, and with the *Transition Region and Coronal Explorer* (*TRACE*). Initial work on this data was presented in Foley *et al.* (2001) who interpret the bulk motion of the erupting structure through the field of view of the CDS to determine that the structure was accelerated ($\sim 3.5 \times$ km s^{-2}) impulsively at around the rise time of the flare. In this short paper we present some complimentary observations which we discuss in the context of the standard flare model.

OBSERVATIONS

An eruptive two-ribbon flare with a *GOES* classification of X2.3 occurred on April 10, 2001 in NOAA Active Region 9415, which was located close to the center of the disk. The region was observed by the *TRACE* and *SOHO* spacecraft to eject a portion of the active region in the form of a flux rope. There was an associated halo CME that was observed with the Large Angle Spectroscopic Coronagraph (LASCO). The *Yohkoh* spacecraft missed the preflare and rise phase because of a transit through the South Atlantic Anomaly. The soft X-ray telescope (SXT) did record some images in the hours preceding. We display a subportion of one of these obtained at 03:38 UT in Figure 1.

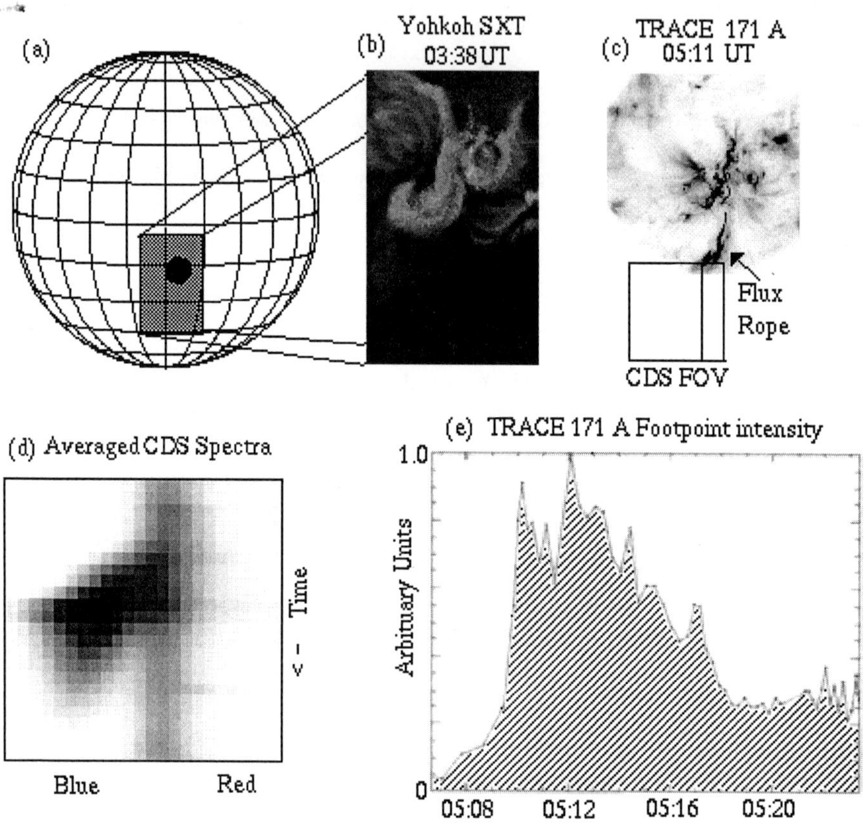

Fig. 1. AR 9415 as observed by (b) the *Yohkoh* SXT and (c) *TRACE*. In (a) we display the position of the partial images. AR 9415 is indicated by the black circle. (c) The *TRACE* image was recorded at 05:11 UT while the eruption was in progress. On this image, we have overplotted the CDS FOV along with the slit position at this time. The erupting flux rope is arrowed. In (d) we have plotted the averaged spectra for each successive CDS step. (e) shows the variation of the intensity measured in the 171 Å images recorded by *TRACE* of the left most footpoint which is observed to recoil, as the right most footpoint centered in AR 9415 is ejected violently.

CDS Doppler Motion

The CDS was rastering approximately 200 arc sec below AR 9415 when the transit of the ejected flux rope (plasmoid) was recorded as part of its synoptic program (see Figure 1c). The signature of the transit was most identifiable in the O V emission line; the average line profile and evolution are illustrated in Figure 1d. The line is observed to shift from a faint redshifted/stationary component to a bright blueshift of around 480 km s^{-1}. This was measured to occur in two phases, an initial short lived (just over a minute) acceleration of $\sim$ 3.5 km s^{-2}, followed by a more modest but sustained acceleration of around 0.7 km s^{-2}.

TRACE EUV Flare Ribbons

The *TRACE* spacecraft recorded images in a range of wavelengths and resolutions. We selected the 171 Å (Fe IX/Fe X) passband images since the flux rope and flare ribbons were most visible in this passband. The tracking of features close to the active region core revealed that the structure was emerging from the active region at a velocity of around 120 km s^{-1}. This was followed after 05:10 UT with an acceleration of around 1.95 km s^{-2}. During this time period the flux rope was observed to increase in intensity as it is observed to move away and apparently detach. This is demonstrated by the intensity variation of the footpoint located outside the active region (*see* Figure 1e).

During the time in which the CDS observations suggest that the flux rope was being accelerated, the EUV flare ribbons were observed to separate in a similarly impulsive fashion. This illustrated in Figure 2, which depicts the time series of the intensity across a portion of the flare ribbons. These appeared to separate between around 05:10 and 05:12 UT with an acceleration of $\sim$ 1.1 km s^{-2}.

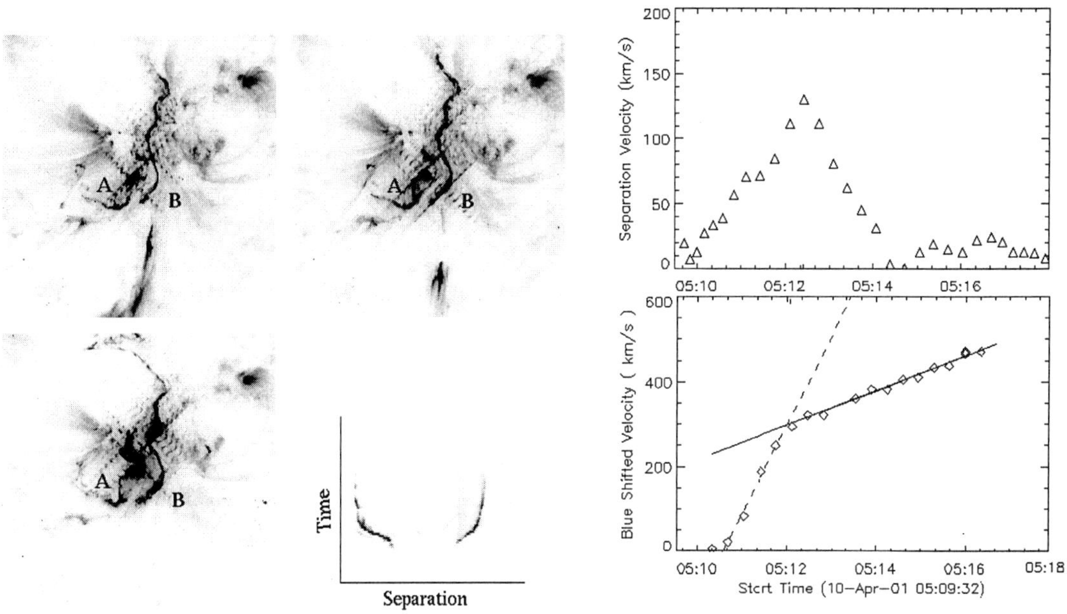

Fig. 2. *Left* Separation of EUV flare ribbons between A and B, which appears to be below the region where the flux rope is observed to be accelerated from. *Right* Velocity evolution of the flare ribbons (top) and flux rope (bottom). The flux rope appears to be driven by the same process which causes the flare ribbons to separate.

DISCUSSION

We have reported on the EUV observations of an X2.3 flare which was associated with a flux rope lifting off the disk of the Sun, using the CDS instrument on board the *SOHO* spacecraft and the *TRACE* EUV imager. We have measured the acceleration of the flux rope as it is ejected. The acceleration was observed to proceed in two phases. A high-acceleration phase (3.5 km s^{-2}) which lasted for just over a minute. This was followed by a more gradual acceleration (0.7 km s^{-2}) for the four minutes until the flux rope left the

CDS FOV. During the preflare phase the flux rope was observed to increase in intensity suggesting that it was being heated prior and during the impulsive phase. The flare ribbons are observed to accelerate apart at 1.1 km s^{-2}, whilst the flux rope was experiencing its initial high-acceleration phase. The observation of these features simultaneously is as predicted by the standard flare model. Observations with coronagraphs of prominence eruptions located at the limb of the Sun have demonstrated that they experience constant or no acceleration. Two possible forms of energy input have been postulated to explain the gradual acceleration of CMEs, based on studies of the velocity evolution of coronal flux ropes by Krall, Chen, & Santoro (2000) within the height range of 0.4-5 solar radii : (1) an increase in the poloidal component of the flux ropes' helical magnetic field and (2) hot plasma injection. Our observations were made early in the impulsive phase and show two phases of acceleration. This may suggest an additional input of energy, due to explosive magnetic reconnection, to explain the initial short-lived higher acceleration phase. This is consistent with the conclusions derived from the observations of flare sprays by Tandberg-Hanssen, Martin, & Hansen (1980). They reported similar morphology and time-height evolution. They attributed the motion observed to an initial acceleration phase that was short-lived and occurred at low altitudes. It was noted from the work of Gosling *et al.* (1976) that the CMEs associated with sprays were generally fast, and those from "disparitions brusques" slow. Observations of CMEs with LASCO (Andrews & Howard 2001) have demonstrated that the CMEs originating from nonflaring/flaring regions display a similar anisotropy. The CMEs associated with flares at their onsets often have accompanying Moreton waves and type II bursts and tend to be faster, brighter, and larger. Observations of broadened line profiles above a flare at the Suns limb have been reported by Innes *et al.* (2001) using SUMER data. The line profiles extend out to 650 km s^{-1} , but without a resolved high-velocity component as reported here. They attributed the motions to shock-driven acceleration and heating of active region loops above the flare location and suggest that the shock provides the energy to drive the CME. Our observations support the idea that explosive reconnection is the underlying CME trigger. However, in our case, instead of a shock front, we see one end of a flux rope being violently disconnected from the Sun.

ACKNOWLEDGEMENTS

C.R.F. acknowledges the support of The Royal Society though provision of a conference grant.

REFERENCES

Andrews, M.D., and Howard, R.A. 2001, *Space Sci. Rev.*, **95**, 147.

Foley, C.R., Harra, L.K., Culhane, J.L., Mason, K.O., 2001, *ApJ Letters*, **560**, L91.

Gosling, J., Hildner, E., MacQueen, R., Munro, R., Poland, A., and Ross, C. 1974, *JGR*, **79**, 4581 532.

Innes, D.E., Curdt, W., Schwenn, R., Solanki, S., Stenborg, G., and McKenzie, D.E. 2001, *ApJ Letters*, **549**, L249.

Krall, J., Chen, J., & Santoro, R. 2000, *ApJ*, **539**, 964.

Masuda, S., Kosugi, T., Tsuneta, S., Hara, H. 1996, *Adv. Space Res.*, *17*, 63.

Tandberg-Hanssen, E., Martin, S.F., and Hansen, R.T. 1980, *Solar Phys.*, **65**, 357.

Ohyama, M., and Shibata, K., **ApJ**, **499**, 934.

Shibata, K., Masuda, S., Shimojo, M., Hara, H., Yokoyama, T., Tsuneta, S., Kosugi, T., Ogawara, Y., *ApJ Letters*, **451**, L83.

PRE-FLARE HEATING AROUND THE TEMPERATURE MINIMUM REGION FOUND RIGHT PRIOR TO AN X-CLASS FLARE

H. Kurokawa[1], T. T. Ishii[1], T. J. Wang[2], and R. Shine[3]

[1]*Kwasan and Hida Observatories, Kyoto University, Yamashina-ku, Kyoto, 607-8471 Japan*
[2]*Max-Planck Institut für Aeronomie, Postfach 20, Katlenburg-Lindau, D-37189, Germany*
[3]*Lockheed Martin Solar & Astrophysics Laboratory, Bldg. 252 Org. L9-41, 3251 Hanover Street, Palo Alto, CA 94304, USA*

ABSTRACT

Flare producing active region NOAA 9026 (June 2000) provided us with a rare opportunity to study a twisted magnetic flux rope and its rapidly-untwisting motions causing a strong flare. From the analyses of the evolution of this region we found for the first time clear evidence of pre-flare heating or energy release from the upper photosphere through the lower chromosphere starting about two hours before an X-class coronal flare. We conclude that the pre-flare energy release was caused by magnetic field reconnection in the lower atmosphere, where the emergence of a twisted magnetic flux rope was abruptly enhanced by the kink instability.

INTRODUCTION

Studies of magnetic shear development and pre-flare activity in flare-producing sunspot regions are of fundamental importance for the study of the flare energy build-up and energy release mechanism. Previous work has demonstrated that the emergence of a twisted magnetic flux rope, which is originally formed in the convection zone, must be the source of the strong magnetic shear development in a sunspot region prior to a large flare (Kurokawa 1987, Tanaka 1991, Ishii *et al.* 2000). However, we are still far from sufficiently understanding how the twisted structure of a magnetic flux rope is formed in the convection zone, and where and how such a twisted magnetic rope untwists and releases its energy as a flare.

OBSERVATIONAL DATA

Flare producing active region NOAA 9026 showed an interesting evolution during coordinated observations between the Domeless Solar Telescope of Hida Observatory, the Swedish Telescope at La Palma, and the *TRACE* satellite, from June 3 through 12 of 2000. This region provided us a rare opportunity to study a twisted magnetic flux rope and its rapidly-untwisting motions causing a strong flare.

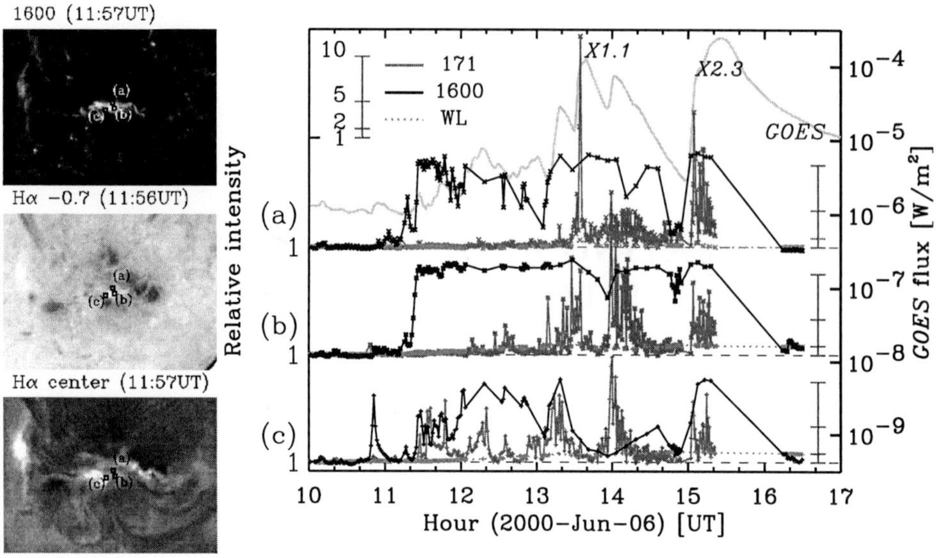

Fig. 1. The light curves of representative locations as a fraction of the intensity at 10 UT.

RESULTS AND DISCUSSION

We studied the evolutional characteristics of the active region NOAA 9026 and constructed a twisted magnetic flux rope model explaining the active region evolution in a previous paper (Kurokawa *et al.* 2002). In this paper we found clear evidence for lower-atmospheric brightenings in *TRACE* 1600 Å and 5000 Å images as well as in Hα wing images shortly prior to the major flares of June 6, 2000.

Figure 1 summarizes the light curves of representative points as a fraction of the intensity at 10 UT for the 171 Å, 1600 Å, and white light images. The representative positions are shown in the left panels. These points are on or near the magnetic field neutral line. Notice conspicuous pre-flare brightenings in 1600 Å as well as in white light from 11 UT through 13 UT, while the 171 Å coronal line only shows strong emission at the time of the major flares. Notice also the 1600 and Hα − 0.7 Å pre-flare emissions at the location of the Hα dark filament in Figure 1. These observations clearly indicate that these pre-flare brightenings originate in the lower atmosphere.

ACKNOWLEDGEMENTS

We thank all members of Solar Observation Team at Kwasan & Hida Observatories. We made extensive use of the *TRACE* Data Center and the *SOHO* MDI Data Service.

REFERENCES

Ishii, T.T., Kurokawa, H., and Takeuchi, T.T., *PASJ*, **52**, 337 (2000).

Kurokawa, H., *Solar Phys.*, **113**, 259 (1987).

Kurokawa, H. *et al.* *ApJ*, in press (2002).

Tanaka, K., *Solar Phys.*, **136**, 133 (1991).

Section IX.

Flare Plasma Dynamics

NON-THERMAL VELOCITIES IN SOLAR FLARES

L. K. Harra

Mullard Space Science Laboratory, University College London, Holmbury St. Mary, Dorking, Surrey RH5 6NT, U.K

ABSTRACT

The high resolution spectroscopic information from the Bragg Crystal Spectrometer on board *Yohkoh* has provided us with new and exciting information about flares. In particular, there has been much work on understanding the excess line broadening above the thermal width (known as non-thermal line broadening). We have been able to look for the first time spectroscopically at the preflare stages in X-rays. The timings of the non-thermal velocity relative to the hard X-ray emission has been investigated. Non-thermal velocities have been observed to increase ten minutes before the main flare begins. Progress has been made to locate the region of dominant non-thermal velocity. This is difficult due to the lack of spatial resolution. A discussion will be made on what can be expected from the EUV Imaging Spectrometer on-board *Solar-B*, which combines both high spatial and spectral resolution simultaneously.

INTRODUCTION

Since *Yohkoh* has been launched enormous progress has been made in understanding the solar flare process. A summary of the Soft X-ray Telescope (SXT) and the hard X-ray telescope (HXT) main achievements will be given elsewhere in these proceedings. This article concentrates on the advances that the observations from the Bragg Crystal Spectrometer has made in the past ten years of *Yohkoh's* successful operation - in particular I will concentrate on the results obtained from *non-thermal velocity* (V_{nt}) which is a critical parameter in understanding energy release in the corona.

Non-thermal velocity is a measurement of the excess line broadening of emission lines and is given by $V_{nt}=[2k(T_D - T_e)/m_i]^{1/2}$, where T_D is the Doppler temperature derived from the total observed line width, T_e is the electron temperature obtained from line ratio diagnostics and m_i is the mass of the ion considered. An example of how lines are broadened is given in Figure 1 which shows a Ca XIX spectrum at the peak and decay phases of a flare. The intensity is lower during the decay phases, but in addition the profile of the spectral lines is much wider at the peak of the flare (135 km/s) whereas during the decay phase it is only 60 km/s in this example.

It has been known for many years that the emission lines are broader than the thermal width during flares, with values ranging between 90–300 kms^{-1} (e.g. Grineva *et al.* 1973, Doschek *et al.* 1983, Antonucci *et al.* 1984) for coronal lines. The non-thermal velocity has been found to reach of order 100 km/s before the flare proper begins in a number of different observations (e.g. McNeice *et al.* 1985). In addition, the transition region also shows broadened lines in the preflare stages (Brueckner *et al.* 1976). This observation was explained by small instabilities in small loops prior to flares. The instabilities were seen at the footpoints.

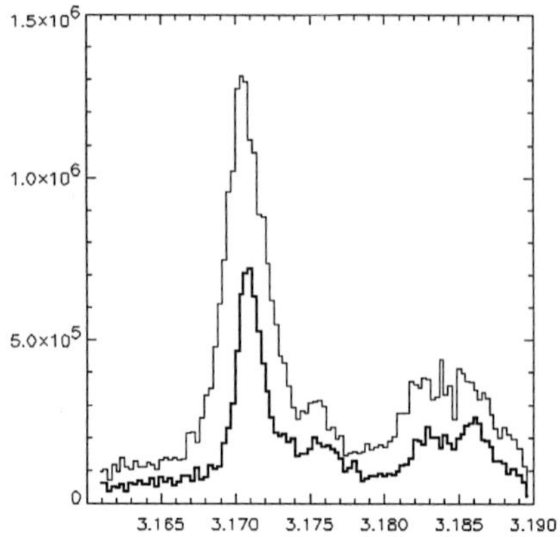

Fig. 1. This shows a Ca XIX spectrum from the BCS on board *Yohkoh* during the peak of a flare (thin line) and during the decay phases (thick line).

There has been some debate on whether the non-thermal velocity observed is a response to chromospheric evaporation. Little or no center-to-limb dependence was found, although Tanaka (1986) found a slight increase for over-the-limb events. This lack of dependence indicates that the broadening is not due to evaporation flows which should not be detectable for limb events (but see Li, these proceedings). However, Doyle, & Bentley (1986) analyzed high time resolution Ca XIX spectra and suggest that the resonance line is composed of many discrete features which are interpreted as mass flows.

Interestingly there has not been a strong relationship found between the intensity of flares and the value of the non-thermal velocity. Harra-Murnion, Akita, & Watanabe (1997) derive the non-thermal velocity for 45 small *GOES* flares observed during the year of 1994 observed by *Yohkoh* concentrating on the peak and rise phase of the non-thermal velocity (V_{nt}). They examined the variation of V_{nt} with electron temperature, *GOES* classification, duration of event, rise time, and the source size of the event. Their results demonstrate that the broadening is independent of the flare size, complexity and intensity of hard X-ray bursts. There is a weak dependence on duration and rise time: as the durations and rise times become longer, the value of V_{nt} decreases.

Large line broadenings are not only observed in flares - they are also apparent in all other phenomena on the Sun. This suggests that understanding non-thermal velocity is critical to understanding the heating and energy release mechanisms.

The BCS instrument is described by Culhane *et al.* (1991). It consists of four spectrometers with wavelength ranges encompassing the resonance lines of helium-like sulphur (5.04 Å), calcium (3.18 Å) and iron (1.85 Å) , along with hydrogen-like iron (1.78 Å). It is a full Sun instrument with a time resolution of 3 s in flare mode. The sensitivity of the system is $\approx$ 10 times that of previous instruments of a similar nature, with an improvement of up to 60 times for the S XV channel. With the higher sensitivity of the Bragg Crystal Spectrometer on board *Yohkoh*, progress has been made on several fundamental aspects of non-thermal velocity;

- *What is the timing relationship between the non-thermal velocity and the start of the flare?.*
- *Where is the location of the largest non-thermal velocity?*
- *What is behavior of the non-thermal velocity in the important preflare stages?*

There has been much progress in understanding the behavior of V_{nt} since the launch of *Yohkoh*, but the process(es) causing this phenomena still remains a mystery. It is caused by at least one of the following;

- Preflare activity
- The elusive flare trigger
- The actual energy release mechanism
- The response to energy release

It is clear is that non-thermal velocity is providing important information for the understanding of the flare trigger and heating mechanisms. A summary of BCS V_{nt} observations over the past decade will be given, followed by a brief discussion on how *Solar-B* can push forward our knowledge even further.

TIMING

It is important to know when during a flare the V_{nt} is dominant in order to distinguish between the different causes. For example if the excess broadening is caused by chromospheric evaporation it would be expected that the main broadening would occur following the impact of the high energy electrons with the chromosphere. Attempts have been made to determine when V_{nt} peaks relative to the Hard X-ray (HXR) bursts. Alexander *et al.* (1998) analyzed 10 flares with *Yohkoh* and found that the V_{nt} measured from S XV spectra exhibits a peak prior to the first significant hard X-ray peak or is already decaying from an earlier unobserved peak. This result suggests that the large values of V_{nt} are related to the actual flare trigger rather than later energy deposition. This observation is consistent with a model developed by Tsuneta (1995) who discusses the possibility that line broadening would be due to turbulence driven by reconnection outflow at the looptop rather than chromospheric evaporation near to the footpoints. However, further work by Mariska & McTiernan (1999) showed that when it is possible to measure the peak of V_{nt} derived from Ca XIX spectra, it occurs after the first significant peak in hard X-rays. This observation differs from the results of Alexander *et al.* (1998), and it was suggested by Mariska & McTiernan (1999) that this was due to a difference in the S XV V_{nt} peak and the Ca XIX V_{nt} peak. Figure 2 shows two examples of temporal variation of non-thermal velocity relative to the hard X-ray emission. In the top case (15-Jan-1992), the non-thermal velocity peaks after the main hard X-ray burst. In the lower case (20-June-1999) the non-thermal velocity peaks at approximately the same time as the hard X-ray peak. In both cases the non-thermal velocity is high before the hard X-ray emission is high.

A larger statistical study has been carried out by Ranns *et al.* (2001). A total of 59 limb flares were analyzed using the Burst and Transient Source Experiments (BATSE) on *CGRO* and *Yohkoh.* A relationship was found between the rise time of the HXRs to a peak value and the delay time between the HXR flux maximum and the V_{nt} maximum, suggesting a relationship between the hard X-ray emission and the observed non-thermal line broadenings (see Figure 3). It has also been found that the time delay between the peak of the HXR and V_{nt} is related to the number of subsidiary HXR peaks present before the main peak. This difference is enhanced when the flares are separated into impulsive and gradual categories. The gradual rise flares show a tendency for V_{nt} to peak before the HXR peak whereas the opposite behavior is observed in impulsive flares.

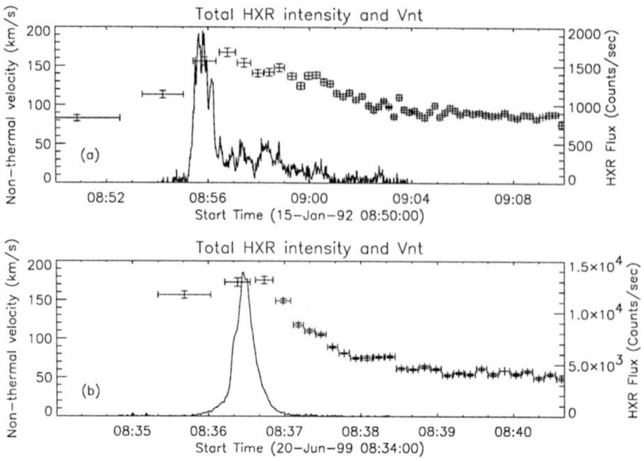

Fig. 2. A figure from Ranns *et al.* (2001) (a) shows the non-thermal velocity derived from Ca XIX spectra (shown with error bars) of a flare on the 15 January 1992, along with the HXR emission determined from the Low channel (25-50 keV) of the Burst and Transient Source Experiment (BATSE) on board the *Compton Gamma Ray Observatory* (*CGRO*). (b) shows the same parameters for a flare on the 20 June 1999.

LOCATION

It is important to determine the location of the V_{nt} as this will provide information on the source. The BCS has no spatial resolution, and hence tricks with occultation have been used in order to obtain information about which region of flaring loops have the largest V_{nt} - the footpoints or the looptops.

Khan *et al.* (1995), and Mariska, Sakao, & Bentley (1996) have presented values of V_{nt} for over-the-limb events where the footpoints of the flaring loops are occulted. Khan *et al.* studied a sequence of 4 flares ranging from C5.4 to C6.9 over a period of 10 hours, which occurred in the same active region. They found values of V_{nt} for the looptop which did not noticeably deviate from the values for the disk events observed by previous missions ($\approx$ 150 km s^{-1}). They did not however directly compare their over-the-limb results with *Yohkoh* BCS observations of limb flares with the footpoints visible. Mariska, Sakao, & Bentley studied eight events, four having the footpoints occulted and four with the footpoints visible. They found that the events happening just in front of the limb, with the footpoints showing, had larger values of V_{nt} (average for Ca XIX is 260 km s^{-1} and for Fe XXV 398 km s^{-1}) than those over-the-limb (average for Ca XIX is 172 km s^{-1} and for Fe XXV is 235 km s^{-1}). The over-the-limb values obtained (with only the looptop showing) were similar to those of Khan *et al.* A further study by Mariska & McTiernan (1999) of a larger statistical study shows none of the difference between the V_{nt} in occulted and non-occulted flares

Ranns *et al.* (2000) have made use of the temperature and emission measure determinations from both BCS Ca XIX and SXT data from *Yohkoh* in order to determine the location of non-thermal velocity. This is illustrated in Figure 4, where the contour levels show the region in the SXT images that the BCS and SXT temperature are similar. Hence the contoured regions show the location of the BCS emission (i.e. the region when the excess broadening is located). At the peak of the non-thermal velocity the location seems to be confined to the looptop. As the flare progresses (Figure 4b) the contoured region becomes larger and hence it can be assumed that the BCS emission is now emitting from a larger spatial location. This indicates a filling of the loops with hot plasma. This is consistent with the work of Khan *et al.* (1995) who show that the dominant source of the non-thermal velocity was at the looptops. It also illustrates that the time period of most interest is the time before the non-thermal velocity peak. Following the peak, the line-broadening seems to be dominated by a response to chromospheric evaporation.

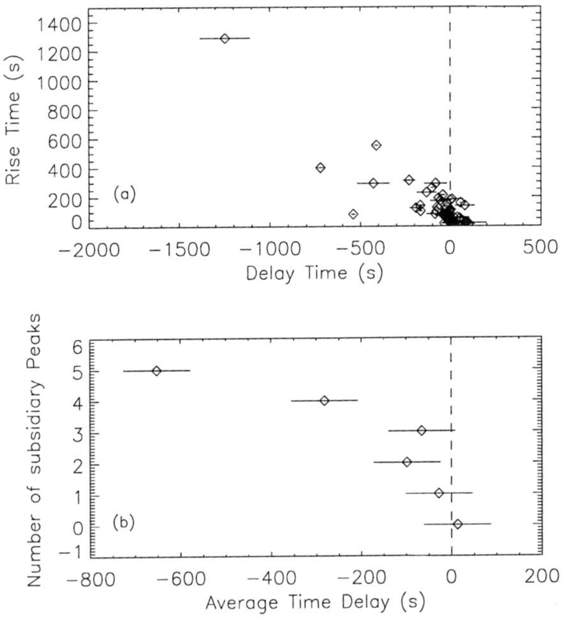

Fig. 3. A figure from Ranns *et al.* (2001) (a) The relationship between the HXR rise time and the delay time between the HXR and non-thermal velocity maxima. (b) The average time delay for flares with a particular number of subsidiary peaks. A negative delay time implies that the non-thermal velocity peak occurred before the hard X-ray peak.

PREFLARE BEHAVIOR

The search for preflare activity has been in progress for decades now. It is the most critical stage of the flare process, and if we are to understand how a flare is triggered we must understand this stage.

Cheng (1990) shows large intensity enhancements in the transition region (TR) and V_{nt} of the order 100 km/s before the HXR bursts. Heating can be as high as 10 MK before the onset of the HXR impulsive phase. The UV emission in the preflare phase is not due to the bombardment of the electrons that produce the HXR emission, but rather the TR is already being excited and heated probably due to some more gradual in situ heating mechanism e.g. joule heating from enhanced current dissipations. The excitation of the plasma creates a favorable environment that prepares the eventual occurrence of a plasma instability resulting in the acceleration of electrons.

There is an extensive summary of preflare signatures in Schmahl *et al.* (1986). This work concluded that there were numerous examples of preflare coronal manifestations, but very few were consistent. The clearest example of preflare activity they found was that of activated filaments, which have long been known to be related to the flare process. More recent work, using *Yohkoh* SXT (Farnik & Savy 1998) analyzed 32 flares which had preflare emission. They found that there was no definite evidence of a special preflare phase when looking at intensity and temperature changes.

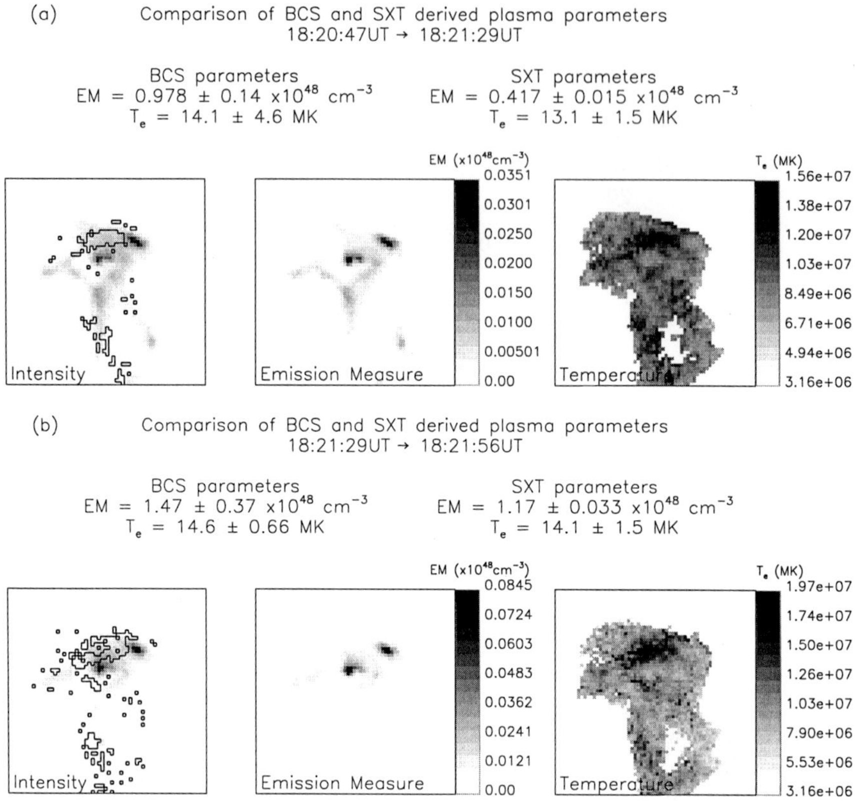

Fig. 4. This figure (Ranns *et al.* 2000) shows the SXT intensity, emission measure and temperature maps of a M1.7 flare that took place on the 22nd June 1999 at the time the non-thermal velocity peaked (18:20 UT) (a) and just after (b). The pixels bound by contours have a temperature within one standard deviation of the BCS temperature for the same interval. This indicates the region of maximum BCS emission and hence the location of the emission producing the non-thermal velocity. The spatial dimensions are 2.5 arcmins^2, and the flare occurred at disk center.

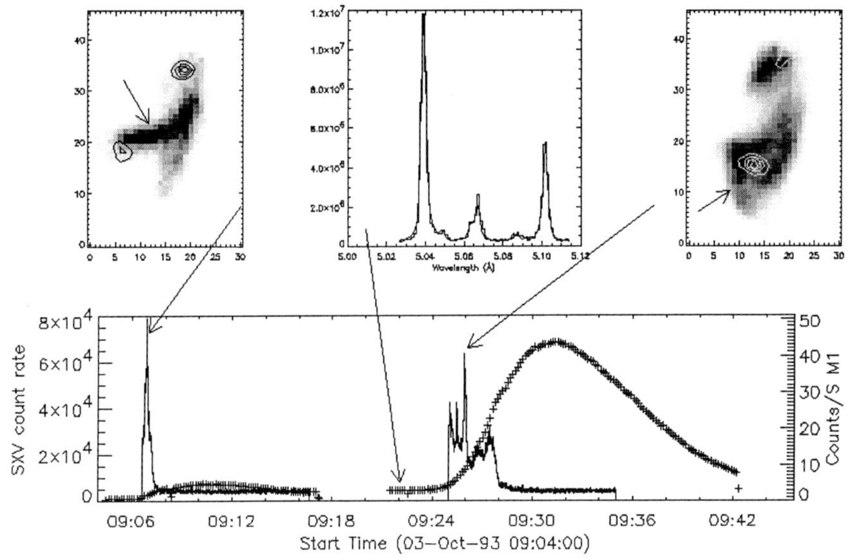

Fig. 5. The BCS and HXT light curves along with one Soft X-ray image (using the Be filter) from each flare (Harra, Matthews, & Culhane, 2001). The Hard X-ray contours from the M1 channel are overlaid. The dimensions of the Soft X-ray images are 75" X 112". A sample BCS S XV spectral fit before Flare B began is shown to demonstrate the high quality of the fits.

One main observational difficulty is following the V_{nt} rise to its peak. To observe this clearly, the V_{nt} needs to be observed at the background active region level and to rise again to the flare level. One way to do this is to observe two flares which have occurred in the same active region in the same *Yohkoh* orbit. Harra, Matthews, & Culhane (2001) discuss one such example - two flares which occurred in October 1993. They discuss the behavior of the V_{nt} in relation to the HXR emission, and show clearly that the V_{nt} starts to rise well before the HXR emission. Figure 5 shows both the flares, along with a BCS light curve, and a sample S XV spectrum during the preflare stages illustrating the excellent quality of the BCS data.

Figure 6 shows that the rise above the V_{nt} active region background level begins 11 minutes before the start of the second flare as defined by the start of the hard X-ray emission. During this extended rise time of V_{nt} there is no increase in the light curves or the electron temperature. Harra, Matthews, & Culhane suggest that this increase is an indicator of turbulent changes in the active region prior to the flare which are related to the flare trigger mechanism. This example is certainly not unique. Warren & Warshall (2001) discuss preflare brightenings in *TRACE* in the UV at 1600 Å. These brightenings occur in the footpoints and loop and happen on the order of minutes before the hard X-ray emission begins. In one example cited in this paper the non-thermal velocity increases from 50 km/s to 150 km/s over a period of over 10 minutes before the hard X-ray emission begins.

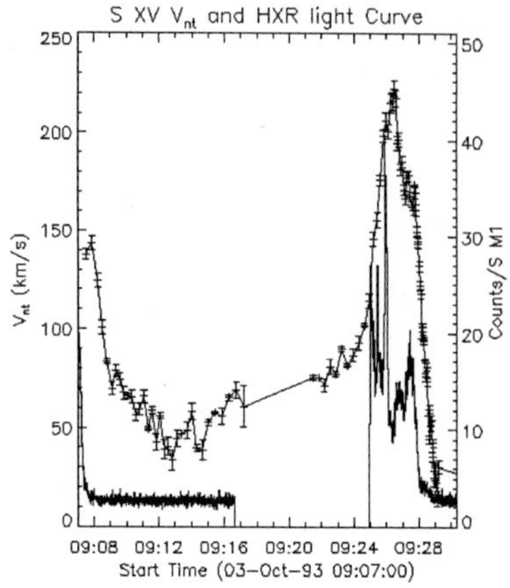

Fig. 6. The non-thermal velocity derived from the S XV spectra (thin line) along with the HXT M1 (23-33 keV) channel light curve (thick line) (Harra, Matthews, & Culhane 2001)

DISCUSSION

There has been huge progress in the pinning down the characteristics of non-thermal velocity. We now know that;

- There is no center to limb dependence on the value of the non-thermal velocity
- The non-thermal velocity reaches at least 100 km/s before the hard X-ray emission begins. This is a rise of at least 50 km/s above the non-flaring active region level in around 10 minutes.
- At later stages the non-thermal velocity appears to be due to chromospheric evaporation
- There are high non-thermal velocities at the loop-top during the flare.
- Flares of all sizes show strong non-thermal velocities.

We have a lot of information now, but still have no definite answers about the physical processes involved. It is clear that after the hard X-ray emission begins the non-thermal velocity is dominated by chromospheric evaporation. It is the period beforehand that is of major interest. It could be that the slow emergence of magnetic field is gradually and slowly reconnecting with the already existing field for a period of time before the main flare begins. Antonucci & Dodero (1995) show evidence for the increase in non-thermal velocity with temperature observed throughout the solar atmosphere and in flare plasma. This evidence coupled with the observations of an enhancement of nonthermal motions in the coronal plasma a few minutes prior to flare onset when the largest non thermal motions are observed strongly supports the idea that nonthermal velocities are a manifestation of the process of energy release in the corona. Ohyama & Shibata (1997) study preflare ejections that can start up to 10 minutes before the flare begins. They suggest that preflare reconnection is taking place. Shibasaki (2001) shows that high beta plasma confined in loops can play important role in solar activity. He suggests that observations of line broadening before and during the impulsive phase can be interpreted as due to enhanced turbulent waves.

The importance of understanding non-thermal velocity is self evident. We are now focusing on the science goals for the EUV Imaging Spectrometer on board *Solar-B*, and how this can make a huge leap in our understanding.

The Future: *Solar-B*

The EUV Imaging Spectrometer on board *Solar-B* is the ideal instrument to achieve the first spatial study of evaporation and turbulence in solar flares. There are two wavelength ranges - 170-210 Å and 240-290 Å which cover a range of coronal active region and flares emission lines as well as some cooler transition region lines. The velocity resolution will be 3 km/s for Doppler velocities and 20 km/s for line widths, and the temporal resolution will be <1 second for flares. In addition we will have high resolution vector magnetograms from the Focal Plane Package (FPP) on the Solar Optical Telescope (SOT) instrument. These will be critical to understanding the changes that we see tens of minutes before the flare begins and will give us an excellent chance to understand the elusive flare trigger. For further information on the EIS instrument look at our website (`www.mssl.ucl.ac.uk/Solar-B`). This also includes a study form to allow one to design potential observations, and to look at studies that the EIS team have developed under our EIS Scientific Planning Guide.

ACKNOWLEDGEMENTS

LKH is grateful to PPARC for the award of an advanced fellowship.

REFERENCES

Alexander, D., Harra-Murnion, L.K., Khan, J.I., & Matthews, S.A., *ApJ Letters*, **494**, L235 (1998).

Antonucci, E., and Dodero, M.A., *ApJ*, **438**, 480 (1995).

Antonucci, E., Gabriel, A. H., and Dennis, B.R., *ApJ*, **287**, 917 (1984).

Culhane, J.L., *et al.*, *Solar Phys.*, **136**, 89 (1991).

Doschek, G., *Solar Phys.*, **86**, 9 (1983).

Doyle, J.G., and Bentley, R.D., *A&A*, **155**, 278 (1986).

Farnik, F., and Savy, S.K., *Solar Phys.*, **183**, 339 (1998).

Farnik, F., Hudson, H.S., and Watanabe, T., *A&A*, **320**, 620 (1997; see also Erratum, *A&A*, **324**, 433 [1997]).

Grineva, Yu. I., Karev, V.I. Korneev, V.V., Krutov, V.V., Mandelstam, S.L., Vainstein, L.A., Vasilyev, B.N., and Zhitnik, I.A., *Solar Phys.*, **29**, 441 (1973).

Harra, L.K., Matthews, S.A., and Culhane, J.L., *ApJ Letters*, **549**, L245 (2001).

Harra-Murnion, L.K., Akita, K., and Watanabe, T., *ApJ*, **479**, 464 (1997).

MacNeice, P., Pallavicini, R., Mason, H.E., Simnett, G.M., Antonucci, E., Shine, R.A., and Dennis, B.R., *Solar Phys.*, **99**, 167 (1985).

Mariska, J.T., Sakao, T., and Bentley, R.D., *ApJ*, **459**, 815 (1996).

Mariska, J.T., and McTiernan, J.M., *ApJ*, **514**, 484 (1999).

Ohyama, M., and Shibata, K., *PASJ*, 49, 249 (1997).

Ranns, N.D.R., Harra, L.K., Matthews, S.A., and Culhane, J.L., *A&A*, **379**, 616 (2001).

Ranns, N.D.R., Matthews, S.A., Harra, L.K., Culhane, J.L, *A&A*, **364**, 859 (2000).

Schmahl, E.J. *et al.* 1986, in NASA Conference Publication 2439, eds. M. Kundu and B. Woodgate.

Tsuneta, S., *PASJ*, **47**, 691 (1995).

Warren, H.P., and Warshall, H.D., *ApJ*, **571**, 999 (2002).

CORRELATED DYNAMICS OF HOT AND COOL PLASMAS IN TWO SOLAR FLARES

B. Kliem[1], I. E. Dammasch[2], W. Curdt[2], and K. Wilhelm[2]

[1]*Astrophysikalisches Institut Potsdam, An der Sternwarte 16, 14482 Potsdam, Germany*
[2]*Max-Planck-Institut für Aeronomie, 37191 Katlenburg-Lindau, Germany*

ABSTRACT

In flare-watch observations using the SUMER spectrometer on board *SOHO*, two limb flares were found to show correlated brightenings and line shifts of hot ($T_e \sim 10^7$ K) and cool ($T_e \sim 10^4$ K) plasmas. The brightenings occurred essentially simultaneously but the cool plasma emission decayed much earlier than the hot plasma emission, which excludes an origin in passive cooling of postflare loops and points to an active role for the cool plasma in the flare dynamics. The observations are consistent with magnetic reconnection that is triggered and sustained by the formation of a coronal condensation through the thermal instability. An MHD simulation supporting this model is presented.

INTRODUCTION

Flares are usually regarded as "hot phenomena": coronal and chromospheric plasma is heated to $\sim$10–30 MK and particles are accelerated to $\sim$10–100 keV or higher (e.g. Sturrock 1980, Hudson 1994, Kosugi 1994). The energy stored in the magnetic field is converted suddenly and locally into kinetic energy of the plasma. Cooling plasma is observed *after* the impulsive energy release of many flares in X rays, UV, and down to temperatures visible in Hα. However, the idea that cooling plasma at temperatures of a few 10^4 K may play a role in *initiating* flares is not commonly included in flare models. It was considered as a possibility to accelerate the tearing mode in the corona by Van Hoven *et al.* (1984), but appears to have not been pursued further since then.

Recently we have obtained spectroscopic observations of a solar limb flare using a number of emission lines in the far-UV range. A strong correlation between the initial brightenings, line shifts, and excess line widths of hot ($\sim 10^7$ K) and cool ($\sim 10^4$ K) plasmas was found. There were no indications of a prominence near the flare site. The correlations in the spectra and the early decay of the cool plasma brightness indicated an active role of cool plasma in the main (and probably also in the impulsive) phase of this flare, different from the usual picture of emergence through (passive) cooling of evaporated chromospheric plasma after the impulsive phase. We have suggested a model in which localized cooling of coronal plasma by the thermal instability triggers magnetic reconnection through the resulting enhanced resistivity, the combined processes leading to the correlated dynamics of hot and cool plasmas in a loop-loop interaction geometry (Dammasch *et al.* 2001, Kliem *et al.* 2002). *Yohkoh*/SXT and *SOHO*/EIT data have been essential in guiding the interpretation by showing that the correlated dynamics did not arise in the loop arcade of a preceding flare but did occur inside this arcade close to the position where a new, presumably very hot loop was formed, and by showing that the event as a whole was not eruptive (Figure 1). *Yohkoh*/HXT provided the timing of the impulsive flare phase with respect to the UV data. UV data and X-ray time profiles of this flare are shown in Figure 2.

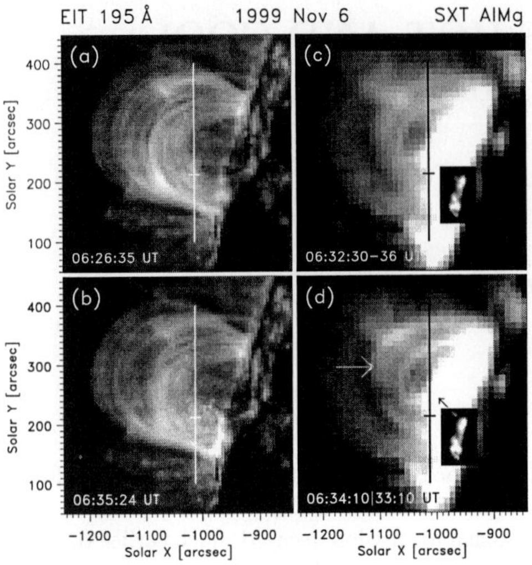

Fig. 1. Morphology of the flare in which the correlated behavior of the hot and cool plasma was discovered. (a, b) *SOHO*/EIT images just before the event and at the time of peak soft X-ray (SXR) emission; (c, d) composites of quarter-resolution and full-resolution *Yohkoh*/SXT images. The fixed position of the SUMER slit and the position of the cool plasma brightening at the slit are indicated. The effects seen by SUMER appeared to be associated with the weak loop-shaped brightening inside the arcade of post-flare loops of a previous flare and with the weaker (the northern) of the two SXT footpoint sources, which gave rise to a weak ejection that faded before the SXR emission peaked (indicated by the small arrow). The lack of any brightening in the SUMER Fe XII observations (Fig. 2) indicates that this brightening actually occurred in the Ca XVII or Fe XXIV lines which fall in the EIT passband at 195 Å. The strong EIT footpoint brightening, the southern SXT source, and the ascending SXT loop (marked by the large arrow) were associated with a separately flaring substructure in the same event. See Kliem *et al.* (2002) for further details.

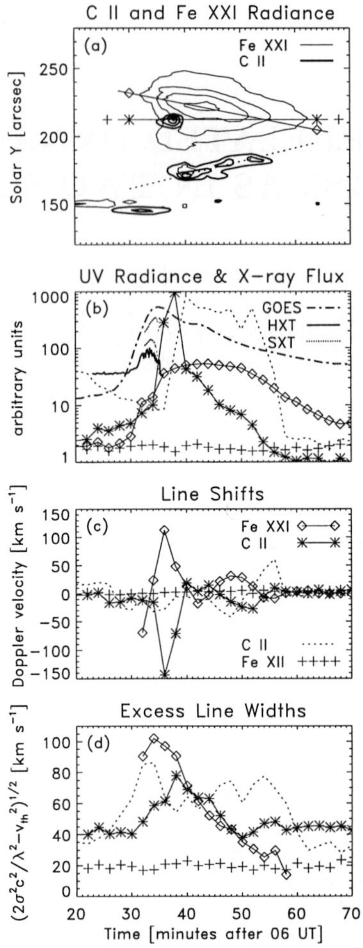

Fig. 2. Summary of SUMER data of the 1999 Nov 6 flare. (a) Equidistant isocontours of line-integrated radiance (linear scale, 20 %–92 % of max.) showing the flare-related brightenings in Fe XXI λ1354.1 ($T_e \approx 10^7$ K; *light contours*) and C II λ1335.7 ($T_e \approx 2 \times 10^4$ K; *heavy contours*). Cut lines, along which the time profiles are plotted in (b)–(d), are indicated. (b) Time profiles of line-integrated UV radiance, of the *GOES* 0.5–4 Å flux, of the *Yohkoh*/HXT 14–23 keV flux, and of the two footpoint sources resolved by *Yohkoh*/SXT. SUMER data points are plotted at the center of each 2 minutes exposure interval. (c) Line shifts relative to the average pre-flare line positions. (d) Excess line widths, where σ is the standard deviation of a Gaussian fit and $v_{\rm th}$ is the thermal velocity of the ion.

NEW OBSERVATIONS

Data of a second event on 2000 Oct 25,that showed a correlated brightening and line shift signature of hot (Fe XIX λ1118.1; $T_e \approx 8$ MK) and cool (S III λ1113.2; $T_e \approx 30\,000$ K) plasmas are displayed in Figure 3. The peak brightenings in this flare coincided exactly (at a spatial resolution along the slit of $1''$ and a temporal resolution of 170 s). As in the event on 1999 Nov 6, the brightening of the cool plasma decayed much earlier than the brightening of the hot plasma. The line shifts showed a simultaneous rise, to the red for both lines, at the onset of the brightenings, the subsequent evolution being different in detail. The main spectral properties suggest a physical relationship between the hot and cool plasmas also for this flare. Again, the cool plasma could not emerge from cooling postflare loops. A series of EIT 195 Å images that includes the time of the flare contains a weak indication that diffuse absorbing (i.e., cool) material was permanently present, extending from the disk out to about the position where the event seen by SUMER occurred. Activity in this cool material that caused a brightening of the S III line – presumably a condensation – appears to have triggered the energy release that led to the Fe XIX brightening. No *Yohkoh* or *TRACE* observations are available for this event.

This flare occurred in the decay phase of a long-duration flare at a remote location and did not produce a signature in the *GOES* light curve, which limits its magnitude to *GOES* class C3. Both cases observed so far are rather small flares. Further observations are required to see whether this is a systematic property.

FIRST SIMULATIONS OF CONDENSATION-DRIVEN MAGNETIC RECONNECTION

To test the hypothesis that a coronal condensation can trigger magnetic reconnection, we have performed two-dimensional MHD simulations of the evolution of a temperature perturbation in a coronal current sheet. The standard resistive and compressional MHD equations were integrated in a numerical setup similar to that used by Kliem *et al.* (2000) but with radiative losses included. The classical Spitzer resistivity was chosen. Its strong temperature dependence, $\eta \propto T^{-3/2}$, is the basis of our hypothesis, since it implies an increase of the resistivity by three orders of magnitude if the temperature drops from coronal to upper chromospheric values. A one-dimensional isothermal Harris current sheet equilibrium, $\mathbf{B} = B_0 \tanh(y/l_0)\mathbf{e}_x$, balanced by a density gradient across the sheet, was initially perturbed by an isobaric temperature drop in the center of the sheet, extending along the sheet by about one sheet width ($2l_0$). The parameters external to the sheet were chosen to be $T_0 = 2$ MK, Alfvén velocity $V_{A0} = 800$ km s^{-1}, Lundquist number $S = 2000$, plasma beta $\beta = 0.1$, and density $n_0 = 5.6 \times 10^{16}(l_0/\mathrm{cm})^{-1}$ cm^{-3}.

The initial temperature perturbation developed into a condensation with the temperature decreasing to about 20 000 K, a value which is determined by the drop of the radiative loss function. The condensation process continued during the whole simulation, with cooling material streaming into the perturbed area. Magnetic reconnection gradually commenced early in the evolution, developing the typical magnetic and velocity field pattern during the first $\approx 10^2$ Alfvén times (i.e., during ≈ 1 min; $\tau_A = l_0/V_{A0} = 0.7$ s for $n_0 = 10^9$ cm^{-3}), and continued also during the whole simulation. The condensation-reconnection process was followed for several $10^2\,\tau_A$ (Figure 4). It turned out to be less vigorous than reconnection driven by anomalous resistivity because the reconnection outflow jets are not accelerated right from the X line but only from the outer boundary of the condensation, but outflow velocities of ≈ 50 km s^{-1} – of the same order as the observed Doppler velocities – were reached. These simulations support our hypothesis that localized condensations, formed by the thermal instability, can trigger and sustain magnetic reconnection in solar flares. To obtain a more complete picture of the condensation-reconnection process, the stabilizing influence of thermal conduction has to be included, which will be the subject of future investigations.

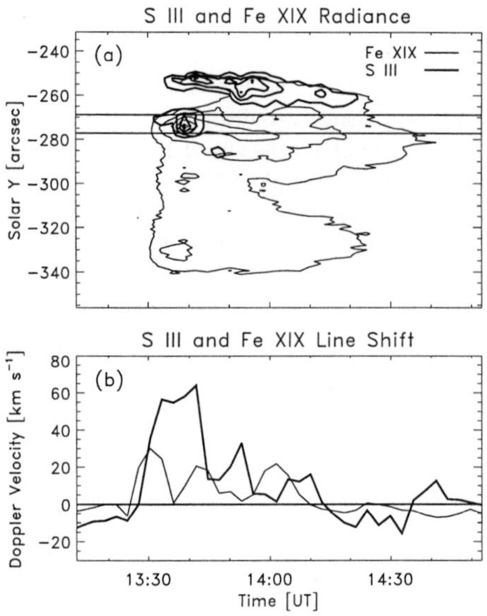

Fig. 3. SUMER data of the 2000 Oct 25 flare taken at a fixed off-limb pointing, $(x, y) = (-1000'', -300'')$, with a cadence of 170 s. (a) Equidistant isocontours of line-integrated radiance (linear scale, 15 %–90 % of maximum) showing the flare-related brightenings in Fe XIX λ1118.1 (*light contours*) and S III λ1113.2 (*heavy contours*). (b) Fe XIX and S III line shifts, averaged over the range $-277'' < y < -269''$ as indicated in (a).

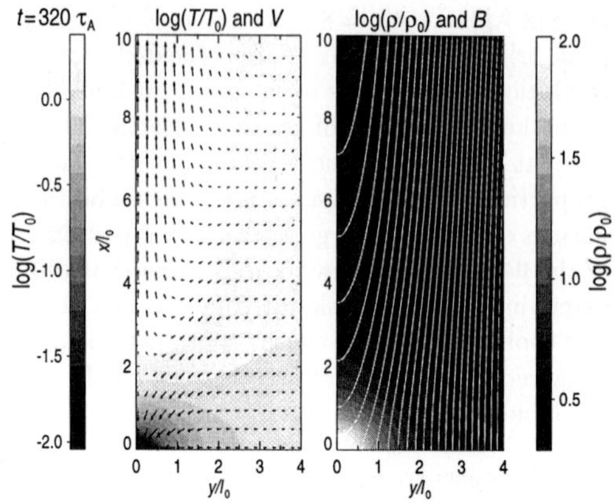

Fig. 4. Snapshot of magnetic reconnection in a two-dimensional resistive, compressible MHD simulation, induced by radiative cooling of an initial isobaric temperature drop at the origin. Displayed are temperature and velocity field (*left*) and density and magnetic field lines (*right*); $T_0 = 2 \times 10^6$ K; l_0 — initial current sheet half width; τ_A — Alfvén time (see text for further parameter values).

ACKNOWLEDGEMENTS

We thank D. E. McKenzie and J. Sato for providing us with reduced *Yohkoh* data. The EIT, SXT, and HXT data are courtesy of the *SOHO*/EIT and the *Yohkoh*/SXT and HXT consortia, respectively. The SUMER project is financially supported by DLR, CNES, NASA, and the ESA PRODEX programme. The work of BK was also supported by the EU under Contract No. HPRN-CT-2000-00153. The John von Neumann-Institut für Computing, Jülich, granted Cray computer time.

REFERENCES

Dammasch, I.E., B. Kliem, W. Curdt, and K. Wilhelm, *Proc. IAU Symp.*, **203**, 264 (2001).

Hudson, H.S., in *New Look at the Sun*, ProcKofu Symp., eds. S. Enome and T. Hirayama, Nobeyama Radio Observatory, NRO Report 360, p. 1 (1994).

Kliem, B., M. Karlický, and A.O. Benz, *A&A*, **360**, 715 (2000).

Kliem, B., I.E. Dammasch, W. Curdt, and K. Wilhelm, *ApJ Letters*, **568**, L61 (2002).

Kosugi, T., in *New Look at the Sun*, Proc. Kofu Symp., eds. S. Enome and T. Hirayama, Nobeyama Radio Observatory, NRO Report 360, p. 11 (1994).

Sturrock, P.A. (ed.) *Solar Flares, A Monograph from the Skylab Solar Workshop II,* Colorado Associated University Press (1980).

Van Hoven, G., T. Tachi, and R.S. Steinolfson, *ApJ*, **280**, 391 (1984).

EARLY RESULTS FROM A MULTI-THERMAL MODEL FOR THE COOLING OF POST-FLARE LOOPS

K. K. Reeves and H. P. Warren

Harvard-Smithsonian Center for Astrophysics, 60 Garden Street, MS 58, Cambridge, MA, 02138, USA

ABSTRACT

We have developed a multi-thermal model for the cooling of post-flare loops. Our model consists of many nested loops that form and cool at offset times to simulate a rising reconnection site. Cooling due to both conductive and radiative processes is taken into account. The free parameters in the model include the initial temperature and density in the loops. The loop width and the initial loop length are constrained by the data. The results from the model are compared to observations from the *Transition Region and Coronal Explorer* and the Soft X-ray Telescope on *Yohkoh.* The many-loop model predicts the observed light curves more accurately than a similar model with a single loop.

INTRODUCTION

The standard model of reconnection states that a rising reconnection site heats loops at successively higher heights (Forbes & Acton 1996). Under this theoretical scenario, there would be many loops consisting of different temperatures and densities coexisting in an arcade. Filter ratios are often used to obtain the temperatures and densities in these loops from observed intensities (e.g. Tsuneta *et al.* 1997). This method assumes a single temperature and density along the line of sight, and for flares on the disk, would produce an emission measure weighted average temperature of a multi-loop structure, rather than the temperatures of the individual loops.

In this paper we present a description of a multi-thermal model of flare cooling. We assume hot flare loops are created at progressively larger heights as the reconnection site rises. The evolution of the temperature and density is determined by the scaling laws derived by Cargill *et al.* (1995). We compare the simulation results to observations of an X6 flare from the *Transition Region and Coronal Explorer* (*TRACE*) and the Soft X-ray Telescope (SXT) on *Yohkoh.*

DESCRIPTION OF THE FORWARD MODEL

The simulation begins with a single loop filled with plasma of an assumed initial temperature and density, uniformly distributed along the loop. As the loop cools, a new loop with the same initial temperature and density but a slightly higher reconnection site is formed above it. This process repeats until the arcade is filled with several hundred loops. The temperature and density in each loop is calculated using an empirical cooling model based on that of Cargill *et al.* (1995), which assumes that conductive and radiative cooling happen in two separate phases, and that there is a single temperature and density along the loop. The conductive cooling time is given by $\tau_c = 4 \times 10^{-10} n_0 L^2 / T_0^{5/2}$. The temperature falls rapidly at first in this phase, and then levels out somewhat. The pressure is assumed constant, which causes the density to rise

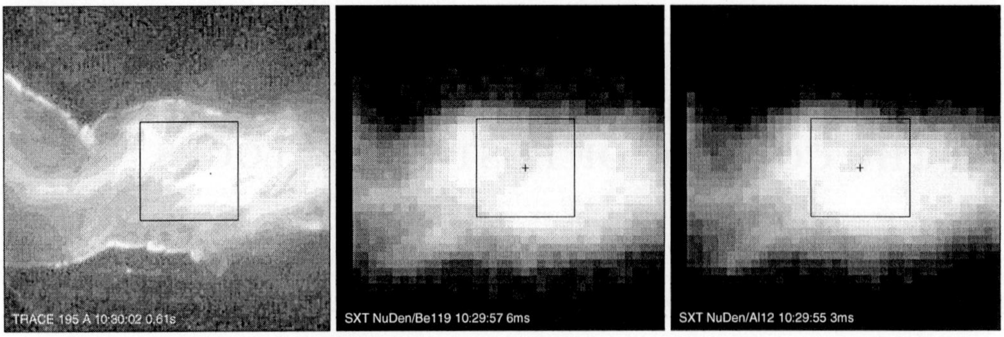

Fig. 1. *TRACE* and SXT images from the June 14 2000 X6 flare. The black box indicates the area used to calculate light curves.

as the temperature falls. The radiative cooling time is given by $\tau_r = 3k_B T_0^{1-\alpha}/n_0\chi$ where L is the loop length, n_0 is the initial density, T_0 is the initial temperature, and α and χ are determined from the radiative loss function, $P_{rad} = \chi T^{\alpha}$. In this phase, the temperature falls rapidly and nearly linearly, while we have assumed that the the density varies as the square root of temperature, based on a scaling law derived from hydrodynamic simulations (Cargill *et al.* 1995). A full description of this model can be found in Reeves & Warren (2002).

The simulation is divided up into time steps on the order of 1s. At the beginning of each time step, the conductive and radiative times are calculated, and the shorter cooling time scale dominates the cooling during that time step. The temperature and density in the loop are calculated at the end of the time step, and the final profiles are passed through the *TRACE* or SXT response function in order to obtain an intensity. The line of sight is taken to be looking down on the loops from above. The intensities of all the loops in the arcade are integrated over this line of sight to find a total simulated light curve.

RESULTS AND DISCUSSION

We compare the results of our simulation to *TRACE* and SXT intensities from an X6 flare that occurred on July 14 2000 at 10:24 UT. The *TRACE* observations for this flare consist of 195 Å images every ~10s with contextual 171 Å and 1600 Å images. SXT took observations with the Be119 and Al12 filters. Representative images of the flare from *TRACE* and SXT are shown in Figure 1.

One Loop Model

We first attempt to fit the observed *TRACE* and SXT light curves with a single loop simulation. To measure loop length, we use the footpoint ribbons visible in the *TRACE* 1600 Å data and assume that the loop is a semicircle extending from the outside edge of one ribbon to the other, measured after the ribbon spreading motion has stopped. The half-length of our single flare loop is measured to be 2.4×10^4 km. For the width of the loop, we measure the width of the more well-defined flare ribbon on the northern side of the flare. This width is measured to be 4.4×10^3 km.

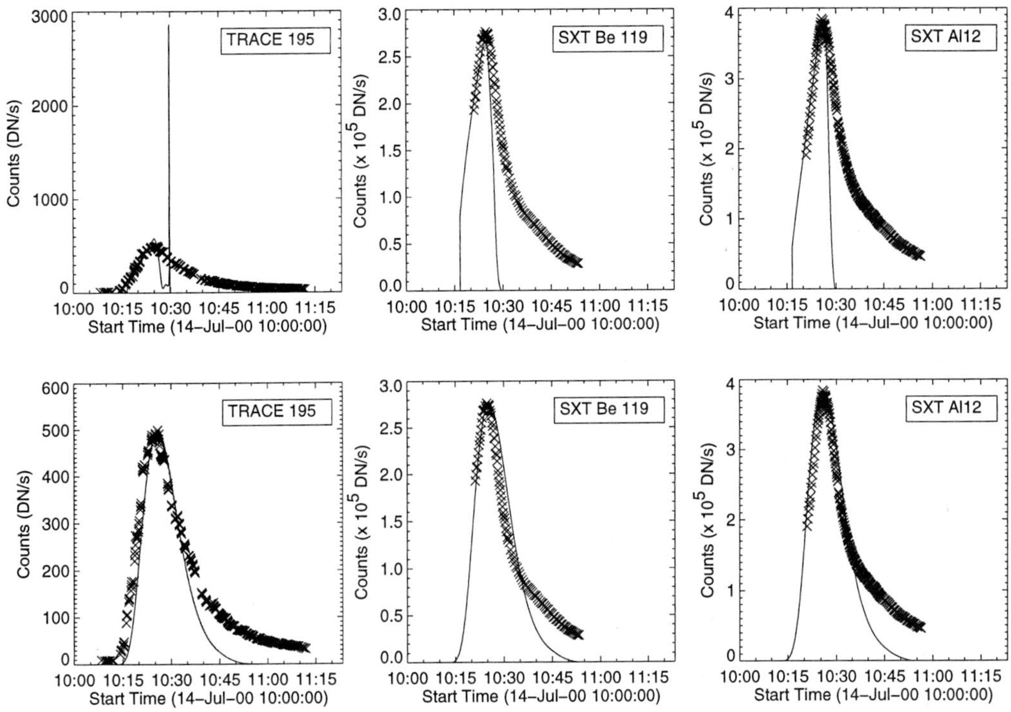

Fig. 2. Observed (crosses) and simulated (solid lines) intensities for *TRACE* 195 Å and SXT Be119 and Al12 filters from the July 14 2000 flare. Shown are simulations for one loop (top) and many loops (bottom). The initial conditions for the single-loop simulation are 35 MK and 2×10^{11} cm^{-3}. For the multi-loop simulation the initial conditions are 25 MK and 4 $\times10^{11}$ cm^{-3}. The simulated intensity is normalized to the maximum of the observed light curve, except for the *TRACE* 195 Å single-loop case, in which the simulated curve before 10:20 UT is normalized to the observed maximum.

We tried a range of initial temperatures (10 – 35 MK) and densities (0.5 – 8 $\times10^{11}$ cm^{-3}) in this model. The parameters that best fit the observed light curves are an initial temperature of 35 MK and an initial density of 2×10^{11} cm^{-3}. The simulated light curve with these initial conditions is plotted on the top row of Figure 2. Also plotted in Figure 2 are the averaged *TRACE* and SXT intensities from inside the black box in Figure 1.

It is clear from Figure 2 that even the best fit curves from the single-loop simulation do not give reasonable fits to the observed intensities. The characteristics of the simulated *TRACE* single loop light curve are due to both Fe XXIV, formed at about 20 MK, and Fe XXIV, formed at about 1.5 MK, which are present in the *TRACE* 195 Å bandpass. There is a broad hump early in the simulation due to the Fe XXIV in the bandpass, and a later, narrow and intense peak where the temperature passes through 1.5 MK, where the 195 Å bandpass is the most sensitive. The observed light curve is markedly different, with a much broader peak and a slow decay in the intensity after the peak. There is no good agreement between the simulated SXT light curves and the observed ones either — the simulated SXT curves drop off much more quickly than their observed counterparts. From this analysis it is clear that a single loop flare model will not work to reproduce the light curves observed by SXT and *TRACE*.

Multi-loop Arcade Model

For the multi-loop simulation, we must determine the reconnection rate in order to determine the loop length as a function of time. The reconnection rate, and thus the time evolution of the loop radius, can be determined from the ribbon spreading velocity, which can be obtained from the distance the ribbons spread with time. The flare ribbons are clearly visible in the *TRACE* 1600 Å images, so we use this data to determine the reconnection rate. The ribbons spread apart quickly at first, and then slow down to a near stop as the flare progresses. This progression was used to determine the loop length at each time in the simulation. The width is the difference in radii between consecutive loops.

We tried several different values for the initial temperature and density, and calculated light curves as described above. We found that the best fit to the data was given by an initial temperature of 25 MK and an initial density of 4×10^{11} cm^{-3}. The simulated and observed light curves are shown in the bottom row of Figure 2. The multi-loop simulation fits the observed light curves much better than the one loop simulation, although the decay phase tail in the observed light curves is not fit very well. Aschwanden and Alexander (2002) used a similar cooling model to fit the timing of the peak intensities in several instruments for this flare, and found average temperatures of 28 – 31.5 MK and densities of 4 – 8 $\times 10^{11}$ cm^{-3}, which are similar to our initial loop values.

Using SXT filter ratios, we calculate the average temperature in the black box in Figure 1 as a function of time. The temperatures found using this method were about 15 MK at the beginning of the flare and fell to 10 MK as the flare progressed. These temperatures are lower than the 25 MK initial loop temperatures needed to obtain a good fit to the light curves using the multi-loop model. We were able to reproduce the SXT temperatures by calculating an emission-measure weighted average of all the loops in the simulation, suggesting that SXT temperatures could represent an ensemble average over a multi-thermal arcade of loops.

CONCLUSIONS

We have developed a model for the cooling of flare plasma. Using this model, we have shown that a representation of a flare as a single, isothermal loop is insufficient to recreate the light curves observed by SXT and *TRACE*. A multi-loop, multi-thermal simulation provides a better fit to the data, although it still does not fit the tail of the observed light curves. This discrepancy could be due to lack of heating terms in the model, or other limitations inherent in the cooling method, such as the assumption that the loops are spatially uniform and the separability of the conductive and radiative cooling phases.

ACKNOWLEDGEMENTS

TRACE is supported by Contract NAS5-38099 from NASA to LMATC.

REFERENCES

Aschwanden, M.J., and D. Alexander, Flare Plasma Cooling From 30 MK Down to 1 MK Modeled from *Yohkoh*, *GOES*, and *TRACE* Observations During the Bastille Day event (2000 July 14), *Sol. Phys.*, in press, (2002).

Cargill, P.J., J.T. Mariska, and S.K. Antiochos, Cooling of Solar Flare Plasmas. 1: Theoretical Considerations, *ApJ*, **439**, 1034 (1995).

Forbes, T.G., and L.W. Acton, Reconnection and Field Line Shrinkage in Solar Flares, *ApJ*, **459**, 330 (1996).

K.K., and H.P. Warren, A model for the Cooling of Post-Flare Loops, *ApJ*, submitted, (2002).

Tsuneta, S., S. Masuda, T. Kosugi, and J. Sato, Hot and Superhot Plasmas Above an Impulsive Flare Loop, *ApJ*, **478**, 787 (1997).

OBSERVATIONS OF MORETON WAVES AND EIT WAVES

K. Shibata [1], S. Eto[1], N. Narukage[1], H. Isobe[1], T. Morimoto[1], H. Kozu[1], A. Asai[1], T. Ishii[1], S. Akiyama[1], S. Ueno[1], R. Kitai[1], H. Kurokawa[1], S. Yashiro[2], B. J. Thompson[2], T. Wang[3], and H. S. Hudson[4]

[1]*Kwasan and Hida Observatories, Kyoto University, Yamashina, Kyoto 607-8471, Japan*
[2]*NASA Goddard Space Flight Center, Greenbelt, MD20771, USA*
[3]*Max-Planck-Institut fur Aeronomie, Max-Planck-Str. 2, D-37191, Katlenburg-Lindau, Germany*
[4]*SPRC, ISAS, Sagamihara, Kanagawa 229-8510, Japan*

ABSTRACT

We study the relationship between Moreton waves and EIT waves by analyzing Hα images taken with the Flare Monitoring Telescope (FMT) at the Hida Observatory of Kyoto University and EUV images taken with *SOHO*/EIT. In the event of November 4, 1997 (Eto *et al.* 2002), the propagation speeds of the Moreton wave and the EIT wave were approximately 780 km/s and 200 km/s, respectively. The data on speed and location suggest that Moreton waves differ physically from EIT waves in these cases. In the event of November 3, 1997 (Narukage *et al.* 2002), the wave is detected in soft X-rays as well as by the Moreton and EIT wave signatures. The propagation speeds of the Moreton wave, X-ray wave, and EIT wave were 490 km/s, 630 km/s, and 170 km/s, respectively, which suggests that we can identify both the X-ray wave and the Moreton wave a coronal fast-mode MHD shock wave, but that the EIT wave is physically different.

INTRODUCTION

Moreton waves are flare-associated waves seen in Hα, propagating within a somewhat restricted solid angle at speeds of 400 – 1100 km/s (Moreton 1960, Smith & Harvey 1971). Uchida (1968) theorized that Moreton waves are the chromospheric manifestation of an MHD fast-mode shock wave propagating through the corona. Recently other wave-like phenomena in the corona have been discovered with *SOHO*/EIT (Extreme ultraviolet Imaging Telescope). These features are now commonly called *EIT waves* (Moses *et al.* 1997, Thompson *et al.* 1998). They are associated with flares and/or coronal mass ejections, and propagate at speeds of 200-300 km/s, sometimes nearly isotropically in direction (Klassen *et al.* 2000). At first it was thought that the EIT wave could be the coronal counterpart of the Moreton wave itself (Thompson *et al.* 1999). However, the marked differences in the propagation pattern and speed are a big puzzle, which raises the question whether EIT waves are invariably identifiable as Moreton waves (Thompson *et al.* 2000, Warmuth *et al.* 2001). More recently another new class of wave-like disturbances in the corona was discovered in soft X-rays with *Yohkoh*/SXT (Khan & Hudson 2000). The first simultaneous observation of a Moreton and an X-ray wave was reported by Khan & Aurass (2002), who suggested that such X-ray waves may be the coronal counterparts of Moreton waves.

In this paper we examine the relation between Moreton waves and EIT waves by analyzing two Moreton waves observed on November 3 and 4, 1997. The November 4 event was fully analyzed by Eto *et al.* (2002). The November 3 (4:36 UT) event is the second example of simultaneous observation of a Moreton wave and

Table 1. Typical Moreton Waves Observed at Hida Observatory in 1997-2000

date (yy-mm-dd)	start (UT)	end (UT)	flare st. (UT)	flare site	GOES class	Moreton wave speed (km/s)	EIT wave speed (km/s)	X-ray wave speed (km/s)
1997-11-03	4:36	4:41	4:32	S20W13	C8.6	490	170	630
1997-11-04	5:56	6:05	5:52	S14W34	X1.2	715	200	no data
1998-08-08	3:14	3:20	3:12	N12E70	M3.0	930	no data	no data
2000-03-03	2:10	2:15	2:08	S17W63	M3.8	1050	200	1300

an X-ray wave, and has been analyzed in detail by Narukage *et al.* (2002a,b). This paper is an initial report of these detailed studies.

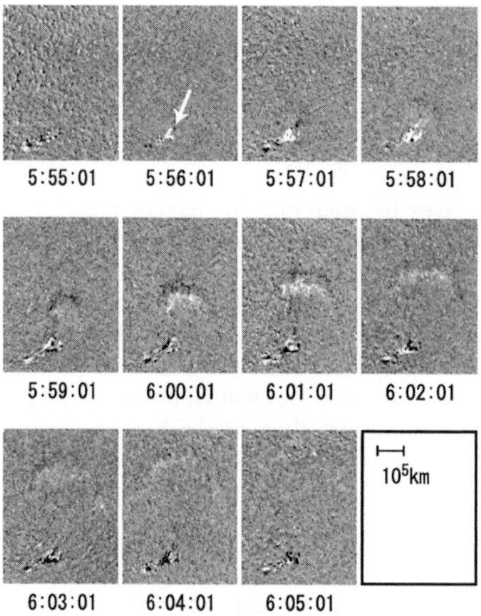

Fig. 1. Running differences of Hα + 0.8 A images of a Moreton wave on November 4, 1997, which are taken at every minute from 05:55:01 UT to 06:05:01 UT. The wave fronts of the Moreton wave are clearly seen as dark and white strips.

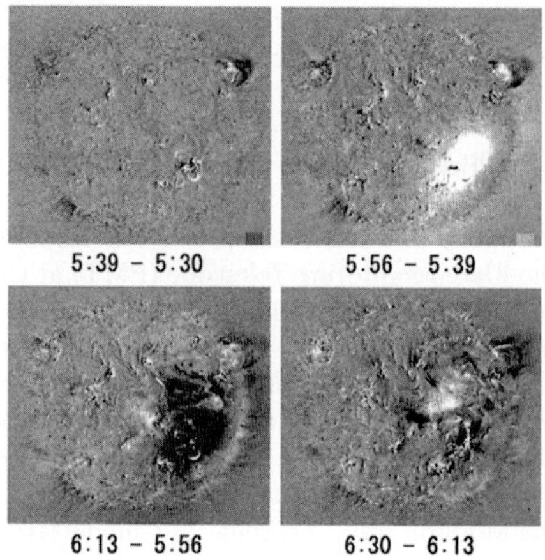

Fig. 2. Running differences of EIT images of an EIT wave associated with the Moreton wave on November 4, 1997. The white oval region in the top right panel is mainly due to the scattered light in the telescope. In the lower panels the disturbance (EIT wave) seems to propagate away from the flare site.

OBSERVATIONS

The Hα images of Moreton waves are taken with the Flare Monitoring Telescope (FMT) at the Hida Observatory of Kyoto University (See `http://www.kwasan.kyoto-u.ac.jp/Hida/FMT/obs-report.html`.) The FMT observes four full-disk images in Hα center, Hα+0.8Å, Hα-0.8Å, the continuum, and one limb image in Hα line center, using a set of five 64 mm aperture telescopes (Kurokawa *et al.* 1995). These full Sun images are taken with 4.2 arcsec pixels at a 2 sec time cadence on video tape and also at a 1 min time cadence on CD-ROM. Typical Moreton waves observed with the FMT at Hida Observatory in 1997-2000 are listed in Table 1. The EUV images of EIT waves are taken with *SOHO*/EIT (Delaboudiniere *et al.* 1995), and the soft X-ray images of X-ray waves are observed with *Yohkoh*/SXT(Soft X-ray Telescope).

THE NOVEMBER 4, 1997 EVENT

A Moreton wave was observed in association with an X-class flare that occurred in NOAA 8100 at 05:52 UT on November 4, 1997. Figure 1 shows the Moreton wave as seen in running differences of the Hα+0.8Å images. The dark small region indicated by an arrow in the image at 05:56:01 UT in Figure 1 seems to be the point of origination of the Moreton wave. The wave front can be identified in the images from 05:58 to 06:04 UT, and measurements show that the Moreton wave speed was about 715 km/s. Figure 2 shows running differences of the EIT images. It is seen that a disturbance propagates nearly isotropically from the flare site; this is the EIT wave. We identify the EIT wave fronts only via the images at 06:13 UT and 06:30 UT. From these two images, we find the wave propagation speed to be about 200 km/s. This is much slower than the propagation speed of the Moreton wave. The wave fronts of both Moreton wave and EIT wave are plotted in Figure 3 with the dimming region seen in the EIT images. Since the observing time of the Moreton wave was 05:58 – 06:05 UT, we cannot directly compare the waves in time. In this case, however, we found a filament oscillation (i.e., winking filament) at 06:12 UT near the north west limb, which is indirect evidence for the further propagation of Moreton wave propagation (see Smith & Harvey 1971) to as late as 06:12 UT. The position of this filament oscillation is also indicated in Figure 3. All these results are summarized in the distance-time diagram as shown in Figure 4 with GOES X-ray flux of the flare associated with these waves. We see that the filament oscillation (asterisk) lies on the extrapolated position of the Moreton wave, and that the Moreton wave and EIT wave are not cospatial.[1] Hence, not only the propagation speed but also the location of the Moreton wave are different from those of the EIT wave, so that we suggest that the Moreton wave is physically different from the EIT wave in this case, in agreement with the Uchida *et al.* (2001) model (see also Chen *et al.* 2002).

THE NOVEMBER 3, 1997 EVENT

A second example of a Moreton wave occurred at 04:32 UT on November 3, 1997, one day before the Moreton wave described in the previous section, and from the same active region. In this case, not only a Moreton wave and an EIT wave were observed, but also an X-ray wave. Interestingly, the wave reported by Khan & Aurass (2002) occurred about four hours later (09:04 UT), again in the same active region. From image analysis (see Figure 1 of Narukage *et al.* 2002b), we found that the propagation speeds of the Moreton wave, EIT wave, and X-ray wave were 490, 170, and 630 km/s, respectively. With the MHD shock theory and the soft X-ray intensities of the X-ray wave, we have confirmed that the X-ray wave is consistent with a coronal fast-mode MHD shock with Mach number of 1.15-1.25. This, as well as the rough cospatiality of the Moreton wave and X-ray wave, suggest that the X-ray wave is indeed the coronal counterpart of the Moreton wave. On the other hand, the EIT wave speed (170 km/s) is much slower than that of other two waves, and the distance-time diagram of the Moreton wave and EIT wave (see Figure 2 of Narukage *et al.* 2002b) is similar to that of the November 4 event, suggesting again that the EIT wave is physically different from the Moreton wave, though in this case there is no other clear evidence (such as a filament oscillation) to support this conclusion.

[1]It is interesting to note that the amplitude of the filament oscillation became maximum at around 06:22 UT when the EIT wave reached the filament (see Figure 8 of Eto *et al.* 2002), though we cannot deny the possibility that the faint front of the diffuse EIT wave reached the filament earlier than 06:22 UT. In relation to this it is interesting to note that Wills-Davey and Thompson (1999) reported coronal loop oscillations excited by EIT waves.

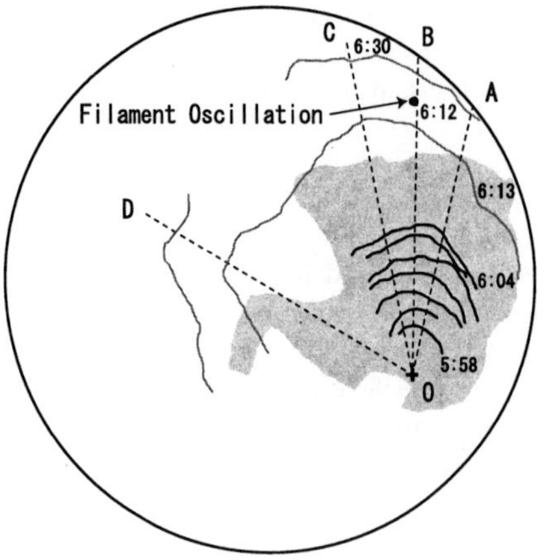

Fig. 3. Observed wave fronts of the EIT wave (gray lines) and the Moreton wave (black lines) and the EIT dimming region (light gray region). The EIT wave fronts are those at 06:13 UT and 06:30 UT, and the Moreton wave fronts are those from 05:58:01 UT to 06:04 UT with intervals of 1 minute. The location of the filament oscillation is indicated by the black filled circle.

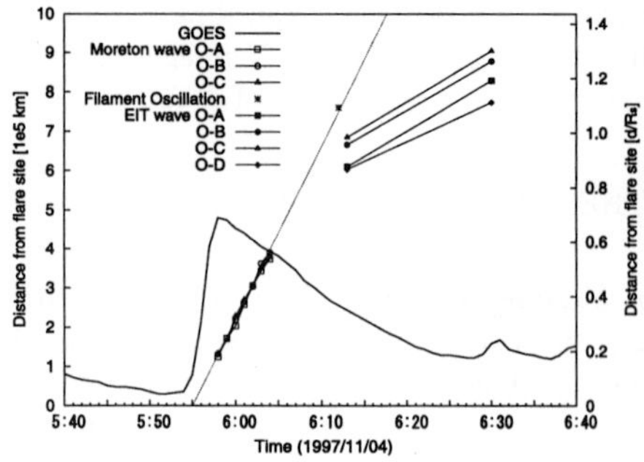

Fig. 4. Time evolutions of the distance of the Moreton wave (open symbols) and the EIT wave (filled symbols) from the flare site (point O in Figure 3) along the lines O-A, O-B, O-C, and O-D in Figure 3. An asterisk at 06:12:00 UT represents the start of the filament oscillation and the distance of the filament from the flare site. *GOES* X-ray flux data from the 1-8Å channel are overplotted. 'Rs' represents the solar radius.

ACKNOWLEDGEMENTS

This work was supported in part by the JSPS Japan-US Cooperation Science Program (Principal Investigators: K. Shibata and K. I. Nishikawa). HSH was supported by NASA under contract NAS 8–37334.

REFERENCES

Chen, P.F., C. Fang, S.T. Wu, and K. Shibata, *ApJ Letters*, in press (2002).

Delaboudiniere, J.P., G. E. Artzner, J. Brunaud, *et al.*, *Solar Phys.*, **162**, 357 (2002).

Eto, S., H. Isobe, N. Narukage, *et al.*, *PASJ*, **54**, No. 3, in press (2002).

Khan, J.I., and H.S. Hudson, *GRL*, **27**, 1083 (2000).

Khan, J.I. and H. Aurass, *A&A*, **383**, 1018 (2002).

Kurokawa, H., K. Ishiura, G. Kimura, *et al.*, *Geomag. Geoelectr.*, **47**, 1043 (1995).

Moreton, G.E., *Astron. J.*, **65**, 494 (1960).

Narukage, N., H.S. Hudson, T. Morimoto, S. Akiyama, R. Kitai, H. Kurokawa, and K. Shibata , *ApJ Letters*, **572**, L109 (2002a).

Narukage, N., K. Shibata, H.S. Hudson, *et al.*, *Adv. Space Res.*, in press (2002b).

Smith, S.F. and K.L. Harvey, in *Physics of the Solar Corona*, ed. C. J. Macris (Reidel), p. 156 (1971).

Thompson, B.J. J.B. Gurman, W.M. Neupert, J. *et al.*, *GRL*,**25**, 246 (1998).

Thompson, B.J., B. Reynolds, H. Aurass, *et al.*, *Solar Phys.*, **193**, 161 (2000).

Uchida, Y., *Solar Phys.*, **4**, 30 (1968).

Uchida, Y., T. Tanaka, M. Hata, and R. Cameron, *Publ. Astr. Soc. Australia*, **18**, 345 (2001).

Warmuth, A., B. Vršnak, H. Aurass, and A. Hanslmeier, *ApJ Letters*, **560**, L105 (2001).

Wills-Davey, M.J., and B.J. Thompson, *Solar Phys.*, **190**, 467 (1999).

SEARCH FOR EVIDENCE OF ALPHA PARTICLE BEAMS DURING A SOLAR FLARE

J. W. Brosius

CUA at NASA's Goddard Space Flight Center, Code 682, Greenbelt, MD 20771, USA

ANALYSIS AND DISCUSSION

We observed NOAA Active Region 9090 (N13 W39) with the Coronal Diagnostic Spectrometer (CDS) and the Extreme-ultraviolet Imaging Telescope (EIT) aboard the Solar and Heliospheric Observatory (*SOHO*) spacecraft between 18:17 and 21:09 UT on 24 July 2000 to search for evidence of alpha particle beams during solar flares. Theoretically, an alpha particle beam will manifest itself during the impulsive phase of a flare through an enhancement in the red wing of the He II Lyα (303.782 Å) emission line, without a corresponding blue wing enhancement. This enhancement is due to downstreaming nonthermal alpha particles undergoing charge exchange with chromospheric neutral hydrogen atoms to form downstreaming nonthermal He II ions. Lyα radiation emitted from these downstreaming ions is Doppler-shifted into the red wing of the Lyα line. Our CDS observing program acquired high time resolution (9.7 s) $4'' \times 4'$ slit spectra between 590 and 630 Å, where we observed He II Lyα in second order (607.564 Å). The CDS and EIT observations (see Figure 1) reveal that AR 9090 underwent significant intensity fluctuations prior to a sudden drastic increase (impulsive phase) around 20:00 UT. The GOES satellite reports a C3.8 event in this region from 19:57 to 20:05 UT. We fit the spectral background and emission line profiles for each CDS spectrum in our observed sequence. Density- and temperature-insensitive intensity ratios of O IV and Mg X lines generally agree with their theoretical values before and after the sudden intensity increase, in support of a reliable relative radiometric calibration for CDS, but differ significantly from their theoretical values during the flare impulsive phase. This may indicate line blending with unknown components, line blending with second order C IV and Fe XV lines, or loss of ionization equilibrium. Most important, however, we find that although the red wing and blue wing backgrounds for He II Lyα remain relatively constant during most of our observation, the blue wing undergoes a more significant enhancement during the impulsive phase than does the red wing. This effect is opposite that expected in the presence of an alpha particle beam. Further, blended spectral line features that mimic the expected nonthermal redshifted He II Lyα beam signal are understood in terms of well known emission line components. Thus we find no evidence for the presence of alpha particle beams in our observations. We estimate an upper limit ~ 250 ergs cm^{-2} s^{-1} sr^{-1} Å^{-1} for the nonthermal redshifted peak spectral intensity due to an alpha particle beam prior to the impulsive phase. See Brosius (2001: *ApJ* **555**, 435) for a much more complete discussion of the observations, data reduction, analysis procedures, and results.

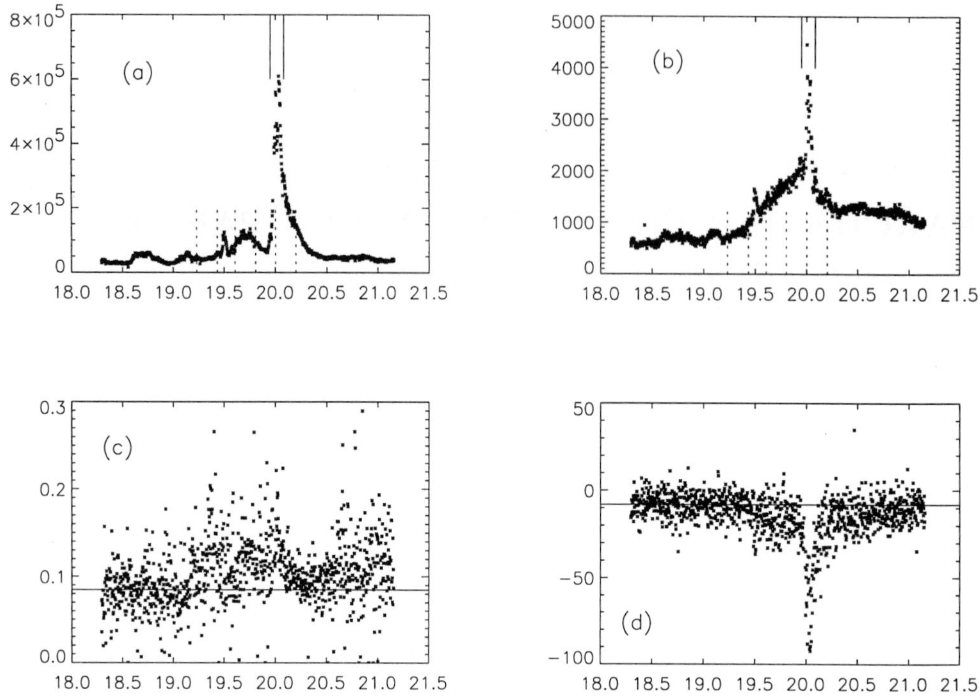

Fig. 1. Time history (hours UT on 24 July 2000) of integrated intensities (ergs cm^{-2} s^{-1} sr^{-1}) of two strong lines observed by CDS: (a) He II 2×303.782, and (b) Mg X 624.939 Å. Dashed vertical lines indicate times at which EIT images were obtained, and solid vertical lines indicate beginning and ending times of the GOES event. Time history of (c) line intensity ratio of O IV 608.3/(553.5 + 554.1 + 554.5 + 555.2), and (d) red minus blue spectral background differences (in ergs cm^{-2} s^{-1} sr^{-1} Å^{-1}) for $\pm$ 10 Å from line center (2×303.782 Å). The solid horizontal line in (c) indicates the theoretical value of the ratio (0.0846), and the solid horizontal line in (d) indicates the average (non-impulsive) red minus blue spectral background difference. Our measured values for O IV 608.3/$\Sigma 554$ are in agreement with theory before the flare, but increase by a factor of 2 - 3 during the flare. This suggests that the O IV line at 608.3 Å may become blended with additional lines that strengthen during the flare, but no clear candidates are evident. The significant increase of the insensitive ratios during the flare could also be a manifestation of the loss of ionization equilibrium. The red and blue wing spectral background intensities are nearly equal during most of our observation, the blue background being slightly greater than the red (as is evident from the negative values of the majority of the red minus blue differences in frame d). However, during the time of the most rapid intensity increase this difference becomes significantly more negative. This is exactly opposite the effect we would expect if an alpha particle beam were present. A similar trend is observed in other spatial locations in which the flare signature is evident, as well as at different wavelength displacements from line center. Thus, based upon our spectral line and background intensity measurements, we find no evidence for the presence of alpha particle beams during the observed flare event.

THE SOLAR CORONAL ORIGIN OF A SLOWLY DRIFTING RADIO PULSATION FEATURE

J. I. Khan[1,2] N. Vilmer[3], P. Saint-Hilaire[4,5], and A. O. Benz[4]

[1]*Mullard Space Science Laboratory, University College London, Holmbury Saint Mary, Surrey, UK*
[2]*Located at: Institute of Space & Astronautical Science, Sagamihara, Kanagawa 229-8510, Japan*
[3]*Observatoire de Paris, Section de Meudon, DASOP, Meudon Cedex, France*
[4]*Paul Scherrer Institute, CH-5232 Villigen PSI, Switzerland*
[5]*Institute of Astronomy, ETH Zentrum, CH-8092 Zürich, Switzerland*

ABSTRACT

In this conference paper we summarize some of the results of our forthcoming paper recently accepted by *Astronomy & Astrophysics* in which we associate drifting radio pulsation emission with plasmoid ejection.

INTRODUCTION

Drifting Pulsation Structures (DPSs) are radio features observed in dynamic radio spectrograms at microwave and dm wavelengths. They are of short duration (generally about 1–2 min), consist of many quasi-periodic pulsations which show a slow drift with time, occur during the impulsive phase of some solar flares, and show no evidence for harmonic emission (Jiřička *et al.* 2001).

Previous work associated the drift of DPSs above 1 GHz with an MHD shock wave ahead of evaporating plasma (Karlický & Odstrčil 1994) and, in a different scenario, the cut-offs in frequency of dm DPSs were associated with a chromospheric evaporation process (Aschwanden & Benz 1995). Recently it has been suggested that DPSs are related to mass ejections based on the temporal coincidence with a few observed cases of soft X-ray (SXR) and microwave ejecta (Hori 1999, Kliem *et al.* 2000, Karlický *et al.* 2001). These mass ejections are termed plasmoid ejections based on 2-D models, but could be considered to be loop ejections in 3-D. Henceforth we use the term plasmoid to refer to such ejecta.

RESULTS

In this work we examined a DPS event observed with the Phoenix-2 Radio Spectrometer (Messmer *et al.* 1999) on 2000 August 25 for which we have imaging observations from the Nançay Radioheliograph (NRH) (Kerdraon & Delouis 1997) and the *Yohkoh* Soft X-ray Telescope (SXT) (Tsuneta *et al.* 1991). These data sets enabled us to identify the radio source of a DPS in relation to coronal features observed in X-rays.

The DPS occurred (from ~14:29:30–14:33:30 UT) during the impulsive phase of an M1.4/1N flare which started at ~14:23 UT on 2000 August 25 and occurred in NOAA sunspot region 9143 near S16E68. The DPS was observed at the NRH frequencies of 236.6, 327, 410.5, and 432 MHz. This DPS consisted of several tens of pulsations whose frequency drift rate was -2.8 ± 0.5 MHz s^{-1} at a mean frequency of 430 MHz.

The *Yohkoh* SXT data show two main features: a small bright flaring kernel region located low in the corona and faint ejecta (loosely termed a plasmoid) above the flaring kernel. We find that none of the NRH observations of the DPS were located near the flaring kernel observed with the SXT. This indicates that the DPS was not associated with chromospheric evaporation. We find that the DPS observed at 327 MHz was located in the vicinity of the SXR plasma ejection above the flaring kernel and it moved outward as the SXR ejection moved outward. At times when the DPS is observed at both 236.6 and 327 MHz we find that the 236.6 MHz sources were located near to, but above, the 327 MHz sources. We find that the projected source speed of the DPS seen at 236.6 MHz was 257 km s^{-1} over the interval 14:31:57–14:32:47 UT, while the speed at 327 MHz was 438 km s^{-1} over 14:29:56–14:31:12 UT and 291 km s^{-1} over 14:31:16–14:32:30 UT. We estimate that the speed of the SXR plasmoid seen in projection in the SXT images was 290 ± 60 km s^{-1} over 14:27:42–14:30:06 UT. These results show that the DPS was closely related to the SXR plasmoid ejection.

There are at least two possible interpretations of the nature of the association of the DPS with the SXR ejection. The first is that the DPS is due to a radio exciter moving through a stationary atmosphere (similar to the interpretation for type II bursts). In that case the frequency drift rate of the DPS is given by $\frac{d\nu}{dt} = \frac{d\nu}{dn}\frac{dn}{dt}$, where ν is the frequency, n is the electron number density and t is the time. The observed negative drift rate would then imply that n decreases with time and that it is related to the speed of the exciter in the solar atmosphere. If this frequency drift is interpreted in terms of a propagating exciter then it implies that the speed of exciter was 1300 km s^{-1}, which is not consistent with the NRH- and SXT-derived speeds. Also we might expect radio emission at a given frequency to occur at a fixed height for a near-limb event, which is not observed (we see motion of the sources at each frequency). The second interpretation is that the DPS is due to an expanding plasmoid of radius R. In that case the drift rate of the DPS is given by $\frac{d\nu}{dt} = -\frac{3}{2}\frac{\nu}{R}\frac{dR}{dt}$. The observed negative drift rate would imply increasing radius of the plasmoid with time (and decreasing density within the plasmoid). In that scenario the radio emission is associated with an emitting region in the vicinity of the plasmoid. If plasmoid expands (and moves outwards) then one would expect the radio sources to show motion. Our observations favor the second interpretation.

Kliem *et al.* (2000) and Karlický *et al.* (2001) suggested that DPS emission might originate from the following: a) a reconnection region between the ejected plasmoid and underlying loops, b) locations where reconnection outflow jets are stopped, c) in the vicinity of a plasmoid (i.e. from particles threading through or trapped within the plasmoid magnetic structure). The observed source locations in this event favor suggestion c).

This work is the first analysis of radio imaging observations of a dm DPS. Based on our results, we associate the DPS emission with particles either trapped within or threading through an erupting plasmoid structure and the observed frequency drift of the DPS with the expansion of the plasmoid as it is ejected.

REFERENCES

Aschwanden, M.J., and Benz, A.O., *ApJ*, **438**, 997 (1995).

Hori, K., in *Solar Physics with Radio Observations*, Proc. Nobeyama Symp., NRO Report 479, p. 267 (1999).

Jiřička, K., Karlický, M., Mészárosová, H., and Snížek, V., *A&A*, **375**, 243 (2001).

Karlický, M., and Odstrčil, D., *Solar Phys.*, **155**, 171 (1994).

Karlický, M., Yan, Y., Fu, Q., *et al.*, *A&A*, **369**, 1104 (2001).

Kerdraon, A., and Delouis, J.-M., in *Coronal Physics from Radio and Space Observations*, ed. G. Trottet, Lecture Notes in Physics, 483, p. 192, Springer-Verlag, Berlin (1997).

Kliem, B., Karlický, M., and Benz, A.O., *A&A*, **360**, 715 (2000).

Messmer, P., Benz, A.O., abd Monstein, C., *Solar Phys.*, **187**, 335 (1999).

Tsuneta, S., Acton, L.W., Bruner, M E., *et al.*, *Solar Phys.*, **136**, 37 (1991).

BROADENING MECHANISMS OF THE Ca XIX RESONANCE LINE IN SOLAR FLARES

Y. P. Li and W. Q. Gan

Purple Mountain Observatory, Nanjing 210008, China, National Astronomical Observatories, Chinese Academy of Sciences

ABSTRACT

There are two explanations for the broadening mechanisms of the Ca XIX resonance line in the rising phase of a solar flare. One theory relates the line broadening to the velocity dispersion of the chromospheric evaporation along the loop. The other theory attributes the line broadening to nonthermal turbulence resulting from the release of flare energy. We distinguish these two possibilities by studying the influence of the loop orientation on the broadening, based on data from SXT and BCS on *Yohkoh*. The results seem to indicate that the broadening of the Ca XIX resonance line is caused by nonthermal turbulence.

INTRODUCTION

The broadening of soft X-ray lines during the impulsive phase of solar flares is one of the two basic characteristics revealed by early observations from the *Solar Maximum Mission*, *P78-1*, and *Hinotori* (e.g. Antonucci 1989). Due to the availability of *Yohkoh* data, studies on line broadening have been greatly refined. At present, it is still not clear if either theory can be ruled out, velocity dispersion or nonthermal turbulence. The strongest evidence against the velocity dispersion theory was thought to be the non-correlation between the line width and the distance from Sun center (Mariska *et al.* 1993). It seems to have been understood for a long time that the evaporation model should predict the line width decreasing from the disk center to the limb. This, however, does not seem to be the case! The line broadening predicted by the evaporation model depends not only on the distance from Sun center, but also on loop orientation. Li *et al.* (1989) showed that there is no line broadening from evaporation if the loop observed is at the solar limb and the plane of the loop is perpendicular to the line of sight. If the plane of the loop is not perpendicular to the line of sight, we may expect to see line broadening even if the flare is at the limb. It is therefore meaningful to check the relationship between line broadening and loop orientation.

RESULTS

Twenty limb events among the samples studied by Gan & Watanabe (1997) were selected. Each event approximates a single ideal loop, the orientation of which can be accurately derived. The loop orientation is defined by the formula given by Li *et al.* (1989)

$$V_{ls} = V_i(\cos\theta\cos\alpha_i \pm \sin\theta\sin\alpha_i\cos\Omega). \qquad (1)$$

where V_i is the velocity along the loop, V_{ls} the velocity along the line of sight, α_i the elevation angle of the segment (see Li *et al.* 1989), θ the angular distance of the loop from disk center, and Ω the orientation of the toroidal axis of the loop with respect to the radius vector between disk center and the midpoint of the loop. When the flare lies at the limb of the Sun, $\theta = 90^o$, and V_{ls} depends only on Ω. If Ω=90°, V_{ls}=0, no line broadening will appear; if Ω=0°, the broadening will be at its maximum. The orientation Ω can be derived from SXT images. Therefore, the line broadening variation with the angle Ω for limb events can provide information on the line broadening mechanism in solar flares.

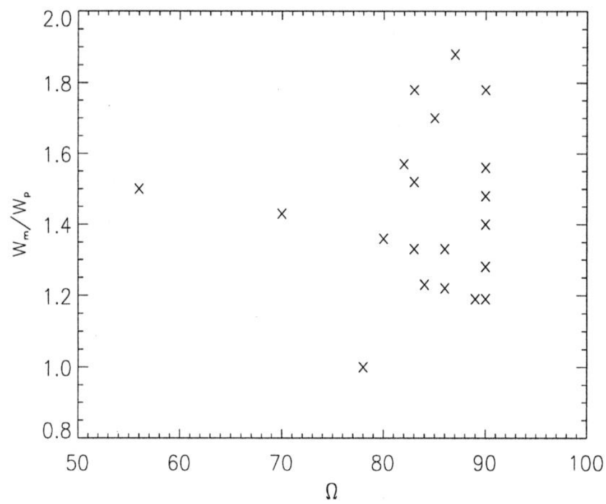

Fig. 1. Line broadening versus orientation of the loop for 20 single loop limb flares.

We measured the maximum Full Width to Half Maximum (FWHM) (W_m) and the FWHM (W_p) at the peak time from the Ca XIX resonance line observed with BCS for each sample flare. We use a relative ratio W_m/W_p to reflect the line broadening. Figure 1 shows W_m/W_p versus Ω for the 20 limb flares observed with SXT and BCS. There are two implications from Figure 1. First, the values of Ω for most samples are close to 90°. From Eq. (1) we know that if $\Omega = 90^o$ then velocity dispersion will not result in line broadening. In other words, the line width for limb events with $\Omega = 90^o$ should be smaller if velocity dispersion is the cause of the line broadening. But from comparing with Figure 6 of Gan & Watanabe (1997) which shows the number distribution of the line widths for all flares (not just limb events), we have found no difference in width values. This provides evidence that velocity dispersion may play an unimportant role in line broadening. Second, from Figure 1 we see that there is no tendency for the line broadening to vary with Ω. This result conflicts with the prediction by the velocity dispersion theory which requires a width distribution decreasing as a function of Ω. Therefore both implications seem to support the theory that line broadening is primarily due to nonthermal turbulence, although more samples are necessary to draw reliable statistical conclusions.

REFERENCES

Antonucci, E., *Solar Phys.*, **121**, 31 (1989).

Gan, W.Q., & T. Watanabe, *Solar Phys.*, **174**, 403 (1997).

Li, P., A.G. Emslie, & J.T. Mariska, *ApJ*, **341**, 1075 (1989).

Mariska, J. T., G.A. Doschek, & R.D. Bentley, *ApJ*, **419**, 418 (1993).

MULTI-WAVELENGTH OBSERVATIONS OF YOHKOH WHITE-LIGHT FLARES

S. A. Matthews[1], L. van Driel-Gesztelyi[2,3,4], H. S. Hudson[5,6], and N. V. Nitta[7]

[1] *Mullard Space Science Lab., University College London, Holmbury St. Mary, Dorking, Surrey, RH5 6NT, UK*
[2] *Center for Plasma Astrophysics, K. U. Leuven, 3001 Leuven, Belgium*
[3] *Observatoire de Paris, LESIA, F-92195 Meudon, France*
[4] *Konkoly Observatory, H-1525 Budapest, Hungary*
[5] *Solar Physics Research Corp., Tucson, AZ, USA*
[6] *Space Science Lab., University of California, Berkeley CA 94720-7450, USA*
[7] *Lockheed Martin Solar & Astrophysics Lab., Bldg. 252 Org. L9-41, 3251 Hanover St., Palo Alto, CA 94304, USA*

ABSTRACT

Most attempts to explain the origin of white-light flares have centered on canonical mechanisms. However, it has become clear that the spatial and temporal correspondence between white-light and hard X-rays (HXR) is not always one-to-one. In this work we attempt to determine the factors that lead to enhancements in the visible continuum by studying the temporal and spatial relationships between emission in the visible, soft X-rays (SXR) and HXR regimes in all of the white-light flares observed by *Yohkoh* (a total of 30 events), and by comparing them with a group of non-white-light flares that produced no white-light emission but had similar other characteristics.

INTRODUCTION

The generally good time coincidence between the hard X-ray burst and white-light flare (WLF) emission has led to the expectation that the electron beam responsible for the HXR is also responsible for the observed WLF emission. While this interpretation holds well for impulsive phase WLF emission, during the gradual phase the WLF is often seen to continue for some time after the HXR emission has fallen off (e.g. Hudson *et al.* 1992, Matthews *et al.* 1999, Sylwester & Sylwester 2000).

OBSERVATIONS

The aspect camera of the Soft X-ray Telescope (SXT) on board *Yohkoh* provided white-light images at 431 nm (the G band) with a bandpass of 3 nm and typical image interval of 10-12 seconds. Simultaneous data from the Hard X-ray Telescope ((HXT), Kosugi *et al.* 1991) and the Be, AlMg, and Al12 filters of SXT (Tsuneta *et al.* 1991) were also used.

In order to identify those flares observed by *Yohkoh* that were WLFs the following criteria were applied: (1) flare mode had been triggered (2) peak count rate in the HXT L-band > 50 or (3) peak count rate in the HXT L-band > 20 and peak count rate in the M1-band > L or $\simeq$ L. Those flares which met the above criteria but did not show white-light emission formed a non-WLF group whose properties are compared with the WLF group. A total of 30 WLFs and 34 non-WL flares were identified for further analysis. A list of the events can be found at `http://www.mssl.ucl.ac.uk/~sam/wlf_cat.html`.

ANALYSIS

In order to compare the WLFs with the non-WLFs we have looked at both unresolved and resolved properties including e.g. *GOES* class; coronal pressure; background subtracted ratios of the count rate in the M2 (33-53 keV) to the M1 (23-33 keV) channels; electron beam energy calculated using a thick target model and a cut-off energy of 33 keV as well as the impulsiveness of the HXR emission in all channels.

The smallest observable WLF in the sample is a C7.8 event and we find that of the combined group the fraction of WLFs increases steadily with GOES class to the point where all 7 X-flares observed showed enhanced WL emission. We also find that WL emission shows a good correlation to coronal over-pressure. This is particularly noticeable in the smaller events where for non-WLFs $\leq$ M2 the pressure is on average $\leq$ 250 dyns cm^{-2} compared to $\geq$ 350 dyns cm^{-2} in all cases except one of the WLFs. Figure 1 shows the variation of pressure with M2/M1 ratio from the peak of the flare, with WLFs having higher pressure on average for a given M2/M1 value. A plot of M2/M1 with peak GOES flux shows no particular tendency for WLFs to have higher M2/M1 ratios than non-WLFs. A full description of the events and their interpretation can be found in Matthews *et al.* (2002).

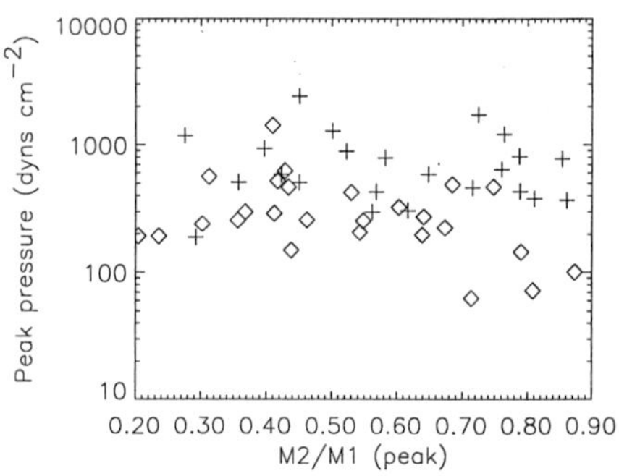

Fig. 1. Peak coronal pressure (dyns cm^{-2}) as a function of M2/M1 ratio at the flare peak. + - WLFs, ◇ - non-WLFs.

ACKNOWLEDGEMENTS

We thank the *Yohkoh* team and the MSSL SURF for providing data for use in this publication. LvDG is supported by Research Fellowship F/01/004 of the K.U. Leuven and by the Hungarian Government grant OTKA T032846 and T-038013. SAM and LvDG are grateful to the Royal Society for a Joint Project award.

REFERENCES

Kosugi, T. *et al.*, *Solar Phys.*, **136**, 37 (1991).

Hudson, H.S., Acton, L.W., Hirayama, T., and Uchida, Y., *PASJ*, **44**, L77 (1992).

Matthews, S.A., van Driel-Gesztelyi, L., Hudson, H.S., and Nitta, N., in *High Energy Solar Physics Workshop - Anticipating HESSI*, ASP Conference Series, Vol. 206, 239, eds. R. Ramaty and N. Mandzhavidze (2000).

Matthews, S.A., van Driel-Gesztelyi, L., Hudson, H.S., and Nitta, N., in preparation (2002).

Sylwester, B., and Sylwester, J., *Solar Phys.*, **194**, 305 (2000).

Tsuneta, S. *et al.*, *Solar Phys.*, **136**, 37 (1991).

ACCELERATION TIME SCALES OF SOLAR DISAPPEARING FILAMENTS

T. Morimoto and H. Kurokawa

Hida and Kwasan Observatories, Kyoto University, 17 Kitakazan Ohminecho, Yamashina-ku, Kyoto, 607-8471, Japan

ABSTRACT

We present an observational result on the acceleration time scale of solar disappearing filaments (SDFs). Measuring the 3D velocity field of 34 SDF events, we found their acceleration time scales vary from one minute to twenty minutes. These time scales are well correlated with the Alfven transit time scales which are calculated from photospheric magnetograms. This result implies that fast and slow prominence eruptions (i.e. "spray prominences" and "eruptive prominences") are merely extreme examples of eruptions with very strong and weak magnetic fields, respectively.

INTRODUCTION

The driving mechanism for solar ejections such as CMEs and prominence eruptions have long been investigated. Though it is still an open question, various models have been proposed and it is believed that magnetic forces play the primary role.

It is well known from observations that there are two types of prominence eruptions. A spray prominence (Warwick 1957) is shot out of the flare region and accelerated to large velocities in a few minutes. An eruptive prominence starts to ascend slowly and is accelerated in many minutes or hours (Valnicek 1964). Since these two types of eruptions are very different from each other not only in acceleration time scale but also in appearance, it has long been believed that they are different phenomena and that we need two different driver mechanisms.

In spite of many recent observations of CMEs and prominence eruptions, no one has yet succeeded in explaining how the magnetic force affects the acceleration time scale of a SDF. In this paper, we report on the quantitative relation between fast and slow solar ejections and the magnetic field strength of their source regions.

DATA ANALYSIS

We analyzed SDF events which were observed by the Flare Monitoring Telescope (FMT) at Hida Observatory. This telescope system can obtain full-disk solar images in both Hα line center and line wings, simultaneously. We selected 34 events using the FMT database from June 1992 through June 2000. In order to investigate the disappearances of filaments more clearly, only filaments longer than 60,000 km were selected.

Making use of the Doppler method, we derived line of sight and therefrom 3D velocity fields for the SDFs. Figure 1 shows the velocity-time plots for the 34 events. Clearly the data points are distributed between the two thick solid lines.

In order to explain what generates such large differences in velocity evolution, we compared the observed acceleration time scales for our SDFs with their Alfven transit time scales. The observed time scales are derived by simply applying exponential fittings to the velocity-time plots. The Alfven transit time scales are computed via the equation $\tau_A = \frac{F\rho_p}{\rho} \cdot \frac{L_p}{V_A}$ (e.g. Shibata & Tanuma 2001) where F is the filling factor, L_p is the length scale of the filament, and ρ_p and ρ are the densities for the filament and corona, respectively. We used photospheric magnetogram data to calculate the Alfven velocity. We also assumed the filling factor (F) and density ratio (ρ_p/ρ) are fixed for all events at 0.1 and 100, respectively.

RESULTS AND DISCUSSION

In Figure 2 one can clearly see that τ_{obs} is proportional or almost equal to τ_A. This means that "spray prominences" and "eruptive prominences" are merely extreme examples of eruptions with very strong and weak magnetic fields, respectively. It is well known that CMEs from active regions are accelerated quickly while CMEs from quiet regions are not. We therefore, conclude that the difference in observed acceleration time scales of solar ejections such as SDFs, prominence eruptions, and CMEs are clearly due to the difference in magnetic field strength of their source region.

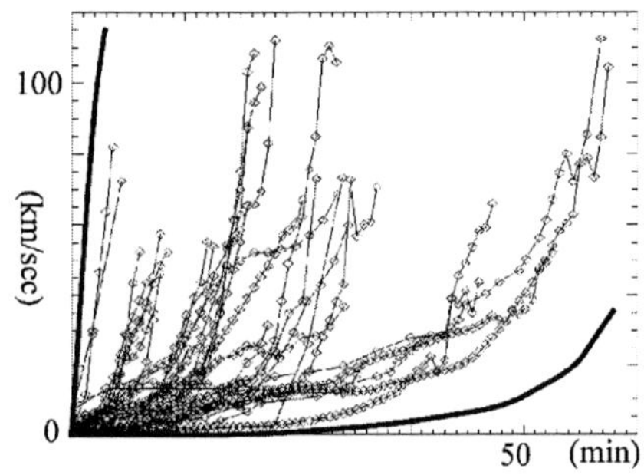

Fig. 1. Velocity-time plots of the 34 SDF events (solid lines with diamond symbols). Two thick solid lines at the bottom and left of the plot window are the typical velocity evolutions for "eruptive prominences" and "spray prominences", respectively.

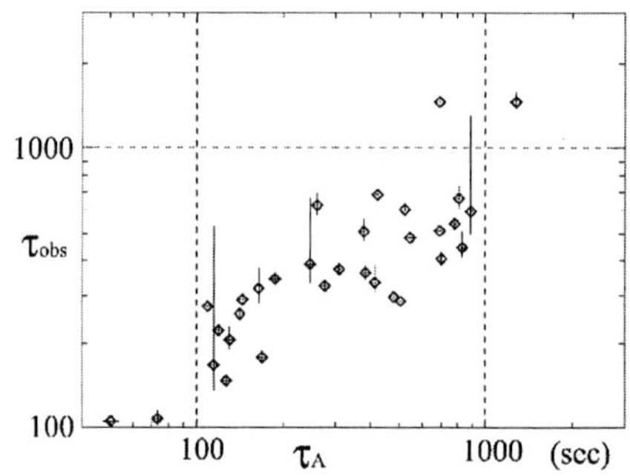

Fig. 2. Observed acceleration time scales (τ_{obs}) against calculated Alfven transit time scales (τ_A) for the 34 SDF events.

REFERENCES

Shibata, K., Tanuma, S., *Earth Planet Spacem* **53**, 473 (2001).

Valnicek, B., *Bulletin of the Astronomical Institute of Czechoslovakia*, **15**, 207 (1964).

Warwick, J.W., *ApJ*, **125**, 811 (1957).

FLARE TEMPERATURES FROM Fe XXV AND Ca XIX: IMPROVED ATOMIC DATA

K. J. H. Phillips[1], J. A. Rainnie[1], L. K. Harra[2], J. Dubau[3], and F. P. Keenan[4]

[1] *Space Science Department, Rutherford Appleton Laboratory, Chilton, Didcot, Oxon. OX11 0QX, UK*
[2] *Mullard Space Science Laboratory, Holmbury St Mary, Dorking, Surrey RH5 6NT, UK*
[3] *Observatoire de Paris - Meudon, Place Jules Janssen, F-92195, Meudon, France*
[4] *Dept. of Physics, Queen's University, Belfast BT7 1NN, N. Ireland, UK*

ABSTRACT

New atomic data are used for analyzing *Yohkoh* BCS Fe XXV and Ca XIX line data. Differences in derived temperatures are shown to be very small.

INTRODUCTION

Temperatures, T_e, and emission measures (N_e^2V, N_e = electron density, V = emitting volume) during flares and from bright active regions have generally been derived from comparison of spectra from the Bragg Crystal Spectrometer (BCS) on *Yohkoh* with spectra calculated from routines in the standard *Yohkoh* software package. The spectra include the resonance lines (w: notation of Gabriel 1972) of Fe XXV, Ca XIX, and S XV (He-like Fe, Ca and S) in the case of three BCS channels together with other He-like lines (x, y, and z) and dielectronic satellites from the Li-like stage, the most intense being j, k, q, and $d13/d15$. The satellites are more prominent in the case of the Fe spectra allowing T_e to be derived from spectral fits that chiefly rely on the j/w and k/w line ratios, though the j and k lines are generally only partly resolved in BCS spectra. For BCS Ca spectra, the wavelength range does not include the j or k lines. In this case, T_e is derived from a single feature due to the $d13/d15$ satellites near the w line. The atomic data needed for comparison with the theoretical spectra are stored in files in the *Yohkoh* software, and in the case of Fe and Ca spectra were compiled some years ago for a similar spectrometer on *Solar Maximum Mission (SMM)*; the collisional rate coefficients describing the theoretical line intensities are from distorted wave (DW) calculations. Close-coupling rates have since become available (Boone *et al.* 1992) from the Queen's University Belfast R-matrix code, a code which is generally regarded as being much more precise than DW formalisms. They are already used in the *Yohkoh* software package for the S XV spectra (Harra-Murnion *et al.* 1996). Here the Boone *et al.* values are substituted for transitions in Ca XIX and Fe XXV (upper levels $n < 4$), but with DW calculations retained for cascades from higher levels.

RESULTS AND CONCLUSIONS

Theoretical spectra were derived with the existing atomic data files and those with the R-matrix rates substituted using the program bcs_synthetic in the *Yohkoh* software and differences examined. The largest differences were in the case of the x, y, and z lines in both Ca XIX and Fe XXV. With the R-matrix data, these lines have theoretical intensities which are up to 5% more than with the previous data. This is

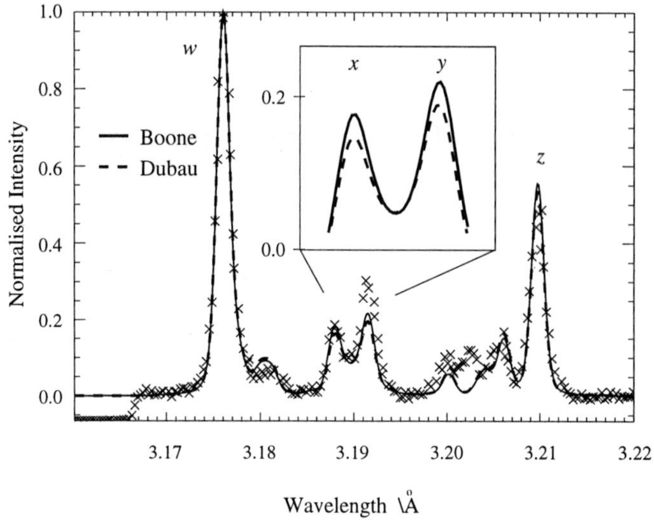

Fig. 1. Ca XIX spectra from existing data in the *Yohkoh* software package (dashed line) and from data based on the R-matrix rate coefficients (solid line) compared with an observed spectrum (crosses) from *SMM.*

illustrated with the Ca XIX spectrum shown in Figure 1 in which theoretical spectra based on the two sets of atomic data are compared, both with $T_e = 13$ MK, and an observed spectrum from the *SMM* spectrometer overlaid. It has been widely known that the observed intensities of the y and z lines in Ca XIX and Fe XXV spectra are more than the theoretical intensities, but the disagreements are about 30%. Evidently the small enhancements with the R-matrix data are inadequate to explain these discrepancies. The R-matrix calculations also add intensity to the w lines of Fe XXV and Ca XIX, but only 1 or 2%. This will have a small effect on derived values of T_e (and therefore emission measures) since the analysis programs use the satellite/w line ratio to obtain T_e; specifically the temperatures will be slightly increased and the emission measures slightly decreased. It was found that the temperature difference ΔT_e between the two sets of atomic data increased with T_e for both Ca and Fe spectra, but was never more than 1MK.

Use of R-matrix Ca and Fe rate coefficients in the the *Yohkoh* analysis software seems not to lead to appreciable differences in derived temperatures and does not resolve a well-known discrepancy between the observed and calculated x and y line intensities. Nevertheless the new data should be an improvement (accuracy $\sim \pm 5\%$: Tako & Nakazaki's (1989) comparison of DW data suggests DW accuracy is $\sim \pm 15\%$) over existing values and will be placed in the atomic data files for general use.

REFERENCES

Boone, A.W., A.E. Kingston, and P.H. Norrington, unpublished Queen's University Belfast paper (1992).

Gabriel, A.H., *MNRAS*, **160**, 99 (1972).

Harra-Murnion, L.K. *et al.*, *A&A*, **306**, 670 (1996).

Kato, T., and Nakazaki, S., *Atomic Data & Nucl. Data Tables*, **42**, 313 (1989).

MULTI-WAVELENGTH OBSERVATION OF A MORETON WAVE ON NOVEMBER 3, 1997

N. Narukage[1], K. Shibata[1], H. S. Hudson[2], S. Eto[3], H. Isobe[1], A. Asai[1], T. Morimoto[1], H. Kozu[1], T. T. Ishii[1], S. Akiyama[1], R. Kitai[1], and H. Kurokawa[1]

[1]*Kwasan and Hida Observatories, Kyoto University, Yamashina, Kyoto 607-8471, Japan*
[2] *Solar Physics Research Corporation, Sagamihara, Kanagawa 229-8510, Japan*
[3]*Department of Astronomy, Kyoto University, Sakyo-ku, Kyoto 606-8502, Japan*

ABSTRACT

We report the observation of a Moreton wave in Hα at 4:37–4:41 UT on November 3, 1997. The same region was simultaneously observed in soft X-rays and EUV, and wave-like disturbances (an "X-ray wave" and "EIT wave") were also found. The propagation speed of the X-ray wave (630 km/s) is almost the same as that of the Moreton wave (490 km/s). The EIT wave (170 km/s) is much slower than the Moreton wave. Assuming that the X-ray wave is an MHD fast shock, we can estimate the propagation speed of the shock, based on the MHD shock theory and the observed soft X-ray intensities. It is found that the estimated fast shock speed is 400–760 km/s, in rough agreement with the observed propagation speed of the X-ray wave. The fast mode Mach number of the X-ray wave is estimated to be about 1.15–1.25. These results suggest that the X-ray wave is an MHD fast shock propagating through the corona and hence is the coronal counterpart of the Moreton wave, but the EIT wave is not.

INTRODUCTION

Moreton waves are flare-associated waves observed to propagate across the solar disk in Hα, especially in the wing of Hα (Moreton 1960). They propagate in a somewhat restricted solid angle at a speed of 500–1500 km/s with arc-like fronts, and are often associated with coronal EIT waves. Moreton waves have been identified as the intersection of a coronal MHD fast-mode shock front with the chromosphere (Uchida 1968). This model predicts the existence of a coronal counterpart to the chromospheric Moreton wave. The purpose of this paper is to examine the basic question whether the EIT wave and the X-ray wave are the coronal counterparts of the Moreton wave, i.e., a coronal MHD fast-mode shock, or not.

OBSERVATIONS AND RESULTS

In this paper we study three kinds of flare-associated waves, a chromospheric Moreton wave, a coronal EIT wave and a coronal X-ray wave accompanied by a C-class flare that occurred in NOAA 8100 at S20, W13 on November 3, 1997 (Figure 1). The flare started at 04:32 UT and peaked at 04:38. The Moreton wave was observed in Hα (line center and +/− 0.8 Å), most clearly in Hα +0.8 Å, with the Flare Monitoring Telescope (FMT) at the Hida Observatory of Kyoto University, the X-ray wave with the Soft X-ray Telescope (SXT) on board *Yohkoh*, and the EIT wave with the *SOHO*/Extreme-ultraviolet Imaging Telescope (EIT).

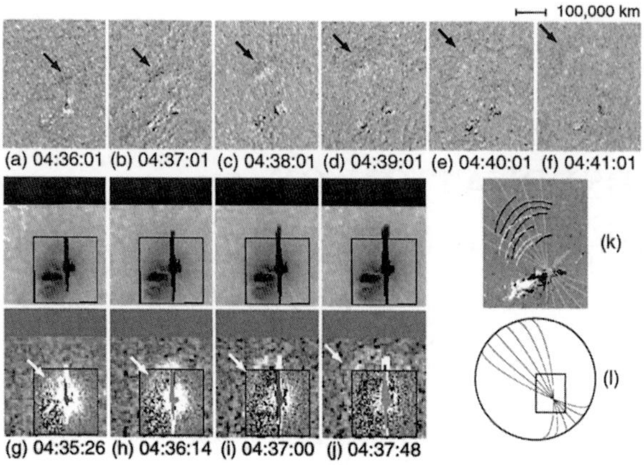

Fig. 1. Observed images on November 3, 1997, at NOAA 8100. Top panels are Hα +0.8 Å "running difference" images of a Moreton wave (black arrows). Middle and bottom panels are soft X-ray and "running difference" images of an X-ray wave (white arrows) taken with Al-Mg filter, respectively.

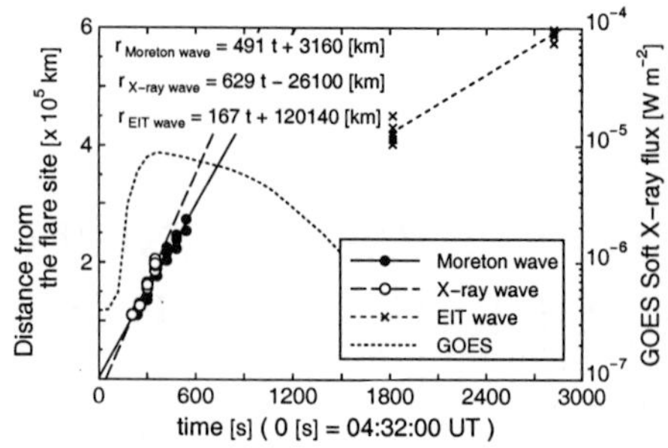

Fig. 2. Propagation of the Moreton wave (closed circles), the X-ray wave (open circles) and the EIT wave (crosses). First-degree polynomial fits of the Moreton wave (solid line), the X-ray wave (dashed line) and EIT wave (short dashed line) are shown. Dotted line shows soft X-ray flux observed by the *GOES* 9 satellite.

Both the position of the X-ray wave front and the direction of propagation of the X-ray wave roughly agree with those of the Moreton wave. The propagation speeds of the Moreton wave and the X-ray wave are about 490 km/s and 630 km/s, respectively. The EIT wave propagates at a speed of 170 km/s (Figure 2).

DISCUSSION

Let us now examine whether the EIT wave and/or the X-ray wave are the coronal counterpart of the Moreton wave or not. The EIT wave propagates at a speed of 170 km/s, which is much slower than the Moreton wave, 490 km/s. Hence, the EIT wave is not the coronal counterpart of the Moreton wave.

The X-ray wave propagates at a speed of 630 km/s, and the position of the wave front as well as the direction of propagation of the X-ray wave roughly agree with those of the Moreton wave. Assuming that the X-ray wave is an MHD fast shock we can estimate the propagation speed of the shock from MHD shock theory (Priest 1982) and the observed soft X-ray intensities. It is found that the estimated fast shock speed is 400–760 km/s, in rough agreement with the observed propagation speed of the X-ray wave. The fast mode Mach number of the X-ray wave is also estimated to be about 1.15–1.25 (Narukage *et al.* 2002). These results suggest that the X-ray wave is an MHD fast shock propagating through the corona and hence the coronal counterpart of the Moreton wave.

REFERENCES

Moreton, G. E., *AJ*, **65**, 494 (1960).

Narukage, N., H.S. Hudson, T. Morimoto, S. Akiyama, R. Kitai, H. Kurokawa, and K. Shibata , *ApJ Letters*, **572**, L109 (2002).

Priest, E. R., in *Solar Magnetohydrodynamics*, D. Reidel Publishing Company, chap. 5.4, Dordrecht (1982).

Uchida, Y., *Solar Phys.*, **4**, 30 (1968).

TIMING AND OCCURRENCE RATE OF X-RAY PLASMA EJECTIONS

M. Ohyama[1] and K. Shibata[2]

[1]*Shiga University, 2-5-1 Hiratsu, Otsu, Shiga 520-0862, Japan*
[2]*Kwasan and Hida Observatories, Kyoto University, Kazan-Oomine, Yamashina, Kyoto 607-8471, Japan*

INTRODUCTION

Yohkoh observations show that solar flares occur through magnetic reconnection above flare loops. It is expected from reconnection models that plasmoid ejections should be observed in all flares (Shibata *et al.* 1995). X-ray plasma ejections are, however, found in only 35–40% of 128 flares from C to X class (Nitta 1996) and in only 20–35% of 71 flares from A to C class (Akiyama 1998). These results seem to be inconsistent with the expectation of the reconnection model. This raises an important question whether X-ray plasma ejections are not general phenomena associated with solar flares. In other words, can many flares occur through mechanisms other than magnetic reconnection above flare loops? To solve this question we study the occurrence rate of X-ray plasma ejections prior to the flare for various flare classes.

TEMPORAL RELATION BETWEEN X-RAY PLASMA EJECTIONS AND FLARES (HXR EMISSION)

We examine the temporal relation between X-ray plasma ejections and flares using 126 limb flares between October 1991 and August 1998, and find that the main X-ray plasmoids already start to rise before the peak of the hard X-ray (HXR) emission (Ohyama & Shibata 2000).

The X-ray intensity of a plasmoid is a few orders of magnitude smaller than that of a flare loop. So it is difficult to determine when an X-ray plasmoid starts to rise. Fortunately, the X-ray plasmoids on December 2 1991 and November 11 1993 were observed from the preflare phase on (Ohyama & Shibata 1997, Tsuneta 1997). Their plasmoids appeared and had already started to rise slowly long before the onset of the impulsive phase, and then were suddenly accelerated just before or at about onset of the impulsive phase (HXR emission).

OCCURRENCE RATE OF X-RAY PLASMA EJECTIONS

To examine the occurrence rate of X-ray plasma ejections we use only 57 flares for which SXT starts to observe before the HXR peak, because X-ray plasmoids already start to rise before the HXR peak. Out of the 57 flares X-ray plasma ejections can be seen in 36 flares, and 4 flares are ambiguous. X-ray plasma ejections occur in both impulsive and LDE flares (Table 1). About 63–70 % of flares are associated with X-ray plasma ejections (Ohyama & Shibata 2000).

Table 1. Number of Flares and X-ray Plasma Ejections

(Life Time)	Flares	Ejections	Rate
Impulsive (< 1h)	50	31(3)	62–68 %
LDE (> 1h)	7	5(1)	71.4–85.7 %
Total	57	36(4)	63.2–70.2 %

Numbers in brackets indicate the ambiguous events.

Table 2. X-ray Plasma Ejections for Each X-Ray Class

X-Ray Class	Flares	Ejections	Rate
X	3	3	100 %
M	38	28 (3)	73.7–81.6 %
C	16	5 (1)	31.3–37.5 %
Total	57	36 (4)	63.2–70.2 %

Numbers in brackets indicate the ambiguous events.

Next, we study the occurrence rate of X-ray plasma ejections for each X-ray class of the flares (Table 2). X-ray plasma ejections can be seen in 100% of X class flares and 74–82% of M class flares, whereas only 31–38% of C class flares have X-ray plasma ejections. Does this result indicate that the mechanism for most C-class flares is different from that of M and X-class flares? In other words, can many small flares occur through mechanisms different from magnetic reconnection above flare loops? If many of the C-class flares occur through a mechanism which is different from that of M and X-class flares, it is expected that there would be a sharp bend around C-class flares in the frequency distribution of flares as a function of total energy. However, it has been found in previous studies that the frequency distribution of flares is well defined by a single power-law function over a broad energy range from 10^{27} ergs for microflares, which are weaker than C-class flares, to 10^{32} ergs for the largest flares (Shimizu 1995, Shimojo & Shibata 1999). Thus we think that small flares occur through the same mechanism: magnetic reconnection above flare loops.

Now we are faced with the following question; "Why is the occurrence rate of X-ray plasma ejections in C-class flares lower than that in larger flares?" The scale size of C-class flares is generally smaller than that of M and X-class flares. Hence it is thought that plasmoids in C-class flares are smaller. Moreover, the lifetime of the plasmoid ejection in the smaller flares is very short, of order of $t \sim$ 10–100 s (Shibata 1997). So we think that it is difficult to detect X-ray plasmoids in the weaker flares, and thus we propose that X-ray plasma ejections are general phenomena associated with solar flares.

REFERENCES

Akiyama, S., Research of X-ray Eruptive Structures Associated with Solar Flares. *Master Thesis*, Department of Aeronautics and Astronautics, Tokai University (in Japanese, 1998).

Nitta, N., A study of major flares observed by *Yohkoh*. In *Magnetic Reconnection in the Solar Atmosphere*, eds. R.D. Bentley and J.T. Mariska, ASP Conference Series No. 111., p. 156 (1996).

Ohyama, M., and Shibata, K., Preflare Heating and Mass Motion in a Solar Flare Associated with Hot Plasma Ejection: 1993 November 11 C9.7 Flare, *PASP*, **49**, 249 (1997).

Ohyama, M., and Shibata, K., Timing and Occurrence Rate of X-ray Plasma Ejections, *Journal of Atmospheric and Solar-Terrestrial Physics*, **62**, 1509 (2000).

Shibata, K., Rapidly Time Variable Phenomena: Jets, Explosive Events, and Flares, in *The Corona and Solar Wind near Minimum Activity*, ed. A. Wilson, ESA SP-404., p. 103 (1997).

Shibata, K., Masuda, S., Shimojo, M., Hara, H., Yokoyama, T., *et. al.*, Hot-plasma Ejections Associated with Compact-loop Solar Flares. *ApJ Letters*, **451**, 83 (1995).

Shimizu, T., Energetics and Occurrence Rate of Active-region Transient Brightenings and Implications for the Heating of the Active-region Corona, *PASP*, **47**, 251 (1995).

Shimojo, M., Shibata, K., Occurrence Rate of Microflares in an X-ray Bright Point within an Active Region. *ApJ*, **516**, 934 (1999).

Tsuneta, S., Moving Plasmoid and Formation of the Neutral Sheet in a Solar Flare, *ApJ*, **483**, 507 (1997).

INTENSITY DYNAMICS OF AN "EIT WAVE" OBSERVED BY TRACE

M. J. Wills-Davey

Physics Department, Montana State University, P.O. Box 173840, Bozeman, MT 59717, USA

ABSTRACT

On 13 June 1998 an "EIT wave" was observed by the *TRACE* satellite, initiated by a flare outside the *TRACE* field of view. This allowed for deep exposure observations, something rare with "EIT waves". To study the wave dynamics we examine a time series of base difference images, taking slices radiating from the flare origin. Graphs of the slices show changes in intensity with respect to a pre-flare image.

We find the "EIT wave" is observable as a propagating crest for 400-500 seconds before it dissipates and/or is disrupted by *in-situ* magnetic structures. We examine eight vectors that can be modelled in part by Gaussian curves. Velocities determined by measuring from peak intensities (the "center" of the wave rather than a visually determined wave front) give a wide range of values; the eastern portion of the wave front moves coherently at $\sim$ 350 km/s, while the western wave front slows dramatically and shows possible deflection. One of the slices appears to map ejecta, rather than a wave front trajectory.

Initial results show that the "EIT wave" crests do not exhibit a monotonic dissipation, but instead seem to maintain or increase in size for hundreds of seconds before dimming and broadening. Similarly, the depleted region following each crest also becomes larger over time. The intensity increase may represent material heating into the *TRACE* 195 Å passband, or it may be evidence of a soliton behavior. The large intensity decrease behind the wave front appears to be the coronal dimming often associated with coronal mass ejections.

DESCRIPTION OF THE DATA

We examine data taken on June 13 1998 at 15:27 UT by the telescope aboard the *Transition Region and Coronal Explorer* (*TRACE*). Between 15:27:11 and 15:45:11, an "EIT wave" was observed in the 195 Å passband which appears associated with a C2.9 *GOES* class flare in AR 8237. Previous studies of this wave event include Wills-Davey & Thompson (1999) and Delannée (2000). We study this wave using base difference images – data normalized with respect to exposure time and subsequently differenced with a pre-flare observation. The resulting images show changes in intensity.

MEASURING INTENSITY DYNAMICS

To study coronal intensity changes due to the "EIT wave", we examine vectors measured from the wave origin, determined in Wills-Davey & Thompson (1999). Each vector is divided into 100 points, and each point is smoothed over a five-pixel radius. Eight vectors show evidence of wave front motion. The positions of the vectors are shown in Figure 1.

The changes in intensity along the vectors are compared over time. An example vector is shown in Figure 2. All eight vectors exhibit propagating pulses which correspond to the wave front, and these pulses are subject to gaussian fits. Radial velocities of the wave are measured between gaussian peaks.

Comparison of the fits in Figure 2 shows an increase in the peak intensity and full-width half-max of the pulses over the first few intervals. All of the measured vectors exhibited similar behavior. This increase in coherence may be due to heating into the *TRACE* 195 Å passband, or is perhaps an effect of shocking.

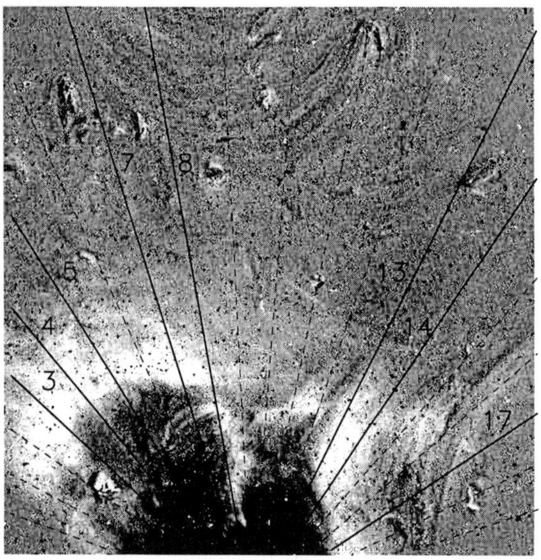

Fig. 1. Base difference image from 13 June 1998, 15:33:20 UT. Vectors (dashed lines) are measured from the wave origin. Eight vectors (solid, numbered lines) are studied in detail.

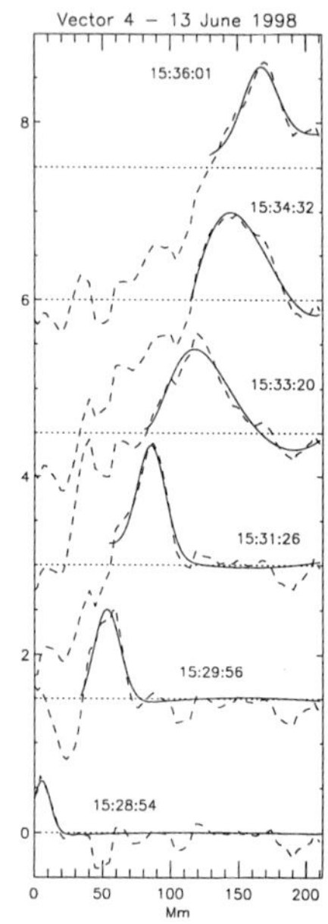

Fig. 2. Change in intensity along Vector 4, derived from base-differenced data. Positive values indicate increased intensity. Each time step has been fitted with a gaussian curve. The original data are shown as dashed lines, the fits are represented by solid lines, and a solid zero reference is shown for each curve.

ACKNOWLEDGEMENTS

This research was supported by an American Fellowship from the American Association of University Women Educational Foundation and by a *TRACE* grant from NASA to Montana State University.

REFERENCES

Delannée, C., Another View of the EIT Wave Phenomenon, *ApJ,* **545**, 545 (2000).

Ofman, L., and Thompson, B.J., Interaction of EIT Waves with Coronal Active Regions, *Solar Phys.,* in press (2002).

Wills-Davey, M.J., and Thompson, B.J., Observations of a Propagating Disturbance in *TRACE*, *Solar Phys.,* **190**, 467 (1999).

Section X.

Coronal Mass Ejections

USE OF YOHKOH SXT IN MEASURING NET CURRENT AND CME PRODUCTIVITY OF ACTIVE REGIONS

D. A. Falconer[1, 2], R. L. Moore[1], and G. A. Gary[1]

[1]*SD 50, NASA/MSFC, Huntsville Al, 35812, USA*
[2]*University of Alabama at Huntsville. Huntsville Al, 35812, USA*

ANALYSIS AND DISCUSSION

Our studies correlating global nonpotentiality of active regions with their CME productivity (Falconer 2001, Falconer, Moore, & Gary 2002) use *Yohkoh*/Soft X-ray Telescope (SXT) images for two purposes: 1) To check for the cusped coronal arcades of long-duration eruptive flares, a proxy for CMEs, as was used by Canfield, Hudson, & McKenzie (1999). 2) To assist in determining an active-region's net electric current, a measure of the region's global nonpotentiality. The net electric current is calculated from Ampere's law ($I_N = \int \mathbf{B_T} \cdot \mathbf{dl}$), integrating the transverse magnetic field around a contour encompassing either the leading or trailing polarity of the active region (Falconer 2001). The problem is that vector magnetographs measure the polarized light of a Zeeman-split spectral line; the transverse field being derived from the linear polarization. This leaves a 180° ambiguity in the direction of the observed transverse magnetic field. The X-ray images show coronal structures that reveal the sense of shear, thus indicating the proper resolution of the ambiguity. Instead of using coronal images for resolving the ambiguity, there are many methods that use only the vector magnetogram, ranging from the simple acute-angle rule (i.e. the observed transverse field is assigned the direction that gives it an acute angle relative to the potential transverse field) to the more sophisticated annealing method (Metcalf 1994). For nearly potential active regions, any of these methods work, but in very nonpotential active regions, in swaths of nearly 90° shear angle (between the observed and potential transverse field), incorrect resolution of the ambiguity can occur, and can be corrected from coronal images (Figure 1). In our sample of 17 magnetograms, the net currents from X-ray-resolved and acute-angle-resolved ambiguity differed by more than a factor of two in 3 cases (Figure 1). Of the three cases: one agreed in magnitude but not sign, one agreed in sign but differed by a factor of four in magnitude, and one (the shown magnetogram) had a negligible acute-angle-resolved net current, whereas it actually had a large net current. As we expand our sample for the CME prediction study, we plan to use coronal images to test various magnetogram-alone ambiguity resolution methods to determine if one of them is as accurate for the purpose of deriving net currents as the present X-ray method.

REFERENCES

Canfield, R. C., Hudson, H. S., & McKenzie D. 1999, *GRL,* **V26, 6,** 627

Falconer, D. A., 2001, *JGR*, **106**, 25185

Falconer, D. A., Moore, R. L., & Gary G. A., 2002, *ApJ*, **569**, 1016

Metcalf, T. R., 1994, *Solar Phys.*, **155**, 235

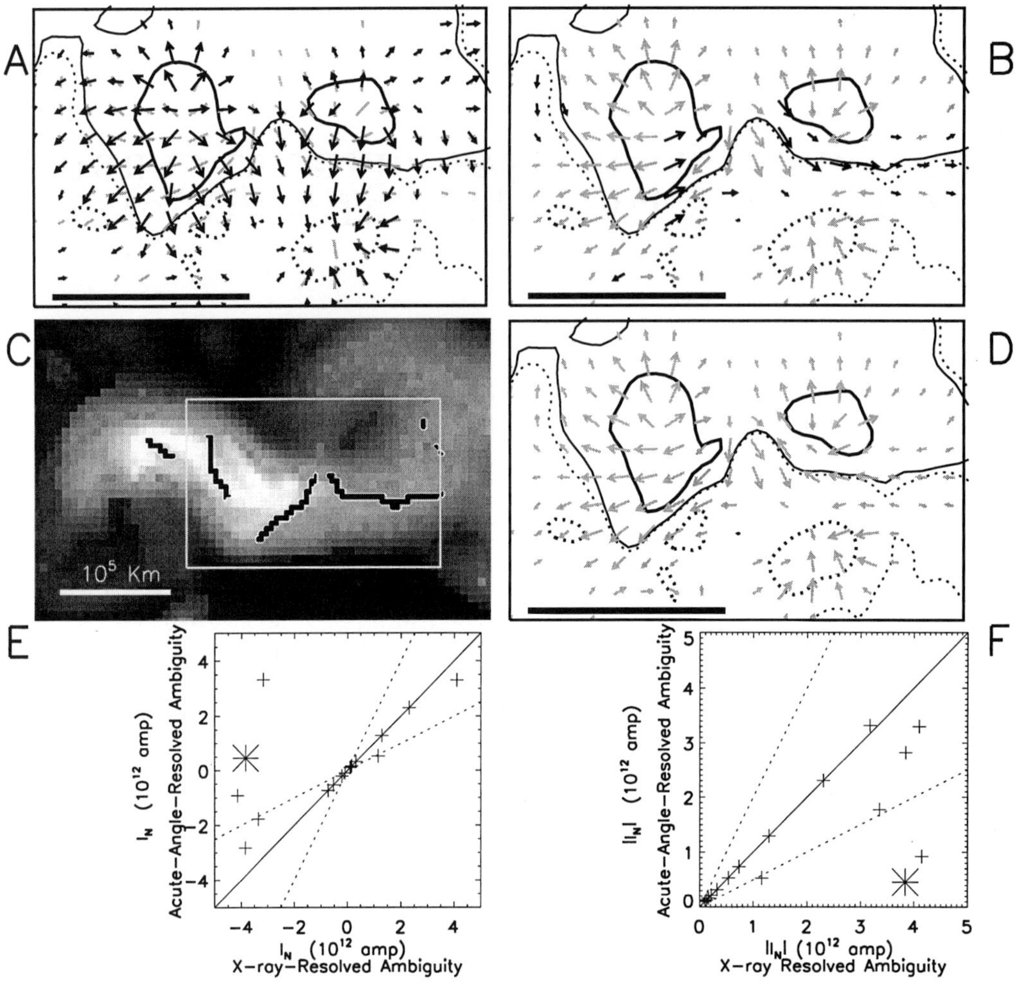

Fig. 1. Example of how SXT images are used to resolve the 180° ambiguity. **A:** Vector magnetogram of AR 9077 July, 14 2000. The contours are of line-of-sight field; the arrows show magnitude and direction of potential transverse field; the gray dashes show magnitude and direction of observed transverse field (with 180° ambiguity). **B:** Acute-angle-resolved magnetogram. Black arrows are ones that are opposite to the X-ray-resolved direction (**D:**). **C:** SXT image superposed with strong-shear sections of the main neutral line and white box showing field-of-view of the magnetogram panels. The X-ray image indicates that the magnetic field overlying the entire main neutral line has a 2-shape (inverse S) topology, with the field originating in the positive northern polarity, sweeping to the left, and rooting in the southern negative polarity, indicating that the observed transverse field points towards the left along the main neutral line, opposite to the acute-angle-resolved field along the west (right) half of the neutral line. **D:** X-ray-resolved magnetogram **E:** Scatter plot of the net current for each of the 17 magnetograms in our sample (*: shown magnetogram; +: the others; vertical axis: acute-angle-resolved current; horizontal axis: X-ray-resolved current; between the dashed lines the acute-angle-resolved current is within a factor of 2 of the X-ray-resolved current.). **F:** Scatter plot of the magnitude of the current. This work is funded by the NSF/Space Weather and NASA/Solar Physics programs.

TRAJECTORIES OF MICROWAVE PROMINENCE ERUPTIONS

K. Hori[1] and J. L. Culhane[2]

[1]*National Space Science and Technology Center, 320 Sparkman Drive, Huntsville, AL 35805, USA*
[2]*Mullard Space Science Laboratory, University College London, Dorking, Surrey, RH5 6NT, UK*

ABSTRACT

The classical coronal mass ejection (CME) scenario describes CMEs as associated with helmet streamer configurations, in which a quiescent prominence over a magnetic neutral line becomes unstable and starts to erupt, resulting in a CME (e.g. Low 1994). Recently, Hori (2000) studied two microwave prominences that were initially not embedded in a streamer but moved toward the streamer latitude and then erupted or disappeared. These prominences seem to be different from those in the classical CME scenario. Further, such prominence behavior suggests that streamers play the role of "mass corridor" through which coronal mass flows out toward the interplanetary space (Crooker *et al.* 1993). In order to address how prominence motions affect or reflect the surrounding coronal structures, we examined trajectories of 50 prominence eruptions that were observed with the Nobeyama Radioheliograph (NoRH) at 17 GHz near solar maximum (1999–2000). By comparing the prominence trajectories with the white-light synoptic maps from the Large Angle and Spectrometric Coronagraph (LASCO) C2 instrument on the *SOHO* spacecraft, we confirmed that coronal mass motions involving eruptive prominences and CMEs are not random but are organized as a bundle of pre-existing streamers. Large scale evolution of coronal features suggests that streamers are a signature of multiple plasma sheets emanating from active regions, arcades, trans-equatorial interconnecting loops, and polar crown filaments, as consistent with Crooker *et al.* (1993). Our study demonstrates that microwave observations can provide useful information on the activity at the base of such "coronal mass corridors".[†]

METHOD

We surveyed 17 GHz prominence eruptions from NoRH daily observations and selected 50 events that had observations from the beginning of the eruption with NoRH, *SOHO* LASCO C2, and *Yohkoh* soft X-ray telescope. For individual events, we made white-light synoptic maps using the LASCO C2 images at 2.5, 3.5, and 4.5 Rs from the Sun center, in which a spline fit of the cubic function was performed along the time axis in order to increase the time resolution (40 s). We marked the heliographic latitude of the top of the prominences as a function of time onto the synoptic maps and compared the latitudinal evolution of the coronal transients observed at different heights. Because of the continuum emission in microwave and white-light, we can examine the continuity in transients without having Doppler effects. This makes our study unique in comparison with other prominence-CME studies using EUV/optical lines.

[†]For details of this work see Hori & Culhane (2002).

RESULTS

This study is weighted towards weak coronal disturbances: Of the 50 events studied, only 12 events occurred at $\geq$ B9.2 flare sites and only 2 events occurred associated with radio type II and IV bursts. However, the association with flare activity will increase if we consider the possibility of behind-the-limb flares. On the other hand, 46 events had a CME association. In spite of the uncertainty due to projection effects, the tops of the eruptive prominences were found to move along preexisting streamers in 82 % of the events. The remaining events started outside of preexisting streamers and apparently moved toward the nearest streamer (Hori 2000). CME material also tends to move in a path parallel to preexisting streamers. The 46 CME structures can be divided into three groups with respect to their relation with preexisting streamers: (a) CMEs that emerged from, and were expelled along, a single streamer, (b) CMEs that encompassed multiple streamers, and (c) CMEs that arose from a sparse mass region outside of, or surrounded by, streamers. The bundle of streamers involved with CMEs in group (b) may suggest that each streamer has a sheet structure, i.e., the CME encompasses "a multiple plasma sheet" (Figure 1, right).

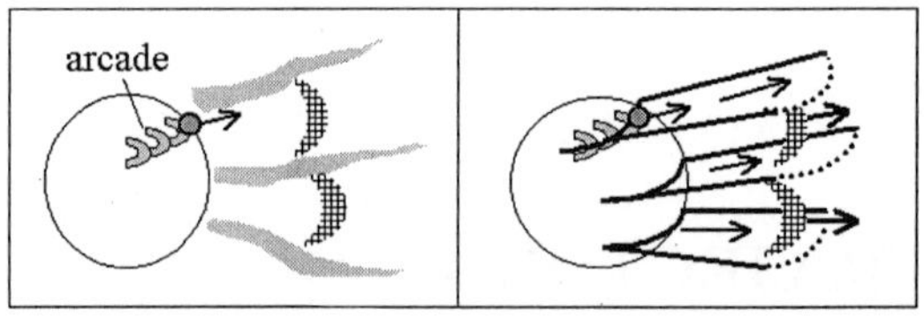

Fig. 1. Multiple sheet corridor, from Hori & Culhane (2002). The circle represents the Sun and the arrows indicate the direction of the mass outflows. Left: Schematic diagram of CMEs. Right: Our interpretation.

DISCUSSION

Since the 1970s it has been recognized that large-scale coronal streamers can be represented by an edge-on view of a single warped heliospheric current sheet (HCS) encircling the Sun. This single-sheet interpretation, however, seems to fail for the CMEs that progressively occur within multiple layers of streamers with a discontinuity across one of the streamers; these CME outflows might be separated by a sheet structure. We verified that the structure of the bundle of streamers can be maintained at least up to 4.5 Rs. While it is possible that these separate structures may finally come into contact at a single HCS far above the solar surface, the multiple layers might still exist as highly structured fields observed at the sector boundaries in the heliosphere (Crooker *et al.* 1993). Most of the prominence eruptions studied here were preceded by low-coronal activities at or near the prominence site, such as flares and the increase of the brightness temperature at 17 GHz. If all streamers are linked as a part of multiple plasma sheets that are encircling the Sun, it is also possible that pressure pulses induced by, e.g., preceding CMEs may travel around the high corona and finally activate the low corona through streamers (e.g. McComas *et al.* 1991).
For further understanding the creation and propagation of coronal transients, the combination of the coming solar mission, *STEREO* (US) and *Solar-B* (Japan/UK/US) may play a key part.

This work was supported by the Particle Physics and Astronomy Research Council (PPARC) in the UK.

REFERENCES

Crooker, N. U., *et al.*, *J. Geophys. Res.*, **98 (A6)**, 9371 (1993).
Hori, K., *ApJ*, **543**, 1011 (2000).
Hori, K. & Culhane, J. L., *A&A* **382**, 666 (2002).
Low, B. C., *Phys. Plasmas*, **1**, 1684 (1994).
McComas, D. J., *et al.*, *J. Geophys. Res. Letters*, **18(1)**, 73 (1991).

NOZOMI OBSERVATION OF INTERPLANETARY TRANSIENTS EJECTED AS LIMB CORONAL MASS EJECTIONS

T. Nakagawa[1], A. Matsuoka[2], and *NOZOMI*/MGF team

[1] *Tohoku Institute of Technology, Taihaku-ku, Sendai, 982-8577, Japan*
[2] *Institute of Space and Astronautical Science, Yoshinodai, Sagamihara, 229-8510, Japan*

ABSTRACT

The magnetic fields of interplanetary objects that were ejected as coronal mass ejections (CMEs) from the limb of the Sun were observed by the *NOZOMI* spacecraft at 1.38 AU above the east limb of the Sun. The solar wind magnetic field whose launch time coincided with ejection of the limb CME was different from that estimated from *ACE* observations near the Earth, suggesting that *NOZOMI* encountered transient magnetic structures. The interplanetary magnetic field showed a complicated structure with a wide range of variations from event to event. The convection speeds of the transients estimated from their travel times were less distributed compared with the initial speeds of CMEs.

INTRODUCTION

The Japanese Mars explorer *NOZOMI*, launched on July 4, 1998, spends more than 5 years in interplanetary space before its insertion into Martian orbit. The magnetometer (Yamamoto & Matsuoka 1998) is operating in its trans-Mars orbit (Nakagawa *et al.* 2002). In 1999, the heliospheric longitude between *NOZOMI* and the Earth continually increased, as they go around the Sun at different rotation periods. During the period from November 1, 1999, to November 30, 1999, *NOZOMI* went as far as 83° – 97° east from the Sun-Earth line at a radial distance of 1.38 AU from the Sun. The large separation from the Earth in heliospheric longitude enabled direct detection of the ejecta launched into interplanetary space in a direction perpendicular to the line of sight of the imagers such as *Yohkoh* and *SOHO* in the vicinity of the Earth. This paper reports *NOZOMI* observations of the magnetic field of the solar wind whose launch time coincided with the time of ejection of limb CMEs in the *SOHO*/LASCO CME Catalog.

INTERPLANETARY MAGNETIC FIELD EJECTED AS CMES

We need to know the date T_{N0} of ejection of the solar wind encountered by *NOZOMI*. However, we had only limited opportunity to measure solar wind speed, so the date was estimated with the aid of *ACE*/SWEPAM measurements. The bottom diagram of Figure 1 shows the relationship of (a) the time T_N of *NOZOMI* observation, (b) the time T_A of *ACE* observation, (c) the time T_{A0} of launch of the plasma toward *ACE*, and (d) the time T_{N0} of launch toward *NOZOMI*. The magnetic field observed by *NOZOMI* agreed well with the result from *ACE* shifted to *NOZOMI* position until November 10, 1999; disagreement started thereafter. Many of the CMEs on the east limb in the *SOHO*/LASCO catalog correspond to the disagreement between *NOZOMI* and *ACE* observations. It is likely that *NOZOMI* encountered transient magnetic structures or that the heliospheric magnetic field re-structured after ejection of the CMEs, although the difference may be due to transients which passed *ACE*, or inaccuracy in solar wind speed determination. The interplanetary magnetic field showed a complicated structure with a wide range of variations from event to event. No common structure has yet been found. The convection speeds (364 – 617 km/s) of the probable transients estimated from their travel times were less distributed compared with initial speeds of the CMEs (153 – 719 km/s), consistent with Figure 5 of Lindsay *et al.* (1999), but the relationship was less clear.

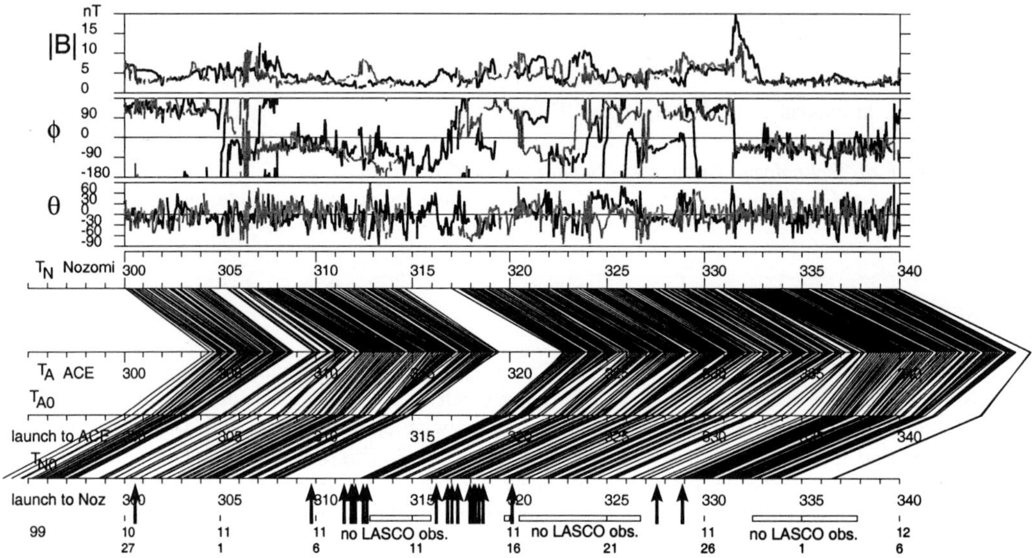

Fig. 1. Hourly averaged magnetic fields obtained by *NOZOMI* (dark) and *ACE* (light), covering the period from November 1, 1999 to November 30, 1999. *ACE* data are shifted to the dates on which the solar wind of the same source was observed by *NOZOMI*. From top to bottom, the magnitude B of the magnetic field, azimuthal angle ϕ and inclination θ. Bottom diagram shows the time T_N of *NOZOMI* observation related with T_A, T_{A0} and T_{N0}. Arrows indicate time of limb CMEs listed in *SOHO*/LASCO CME catalog (http://cdaw.gsfc.nasa.gov/CME_list/).

ACKNOWLEDGEMENTS

The authors are indebted to N. Gopalswamy, S. Yashiro and G. Michalek for the *SOHO*/LASCO CME Catalog, Charles W. Smith for *ACE*/MAG data and David J. McComas for *ACE*/SWEPAM data.

REFERENCES

Lindsay, G. M., J. G. Luhmann, C. T. Russell, & J. T. Gosling, Relationship between coronal mass ejection speeds from coronagraph images and interplanetary characteristics of associated interplanetary coronal mass ejections, *J. Geophys. Res.*, **104**, 12515 (1999).

Nakagawa, T., A. Matsuoka, & *NOZOMI*/MGF team, *NOZOMI* observation of the interplanetary magnetic field in 1998, *Adv. Space Res.*, **29**-3, 427 (2002).

Yamamoto, T., & A. Matsuoka, Planet-B magnetic fields investigation, *Earth,Planets.Space*, **50**, 189 (1998).

ON THE RELATION BETWEEN FLARES AND CMES

N. V. Nitta

Lockheed Martin Advanced Technology Center, Org. L9-41, B. 252, 3251 Hanover Street, Palo Alto, CA 94304, USA

ABSTRACT

Many spectacular coronal mass ejections (CMEs) are known to occur without major flares. This has created a trend in which flares are considered to be unimportant in the processes of CME initiation. We point out that energetic CMEs are often associated with intense short-duration flares, and that there may even be a class of CMEs caused by flares. In order to prove or disprove the latter possibility, we need to go back to flare studies in a more systematic manner. Distinguishing waves from plasma ejections in flares seems to be important.

FILAMENT ERUPTIONS AND FLARES

Historically, a CME has been considered to be associated with either a flare or a filament eruption. Indeed, height-time plots of CMEs indicate that those associated with flares tend to attain large acceleration close to the Sun and that those associated with filament eruptions are accelerated over longer time and distance (cf. MacQueen & Fisher 1983, Sheeley *et al.* 1999). However, we would like to ask what a flare stands for (cf. Cliver 1995), since many filament eruptions are accompanied by what we call long decay events (LDEs). It is well known that CMEs are much better correlated with LDEs than with short-duration flares. Therefore, as indicated in Figure 2 of Forbes (2000), we may understand a CME and the associated LDE in terms of a single scenario.

IMPULSIVE FLARES AND CMES

As we have shown elsewhere (Nitta & Hudson 2001), recurrence of six halo CMEs within a 2.5 day period was found to be associated with short-duration X- and M-class flares from the same active region. As shown below, we now find that the association of X-class flares with halo CMEs is quite high, irrespective of flare durations. In the present cycle, a total of 56 X-class flares have occurred, of which only a few are in the LDE category. For the 49 X-class flares with LASCO coverage, 73% are associated with halo CMEs (including partial halos), and only 10% lack CME association. In contrast, out of the 51 M-class flares between July 1996 and May 1998 that occurred within LASCO observing times, 67% are not associated with any CMEs, and only 16% are associated with halo CMEs.

These numbers indicate that some extended CMEs may be the result of the so-called "Big Flare Syndrome" (Kahler 1982), meaning that these CMEs may be *caused by* flares. Although such a concept is now considered by many to be outdated, Uchida *et al.* (2001) propose that their "bubble type" CMEs signify flare blasts. Non-LDEs may be associated with CMEs through flare-produced waves.

FLARE EJECTA AS AN IMPORTANT OBSERVABLE

But there may be a caveat in the distinction of flares in terms of durations. Indeed, almost all flares have X-ray light curves with a longer decay time than a rise time. Moreover, Shibata *et al.* (1995) reported that hot plasmoid ejections, an X-ray analog of Hα filament eruptions, were observed in all of their eight short-duration flares at the limb, arguing that these flares are not fundamentally different from LDEs. As shown in Figure 1, these ejections are typically observed in SXT over-exposed partial-frame images with a $\sim 10'$ field-of-view (FOV). It was shown later (Nitta & Akiyama 1999) that flares with X-ray ejections are always associated with CMEs, but that there are non-ejective flares too. We also find "confined" ejections, in which X-ray ejections are seen only in a smaller FOV.

Here, we point out that we are far from understanding the nature of these flare-associated ejections especially in the context of CME initiation, because the number of events closely looked at is still small. More importantly, there seems to be no objective definition of the ejections. For example, the two arrows in Figure 1 indicate two moving features separate from the central flare loop, but one of them is considered to signify blast waves and the other to signify flare ejecta (Hudson *et al.* 2002). Distinguishing waves from ejecta is important in order to address the role of flare processes in launching CMEs, especially since the rare Moreton wave phenomenon as observed in Hα tends to be seen in a similar configuration of an intense flare located close to one leg of the associated CME.

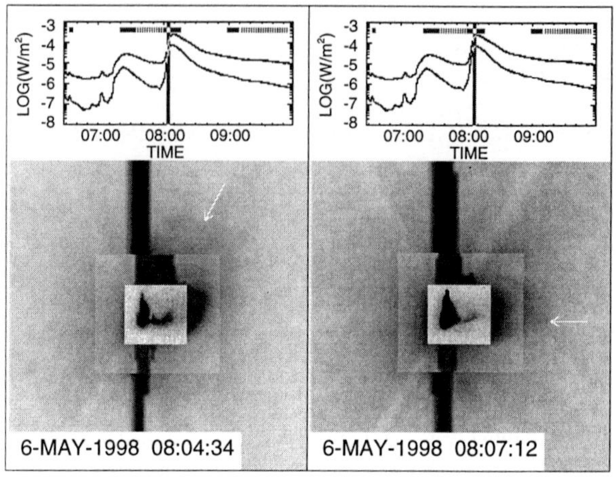

Fig. 1. The flare of 1998 May 6 (cf. Khan & Hudson 2000, Hudson *et al.* 2002). Arrows indicate two "ejecta".

CONCLUDING REMARKS

- We cannot rule out from coronagraphic observations alone the possibility that different processes exist that explain the launch of CMEs.
- Presently, flare ejections seem to be the only direct observations of at least part of CMEs. But they mean different processes to different scientists. This situation has to be ameliorated.

REFERENCES

Cliver, E. W., *Sol. Phys.*, **157**, 285 (1995).

Forbes, T. G., *JGR*, **105**, 23153 (2000).

Hudson, H. S., Khan, J. I., Lemen, J. R., Nitta, N. V., & Uchida, Y., *Sol. Phys.*, submitted (2002).

Kahler, S. W. *JGR*, **87**, 3439 (1982).

Khan, J. I., & Hudson, H. S., *GRL*, **27**, 1083 (2000).

MacQueen, R. M., & Fisher, R. R., *Sol. Phys.*, **89**, 89 (1983).

Nitta, N. & Akiyama, S., *ApJ*, **525**, L57 (1999).

Nitta, N. V., & Hudson, H. S., *GRL*, **28**, 3801 (2001).

Sheeley, N. R., Jr., Walters, J. H., Wang, Y.-M., & Howard, R. A., *JGR*, **104**, 24739 (1999).

Shibata, K., Masuda, S., Shimojo, M., Hara, H., Yokoyama, T. *et al.*, *ApJ*, **451**, L83 (1995).

Uchida, Y., Tanaka, T., Hata, M., & Cameron, R., *PASA*, **18**, 345 (2001).

THE FORCE FREE MAGNETIC STRUCTURE INSIDE A TOROID

E. P. Romashets

Institute of Terrestrial Magnetism, Ionosphere, and Radio Wave Propagation of Academy of Sciences (IZMIRAN), Troitsk, Moscow Region, 142190, Russia

ABSTRACT

We present a general form of a solution of the equation for the force free magnetic field $curlB = \lambda B$ inside toroid with arbitrary large and small radii. The solution consists of a set of harmonics. To determine the real magnetic structure corresponding to a passage of a magnetic cloud of toroidal shape, the field inside it is assumed to be nearly force-free, since the ratio of plasma and magnetic pressure (beta) is less than 0.1. As a rule, for such objects one must determine coefficients for each harmonic in the set.

INTRODUCTION

There is much evidence that some of the solar and interplanetary structures which can trigger strong geomagnetic storms when interacting with the Earth's magnetosphere, contain force free magnetic fields in their interior.

METHODS

One of the possible geometries for such structures is toroidal. In Miller & Turner (1981) a solution for a large aspect ratio torus was found using curved cylindrical coordinates r, θ and z. We use here toroidal coordinates μ, η and φ, described in Morse & Feshbach (1953), and used recently for calculation of fields outside the body in Romashets & Vandas (2001). These coordinates are related to cartesian ones by

$$x = \frac{a \sinh \mu \cos \varphi}{\cosh \mu - \cos \eta}; \quad y = \frac{a \sinh \mu \sin \varphi}{\cosh \mu - \cos \eta}; \quad z = \frac{a \sin \eta}{\cosh \mu - \cos \eta} \quad . \tag{1}$$

Here $a = \sqrt{R_0^2 - r_0^2}, and$ R_0 and r_0 are the large and small radii of the torus.

As it has been shown in Chandrasekhar and Kendall (1957) the field distribution

$$\mathbf{B} = B_0 \nabla \times (\mathbf{b}\Psi) \tag{2}$$

is force free. Here $\mathbf{b}$ is a fixed vector and Ψ must satisfy $\Delta\Psi = \lambda^2\Psi$.

In order to find Ψ in toroidal coordinates we can express it as

$$\Psi = \sqrt{\cosh\mu - \cos\eta}M(\mu)N(\eta)F(\varphi). \tag{3}$$

And we have for M the following equation from $\Delta\Psi = \lambda^2\Psi$ if it is assumed that $N(\eta) = e^{in\eta}$ and $F(\varphi) = e^{im\varphi}$

$$\frac{1}{\sinh\mu}\frac{d}{d\mu}\left(\sinh\mu\frac{dM}{d\mu}\right) - \frac{m^2M}{\sinh^2\mu} - \left(n^2 - \frac{1}{4}\right)M = \frac{\lambda^2a^2}{(\cosh\mu - \cos\eta)^2} \quad . \tag{4}$$

If $\lambda = \frac{1}{a}\frac{\cosh\mu - \cos\eta}{\sinh\mu}m\sqrt{3}$ inside the torus, $\mu > \mu_0$, one can see that the solution (3) is a linear combination of

$$\Psi_n^m = Re\left(\sqrt{\cosh\mu - \cos\eta}B_0aQ_{n-1/2}^{2m}(\cosh\mu)\, e^{in\eta}e^{im\varphi}\right) \quad , \tag{5}$$

where B_0 is a complex constant, $B_0 = B_{0r} + B_{0i}$, and $Q_{n-1/2}^m$ is a Legendre function of a second kind.

Now, if $\mathbf{b} = \mathbf{e}_\mu$ in (2) we have the following force free magnetic field distribution inside toroid:

$$B_\mu = 0 \quad , \tag{6}$$

$$B_\eta = -Re\left(\frac{imB_0}{\sinh\mu}(\cosh\mu - \cos\eta)^{3/2}\,Q_{n-1/2}^{2m}(\cosh\mu)\,e^{in\eta}e^{im\varphi}\right) \quad , \tag{7}$$

$$B_\varphi = Re\left(B_0\left[\frac{\sin\eta}{2} - in\sqrt{\cosh\mu - \cos\eta}\right]\sinh\mu\sqrt{\cosh\mu - \cos\eta}Q_{n-1/2}^{2m}(\cosh\mu)\,e^{in\eta}e^{im\varphi}\right) \quad . \tag{8}$$

CONCLUSIONS

The distribution of force free magnetic field within a torus with a superconductive wall has been found by solving the problem in toroidal coordinates. The solution obtained can be used to describe magnetic configurations inside interplanetary clouds as well as filaments and other regions with strong fields on the Sun.

ACKNOWLEDGEMENTS

The author thanks numerous colleagues for their encouragement of the work on force free magnetic fields. Especially helpful comments were made by Kim Ivanov. This work was supported by EU-INTAS-ESA grant 99-00727.

REFERENCES

Chandrasekhar, S., and P. C. Kendall, *Astrophys. J*, **126**, 457 - 460, (1957).

Miller, G., and L. Turner, *Phys. Fluids*, **24**, 363 - 365, (1981).

Morse, P. M., and H Feshbach, *Methods of Theoretical Physics, McGraw-Hill, New York*, (1953).

Romashets, E. P., and M. Vandas,*J. Geophys. Res.*, **106**, 10615 - 10624, (2001).

DECIMETRIC REVERSE DRIFT AND U-TYPE BURSTS IN THE APRIL 9, 2001 FLARE

J. R. Cecatto[1], H. S. Sawant[1], F. C. R. Fernandes[1], V. Krishan[1,2], R. R. Rosa[1], and M. Karlický[3]

[1]*National Institute for Space Research - INPE, São José dos Campos, SP - Brazil*
[2]*Indian Institute of Astrophysics, Bangalore, India*
[3]*Ondřejov Observatory, 25165 Ondřejov, Czech Republic*

ABSTRACT

The Brazilian Solar Spectroscope (BSS) as well as the radio spectrographs of the Ondřejov Observatory (OO) observed, in decimeter wavelengths, spectra consisting of groups of type III reverse drift (RS) bursts with time resolution of 50 ms and also observed a partial inverted long duration (about 25 s) U-type burst. The speeds of their exciting agents have been determined. The estimated speeds suggest that distinct exciters are responsible for the emissions of the inverted U-type and type-III RS bursts.

OBSERVATIONS AND DISCUSSION

High-resolution observations of the solar decimetric radio emission are of a great diagnostic value for understanding particle acceleration and the mechanism of energy release in flares. Time variations, periodic or otherwise, are related to the process of energy release. Here, we report for the first time high resolution (50 ms) observations of decimetric type-III RS bursts. Also, a partial inverted U-type burst of long duration (25 s) is being reported for the first time in the frequency range of 900-2000 MHz. This flare has also been observed by the *SOHO* spacecraft, and the SXT instrument on the *Yohkoh* satellite. Details of the spacecraft observations will be published elsewhere.

The spectra recorded at the OO show a U-type like burst starting at 15:25:23 UT around 1000 MHz. This burst attains its minimum mean frequency of 900 MHz at 15:25:32 UT. Then the burst drifts to higher frequencies ending at 15:25:48 UT as shown in Figure 1. The BSS (Sawant *et al.* 2001) recorded a group of type-III RS bursts around 2000 MHz having bandwidth of about 500 MHz at 15:33 UT which are superimposed on a continuum as shown in Figure 2.

U-type burst: The speed of the exciting agency for the $\sim$ 6 s duration descending part of the partial inverted U-type burst can be estimated assuming the original density model of Aschwanden and Benz (1995) suitably modified for high frequencies above 1000 MHz by Meléndez (1997) as given by: $n_e = n_1(\frac{h}{h_1})^{-p}$, with p=5, $h_1 = 3.5 \times 10^5$ km, and $n_1 = 4.3 \times 10^7$ cm^{-3}. Here h_1 and n_1 are the parameters in the transition region from low to high corona where the density law changes from power-law to exponential. The heights of these emissions are estimated for the frequency range of 1000-2000 MHz. Assuming time of flight for the descending part of the partial inverted U-type burst to be 6 sec, estimated speeds are $\simeq 4.0 \times 10^3$ km/s and 6.5×10^3 km/s at the fundamental and second harmonic, respectively. These speeds are about one order of magnitude less than those speeds of an exciter required to produce type-III bursts and higher than those of an exciter required to produce type-II bursts. Efforts are being made to use *SOHO* and SXT data to determine the separation between the footpoints of the flaring loop which will lead us to a better understanding of the type of exciter. Details of this analysis will be published elsewhere.

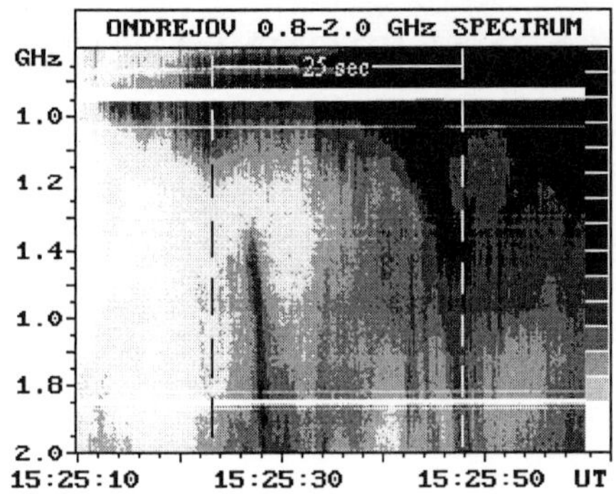

Fig. 1. Radio spectrum (800-2000 MHz) recorded at the Ondřejov Observatory for the April 9th, 2001 event showing a partial inverted U-type burst limited by the dashed lines.

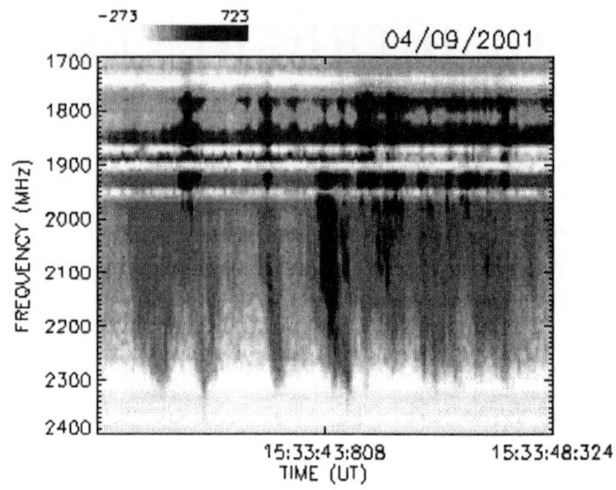

Fig. 2. BSS high resolution spectrum (1700-2400 MHz) for the April 9th, 2001 event showing a group of type-III RS bursts.

Type-III RS bursts: The frequency drift rate D_f (MHz/s) is related to the observed frequency f_{obs} (MHz) as $D_f = \frac{f_{obs}}{2}\frac{V}{H}$. Here V is the speed of the exciting agency in km/s. For determining the scale height of density variation H we can make use of the same density model as described above. High resolution observations permits us to determine more accurately the drift rates. By using the density model above, the drift rate at high frequencies is given by: $|df/dt| = (0.09 \pm 0.03) \times f^{(1.35\pm0.10)}$ where df/dt is in MHz/s, and f in MHz (Melendez *et al.* 1999). Observed drift rates of a sample (15:33:40-15:33:48 UT) of these bursts is in the range of 980 - 1720 MHz/s, in agreement with the above equation. Thus for D_f = 1300 MHz/s at $f_{obs} = 2 \times 10^3$ MHz, we find $V \simeq 2.9 \times 10^4$ km/s for type-III RS bursts, which is about one order of magnitude higher than the speed of the exciter determined above for the descending part of the inverted U-type burst.

The decimetric (1000-3000 MHz) observations and preliminary analysis of the first reported partial inverted long duration U-type, and high time resolution type-III RS bursts indicate the presence of different exciters. Thus distinct acceleration mechanisms are operating in the flaring region. A more detailed modeling will give more insight into the physics of the flaring region. F.C.R.F., V.K. and M.K. acknowledge scholarship from FAPESP.

REFERENCES

Aschwanden, M.; A.O. Benz, Chromospheric evaporation and decimetric radio emission in solar flares, *Astrophys. J.*, **438**, 997 (1995).

Meléndez, J.L.M. Decimetric solar type-III bursts in association with impulsive phase of solar flares. *MSc. Thesis.* INPE, São José dos Campos, 1997. 115p. (INPE-6382-TDI/601).

Meléndez, J.L.M; H.S. Sawant, F.C.R. Fernandes, A.O. Benz, Statistical analysis of high-frequency decimetric type III bursts, *Solar Phys.*, **187**, 77 (1999).

Sawant, H.S., K.R. Subramanian, C. Faria, F.C.R. Fernandes, J.H.A. Sobral, J.R. Cecatto, R.R. Rosa, H.O. Vats, J.A.C.F. Neri, E.M.B. Alonso, F.P.V. Mesquita, V.A. Portezani, A.R.F. Martinon, Brazilian Solar Spectroscope (BSS), *Solar Phys.*, **200**, 167 (2001).

THE 1.0-4.5 GHz ZEBRAS IN THE JUNE 6, 2000 FLARE

H. S. Sawant[1], M. Karlický[2], F. C. R. Fernandes[1], and J. R. Cecatto[1]

[1]*Instituto Nacional de Pesquisas Espaciais, INPE, C.P. 515, 12201-970, São José dos Campos, SP, Brazil*
[2]*Astronomical Institute of the Academy of Sciences of the Czech Republic, 25165, Ondřejov, Czech Republic*

ABSTRACT

A long lasting X2.3 flare was recorded in radio waves during 15:00-17:00 UT on June 6, 2000. A full-halo coronal mass ejection was associated with this flare. The flare was unusually rich in the high-frequency radio zebras above 1 GHz. A unique case of zebra branches in a harmonic ratio of 1:2 was observed. These zebras were analyzed and interpreted in the model based on the double plasma resonance instability. Longitudinal upper-hybrid waves are excited at positions of cyclotron resonances and then transformed into electromagnetic waves. Using this model, the magnetic field strength in this flare is estimated.

INTRODUCTION

The metric radio zebras have been known for a long time (Slottje 1971). Recently, zebras have been also reported in the frequency range of 1.0-4.5 GHz. All these zebras patterns can be used for magnetic field estimates in the corona. Recently, a new model for zebra emissions was suggested by Ledenev *et al.* (2001), which gives more realistic values of the magnetic fields in the lower heights of the solar flare atmosphere. In the present paper, we apply this model of zebras to the case of the June 6, 2000 flare, which was very rich in high-frequency zebra patterns. Here, we described briefly the observations and the magnetic field estimates.

OBSERVATIONS

The June 6, 2000 flare, classified as X2.3, was observed during 15:00-17:00 UT in the active region NOAA AR 9026. This flare was unusually rich in the high-frequency zebras. Several examples were recorded in the 1.2-1.7 GHz and 2.0-4.5 GHz frequency ranges, respectively, by the Brazilian Solar Spectroscope (BSS) (Sawant *et al.* 2001) and by the Ondřejov Observatory and four of them are shown in the Table 1. The frequency ratios of the neighboring zebra lines are below the value of 1.024 and there is a tendency to decrease these ratios towards lower frequencies. The number of observed zebra lines is in the range of 2-6 lines. In one case unique zebra patterns in a harmonic ratio of 1:2 were observed. While in the low-frequency band 3 zebra lines were recognized, on the double frequency only 2 zebra lines were recorded.

MAGNETIC FIELD ESTIMATES

In agreement with the model of Ledenev *et al.* (2001) we assume that the zebra pattern lines are generated at positions in the solar atmosphere where the following resonance condition is fulfilled: $\omega_{UH} = (\omega_{pe}^2 + \omega_{Be}^2)^{1/2} = s\,\omega_{Be}$, where ω_{UH}, ω_{pe}, and ω_{Be} are the upper hybrid, electron plasma and cyclotron frequencies, respectively and s is the integer harmonic number. The characteristic space scale of the electron density is assumed to

Table 1. Zebra patterns in the 1.0-4.5 GHz range.

#	Start (UT)	End (UT)	Frequency Range (MHz)	No. Zebra	Zebra line frequency	Ratio of succeeding frequencies	Magnetic field (G)
1	15:37:42	15:37:50	1590-1685	3	1620/1635/1675	1.009/1.024	150
2	15:42:51	15:42:52	1220-1265	3	1220/1240/1265	1.016/1.020	113
1	15:37:44	15:37:49	3300-3400	2	3290/3355	1.019	150
2	15:43:12	15:43:13	3500-4000	6	3639/3683/3761 3818/3897/3977	1.015/1.018/1.015 1.020/1.020	178

be much greater than that of the magnetic field. Then, the ratio of the neighboring zebra line frequencies is:

$$\omega_s/\omega_{s+1} = [\omega_{pes}/\omega_{pes+1}] \left[s^3(s+2)/(s+1)^3(s-1)\right]^{1/2}. \quad (1)$$

If we take roughly $\omega_{pes} \approx \omega_{pes+1}$ then $\omega_2/\omega_3 \approx 1.088$, $\omega_3/\omega_4 \approx 1.027$, $\omega_4/\omega_5 \approx 1.012$, $\omega_5/\omega_6 \approx 1.006$ and so on.

For magnetic field estimation the most important aspect is the determination of the s parameter for the zebra line with the highest frequency. This can be done by a comparison of the highest frequency ratio in the specific zebra pattern with the theoretical minimum values. In our case, the fourth harmonic ($s = 4$) is the most probable. Excepting the zebra patterns in harmonic ratio (1:2) we assume that the zebras are observed on the frequency of the upper hybrid waves (fundamental branch). Then the magnetic field strength from different zebra patterns can be determined as $B = f_{max}/(2.8s)$, where $s = 4$ in our case, B is in Gauss and f_{max} is the highest frequency of zebra lines in MHz. Thus we estimate the magnetic field to be in the range of 110-180 G.

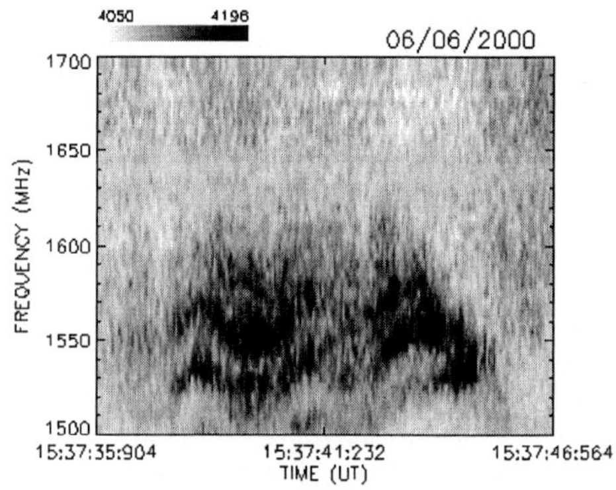

Fig. 1. *Dynamic spectrum of zebra recorded by the BSS, simultaneously with Ondřejov Observatory.*

ACKNOWLEDGEMENTS

M.K. thanks to FAPESP (01/00144-5) authorities for supporting his visit at INPE. This work was also supported by the grant S1003006, AS CR. F.R.C.F. thanks FAPESP for scholarship (99/10529-1).

REFERENCES

Ledenev, V.G., M. Karlický, Y. Yan, and Q. Fu, *Solar Phys.*, **202**, 71 (2001).

Sawant, H. S., K. R. Subramanian, C. Faria, F. C. R. Fernandes, J. H. A. Sobral et al., *Solar Phys.*, **200**, 167 (2001).

Slottje, C., *Proceeding of the CESRA-2 Meeting*, (Trieste), 88 (1971).

CORONAL MASS EJECTIONS AND INTERPLANETARY SCINTILLATION

Hari Om Vats[1], R. M. Jadhav[2], K. N. Iyer[2,3], and H. S. Sawant[3]

[1]*Physical Research Laboratory, Ahmedabad-380009, Bharat*
[2]*Saurashtra University, Rajkot-360005, Bharat*
[3]*Instituto Nactional de Pesquisas Espaciais, INPE, C.P. 515, 12201-970, Sao Jose dos Campos, SP, Brazil*

ABSTRACT

Here we report the Interplanetary Scintillation (IPS) observations of three events on April 2000, May 1999 and May 1997. The April 2000 observations are the interplanetary consequences of a halo Coronal Mass Ejection (CME). The May 1999 event is termed a solar wind disappearance event; it appears to be due the passage of a large void. The May 1997 event is that due to an Earth directed CME and the IPS observations of a radio source 3C48 show that the interplanetry disturbance due to this CME had density ~4 times more than the ambient, but moved slower than the ambient medium.

INTRODUCTION

An excellent set of observations of coronal mass ejections are now available from *Yohkoh* and *SOHO*. It has become possible to investigate the initial ejection and acceleration of these CMEs from these X-ray images. The monitoring of propagation of the coronal mass ejections away from the Sun in the heliosphere is possible by an indirect method of interplaneatry scintillation (IPS). In principle this phenomena is similar to the twinkling of stars and senses the presence and propagation of plasma irregularities crossing the line of sight to a compact radio source from the observing radio telescope. There are two radio telescopes operating at 327 and 103 MHz in Japan and Bharat, respectively, to observe this phenomena on regular basis. The passage of a coronal mass ejection through the heliosphere is usually termed an interplanetary disturbance (IPD). The statistical analysis of IPDs has shown that there are about three times more IPD events around the solar maximum phase than those around the solar minimum phase. Here we present the case studies of a few selected coronal mass ejections and their associated IPDs namely, April 2000, May 1999 and May 1997.

April 2000 event: One halo CME with a bright front (as seen by *SOHO*/LASCO) began on April 04, 2000 at about 1632 UT. This appeared to be associated with a C9 flare in AR 8933. With IPS observations at 103 MHz we detected the effect of this CME two days later at the line of sight of 3C459 and three days later at the line of sight of 3C2, 3C119 and 3C122. The observations of Jadhav *et al.* (2001) are shown in Figure 1. At the line of sight of 3C48 there appeared a very feeble effect of the passage of this CME. This could be due to the projection effect. The CME of April 4, 2000 produced a shock which was detected by *ACE* solar wind velocity detectors (the radial velocity increased from 375 km s^{-1} to 575 km s^{-1} at 16 UT on April 6, 2000). This shock led to a very large drop (~ 300 NT) in equatorial Dst and produced one of the largest geomagnetic storms on record.

May 1999 event: This event was popularly termed as "solar wind disappearance event". The in situ observations of several satellites showed that during this event the bow shock of the geomagnetic field was highly extended and had crossed 60 RE. In fact the IPS observations of Vats *et al.* (2001a) of two radio sources; (one sensitive to the IPM close to the Earth and the other one sensitive to the IPM away from the Earth) clearly showed that this was due to the passage of a large void, 150 R_E wide by 4000 R_E long..

May 1997 event

The IPS observations at 103 and 327 MHz showed an Earth-directed coronal mass ejection (CME), which occurred, near the center of the solar disk at 0435 UT on May 12, 1997. This event was associated with a two-ribbon flare. The ionospheric effects of soft X rays from this solar flare were observed by a digital ionosonde at Ahmedabad in the form of excess ionization (~1200 el cm^{-3}) in the *D* region of the ionosphere. The associated IPD was found to have plasma density ~ 4 times more than that of the ambient plasma at a distance of ~ 0.5 AU from the Sun. The most peculiar aspect of this CME, (Vats *et al.* 2001b), is that it appears that the disturbance moved slightly slower than the ambient medium (Figure 2). The *Solar and Heliospheric Observatory (SOHO)* and interplanetary scintillation (IPS) estimates of solar wind are quite different; it appears that the difference could be due to the projection effect of the *SOHO* image. Though the disturbance was not very severe, its impact on Earth's environment produced a geomagnetic storm.

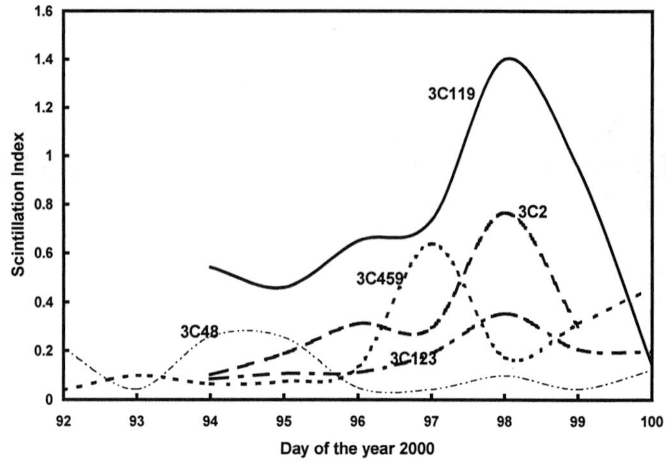

Fig. 1. Temporal variation of scintillation index at 103 Mhz.

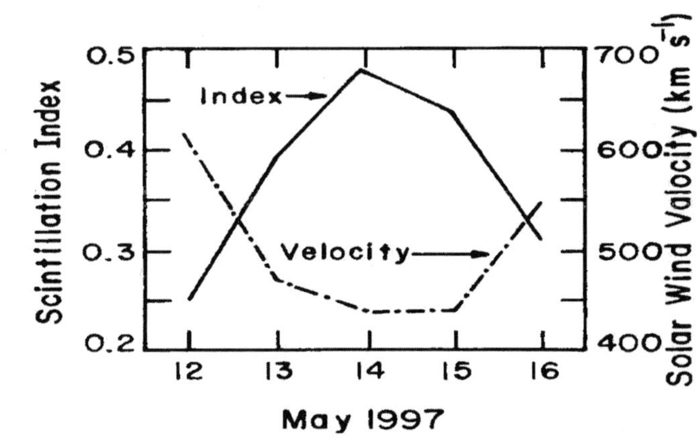

Fig. 2. Variation of scintillation index at 103 MHz and solar wind velocity at 327 MHz for the radio source 3C48

These events are a few of the many CME events that occur. Their interaction with the terrestrial environment produces several effects that are very important for space weather and for understanding the solar-terrestrial relationship.

REFERENCES

Jadhav, R.M. H.O. Vats, & K.N. Iyer, Interplanetary disturbance of 4 April 2000 and associated geomagnetic effects, in *Plasma 2001*, SA-07, CPP Guwahati, Bharat, December 17-20 (2001).

Vats Hari Om, H.S. Sawant, Rupal Oza, K.N. Iyer and Ravi Jadhav, Interplanetary scintillation observations for the solar wind disappearance event of May 1999, *J. Geophys. Res. (Space Physics)*, **106**, 25121 (2001a).

Vats Hari Om, Som Sharma, R. Oza, K.N. Iyer, H. Chandra, H.S. Sawant and M.R. Deshpande, Interplanetary and terrestrial observations of an earth directed coronal mass ejection, *Radio Science*, **36**, 1769 (2001b).

CORONAL MASS EJECTIONS: RELATIONSHIP WITH SOLAR FLARES AND CORONAL HOLES

V. K. Verma

State Observatory, Naini Tal-263129, India

ABSTRACT

The Coronal Mass Ejections (CMEs) observed by the LASCO coronagraph and associated solar activity phenomena whose locations were identified by EIT and solar $H\alpha$ flare observations during year 2000 indicate that about 40%, 26% and 30% of CMEs were observed when there were coronal holes (CHs) within 1-10, 11-20 and 21-40 degrees, respectively from the location of solar $H\alpha$ flares. We are of the view, as also earlier suggested by Verma & Pande (1989), that CMEs might have been produced by the same mechanism by which the mass ejected by the same solar flares or active prominences connects with open magnetic lines of CHs (source of high speed solar wind streams) and moves along them to appear as CMEs.

OBSERVATIONAL DATA, ANALYSIS AND RESULTS

To study the relationship of CMEs with solar active phenomena and CHs for year 2000, we have taken data for CMEs, EIT, $H\alpha$ flares and CHs from websites of the LASCO , EIT , Solar Geophysical Data and KPNO. Once we know the locations of CME-related solar flares, we calculate their locations in heliographic longitude and latitude and plot them on a synoptic coronal map of CHs.

For example, in Figure 1 we have shown locations of three solar flares which were associated with CMEs. The first flare event observed on December 12, 2000 at 14:48 UT is associated with a CME that occurred in AR 9267, It has a heliographic location N08 081 and the distance between nearest coronal hole (NCH) boundary and solar flare was in the range of 1-10 degrees. The second flare event observed on December 14, 2000 at 09:48 UT is associated with a CME that occurred in AR 9264. It has a heliographic location S24 060 and the distance between the NCH boundary and the solar flare was in the range of 21-30 degrees. The third flare event, observed on December 18, 2000 at 11:12 UT, is associated with CMEs that occurred in AR 9269. It has a heliographic location N15 027 and the distance between the NCH boundary and solar flares was in the range of 41-50 degrees.

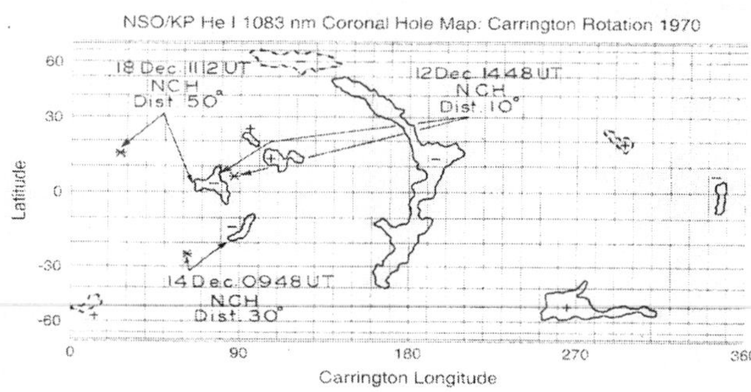

Fig. 1. Synoptic map showing locations of three flares that were associated with CMEs and coronal holes.

In this way, we found that during year 2000, a total number of 196 CMEs were simultaneously observed by LASCO and EIT instruments or by the solar $H\alpha$ flare instruments of various ground observatories . Out of 196 CMEs 79(40.5%), 51(26%), 47(24%), 12(06%) and 07(03.5%) CMEs were observed when there were coronal holes in the range of 1-10, 11-20, 21-30, 31-40 and 41-70 degrees, respectively, from the location of solar $H\alpha$ flares. The plot of the minimum distance between the location of flares which were associated with CMEs and the boundary of a coronal hole, versus number of CME events, is shown in Figure 2.

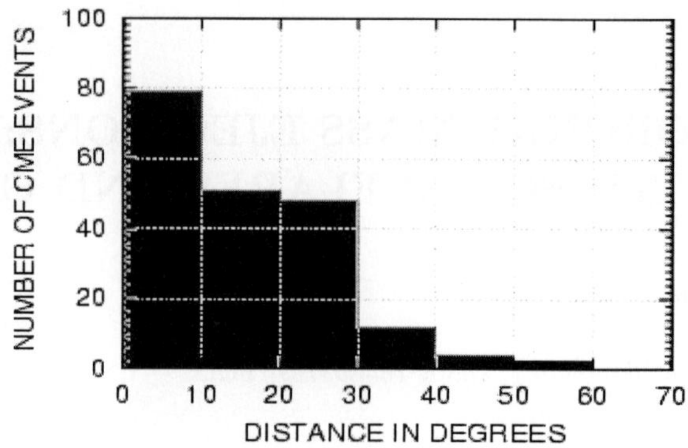

Fig. 2. Shows number of CMEs versus minimum distance between location of solar phenomena and boundary of CHs.

DISCUSSION AND CONCLUSIONS

In the preceding section we identified the nearest coronal hole to each CME observed by LASCO that had solar flares seen by EIT. The CHs are said to be sources of high speed solar wind (HSSW). It is well known that the separation of foot points of X-ray loops ranges up to 250,000-500,000 km or 33-66 degree of sun (1 arcsec=750 km) and separation between $H\alpha$ footpoints of loops ranges up to 35,000-50,000 km or 4.6-6.7 degree, and most or all the loops are anchored to sunspots (Bray *et al.* 1991). Therefore reconnection between magnetic structures with separations in the range of 01-70 degrees to produce CMEs and flares is possible. Magnetic reconnection scenarios are detailed by McKenzie (2002) in these proceedings. We are of the view that CMEs may originate through the combined action of solar activity phenomena and coronal holes, as earlier suggested by Verma & Pande (1989). Such a possible scenario of a CME originating in the presence of a CH is shown in Figure 3.

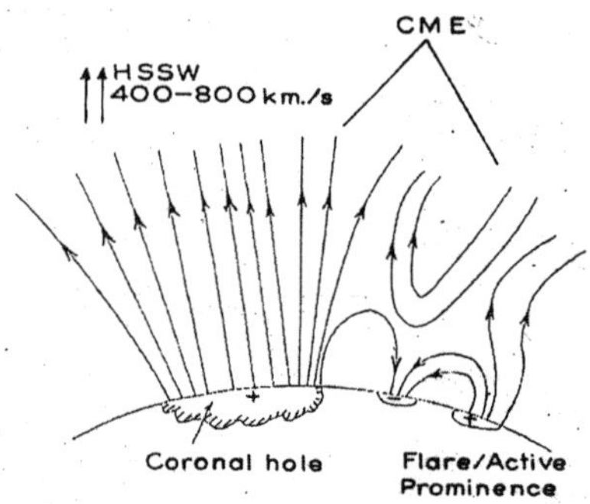

Fig. 3. Shows possible scenario of CME originating in the presence of Coronal Holes.

ACKNOWLEDGEMENTS

The author is thankful to the organizers of the *Yohkoh* 10th Anniversary Meeting, Hawaii, USA, for providing financial assistance for travel and local support, and to the anonymous referee for useful comments. The author also thanks the scientists of LASCO, EIT and KPNO whose data is used here.

REFERENCES

Bray, R. J., *et al.*, *Plasma Loops in the Solar Corona,* Cambridge Univ. Press, Cambridge, p.29, 156, 246 (1991).

McKenzie, D. E. these proceedings (2002).

Verma, V. K. & Pande, M. C, *Proc. IAU Colloq. 104 " Solar and Stellar Flares"* Poster Paper (Eds. B. M. Haisch and M. Rodono), Stanford University, Stanford, USA, p. 239 (1989).

Section XI.

Solar Cycle Studies

COMPARATIVE ANALYSIS OF SOLAR NEUTRINO DATA AND SXT X-RAY DATA

P. A. Sturrock and M. A. Weber

Center for Space Science and Astrophysics, Stanford University, Stanford, CA 94305

ABSTRACT

We compare power spectra derived from the Homestake and GALLEX-GNO solar neutrino experiments with power spectra derived from equatorial SXT data and from high-latitude SXT data, and find a remarkable agreement. In the range 10 – 16 y^{-1}, the principal peak in the Homestake spectrum is at 12.88 y^{-1}, whereas the principal peak in the high-latitude SXT spectrum is at 12.86 y^{-1}; the principal peak in the GALLEX-GNO spectrum is at 13.59 y^{-1}, whereas the principal peak in the equatorial SXT spectrum is at 13.55 y^{-1}. We estimate the correspondence to be significant at the 0.04% level. We speculate on the implications of this result.

INTRODUCTION

When solar neutrino data were first obtained by the Homestake experiment, there was great interest in the possibility that the flux may be variable. Sakurai (1981) suggested that the flux varies with a 2 y period. There were also claims that the flux varies with the solar cycle - usually based on claimed correlations between the neutrino flux and some solar activity index such as the sunspot number (see, for instance, Massetti & Storini 1996). Walther (1997) has reviewed these claims and found reasons to discount them. Although there is no convincing evidence that the solar neutrino flux varies with the solar cycle, evidence is now accumulating that the flux varies on a much shorter time scale. Power spectrum analysis of the Homestake data (Cleveland *et al.* 1998) yields a significant peak at 12.88 y^{-1} (period 28.4 days, corresponding to rotation near the tachocline; Sturrock, Walther, & Wheatland 1997). Similar analysis of the GALLEX-GNO data (Hampel *et al.* 1997; Altmann *et al.* 2000) yields a significant periodicity at 13.59 y^{-1} (26.9 days, corresponding to the deep convection zone; Sturrock & Weber 2002). We find additional evidence for variability in the fact that the variance of the Homestake data is larger than expected (Sturrock, Walther, & Wheatland 1997), and the histogram of GALLEX-GNO data is bimodal (Sturrock & Scargle 2001). We here compare these results with the power spectrum of SXT data.

POWER SPECTRUM ANALYSIS

We have analyzed Homestake and GALLEX-GNO data by the method developed by Lomb (1976) and Scargle (1982) for the analysis of irregular time series. The spectrum obtained from Homestake data, over the frequency range 10-16 y^{-1}, is shown as Figure 1a; the biggest peak occurs at 12.88 +/-0.02 y^{-1} (period 28.36 days). The spectrum obtained from GALLEX-GNO data is shown in Figure 1b; the biggest peak occurs at 13.59 +/-0.03 y^{-1} (period 26.88 days). We have carried out a similar spectrum analysis of SXT X-ray data compiled over the six years 1992 through 1997, divided into nine latitude ranges centered on 60N, 45N, 30N, 15N, Equator, 15S, 30S, 45S and 60S (Weber & Sturrock 2002). We formed the logarithm of the flux measurements, in order to de-emphasize singular events such as flares, and de-trended the data. Figure 2a shows the mean of the spectra formed from 60N and 60S data; the principal peak is at 12.86 +/- 0.02 y^{-1}. Figure 2b shows the spectrum of the SXT equatorial flux; the principal peak occurs at 13.55 +/- 0.02 y^{-1}.

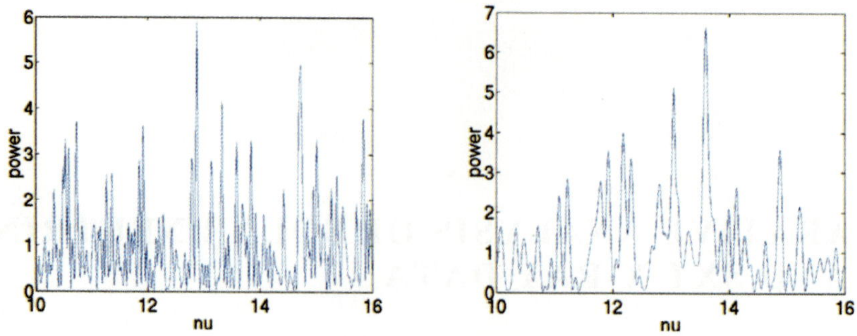

Fig. 1. Power spectra for Homestake (left) and GALLEX-GNO (right).

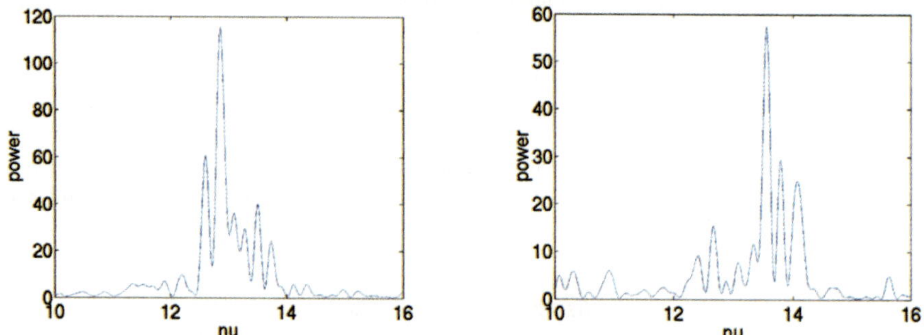

Fig. 2. Power spectra for SXT Latitudes N60 and S60 (left) and Equator (right).

To simplify the comparison of the spectra, we show in Figure 3 the corresponding probability distribution functions (pdf's; see Bretthorst 1988). Each pdf of the power spectrum $P(\nu \mid sp)$ is related to the power S by

$$P(\nu \mid sp) \propto \exp(S). \tag{1}$$

There is excellent agreement between the principal Homestake peak and the equatorial SXT peak, and between the principal GALLEX-GNO peak and the high-latitude SXT peak.

DISCUSSION

If B is the width of the search band (6 y^{-1}) in Figures 1 and 2, $\delta_H \nu$ the separation (0.02 y^{-1}) between the Homestake peak and high-latitude SXT peak, and $\delta_G \nu$ the separation (0.04 y^{-1}) between the GALLEX-GNO peak and equatorial SXT peak, the probability of obtaining both correspondences by chance is the product of the probability of finding the Homestake peak within $\delta_H \nu$ of one of the two SXT peaks times the probability of finding the GALLEX-GNO peak within $\delta_G \nu$ of one of the two SXT peaks.
This is given by

$$p = \left(\frac{4\delta_H \nu}{B} \right) \left(\frac{4\delta_G \nu}{B} \right). \tag{2}$$

We find that $p = 0.0004$, so that the correspondence is significant at the 0.04% level.

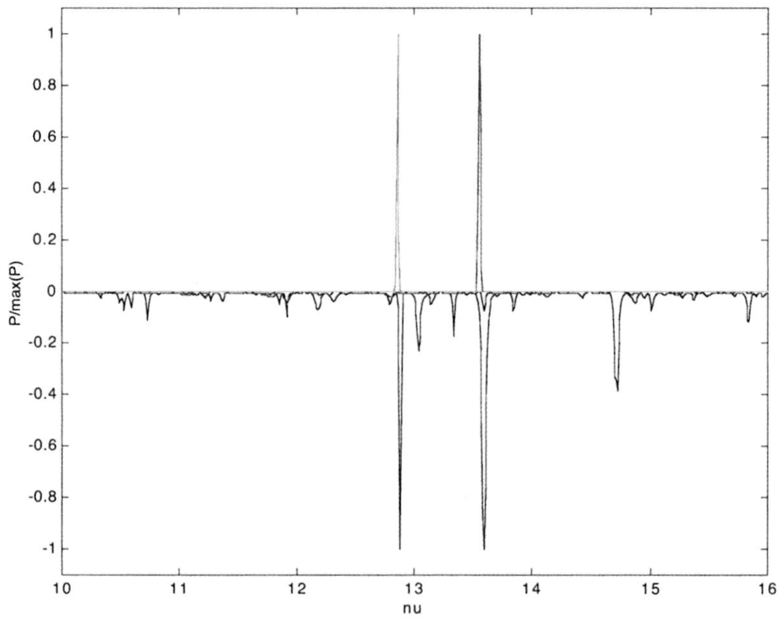

Fig. 3. Comparison of normalized probability distribution functions formed from spectra of data from SXT Equator (red), SXT N60-S60 (green), Homestake (black) and GALLEX-GNO (blue).

The most likely interpretation of the variation of the solar neutrino flux is that neutrinos have nonzero mass and nonzero magnetic moment, so that they are subject to Resonant Spin-Flavor Precession, whereby neutrinos change both flavor and spin as they travel through matter permeated by a magnetic field (Akhmedov 1997). According to this scenario, the neutrino flux is probably influenced by two distinct magnetic structures within the Sun: one located where the rotation rate is 13.57 +/- 0.04 y^{-1}, influencing primarily the equatorial SXT flux and the GALLEX-GNO neutrino flux; and the other located where the rotation rate is 12.87 +/- 0.02 y^{-1}, influencing primarily the high latitude X-ray flux and the Homestake neutrino flux. In order to locate these two regions, we have examined the "rotational modulation index" (Sturrock & Weber 2002),

$$\Xi(r,\lambda) = \int_{\nu_l}^{\nu_u} d\nu P(\nu \mid sp) P(\nu \mid r,\lambda), \tag{3}$$

where $P(\nu \mid sp)$ is the pdf of the power spectrum, and $P(\nu \mid r,\lambda)$ is the pdf of the rotation rate at each radius and latitude, as determined by the MDI instrument on the *SOHO* spacecraft (Schou *et al.* 1998). We map this statistic on a meridional section of the Sun. Since the SXT spectra are simpler than those derived from neutrino data, we adopt the former for display purposes. The rotational-modulation map of the equatorial SXT data (and GALLEX-GNO data) is shown in Figure 4. That of the high-latitude SXT data (and Homestake data) is shown in Figure 5.

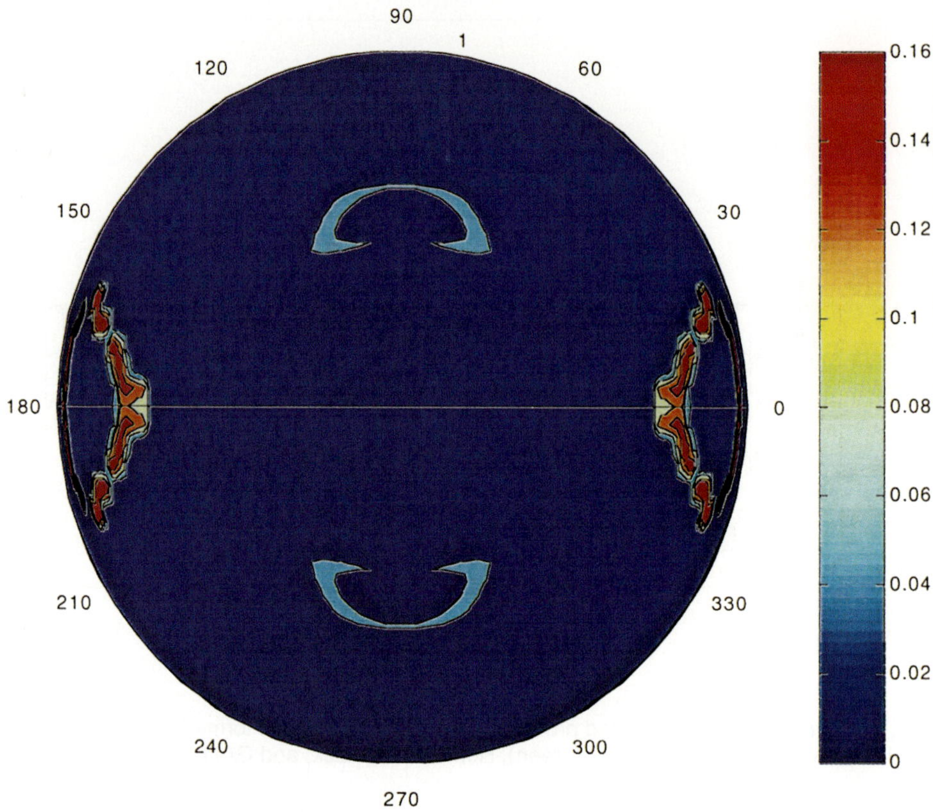

Fig. 4. Rotational modulation statistic formed from the spectrum of the equatorial SXT flux.

Since the neutrino flux is produced in the core of the Sun, modulation must occur at or close to the equator. Hence we see that the region responsible for modulation of the GALLEX-GNO flux and of the equatorial SXT flux appears to be located in the convection zone at normalized radius 0.8. The region responsible for modulation of the Homestake neutrino flux and the high-latitude X-ray flux is probably located near the tachocline, but possibly in the radiative zone at normalized radius 0.6.

Since the above spectra were formed from many-year data sets, it appears that the structures responsible for the above rotational modulation have lifetimes of at least several years and perhaps many years. The above results are, therefore, a challenge to the conventional model of the solar dynamo, in which the magnetic structure changes completely every eleven years.

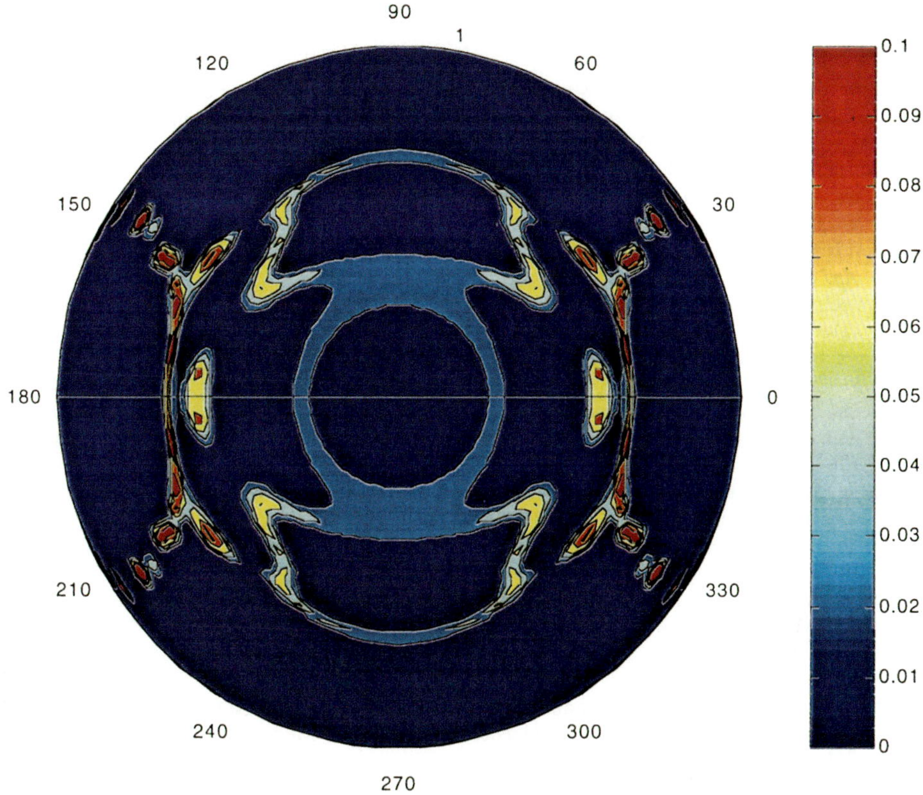

Fig. 5. Rotational modulation statistic formed from the spectrum of the N60 and S60 SXT flux.

ACKNOWLEDGEMENTS

This article is based on work supported in part by NASA grant NAS 8-37334 and NSF grant AST-0097128.

REFERENCES

Akhmedov, E.K., The Neutrino Magnetic Moment and Time Variations of the Solar Neutrino Flux, in *Fourth International Solar Neutrino Conference*, ed. W. Hampel, Max-Planck-Institut fur Kernphysik Heidelberg, Germany, p. 388 (1997).

Altmann, M., *et al.* GNO Solar Neutrino Observations: Results for GNO I, *Physics Letters B*, **490**, 16 (2000).

Bretthorst, G.L. *Bayesian Spectrum Analysis and Parameter Estimation*, Vol. 48 of Lecture Notes in Statistics, ed. J. Berger *et al.*, Springer-Verlag, Berlin, Germany (1988).

Cleveland, B.T., *et al.*, Measurement of the Solar Electron Neutrino Flux with the Homestake Chlorine Detector, *ApJ* **496**, 505 (1998).

Hampel, W., *et al.*, GALLEX Solar Neutrino Observations: Results for GALLEX IV, *Phys. Lett. B*, **447**, 127 (1999).

Lomb, N.R., Least-Squares Frequency Analysis of Unequally Spaced Data, *Astrophys. Space Sci.*, **39**, 447 (1976).

Massetti, S., & M. Storini, Spacetime Modulation of Solar Neutrino Flux: 1970 – 1992, *ApJ*, **472**, 827 (1996).

Sakurai, K., Quasi-Biennial Periodicity in the Solar Neutrino Flux and its Relation to the Solar Structure, *Solar Phys.*, **74**, 35 (1981).

Scargle, J.D., Studies in Astronomical Time Series Analysis. II. Statistical Aspects of Spectral Analysis of Unevenly Spaced Data, *ApJ*, **263**, 835 (1982).

Schou, J., *et al.*, Helioseismic Studies of Differential Rotation in the Solar Envelope by the Solar Oscillations Investigation using the Michelson Doppler Interferometer, *ApJ*, **505**, 390 (1998).

Sturrock P.A., & J.N. Scargle, Histogram Analysis of GALLEX, GNO and SAGE Neutrino Data: Further Evidence for Variability of the Solar Neutrino Flux, *ApJ (Letters)*, **550**, L101 (2001).

Sturrock, P.A., G. Walther, & M.S. Wheatland, Search for Periodicities in the Homestake Solar Neutrino Data, *ApJ*, **491**, 409 (1997).

Sturrock, P.A., & M.A. Weber, Comparative Analysis of GALLEX-GNO Solar Neutrino Data and SOHO/MDI Helioseismology Data: Further Evidence for Rotational Modulation of the Solar Neutrino Flux, *Ap.J.*, **565**,1366 (2002).

Walther, G., Absence of Correlation between the Solar Neutrino Flux and the Sunspot Number, *Phys. Rev. Lett.*, **79**, 4522 (1997).

Weber, M.A., & P.A. Sturrock, Differential Rotation of the Soft X-Ray Corona over a Solar Cycle, these proceedings (2002).

CORONAL PATTERNS OF ACTIVITY FROM YOHKOH AND SOHO/EIT DATA

E. E. Benevolenskaya[1,2], A. G. Kosovichev[2], P. H. Scherrer[2], J. R. Lemen[3], and G. L. Slater[3]

[1]*Pulkovo Astronomical Observatory, St. Petersburg, Russia*
[2]*W.W.Hansen Experimental Physics Laboratory, Stanford University, Stanford, USA*
[3]*Lockheed Martin Solar and Astrophysics Laboratory, Palo Alto, USA*

ABSTRACT

We have studied the evolution of large-scale coronal structures using soft X-ray data from *Yohkoh* and EUV data from *SOHO*/EIT during the rising phase of the current solar cycle 23, and compared with the evolution of the photospheric magnetic field. The poleward-migrating high-latitude coronal structures ("coronal activity waves") that received significant attention in the past and are represented by the bright polar footpoints of the giant loops in the EUV/EIT data, are visible as the whole loop structures on the solar disk in the soft X-ray *Yohkoh* data. The *Yohkoh* data show that these large-scale magnetic connections appear mostly during the rising phase of the solar cycle and its maximum. These connections did not appear or were very weak during the declining phase of the solar cycle. The high-latitude coronal activity waves in the SXT data are closely related to the low-latitude magnetic flux and sunspot numbers and display intriguing quasiperiodic 'impulses' with the characteristic period of 1-1.5 years. The relation between the soft X-ray flux, which is considered to be a proxy of the coronal energy flux, and the magnetic flux, can be approximated by a power law with an average index close to 2, and varies with solar cycle.

INTRODUCTION

Coronal structures play a very important role for understanding the evolution of the solar magnetic field during the solar cycle. They have a long history of investigation (Secchi 1877, d'Azambuja 1945, Waldmeier 1957, Leroy & Trellis 1974, Rŭsin *et al.* 1990, Altrock 1997). These studies reveal equatorward and poleward migrating structures in coronal emissions, that sometimes are called "coronal activity waves". These observations led to a suggestion that the polar branch of activity might be associated with an additional dynamo wave in the convection zone (Makarov *et al.* 1987, Belvedere *et al.* 1991). It has also been found that both low- and high-latitude waves of solar activity may last longer than the 11-year sunspot cycle, and be associated with the so-called "extended solar cycle" (e.g. Wilson *et al.* 1988). We have recently identified the coronal activity waves in the extreme ultraviolet data from *SOHO*/EIT (Benevolenskaya *et al.* 2001). It is found that the bright coronal structures, which migrate to the poles during the rising phase of the solar cycle, are formed by density enhancements in the poleward footpoints of magnetic field lines. These magnetic lines connecting the magnetic fields of the following parts of active regions with the polar field are represented by giant polar coronal loops. In other words, these giant magnetic loops connect the toroidal field of the new solar cycle with the polar poloidal field formed during the previous cycle.

In this paper, for further investigation of large-scale coronal activity, we use synoptic observations of the solar corona during 1991-2001 in soft X-rays from the SXT instrument on Yohkoh (Tsuneta *et al.* 1991).

This period covers the declining phase of solar cycle 22 and the rising phase of the current cycle, 23.

SYNOPTIC STRUCTURE OF THE CORONA

The image sets (from which the SXT synoptic maps were derived) consist of full sun images taken in two different filters of the Yohkoh Soft X-Ray Telescope: a thin Al filter with an approximate pass band of $\approx 6-13$ Å (Al), and a composite Al/Mg/Mn filter with a slightly shorter passband of $\approx 5-12$ Å (AlMg) from November 11, 1991 to March 13, 2001. The full SXT image archive contains images taken at different resolutions (2.45 arcsec, 4.91 arcsec, and 9.8 arcsec). Full disk images were chosen at a 6-hour cadence at the highest available resolution available at each sample time. Most of the images are 4.91 arcsec resolution images. Each raw image was processed in the following way: the instrumental background noise was subtracted. Additionally, contamination from white light leakage onto the CCD was subtracted using a sophisticated model of the instrumental white light leakage pattern.

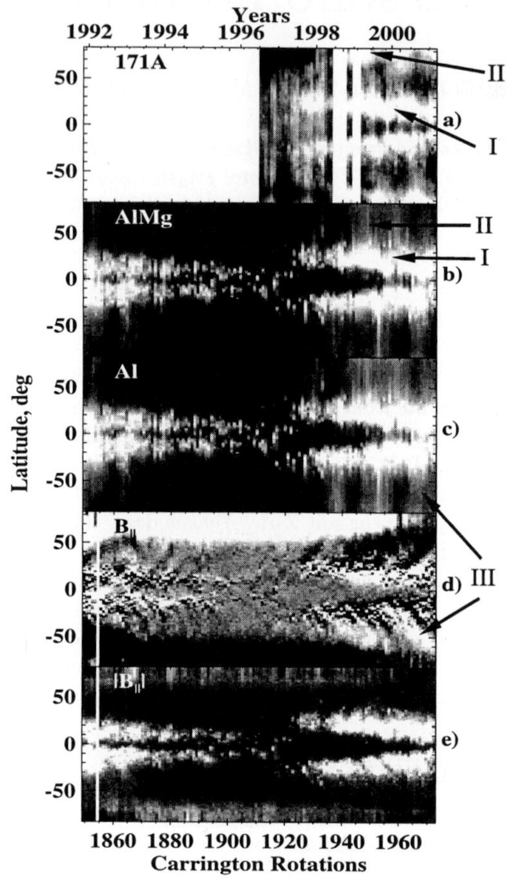

Fig. 1. Axisymmetrical distributions as a function of latitude (from $-83°$ to $83°$) and time from 11 November 1991 to 13 March 2001 (Carrington rotations from 1849 to 1973) for: a) EUV flux in Fe IX,X lines (available only since CR1911, 28 June 1996); b) X-ray flux in the AlMg filter; c) X-ray flux in the Al filter; d) $B_{||}$-component of the magnetic field, the grayscale range is [-1G 1G]; e) unsigned magnetic flux $|B_{||}|$, in [0 10G] range.

The Carrington synoptic maps of the corona are represented by values of the intensity centered on the central meridian. The resolution of these maps is 1° in both longitude from 0° to 360° and latitude from $-83°$ to $83°$. To obtain the latitudinal distribution we averaged the synoptic maps over longitude, and plotted as a function of latitude and time. Examples of the latitudinal distribution of the EUV and soft X-ray intensity are shown in Figure 1a-c. For comparison, in Figure 1d-e we plotted the azimuthally averaged distributions of the line-of-sight magnetic flux, $B_{||}$, and the corresponding unsigned (absolute) magnetic flux, $|B_{||}|$, obtained from the Kitt Peak Observatory synoptic magnetic maps. The zonal magnetic neutral lines separating magnetic polarities at high latitudes are seen as the contrast lines between white and black colors in Figure 1d.

The neutral lines reflect the position of $B_{||}$=0 for magnetic field averaged over longitude and separating the polar field formed during the previous solar cycle from the field emerged during the current cycle. The $|B_{||}|$ map, which shows the location of active regions, reveals the "butterfly" diagram: the sunspot zones of the current solar cycle start at about 30° latitude in mid 1997, and then gradually migrate towards the equator as the cycle progresses. The coronal EUV map (Figure 1a) shows in each hemisphere two sets of migrating structures: low-latitude structures (marked I) that migrate toward the equator following the evolution of $|B_{||}|$ (low-latitude coronal activity waves) and high-latitude structures (II), or high-latitude waves, that migrate toward the poles parallel to the magnetic neutral lines. In the SXT coronal maps (Figures 1b and 1c), the low-latitude migrating structures are similar to those in the EUV map.

However, the high-latitude structures look different, without pronounced brightening in the polar regions, and more uniform latitudinally, connecting the low-latitude bands with the polar regions. The high-latitude structures appear mostly during the rising phase of the solar cycle. The axisymmetrical distributions of the coronal intensity shown in Figure 1 also reveal that the high-latitude coronal structures (visible as bright connections between the low-latitude coronal structures and polar regions) tend to appear quasi-periodically with $\sim$ 1-1.5-year period, correlating with "impulses" of magnetic flux. An example of such impulse in the soft X-ray and magnetic fluxes indicated by arrows and symbol 'III' in Figure 1c-d.

The poleward-migrating high-latitude coronal structures ("coronal activity waves") that received significant attention in the past and are represented by the bright polar footpoints of the giant loops in the EUV/EIT data, are visible as the whole loop structures on the solar disk in the soft X-ray *Yohkoh*/SXT data. Therefore, the temperature of these loops is in the range of 2-3$\times 10^6$ K. The giant magnetic loops connect the magnetic flux of the following parts of the active regions with the magnetic flux of the polar regions which have the opposite polarity.

RELATIONSHIP BETWEEN THE CORONAL X-RAY INTENSITY AND MAGNETIC FLUX

Using the axisymmetrical distributions presented in Figure 1 we have investigated the relationship between the soft X-ray emission and the magnetic flux. We find a good correlation between the intensity of the soft X-ray emission and the magnetic flux for latitudes less than 55° (correlation coefficient $r \simeq 0.9$).

It is emphasized by Sturrock (1999) that the relationship between the energy flux into the corona (F) and the magnetic flux at the photospheric level is most likely represented by the power law $F = k\langle|B|\rangle^n$. Soft X-ray measurements are used as a proxy for energy flux.

We have calculated the power-law index for 12 individual subintervals of 10 rotations, using the whole magnetic flux range (Figure 2). The straight lines show the power-law fits.

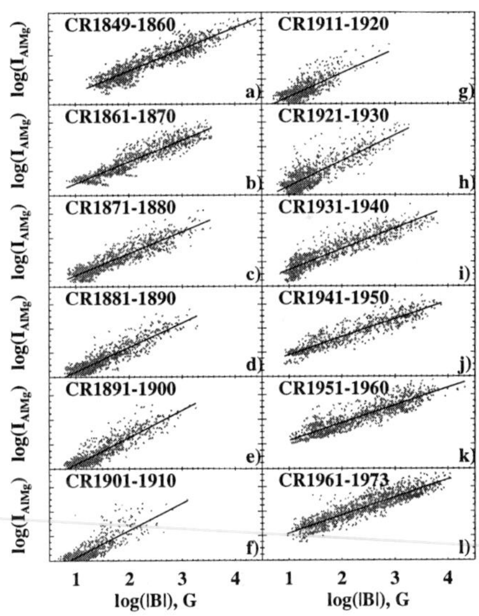

Fig. 2. Scatter plots of the soft X-ray intensity from SXT (*Yohkoh*) data in the AlMg filter as a function of the natural logarithm of magnetic flux (from Kitt Peak Observatory) for 12 periods from November 11, 1991 to March 13, 2001.

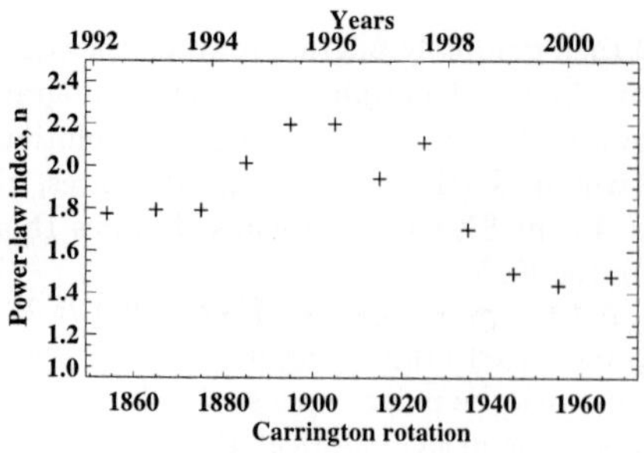

Fig. 3. Power-law index, n, of the $I_{\rm AlMg} \sim \left\langle |B_{\|}| \right\rangle^n$ dependence as a function of Carrington rotation number and time.

The results shown in Figure 3 indicate that the power index varies with the solar cycle between 1.5 in the solar maximum and 2.2 in the minimum. This suggests that the soft X-ray flux depends not only on the amount of the photospheric magnetic flux but also on some magnetic field structural properties that vary with the solar cycle. Our estimates of the power-law index are generally consistent with the previous results by Golub *et al.* (1980), 1.5-1.9, and Wolfson *et al.* (2000), 1.86-1.88, but significantly higher than 1.19 estimated by Fisher *et al.* (1998).

Wolfson *et al.* have found that at 45° and higher, the coronal emissions correlate with the lower-latitude (30°) magnetic flux. Our results agree with this conclusion and provide an explanation. We can conclude that the correlation between the high-latitude coronal emission and the low-latitude magnetic flux is related to the giant coronal loops connecting the following parts of the active regions with the polar regions.

The giant loops are most likely formed by magnetic reconnections in the corona and provide an important topological connection between the poloidal magnetic field formed during the previous solar cycle in the polar regions with the toroidal low-latitude field of the new cycle. In the axisymmetrical distributions these loops are visible as diffuse structures extended from middle-latitude to high-latitude and represent impulses of coronal activity. These structures are similar to 'ghosts' noticed on stacks of coronal maps by Pevtsov & Acton (2001), which we can also explain as caused by the giant coronal loops. The *Yohkoh* data show that these large-scale magnetic connections appear mostly during the rising phase of the solar cycle and its maximum. The high-latitude coronal structures in the SXT data are closely related to the low-latitude magnetic flux and display intriguing quasiperiodic 'impulses' with the characteristic period of 1-1.5 years.

This work was partly supported by JURRISS Program NASA NRA 98-OSS-08 and the SOI/*SOHO* NASA contract NAG5-8878 to Stanford University, and by the Russian Federal Program 'Astronomy', Grant 1.5.3.4.

REFERENCES

Altrock, R. C. 1997, *Sol. Phys.*, **170**, 411

Belvedere, G., Lanzafame, G. & Proctor, M. R. E. 1991, *Nature*, **350**, 481

Benevolenskaya, E. E., Kosovichev, A. G., Scherrer, P. H., 2001, *ApJ*, **554**, L107

D'Azambuja, L. 1945, *ApJ*, **101**, 260

Fisher, G. H., Longcope, D. W., Metcalf, T. R., & Pevtsov, A. A. 1998, *ApJ*, **508**, 885

Golub, L., Maxson, C., Rosner, R., Vaiana, G. S. & Serio, S. 1980, *ApJ*, **238**, 343

Leroy, J. L., Trellis, M. 1974, *A&A*, **35**, 283

Makarov, V. I., Ruzmaikin, A. A., & Starchenko, S. V. 1987, *Sol. Phys.*, **111**, 267

Pevtsov, A. A., & Acton, L. W. 2001, *ApJ*, **554**, 416

Rŭsin, V., Rýbansky, M., & Minarovjech, M. 1998, in: *Synoptic Solar Physics*, eds. K. S. Balasubramaniam, J. W. Harvey, & D. M. Rabin, ASP Conf. Ser., v. 140, 353

Secchi, P.A. 1877, *Le Soleil*, **v. 2**, Gauthier-Villars, Paris

Sturrock, P.A. 1999, *ApJ*, **521**, 451

Tsuneta, S. 1991, *Solar Physics*, **136**, 37

Waldmeier, M. 1957, *Die Sonnenkorona*, **Vol. II**, Birkäuser, Basel

Wilson, P. R., Altrock, R. C., Harvey, K. L., Martin, S. F., Snodgrass, H. B., 1988, *Nature*, **333**, 748

Wolfson, R., Roald, C. B., Sturrock, P. A., & Weber, M. A. 2000, *ApJ*, **539**, 995

LARGE-SCALE AND LONG-LIVED CORONAL STRUCTURES DETECTED IN LIMB SYNOPTIC MAPS

J. Li[1], B. LaBonte[1], L. Acton[2], and G. Slater[3]

[1]*Institute for Astronomy, University of Hawaii, 2680 Woodlawn Drive, Honolulu HI 96822, USA*
[2]*Montana State University, P.O. Box 173840, Bozeman, MT 59717, USA*
[3]*Lockheed-Martin Advanced Technology Center, Bldg. 252 Org. L9-41, 3251 Hanover Street, Palo Alto, CA 94304*

ABSTRACT

We summarize our recent studies of large-scale, long-lived coronal structures detected in limb synoptic maps using *Yohkoh*/SXT data. These structures appear as "polar sinusoids" on full limb synoptic maps made from data. A complete limb synoptic map is displayed, which covers from October 1991 to August 2001 (almost the entire *Yohkoh* mission). Several properties of the coronal polar sinusoids have been revealed. Comparison with optical data shows that these vast coronal structures are associated with underlying, non-contemporaneous sunspot clusters. Individual spots in the clusters are short-lived, but contribute magnetic flux collectively to sustain the long-lived coronal features. This implies the existence of a subphotospheric magnetic region from which the sunspots erupt.

INTRODUCTION

We are studying the long-term variation of the solar corona using full limb synoptic maps. The advantages of these maps over the traditional Carrington synoptic maps are 1) coronal structures are displayed continuously over long time intervals instead of being segmented by a single Carrington rotation; and 2) coronal structures around the full limb are displayed simultaneously including entire coronal holes. In addition, limb synoptic maps allow us to condense a vast amount of observational data (e.g. 30 Gbytes of *Yohkoh* data) into compact and easily studied maps. These advantages allow us to simultaneously study the latitude distribution of the corona as well as temporal variations on timescales comparable to the solar cycle.

Some of the most prominent features on full limb synoptic maps are sinusoid-like structures that cross both polar holes and connect to more equatorial active regions (Li, Jewitt, & LaBonte 2000). Close inspection shows that these polar sinusoids are formed by projection when coronal gas above the active regions is very extended in both latitude and altitude. Most recently, we found that these long-lived, large-scale coronal streamers are sustained by non-contemporaneous sunspot clusters which emerge from a narrowly-defined area on the sun. "While individual sunspots are short-lived, the clusters of sunspots provide magnetic field to reinforce the large-scale, long-lived coronal streamers." (Li *et al.* 2002). The implication is that a large, sub-surface magnetic region lives for many solar rotations, from which sunspots repeatedly erupt.

In this paper, we will demonstrate the technique for making full limb synoptic maps and give a brief summary of recent results based on these maps. Readers are referred to the recent papers on this work for details (Li, Jewitt, & LaBonte 2000, Li *et al.* 2002).

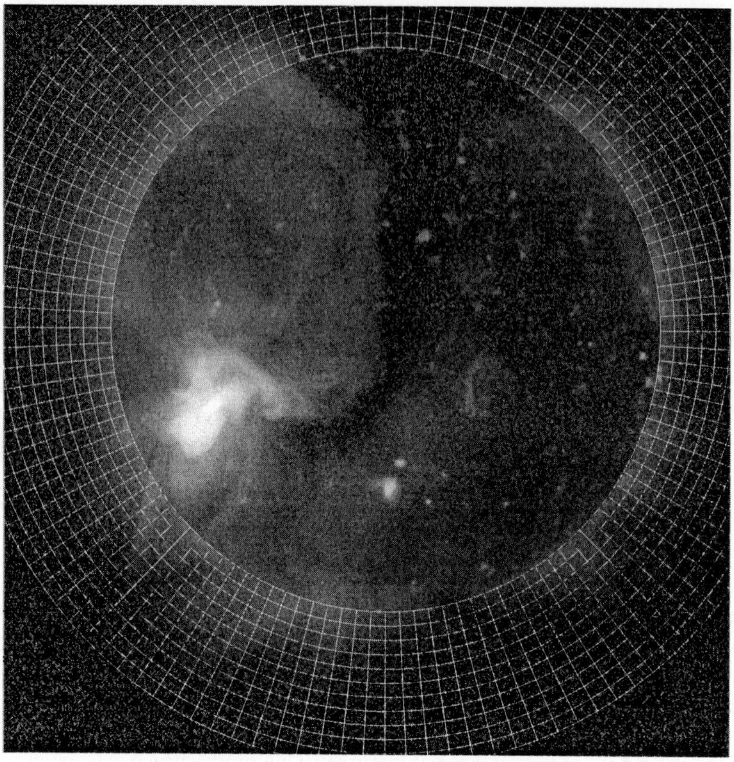

Fig. 1. Example of limb signal extraction. This *Yohkoh*/SXT image was taken on August 27, 1996 05:47:14 UT using the AlMg filter. Each cell covers 3° in azimuth and $0.05R_\odot$ in altitude.

THE LIMB SYNOPTIC MAP

We divide the solar limb into cells in azimuth and in the radial direction (see Figure 1). The signals extracted from each cell are functions of polar angle (north pole is 0° and the east is 90°) and radial distance. They are aligned in time and used to create limb synoptic maps like that shown in Figure 2.

Three significant features are immediately seen in the figure:

Coronal Butterfly Pattern. The two bright horizontal belts near the equatorial regions change their distance to the equator with time. They represent the active regions of the northern and southern hemispheres. As the solar cycle progresses, the coronal belts approach the equator, just like the sunspots underlying them. When the new solar cycle starts, the belts shift to high latitudes leaving the equatorial regions as a dark channel (Figure 2). This coronal pattern resembles the sunspot butterfly diagram.

Coronal Polar Sinusoids. Dense vertical finger features emanate from active regions and pass across both polar holes. Because they show a quasi-sinusoidal pattern above the polar holes, they are called "coronal polar sinusoids" (Li, Jewitt, & LaBonte 2000). They are produced by projection of high altitude plasma above active regions across the polar regions. Solar rotation creates the sinusoidal appearance in the limb synoptic maps. Individual polar sinusoids can live as long as a year.

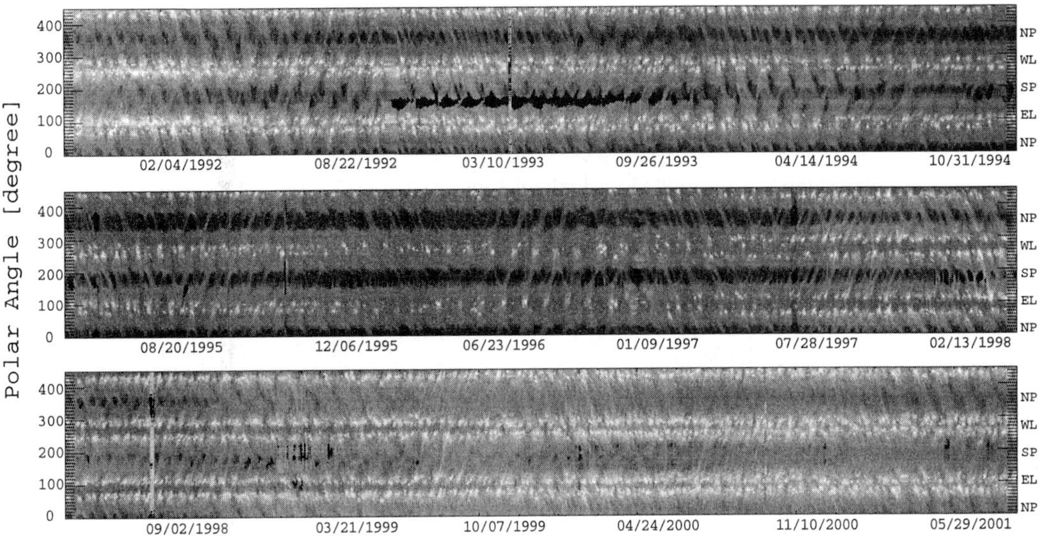

Fig. 2. The coronal limb synoptic map at 1.000 to 1.015 $R_\odot$ from *Yohkoh*/SXT images with Al.1 filter taken between October 1991 to August 2001. The horizontal axis is time and vertical axis is polar angle. Zero limb angle corresponds to the north pole (NP), while the east limb is at 90° (EL). To display the entire northern polar hole (0° and 360°), the polar angle is extended by 90° beyond 360°. To clearly demonstrate the coronal features described in the text, the map is equally divided into 3 portions in time sequence. The imperfections evident in this time series, e.g., the long dark gap in 1993 at polar angle 160°, result from imperfect correction for visible stray light in the SXT images. This problem will be corrected in the permanent SXT data archive.

Coronal Holes. The visibility of the coronal holes is modulated by the coronal polar sinusoids, i.e., the high temperature plasma above the active regions. This may explain why polar holes are well defined near the solar minimum, and hardly seen near the solar maximum.

Sunspot clustering phenomena have been previously reported in terms of "complexes of activity" (Bumba & Howard 1965, 1969; Gaizauskas *et al.* 1983), "active longitudes" (e.g. Bogart 1982), "active/sunspot nests" (Castenmiller, Zwaan, & Van der Zalm 1986, Brouwer & Zwaan 1990) and "hot spots" (Bai 1990). Although it is not always obvious that these previous works refer to the long-lived structures we have identified in the corona, the "complexes of activity" by Gaizauskas *et al.* (1983) and "sunspot nests" by Brouwer & Zwaan (1990) appear to be relevant. The former study used synoptic magnetograms in the photosphere for a 2 year period. The "complexes of activity" were described as "they are maintained for typically 3 to 6 solar rotations by fresh injections of magnetic flux seen as a sequence of many active regions which emerge in the same belt of activity and within well-defined zones of longitude." The latter study followed the appearance of sunspot groups for several years and wrote "A sunspot nest is a relatively small space on the surface within which a succession of spot groups appears."

Unlike the previous studies which focused on the photospheric sunspots and magnetogram observations, we study large-scale and long-lived coronal features using the limb synoptic maps. The surprise of our work is that these features persist for long times and are clearly tied the non-contemporaneous sunspot clusters.

SUMMARY

1. Limb synoptic maps allow us to condense a huge amount of observational data into compact, easily interpreted images. They reveal long-lived, large coronal structures that are less well seen in other forms of data display.

2. The coronal polar sinusoids are formed by high altitude plasma associated with distant active regions. The lifetimes of the coronal sinusoids are much longer than the lifetimes of the individual active regions which create them, up to a year in some cases.

3. The sinusoids are produced by non-contemporaneous sunspot clusters in the photosphere. The spot clusters are spatially localized regions from which sunspots erupt and decay. This process can continue for up to a year.

4. Continuously re-emerging sunspots imply long-lived, subsurface magnetic structures from which sunspots erupt. To our knowledge, such extended buried structures have yet to be modeled.

ACKNOWLEDGEMENTS

JL thanks Alisdair Davey for his support on the MSU computer facility. It is not possible to complete the SXT limb synoptic maps covering 8 years of data without the reliable access to the *Yohkoh* database in MSU from Hawaii. This work was supported by NASA grant NAG 5-4941 and *Yohkoh*-SXT subcontract LR01 M8801R.

REFERENCES

Bai, Taeil, Solar 'hot spots' are still hot, *ApJ*, **364**, L17 (1990)

Bogart, R. S., Recurrence of solar activity - Evidence for active longitudes, *Solar Phys.*, **76**, 155 (1982)

Brouwer, M. P., Zwaan, C., Sunspot nests as traced by a cluster analysis, *Solar Phys.*, **129**, 221 (1990)

Bumba, V., & Howard, R., A Study of the Development of Active Regions on the Sun, *ApJ*, **141**, 1502 (1965)

Bumba, V., & Howard, R., Solar Activity and Recurrences in Magnetic-Field Distribution, *Solar Phys.*, **7**, 28 (1969)

Castenmiller, M. J. M., Zwaan, C., van der Zalm, E. B. J., Sunspot nests - Manifestations of sequences in magnetic activity, *Solar Phys.*, **105**, 237 (1986)

Gaizauskas, V., Harvey, K. L., Harvey, J. W., and Zwaan, C., Large-scale Pattern Formed by Solar Active Regions During the Ascending Phase of Cycle 21, *ApJ*, **265**, 1056 (1983)

Li, J., Jewitt, D., & LaBonte, B., The Nature of Solar Polar Rays, *ApJ*, **539**, L67 (2000)

Li, J., LaBonte, B., Acton, L., & Slater, G., Persistent Coronal Streamers and the Identification of Sunspot Clusters, *ApJ*, **565**, 1289 (2002)

LONG-TERM VARIATION OF THE ROTATION OF THE SOLAR CORONA

R. C. Altrock

Air Force Research Laboratory, Space Vehicles Directorate, National Solar Observatory at Sacramento Peak, Sunspot, NM 88349, USA

ABSTRACT

Synoptic photoelectric observations of the coronal Fe XIV and Fe X emission lines at 530.3 nm and 637.4 nm, respectively, are analyzed to study the rotational behavior of the solar corona as a function of latitude, height and time. The data used are measurements made with the Sacramento Peak 40-cm coronagraph and Emission-Line Coronal Photometer of the intensity of these lines observed at 1.15 to 1.45 solar radii ($R_\odot$) between 1973 (1984 for Fe X) and 2000. An earlier similar temporal-correlation analysis of the Fe XIV data at 1.15 $R_\odot$ over only one 11-year solar activity cycle (Sime, Fisher, & Altrock 1989) found suggestions of solar-cycle variations in the differential-rotation and latitude-averaged-rotation patterns that combined the effects of large-scale patterns seen in the white-light corona and smaller-scale patterns seen in chromospheric and photospheric rotation. These results are tested over the longer epoch now available. In addition, the new 1.15 $R_\odot$ Fe XIV results are compared with those at greater heights and with results from the Fe X line to form a global picture of solar rotation throughout the corona and over more than two solar cycles.

INTRODUCTION

Previous observations of the rotation period of the solar corona have varied between rigid and latitudinally-dependent results. A previous study of Fe XIV observations during a 10-year period by Sime, Fisher, & Altrock (1989) indicated the possibility of temporal variation between rigid and latitudinally-dependent rates. This study, using up to 26 years of data, will attempt to clarify the question of a temporal variation in the rates and take a first look at the possibility of height variation in the rate.

OBSERVATIONS

Coronal scans have been taken daily at the John W. Evans Solar Facility of the National Solar Observatory at Sacramento Peak in Fe XIV and Fe X since 1973 and 1983, respectively. The instrument used to record the intensities is the Emission-Line Coronal Photometer, which removes the sky background in real time. Data are taken at 3° intervals around the disk at 1.15 $R_\odot$ (solar radii) and greater heights. For further information on the observations, see Altrock (1997).

Figure 1 shows a grey-scale plot of the intensities at 1.15 $R_\odot$ in Fe XIV as a function of position angle from 1975 through 2000. Labels along the right side indicate the poles (Np, Sp) and East- and West-limb equator (E, W). The "butterfly diagrams", similar to those for sunspots, of solar cycles 21, 22 and part of 23 are clearly seen in the coronal intensity. The 1.25 $R_\odot$ data are only available from 1985 onwards, and no major differences are seen with Figure 2. The Fe X data are only available from 1983 onwards. An apparent

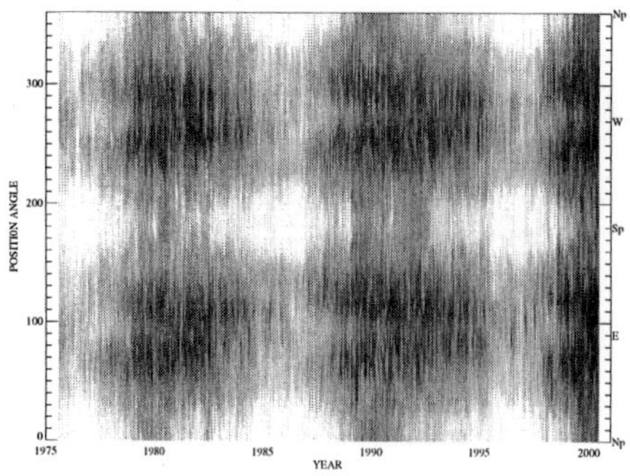

Fig. 1. Intensities at 1.15 $R_\odot$ in Fe XIV.

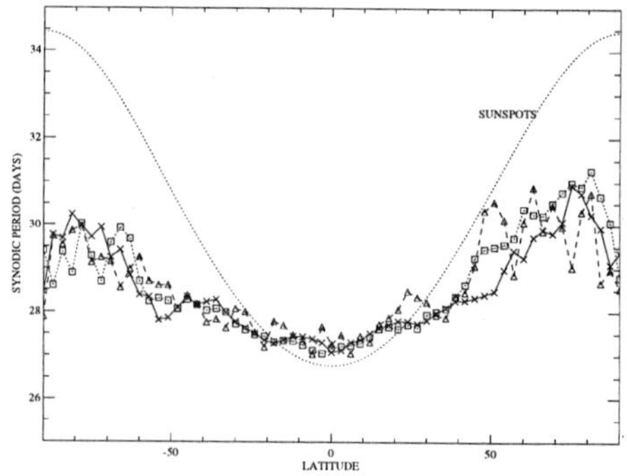

Fig. 2. Long-term-averaged rotation period vs. latitude in Fe XIV 1.15 $R_\odot$ (X), Fe XIV 1.25 $R_\odot$ (□) and Fe X 1.15 $R_\odot$ (△).

discontinuity in the intensity scale in 1988 may indicate possible problems with determining Fe X rotation periods.

INTERPOLATION

In order to determine rotation periods, I linearly interpolated over missing data. Cubic interpolation was tried but did not produce physically-reasonable results. The linearly-interpolated data from Figure 1 appear very similar to Figure 1, indicating that interpolating the data does not introduce any specious features. The interpolated Fe XIV 1.25 $R_\odot$ data, and especially the Fe X 1.15 $R_\odot$ data, show some differences with respect to the raw intensities, perhaps indicating some problems with the interpolation method due to sparser data and lower signal-to-noise ratio from lower intensities, which may lead to some difficulty in determining rotation periods. Over the epoch 1986 through 1999, during which we have data for Fe XIV 1.15 $R_\odot$, Fe XIV 1.25 $R_\odot$ and Fe X 1.15 $R_\odot$, there were 2939 days of observations for Fe XIV 1.15 $R_\odot$ and 2560 days for Fe XIV 1.25 $R_\odot$, or 13% fewer, and there were 2349 days of observations for Fe X 1.15 $R_\odot$, or 20% fewer than for Fe XIV 1.15 $R_\odot$.

VARIATION OF ROTATION WITH LATITUDE

Rotation periods are determined by performing the cross-correlation between the East- and West-limb intensity traces on a yearly basis. A lower limit of 0.1 has been adopted for usable cross-correlation coefficients (CCC). Rotation periods with CCC < 0.1 are not used in the analysis. In addition, data yielding synodic periods outside the range $25 < P < 35$ days are not used in the analysis. Periods obtained from negative lags were more robust than those obtained from positive lags, so only negative lags were used.

Figure 2 shows the synodic rotation period vs. latitude in Fe XIV 1.15 $R_\odot$ (26-year-average), Fe XIV 1.25 $R_\odot$ (17-year-average) and Fe X 1.15 $R_\odot$ (18-year-average), compared with the sunspot period. As found in the earlier Fe XIV study, the Fe XIV 1.15 $R_\odot$ rotation period is a combination of rigid and latitudinally-dependent (differential) rotation. The Fe X 1.15 $R_\odot$ and Fe XIV 1.25 $R_\odot$ periods are similar to Fe XIV 1.15 $R_\odot$, and there is an indication of a possible maximum in the period near 80° in all three data sets. There is some indication that the periods in Fe XIV 1.25 $R_\odot$ at middle latitudes may be greater than those in Fe XIV 1.15 $R_\odot$.

If the epoch of the data set for Fe XIV 1.15 $R_\odot$ is reduced to the same epoch as that of Fe XIV 1.25 $R_\odot$,

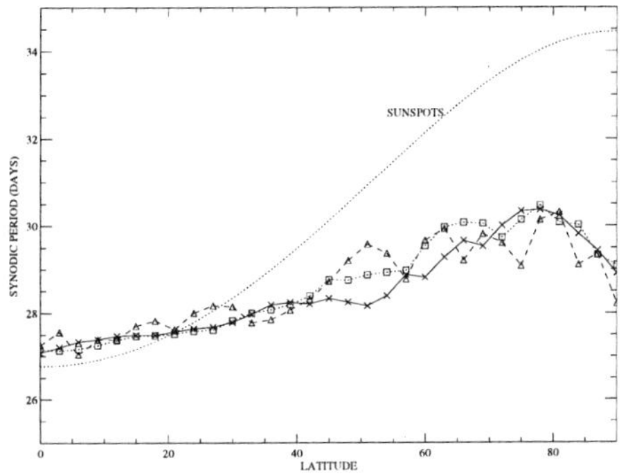

Fig. 3. The values of Fig. 2 averaged over the north and south hemispheres.

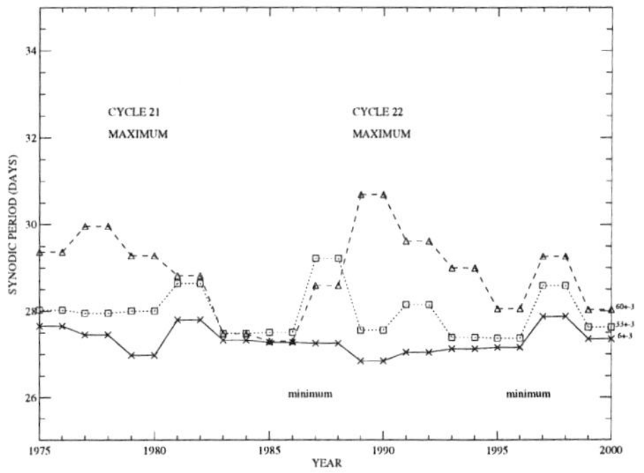

Fig. 4. Rotation period of three latitude bands given in degrees on the right side in 1.15 $R_\odot$ Fe XIV as a function of time (N+S 2-year averages).

the 1.15 $R_\odot$ noise level approximates that of 1.25 $R_\odot$ seen in Figure 2. However, the general properties of the two curves remain similar to those seen in Figure 2. The same is true for Figure 3. However, the effect of reducing the size of the 1.15 $R_\odot$ set for the 6°-bin data is insignificant.

Since the rotation curves for the north and south hemispheres appear symmetric within the noise level, it seems reasonable to average them, in order to increase the signal-to-noise ratio. Figure 3 shows these averaged curves as in Figure 2. The most interesting feature in both ions is a more-definite indication of a maximum in the period near 80°, with the period decreasing near the poles. Fe X may be rotating slower than Fe XIV in the mid-latitude range. There is also an indication of an increase in the Fe XIV period with height in the mid-latitude range. Introducing 6° bins smoothes the Fe X data considerably in Figure 3 but has little effect on Fe XIV 1.15 and 1.25 $R_\odot$ data and does not change the above results.

However, since the bases of prominences and the overlying helmet streamers only approach the poles during a short epoch of the solar cycle when the polar crown reaches the poles, it is puzzling that these data appear to show with high precision a spinup or polar vortex above 80° latitude. One possible explanation is that streamers rooted at lower latitudes may be seen in the polar region as they rotate in front of and behind the disk. Since these streamers have the higher rotation rate appropriate for the latitude of their bases, this higher rate would be *apparent* in the observations near the poles, and one might expect that this phenomenon would become more effective as one approaches the poles (hence the decrease of the period towards the poles). One could infer from the apparent polar rate that the streamers responsible for the results seen at 90° latitude are rooted near 60° latitude (the location where the polar rate matches the lower-latitude rate).

VARIATION OF ROTATION WITH HEIGHT

The average Fe XIV periods for latitudes above 40° and below 70° for 6°-binned data similar to Figure 3 are 28.75 ± 0.58 for 1.15 $R_\odot$ and 29.23 ± 0.64 for 1.25 $R_\odot$, where the uncertainties are the standard deviations. If equal-epoch data sets are used (see discussion in the preceding section), these values become 28.74 ± 0.67 for 1.15 $R_\odot$ and 29.15 ± 0.61 for 1.25 $R_\odot$. For 9° bins, these values become 28.81 ± 0.63 for 1.15 $R_\odot$ and 29.33 ± 0.63 for 1.25 $R_\odot$. Thus, the Fe XIV periods for latitudes above 40 deg and below 70 deg are consistently higher at 1.25 $R_\odot$ than at 1.15 $R_\odot$, although the standard-deviation error bars overlap. A future analysis of data at 1.35 $R_\odot$ may help clarify this.

The results of Figures 2 and 3 are at odds with those of Vats *et al.* (2001), who looked at rotation periods from solar radio fluxes from 1997 to 1999. Their results show synodic periods that decrease from 25.81 days at 1.084 $R_\odot$ to 25.38 days at 1.180 $R_\odot$. Although there is little overlap in the height range of the two studies, their values are significantly lower than the values in this study, as well as those from earlier optical studies (Sime, Fisher and Altrock, 1989, and Fisher and Sime, 1984). It must be said that the radio and optical studies are in disagreement. Perhaps a contributing factor is that the radio data are full-disk fluxes, whereas the optical data isolate a small range in latitudes near the limb.

VARIATION OF ROTATION WITH PHASE OF THE SOLAR CYCLE

Figure 4 shows the rotation period of three latitude bands (3°-9°, 30°-36° and 57°-63°) in 1.15 $R_\odot$ Fe XIV as a function of time. The temporal development is consistent with the earlier study: there is rigid rotation near solar minimum (curves converge toward a single period) and differential rotation (curves diverge) near solar maximum (although cycle 23 is uncertain). The results for Fe XIV 1.25 $R_\odot$ are somewhat less certain for these noisier data, but there is an indication of a divergence toward solar maximum in 2001 (the Fe XIV 1.15 $R_\odot$ data are shown only through 2000). The Fe X data are much noisier. It is difficult to draw any conclusion about the temporal behavior of Fe X.

CONCLUSIONS

These data contain new information about the variation of coronal rotation with time and latitude. On average, the corona rotates more rigidly than sunspots, as seen in two different ions. The coronal equatorial rate is perhaps slightly slower than for sunspots. Both ions show a partially-differential rotation period that may peak near 80° latitude and then decrease to the poles. There is an indication that the rotation period may increase with height between 40° and 70° latitude: the data in Fe XIV 1.25 $R_\odot$ appear to show slower rotation in that range as compared with Fe XIV 1.15 $R_\odot$. The temporal evolution of coronal rotation as seen in Fe XIV at 1.15 $R_\odot$ is consistent with the earlier study: there is rigid rotation near solar minimum and differential rotation near solar maximum. The Fe X 1.15 $R_\odot$ and Fe XIV 1.25 $R_\odot$ data are too noisy to determine the temporal development. The periods found here and in earlier optical studies of the corona are not in agreement with those found by Vats *et al.* (2001) from radio data. These results have implications for torsional oscillations in the corona and for models of the solar activity cycle.

ACKNOWLEDGEMENTS

This paper was supported by the Air Force Office of Scientific Research.

REFERENCES

Altrock, R.C., An "Extended Solar Cycle" as Observed in Fe XIV, *Solar Phys.*, **170**, 411 (1997).

Fisher, R.R., & D.G. Sime, Rotational Characteristics of the White-Light Solar Corona 1965-1983, *Astrophys. J.*, **287**, 959 (1984).

Sime, D.G., R.R. Fisher, & R.C. Altrock, Rotation Characteristics of the Fe XIV (5303 Å) Solar Corona, *Astrophys. J.*, **336**, 454 (1989).

Vats, H.O., J. R. Cecatto, M. Mehta, H. S. Sawant, & J.A.C.F. Neri, Discovery of Variation in Solar Coronal Rotation with Altitude, *Astrophys. J.*, **548**, L87 (2001).

WHAT ARE THE ORIGINS OF QUIESCENT CORONAL SOFT X-RAYS?

C. R. Foley,[1], J. L. Culhane[1], S. Patsourakos[1], R. Yurow,[2], C. Moroney[2], and D. MacKay[3]

[1]*Mullard Space Science Laboratory, University College London, Dorking, Surrey RH5 6NT, UK*
[2]*Emergent Information Technologies, Inc., 2000 Park Towers Dr., Vienna, VA 22180 United States*
[3]*Department of Mathematical and Physical Sciences, St. Andrews University, Scotland, UK*

ABSTRACT

Long term observations performed with instruments on the *SOHO* and *Yohkoh* spacecraft, obtained over the rising phase of the current solar cycle 23, provide new data for investigating the solar cycle variation in coronal heating and solar activity. We have used the Coronal Diagnostic Spectrometer (CDS) onboard *SOHO* to analyze conditions in coronal streamer structures observed close to solar minimum (1996 July 8) and near solar maximum (1999 August 5).

INTRODUCTION

Long term, synoptic observations of the sun with instruments like the *Yohkoh* Soft X–ray Telescope (SXT) and *SOHO* CDS allow us to investigate how the heating requirements vary through the solar cycle and ultimately with the strength of the ambient magnetic field. In Figure 1, we present the solar cycle evolution of the quiet sun EUV flux measured by the *SOHO* CDS/GIS. The increase over the cycle in coronal line Fe XIV is of the same magnitude as that for the full sun, which is dominated by the emission from active regions. Emission from a transition region line, such as Fe IX, appears flat. This clearly demonstrates that the principal reason why the quiet sun coronal flux increases by almost two orders of magnitude is because the quiet corona becomes hotter into the solar cycle, an effect of the increasing magnetic field.

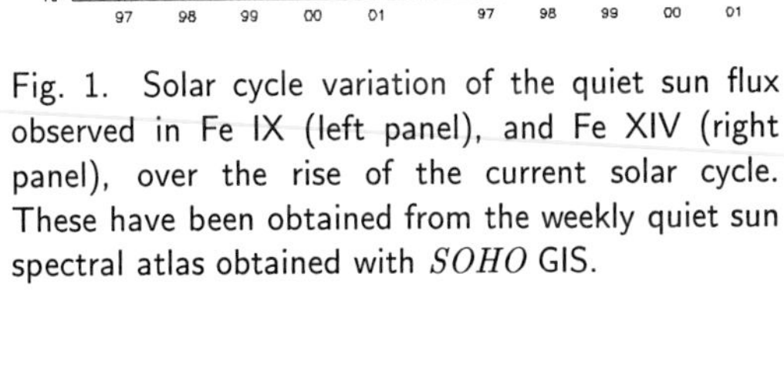

Fig. 1. Solar cycle variation of the quiet sun flux observed in Fe IX (left panel), and Fe XIV (right panel), over the rise of the current solar cycle. These have been obtained from the weekly quiet sun spectral atlas obtained with *SOHO* GIS.

To be able to determine how much extra heating is required to produce this hotter corona, we need to know the density and temperature structure in the quiet sun and to solve the energy equation. We have made measurements of the latter at the solar min (July 8, 1996) and at the current solar max (August 4, 1999). We present these measurements in the next section, along with an estimate for the heating.

ANALYSIS AND CONCLUSIONS

The role of the coronal magnetic field in coronal heating is well established with many different heating mechanisms being distinguished by their scaling with magnetic field strength. In such models the coronal heating rate is proportional to B^{α}, with α being a function of the specific mechanism of magnetic energy build-up, release and dissipation. In Figure 2, we display the *SOHO* CDS results of the solar minimum and solar maximum coronal streamers from Foley *et al.,* (2002). In that paper we made an estimate of the dependency of the heating rate with the strength of the magnetic field. We can do this since the temperature scale heights for the solar minimum and maximum coronae are more or less identical (cf. Figure 2). This means that the ratio of conductive flux is equal to the ratio of magnetic heating at solar maximum to minimum: $(\text{Tmax}/\ \text{Tmin})^{(5/2)} = (\text{Bmax}/\ \text{Bmin})^{\alpha}$, where Tmax and Tmin are the maximum coronal temperature for solar maximum and minimum. To do this we are also assuming that the dominant energy loss process within coronal streamers is conduction. A typical value for the ratio of the magnetic field strength (Bmax/ Bmin) is 2 (see e.g. Pevtsov & Acton 2001). This value along with our finding that (Tmax/ Tmin) = 1.5 would indicate that a reasonable value for α is 1.5. This value of α compares well with the α predicted by 'direct current' stressing models based on reconnection and turbulence with constant dissipation coefficients (cf. Mandrini *et al.* 2000).

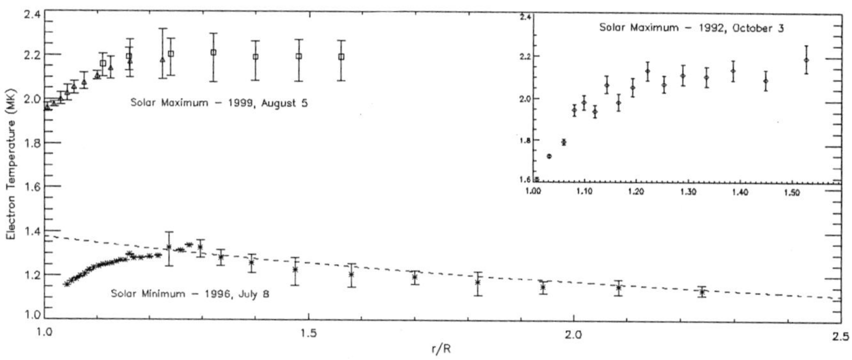

Fig. 2. In this figure we plot the radial temperature structure determined for solar maximum (August 5, 1999, triangles - NIS; squares - GIS) and solar minimum (July 8, 1996, asterisks). The values from Foley (1998) are over plotted as diamonds in the inset panel and were determined through a similar region at the last solar maximum.

ACKNOWLEDGEMENTS

C.R.F. acknowledges the support of The Royal Society though provision of a conference grant.

REFERENCES

Foley, C. R., 1998, PhD Thesis, `http://www.mssl.ucl.ac.uk/www_solar/publications/foleydiss.zip`.

Foley, C. R., Patsourakos, S., Culhane, J. L., & MacKay, D., 2002, *A&A*, **381**, 1049.

Mandrini *et al.*, 2000, *ApJ*, **530**, 999.

Pevtsov, A. A., & Acton, L. W., 2001, *ApJ*, **554**, 416.

EVOLUTION OF THE 'GORGEOUS' CORONAL HOLE

A. Takeda[1,2] and S. Kubo[2]

[1]*Solar Physics Research Corporation, 4720 Calle Desecada Tucson, AZ 85718, USA*
[2]*Institute of Space and Astronautical Science, 3-1-1 Yoshinodai, Sagamihara, Kanagawa, 229-8510, Japan*

ABSTRACT

A distinct (or 'gorgeous') coronal hole was observed from October to December, 2000, at around 180° in Carrington longitude. Its formation, evolution, and rotation rates are discussed based on an analysis of *Yohkoh* Soft X-ray Telescope (SXT) images and synoptic maps of the Kitt Peak magnetic field.

INTRODUCTION

Extensive studies on coronal holes have been done since the mid-1970s, after the advent of *Skylab*. The famous boot-shaped coronal hole, 'CH1', observed during the *Skylab* mission (June – September, 1973), was analyzed in detail by Timothy *et al.* (1975), later explained by the model of Wang & Sheeley (1990, 1993), and is still referred to as a prototype for coronal holes. Compared with the 1970s when CH1 was observed, we now obtain images with much better temporal and spatial resolutions and with a greater variety of wavelengths. It is therefore worthwhile to revisit detailed analysis of an individual coronal hole to verify the current understanding of coronal holes. As a first step, in this short paper we will present some results from a morphological study of a coronal hole observed from October to December, 2000. This coronal hole had an elongated shape in the north-south direction across the equator and maintained its outline for at least three Carrington Rotations (C.R.) (1968–1970, see Figure 1). One may describe it as 'gorgeous' in the sense that it took a boot-like shape, reminiscent of CH1, during its October rotation, and transformed into a chevron-like shape in its December rotation. We generated a series of synoptic maps of the soft X-ray intensities obtained with the *Yohkoh* SXT (for C.R. 1964–1972) and closely compared with the corresponding maps of the photospheric magnetic fields obtained at Kitt Peak.

RESULTS AND DISCUSSION

Comparison of SXT and Kitt Peak synoptic maps during the same rotation shows that the coronal hole area defined in SXT was located roughly over the central part of a large negative unipolar region in the magnetic field image. Looking over consecutive magnetic field maps for a few rotations before and after the period shown in Figure 1, it turns out that this large unipolar region was comprised of three parts which originated from different groups of active regions which emerged during C.R. 1964–1965. The first group of active regions were those which emerged in the range of 180°–230° in Carrington longitude and 0°–20°S in heliographic latitude. The negative polarity part of their remnants were merged together, moved westward with a velocity near to differential rotation rate at their latitude, and came to form the equatorial part of the coronal hole in C.R. 1968. The two other groups of active regions were located at latitudes 25°N and 15°S and were in the range 90°–180° in longitude. The latter two remnants moved eastward on synoptic maps due to their differential rotation and formed the high-latitude parts of the coronal hole in each hemisphere.

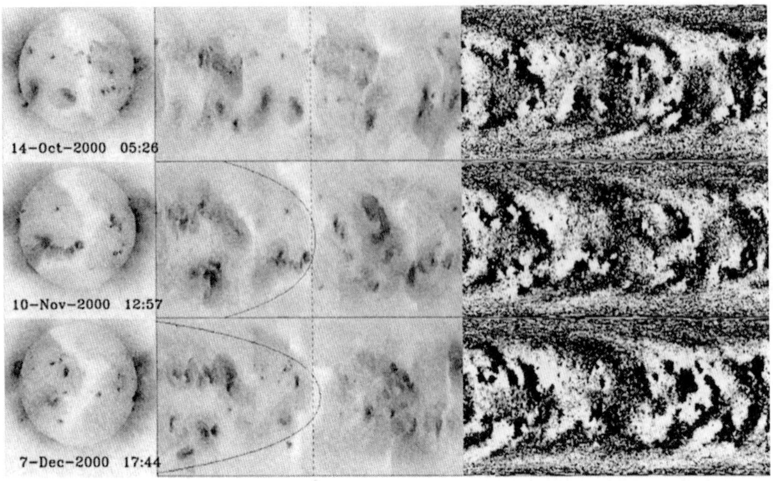

Fig. 1. Left column : *Yohkoh* SXT images of the 'gorgeous' coronal hole for three consecutive central meridian passages. Middle column : SXT synoptic maps for C.R. 1968–1970 (from top to bottom). Dashed lines show the approximate axis of the coronal hole as of C.R. 1968, and the solid curves represent the predicted location of the axis after one or two Carrington rotation periods assuming the differential rates given in *Astrophysical Quantities* (Allen 1973). Right column : synoptic maps of the Kitt Peak magnetic field intensities.

Since C.R. 1968, when the outline of the coronal hole first became clear in the SXT images, the coronal hole rotated with velocities particular to their latitude and finally broke into three pieces, disappearing in early January, 2001. The rotation rate of the SXT boundaries at low latitudes ($\leq 15°$) was near to the differential rate, in accordance with the equatorial part of the underlying magnetic unipolar region. However, at higher latitudes, the rotation rate was significantly faster than the differential rate, as indicated in the SXT synoptic maps shown in Figure 1. It is known that CH1 rotated quasi-rigidly, i.e., its rotation rate at high latitude was faster than the differential rate (Timothy *et al.* 1975). According to Wang & Sheeley (1993), this property is typical for a coronal hole, called 'polar extension', and is often observed in the declining phase of the solar cycle. However, the hole analyzed here appeared around solar maximum, when positive polarities were still maintained at the north pole. Thus it has a different origin from CH1. The same rotational property observed here may require a different mechanism from the quasi-rigid rotation of coronal holes. In the case of the 'gorgeous' hole, the western side of the underlying unipolar region appeared to be more effectively reinforced by the nearby active region remnants, and this seemed to cause a drift of the coronal hole boundary away from the unipolar regions above which the hole had initially been located.

REFERENCES

Allen, C.W., *Astrophysical Quantities*, 3rd ed, Athlone, London, (1973).

Timothy, A.R., Krieger, A.S., & Vaiana, G.S., Solar Physics **42**, 135 (1975).

Wang, Y.-M., & Sheeley, N.R.Jr., Astrophysical Journal **365**, 372 (1990).

Wang, Y.-M., & Sheeley, N.R.Jr., Astrophysical Journal **414**, 916 (1993).

EXCITATION OF THE MID- AND LOW-LATITUDE ROSSBY VORTICES AT THE BASE OF THE SOLAR CONVECTION ZONE AND FORMATION OF THE COMPLEXES OF ACTIVITY

E. Tikhomolov

TRIUMF, Canada's National Laboratory, 4004 Wesbroo k Mall, Vancouver, BC, V6T 2A3, Canada

ABSTRACT

Rossby vortices excited near the bottom of the solar convection zone are very effective sources of the global magnetic structures. We show how such vortices can be forced by deformational long-wave instability that is realized near the interface between the convective and radiative zones.

ROSSBY VORTICES AT THE BOTTOM OF THE SOLAR CONVECTION ZONE

Several years ago we suggested a new interpretation of global magnetic structures (Tikhomolov 1995). The main idea is that global magnetic structures are generated by the large-scale vortices that appear in a thin layer near the base of the solar convection zone. We developed a shallow-water model of this thin interfacial layer that occupies part of the penetrative convection layer. We called the layer under consideration the Active Transition Layer (ATL). Recently, Dikpati & Gilman (2001) continued investigations in the same direction. One of the properties of the ATL is that the effective turbulent diffusion coefficient in it is much smaller than that in the convection zone. Both this small diffusion coefficient and the thinness of the layer permit the large-scale (on the order of several helio-degrees) and long-lived (with a lifetime on the order on several Carrington rotation periods) vortices to exist. Usually, such vortices are referred to as Rossby vortices (Pedlosky 1982).

Deformational long-wave instability can play a key role in the excitation and forcing of Rossby vortices near the base of the solar convection zone (Tikhomolov 1996, 1998). Numerical simulations show that the main feature of an established stationary flow is the existence of giant cyclones in each hemisphere (Figure 1a).

Another process that can give rise to the excitation of Rossby vortices is the shear instability (Dikpati & Gilman 2001). The forcing of Rossby vortices by shear instability in our model requires rather strong shear in latitude. Possibly, a two-step scenario is more realistic: first, the giant Rossby vortices (or Rossby waves) are forced by deformational long-wave instability, and then smaller-scale vortices are excited in the areas of formed strong shear.

Rossby vortices are the effective sources of the giant magnetic structures one can see in Figures 1b and 1c. After excitation, Rossby vortices twist magnetic field lines, which leads to the formation of closed configurations. The maximum value of the magnetic field strength significantly increases in the regions affected by the actions of the Rossby vortices. The strong magnetic field flows up from the bottom of the solar convection zone to the solar surface and gives rise to the formation of the complexes of activity.

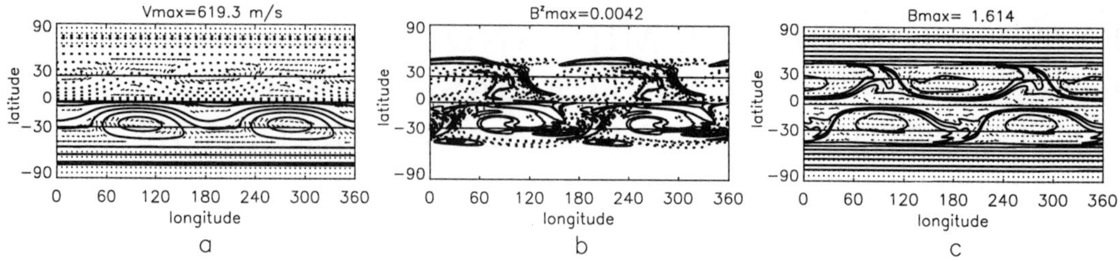

Fig. 1. Numerical simulations of the generation of the magnetic field by Rossby cyclones near the base of the convection zone. Solid and dashed lines show, respectively, positive and negative values of the specified quantities. All values for each figure are normalized by their maximum value. The maximum values of velocity (Vmax), vertical (B^zmax) and toroidal magnetic field (Bmax) components are indicated at the top of the figures. (a) The evolution of the Rossby vortices. Shown are the contours of stream function ψ: ± 0.05, ± 0.1, ± 0.3, ± 0.5, ± 0.7, and ± 0.9. Arrows show distribution of the velocity V. (b) The evolution of the vertical component B^z for flows depicted in (a). (c) The evolution of the toroidal component for flows depicted in (a). Shown are the contours of stream function T: 0.1, 0.2, 0.3, 0.4, 0.5, 0.6, 0.7, 0.8, and 0.9. Arrows show distribution of the toroidal field B.

REFERENCES

Dikpati, M., Gilman P. A, Analysis of hydrodynamic stability of solar tachocline latitudinal differential rotation using a shallow-water model, *Astrophys. J.*, **551**, 536 (2001).

Pedlosky, J. *Geophysical Fluid Dynamics*, Springer-Verlag, Berlin, (1982).

Tikhomolov, E., Rossby Vortices as Sources of Global Magnetic Structures on the Sun, *Solar Phys.*, **156**, 205 (1995).

Tikhomolov, E., Short-scale convection and long-scale deformationally-unstable Rossby wave in a rotating fluid layer heated from below, *Physics of Fluids.*, **8(12)**, 3329 (1996).

Tikhomolov, E., . Forcing of differential rotation and Rossby waves at the interface between the convectively stable and unstable layers, *Astrophys. J.*, **499**, 905 (1998).

DIFFERENTIAL ROTATION OF THE SOFT X-RAY CORONA OVER A SOLAR CYCLE

M. A. Weber and P. A. Sturrock

Center for Space Science and Astrophysics, Stanford University, Stanford, CA 94305, USA

ABSTRACT

The *Yohkoh* Soft X-ray Telescope has provided unprecedented high spatial and temporal resolution images of the solar corona in X-rays for a complete activity cycle. Building upon earlier work, we perform time-series analysis on latitude bins of the full-disk images (SFDs) to describe the differential rotation of the corona over latitude and cycle phase. The bins are formed by integrating over localized regions in heliographic longitude and latitude. Using techniques of Bayesian harmonic analysis, we find that the rotation signal in the data comprises several components, of which the solar activity cycle is dominant. Typically, there are several rotation components, but one is usually preeminent. We find evidence for rotation power in the corona that matches signals found in solar neutrino experiments.

INTRODUCTION

The differential rotation of the solar corona is an interesting subject, in part because the latitudinal profile of the rotation rate seems to vary less (i.e. is more "rigid") than that of the underlying photosphere. Although there has been a history of studies on the coronal rotation rate, including some of Soft X-Ray (SXR) data, today we benefit from resolved SXR data spanning an entire solar activity cycle, acquired by the *Yohkoh* Soft X-Ray Telescope (SXT; Tsuneta *et al.* 1991). This study builds on earlier work (Weber *et al.* 1999) by extending the dataset to ten years, and by using a fully Bayesian methodology instead of the earlier, more *ad hoc* approach.

DATA REDUCTION, BAYESIAN HARMONIC ANALYSIS, AND RESULTS

The data used are derived from "SFD" images (the full-frame images of the sun which have been used as the frames of the famous SXT movie). Time-series are formed from the integrated flux from each of nine latitude bins located on the central meridian at {60S, 45S, 30S, 15S, equator, 15N, 30N, 45N, 60N}. (The data reduction techniques are discussed in more detail in Weber *et al.* 1999.) For the results presented here, we only consider the data from the AlMg diagnostic filter. The total time studied runs from September 1991 to December 2000 (about 9.2 years). The dataset included 215,828 images, for a mean sampling rate of about 64 images per day.

Bretthorst (1988) has developed the groundwork for doing harmonic analysis of a time-series using Bayesian techniques. For this presentation, we utilized his approach. We presume that the time-series $d_i = d(t_i)$ (where $i = 1, \cdots, N$) can be described by the model

$$d_i = f_i + e_i = \sum_{j=1}^{m} [A_j \sin(\omega_j t_i) + B_j \cos(\omega_j t_i)] + e_i, \tag{1}$$

where $f = f(\{A_j\}, \{B_j\}, \{\omega_j\})$ and e_i is normally distributed noise with variance σ^2. The periodogram of the data $\overline{h^2}$ can be shown to be the "sufficient statistic" for optimally estimating the frequencies of the model. For a model with m harmonics, $\overline{h^2}$ is an m-dimensional periodogram.

Bretthorst (1988) also provides formulae and procedures for determining the moments of the frequency estimations (and hence, their uncertainties), as well as for the amplitudes and their uncertainties, the noise variance σ^2, and the relative likelihood of models with differing numbers of harmonic components.

Our procedure was to (1) locate the first component using a single-harmonic model; (2) subtract the estimated component(s) from the data; and (3) analyze the residual with a single-harmonic model to initially estimate the next component. (4) Starting from previous estimates, analyze the *total* data set with the multi-harmonic model. [This amounts to searching for $\max(\overline{h^2})$. To locate the maximum of $\overline{h^2}(\{\omega\})$, we used a "pattern-search" algorithm (Hooke & Jeeves 1961).] (5) Repeat steps 2–5.

We have just begun to get results from this analysis, but there are a few things that can be said at this time.

Dominance of solar cycle modulation and rectification. From the harmonic analysis, we find a solar cycle component with a frequency of about 0.106 cyc/yr (~9.5 yr period). This component typically carries about 5× as much power as the dominant rotation component. On a smaller note, we observe that the time-series is rectified at solar minimum, as happens with the sunspot number.

Relatively rigid rotation of corona. We examined the differential rotation year by year of the dominant rotation components. A common characteristic is that the coronal rotation rate only matches the underlying photospheric rate near the equator, and rotates faster than the photosphere at higher latitudes; thus the rotation profile of the corona across latitude is shallower, i.e., more rigid than that of the photosphere.

Indication of connection to solar interior. We examined the power spectra across latitude, for the period 1991–2000. Low latitudes displayed a component at 13.6 cycles/yr, and high latitudes displayed one at 12.9 cycles/yr. These frequencies are noteworthy because they correspond to the dominant frequencies in the power spectra of the Homestake and GALLEX/GNO solar neutrino data. This correspondence is elaborated upon further by Sturrock & Weber (2002).

The Bayesian approach appears to be quite promising for time-series analysis of solar data. Even better would be to reimplement Bretthorst's work for wavelets in place of harmonic components.

We gratefully acknowledge that this research was supported by NASA grants NAS 8-37334 and NAG 5-9784.

REFERENCES

Bretthorst, G.L., *Bayesian Spectrum Analysis and Parameter Estimation*, Springer-Verlag, Germany (1988).

Hooke, R., and Jeeves, T.A., *Journal of the Assoc. for Computing Machinery,* **8(2)**, 212 (1961).

Sturrock, P.A., and Weber, M.A., these Proceedings (2002).

Tsuneta, S., *et al., Solar Phys.,* **136**, 37 (1991).

Weber, M.A., Acton, L.W., Alexander, D., Kubo, S., and Hara, H., *Solar Phys.,* **189**, 271 (1999).

Section XII.

High Energy Emission in Flares

HARD X-RAY SOLAR FLARES REVEALED WITH YOHKOH HXT – A REVIEW

S. Masuda

Solar-Terrestrial Environment Laboratory, Nagoya University, Toyokawa, Aichi 442-8507, Japan

ABSTRACT

Yohkoh/HXT has detected more than 2500 solar flares over its 10 years of operation. In the first five years of operation HXT has provided us with a lot of new information on solar flares. Some of its major achievements are providing both statistical and detailed event studies of double footpoint sources, the discovery of the above-the-looptop source as evidence for magnetic reconnection, and revealing characteristics of a super-hot (> 30 MK) thermal component. These results clearly indicate that many types of hard X-ray sources exist simultaneously even in a single flare. Flares, especially intense flares, which occurred during the current solar cycle, have been analyzed and many new results have been reported recently. They are summarized in this paper.

INTRODUCTION

The hard X-ray telescope (HXT, Kosugi *et al.* 1991) onboard *Yohkoh* (Ogawara *et al.* 1991) has detected more than 2,500 solar flares during its ~ 10-years of operations. HXT was a unique hard X-ray imager in the 1990s, although hard X-ray imaging observations started in the early 1980s. HXT has several advantages compared with the previous imagers, such as a wide effective area (~ 60 cm^2), a high time (0.5 s) and a spatial resolution (~ 5 arcsec), and a wide field-of-view (full sun). From the scientific view point the following two points are crucially important among the many advantages.

1. HXT is not a simple imager, but an imaging spectrometer. HXT has four energy bands, the L-, M1-, M2-, and H-band (14 – 23 – 33 – 53 – 93 keV). *Four* is a small number, but *not one*. This is quite important. We can derive spatially-resolved spectral information. This information is very helpful to specify the type of each hard X-ray source. Actually it was found that there were several types of hard X-ray sources even at one moment in a single flare. Those sources show different spectra and different time behaviors. The characteristics of these sources, such as the double footpoint sources, the looptop impulsive source, and looptop gradual source, are mentioned in the following sections.

2. The *Yohkoh* satellite has another X-ray telescope, the soft X-ray telescope (SXT; Tsuneta *et al.* 1991). The observational objects of SXT are mainly thermal plasmas, whose temperatures are ~ 2 – 20 MK, contained in magnetic loops, while the observational targets of HXT are higher (> 20 MK) temperature plasmas and non-thermal particles. These two telescopes provide us with complementary data for solar flare studies. Co-alignment between these two telescopes has been achieved with an accuracy of ~ 1 arcsec (Masuda 1994). Thanks to this, simultaneous observations with SXT give us the spatial relationship between magnetic field lines involved in a solar flare and hard X-ray sources. A lot of important results from HXT, such as the discovery of the above-the-looptop source, originated from comparing with SXT data. The existence of SXT is a great advantage for HXT.

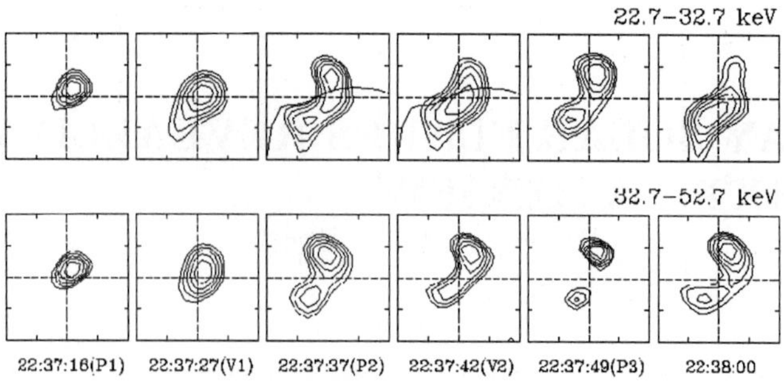

Fig. 1. HXT images at the peaks and minima of the flux profile during the impulsive phase ($37'' \times 37''$) of the November 15 1991 flare. The pointing jitter of the spacecraft was corrected. Images in the M1-band are shown in the upper row, while those in the M2-band are in the lower row. Magnetic neutral lines are shown in P2 and V2 images in the M1-band. The contour levels are 18, 25, 35, 50, and 71% of the maximum brightness for each image. The averaging times are 1, 1.5, 0.5, 0.5, 0.5, and 1 s for the M1-band images, and 1.5, 3, 0.5, 1.5, 1, and 2 s for the M2-band images, respectively. (after Sakao *et al.* 1992)

HARD X-RAY FEATURES DURING THE IMPULSIVE PHASE

The hard X-ray intensity varies very rapidly in the impulsive phase. It is believed that this temporal behavior corresponds to non-thermal energy release and energy deposition. During the impulsive phase various types of hard X-ray sources are observed with *Yohkoh*/HXT. In this section some of them are introduced.

Double Footpoint Sources

A double source structure is often observed during the impulsive phase, especially in higher energy hard X-rays (Hoyng *et al.* 1981, Duijveman *et al.* 1982, Sakao *et al.* 1992). According to Sakao's statistical analysis of 28 flares selected with a criterion only on the hard X-ray flux, 39% (11/28) of the events showed a 'double source' structure (Sakao 1994). This structure was the most frequent type. The fractions of 'single source' and 'multiple source' flares were 29% and 32%, respectively. Incorporating other observational results, Sakao inferred that the double-source structure corresponded to the two footpoints of a flare loop, and that it is a fundamental structure in the energy range above 30 keV (Sakao 1994).

In the case of an X-class flare occurring on November 15 1991 the double footpoint sources show an impressive evolution (Sakao *et al.* 1992). In this flare there are three outstanding spikes in the time profile of the hard X-ray intensity observed with HXT. Comparing the M2-band images at the three peak times, the separation of the double sources monotonically increases with time (Figure 1). Also, a significant difference is seen between the two images taken at a peak time and a minimum of the flux profile. At the peak time the double source structure is clearly observed. However, at the minimum this structure is not so clear because another emission appears between the double sources (Figure 1). Sakao *et al.* also analyzed the temporal variations of the two sources in hard X-ray intensity. Both variations are quite similar, and the time-lag is less than 0.2 s. This result suggests that the two sources correspond to the two ends of a magnetic loop and that high energy electrons precipitate into the regions where the two sources are located.

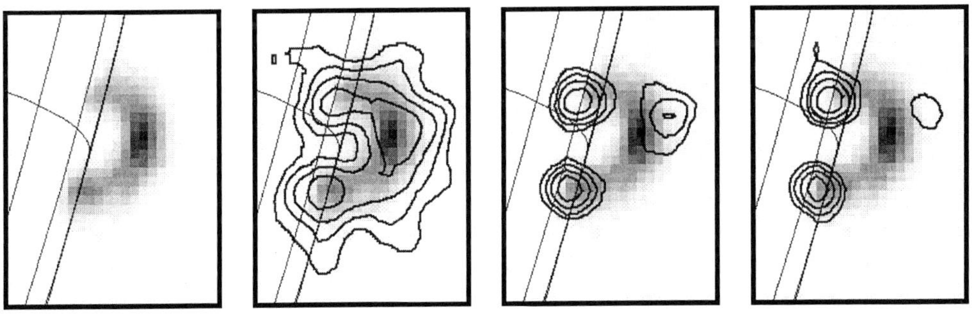

Fig. 2. Hard X-ray and soft X-ray images of the January 13 1992 flare. The left panel shows a soft X-ray image taken with SXT/Be filter at 17:28:07 UT. From left to right the remaining three panels are the L-, M1, and M2-band images (contours) overlaid on the same soft X-ray image (grey scale). The photon accumulation time is 17:27:35 – 17:28:15 UT for these three images. The contour levels are 6.25, 12.5, 25.0 and 50.0% of the peak value in each image. The field of view is $59'' \times 79''$ for all panels.

Looptop Impulsive Sources (Above-the-Looptop Sources)

Figure 2 shows the soft and hard X-ray images of an impulsive flare which occurred at the west limb on January 13 1992. In this figure it is clearly seen that a hard X-ray source is located well above the corresponding soft X-ray loop (Masuda 1994, Masuda *et al.* 1994, Masuda *et al.* 1995). This hard X-ray source was observed even in the M2-band (33 – 53 keV) and has a relatively hard spectrum ($\gamma \sim 4$). It also shows impulsive time behavior, similar to that of the footpoint sources (Masuda *et al.* 1994). This discovery indicates that the flare energy-release, probably magnetic reconnection, takes place *not in* the soft X-ray loop, *but above* the loop, even in impulsive solar flares.

After this discovery further studies of this flare have been carried out by many solar physicists. Analyzing soft X-ray images taken with wide field of views, soft X-ray hot-plasma ejections were found during this flare (Shibata *et al.* 1995). This supported the model that magnetic reconnection took place above the soft X-ray loop. A detailed comparison with temperature map derived from SXT was made by Tsuneta *et al.* (1996). Other observational data, taken at ground-based observatories and other satellites, are also analyzed to understand the above-the-looptop hard X-ray source (e.g. Wang *et al.* 1995, Aschwanden *et al.* 1996*a*).

Looptop Gradual Sources

Another coronal hard X-ray source is observed in the impulsive phase. In Figure 2 the L-band image is similar to the soft X-ray image. This hard X-ray source has a very soft spectrum and is regarded as thermal emission from a high temperature plasma. This source becomes dominant in the gradual phase (see the next section). However, it is clear that the emission starts in the impulsive phase.

Also, in some flares other types of hard X-ray sources are very weak and only this component is observed during almost the whole flare period, for example, the February 6 1992 flare (Kosugi *et al.* 1994). In this case the hard X-ray spectrum is well fitted by a thermal (T $\sim$ 40 MK) spectrum.

Diffuse Coronal Sources with a Hard Spectrum

Sato reported recently that, during the April 23 1998 series of flares an extended hard X-ray source with a very hard spectrum was located in the high corona ($\sim 5 \times 10^4$ km) (Sato *et al.* 2001). This source is regarded as a non-thermal source. In the same series another non-thermal source is observed in the lower corona ($\sim 2.7 \times 10^3$ km). This lower source reaches its peak first, about two minutes before the higher extended source. These observations suggest that the energization of electrons occurred first at the lower altitude and then those electrons are injected into the high coronal region. No similar event has been detected by HXT so far. The intense footpoint sources were occulted by the solar limb since this flare took place behind the limb. Such a special situation might be needed to detect this kind of hard X-ray source.

In the 1980s so-called Type-C flares were observed with *Hinotori* (Tsuneta *et al.* 1984, Tanaka 1987). Their characteristics are a hard X-ray source in the high-altitude corona and spectral hardening with time. This high-altitude source is interpreted as the emission from high energy electrons trapped in a magnetic bottle. The hard X-ray source in the April 23 1998 flare seems to be slightly different from this type-C hard X-ray source. The relationship between them is not clear yet.

Moving Coronal Sources

Very recently a surprising hard X-ray feature was reported (Hudson *et al.* 2001), a high-speed ($\sim$ 1000 km/s) coronal ejection. It was observed in the 23 – 33 keV energy band during the April 18 2001 flare. Simultaneous imaging at 17 and 34 GHz by the Nobeyama radio heliograph shows complex moving features with the same speed, including a compact moving feature observed in hard X-rays. Also observed was a bright CME about 15 minutes later with the same position angle as that of the hard X-ray ejective feature.

HARD X-RAY FEATURES DURING THE GRADUAL PHASE

After the impulsive phase a coronal source located at the apex portion of the corresponding soft X-ray loop becomes dominant in the L-band. In the M1-band the loop-top source is not dominant, but shows comparable intensity to the footpoint sources. As mentioned in the previous section, this loop-top source begins to brighten in the impulsive phase and reaches its maximum in the gradual phase. This source shows a very soft spectrum, which is well explained by an emission from a super-hot thermal plasma with a temperature of $\sim 30 - 50$ MK (Masuda 1994).

These characteristics are similar to the hard X-ray component that was first detected by Lin's balloon experiment (Lin *et al.* 1981). This was only a spectroscopic, not an imaging observation. Now HXT can clearly show where this super-hot thermal component exists, and how it evolves through the whole duration of a flare.

A similar component is observed during so-called long-duration events (LDEs). The size of LDEs is generally larger than that of impulsive flares. For HXT it is more difficult to synthesize such an extended hard X-ray source. However, after a new calibration and improvement of the software (Sato *et al.* 1999), good quality images of some LDEs were finally obtained. In the case of a famous LDE which occurred on February 21 1992, SXT observations clearly show a cusp-shaped loop (Tsuneta *et al.* 1992). In hard X-rays (14 – 23 keV) an extended source seems to fill the whole of a bright soft X-ray loop. There are a few bright patches inside the extended source, but no hard X-ray sources in the footpoint region. Since the hard X-ray spectrum of this extended source is very soft and shows a very gradual temporal behavior, it might be the same type of hard X-ray source as the looptop source observed during the gradual phase of impulsive flares (Masuda 1998).

HARD X-RAY SOURCES IN THE INITIAL PHASE

What kind of mechanism triggers a solar flare? To answer this question it is important to observe hard X-ray features in the very early phase (initial phase). Several authors have taken up the challenge to analyze HXT data during the initial phase, in spite of the fact that the number of photons is not large enough to synthesize a hard X-ray image with a good quality.

Takakura *et al.* (1993) reported that in several flares a hard X-ray source first appears at the apex of a flare loop and then the (double) footpoint sources appear in the impulsive phase. This suggests that the energy-release takes place high in the corona. On the other hand, recently it was reported that initially faint double footpoint sources appear, then a hard X-ray looptop source with a very high temperature ($\sim$ 80 MK), and finally again double footpoint sources brighten in the impulsive phase (Uchida *et al.* 2001). In this case it seems that the initial energy is released at the footpoints. Generally it's not easy to analyze the initial phase because the number of photons is very small compared to that in the impulsive and gradual phases. However, its importance is clear and we need further studies on this topic.

PARTICLE ACCELERATION

Acceleration Sites

A hard X-ray source exists above the corresponding soft X-ray loop during the impulsive phase (Masuda 1994, Masuda *et al.* 1994). This is evidence that high-energy electrons are energized outside of the loop. Aschwanden *et al.* (1996*a*) analyzed the January 13 1992 flare using BATSE data from the *Compton Gamma Ray Observatory (CGRO)* They claim that for the thick target model the distance between the coronal acceleration site and the chromospheric hard X-ray emission site can be determined from velocity-dependent electron time-of-flight (TOF) differences (Aschwanden *et al.* 1996*b*). The TOF distance they find is $44,000 \pm 6,000$ km, which corresponds to a higher altitude than that of the soft X-ray loop ($\sim 12,500$ km) and the above-the-looptop hard X-ray source ($\sim 22,100$ km). This result suggests that the particle acceleration takes place above the hard X-ray source. Later a statistical analysis of TOF distance was done by the same group (Aschwanden *et al.* 1996*c*), in which they analyzed 42 events, including five flares which show a hard X-ray source above the soft X-ray loop. They find that there is a relationship between the TOF distance, l', and the flare loop half-length, s, $l'/s = 1.4 \pm 0.3$. The heights of the hard X-ray sources in these five flares are consistent with the electron TOF distance to the footpoints. They conclude that particle acceleration in solar flares occurs in the cusp region above the flare loop and that the above-the-looptop hard X-ray source is a signature of the acceleration site, probably controlled by the magnetic reconnection process.

On the other hand, another acceleration site may exist. Nishio *et al.* (1997) analyzed 14 impulsive flares, simultaneously observed with *Yohkoh* and the Nobeyama radio heliograph. At least 10 out of the 14 flares involve two loops, one short ($< 20''$) and one long ($30'' - 80''$). Microwave emission is detected from both loops, while hard X-ray emission originates from the shorter loop which is also brighter in soft X-rays. However, the intensity variations of the microwaves from both loops are similar. Nishio *et al.* conclude that this type of flare is caused by interaction between two loops, that high-energy electrons are injected into both loops, and that the lack of hard X-ray emission at one end of the larger loop might be due to the magnetic mirror effect. Hanaoka analyzed three C-class flares, observed with *Yohkoh*/HXT, *CGRO*/BATSE, and the Nobeyama radio heliograph (Hanaoka 1999). All of the these flares show a double-loop structure. The main radio/hard X-ray source is located near one of the footpoints of a large overlying loop, where a newly loop emerges. The main and the remote source show a correlated brightness fluctuation, but the fluctuation of the remote source lags behind that of the main source by about 500 ms. This suggests that the particles are accelerated near the main source by the interaction between the small emerging loop and the overlying large loop, while the time-lag corresponds to the travel time of high-energy electrons from the interaction site to the far end of the large loop.

Acceleration Mechanisms

Although the most important goal of HXT was to reveal the particle acceleration mechanism in solar flares, we have not yet determined what kind of acceleration mechanism dominates. There are many acceleration models such as stochastic acceleration, shock acceleration, DC electric field acceleration and so on (Miller *et al.* 1997 and references therein).

The pitch angle distribution of accelerated electrons gives us information about the mechanism. However, we can not directly measure it. Hard X-ray observations could contribute to solve this problem. Sakao (1994) and Sakao *et al.* (1994) studied seven impulsive flares which show a double footpoint structure in the energy range above 30 keV during the impulsive phase. They compared the hard X-ray intensities of the double sources, their spectral hardness, and photospheric magnetic field strengths where they are located. They found that the brighter source has a harder spectrum and that it is located at the weaker magnetic field region. These results suggest that electrons precipitating to the less bright footpoint are mirrored back at a relatively higher altitude due to the stronger magnetic field convergence. A mechanism that accelerates electrons perpendicular to the magnetic field, is preferred by these results, but other mechanisms are not rejected.

This is a statistical approach. Another approach was taken by Masuda *et al.* (2001). They studied in detail a large arcade-type flare which occurred on July 14 2000. For the first time a clear two-ribbon structure is observed in the energy range above 30 keV during this flare. That structure is very similar to the two-ribbons observed in Hα and EUV. This suggests that electrons are in fact accelerated in the whole of this arcade, not merely in a particular dominant loop. Analyzing the spectral distribution of the hard X-ray ribbons outer edge regions shows harder spectra. This means that higher energy electrons precipitate more at the outer edge than at the inner area of the ribbons. There are several interpretations. One of them is as follows. In the cusp-type magnetic reconnection scenario the outer loops are newly reconnected loops. According to this, higher-energy electrons dominate in the newer loops. This might be caused by energy-dependent pitch-angle distribution, i.e. higher-energy electrons tend to have smaller pitch angles. In this case the acceleration should occur along the magnetic field. Of course, this is just one of several interpretations. In any case it is important to find small differences in hard X-ray images taken in different energy bands, especially in the M2- and H-bands. Even a small difference may give us a hint towards revealing the particle acceleration mechanism in solar flares.

Lower-Cutoff Energy of Accelerated Electrons

It is very important to determine the lower cut-off energy of non-thermal electrons because the total energy of non-thermal electrons strongly depends on it in the thick-target model (Brown 1971). Though 20 keV or 25 keV is often used for the calculation of the total energy, those values are not confirmed by observations. Required are spectroscopic observations in the energy range of 10 – 40 keV with a high energy and time resolution, such as those by *RHESSI* and SoXS (Lin *et al.* 2000, Jain *et al.* 2002). Of course there are several simultaneous components in a flare. Imaging observations like those from HXT can help us to interpret the results derived from such spectroscopic observations.

PLASMA HEATING

As mentioned above, a super-hot thermal source is observed at the looptop portion in many flares. How is such a super-hot plasma created?

A statistical result derived from HXT gives a hint towards the answer of this question. Sakao *et al.* (1998) found that flares which have a significant super-hot component tend to increase their footpoint separations with time, while the ones which show only a non-thermal component, don't show this increase, sometimes

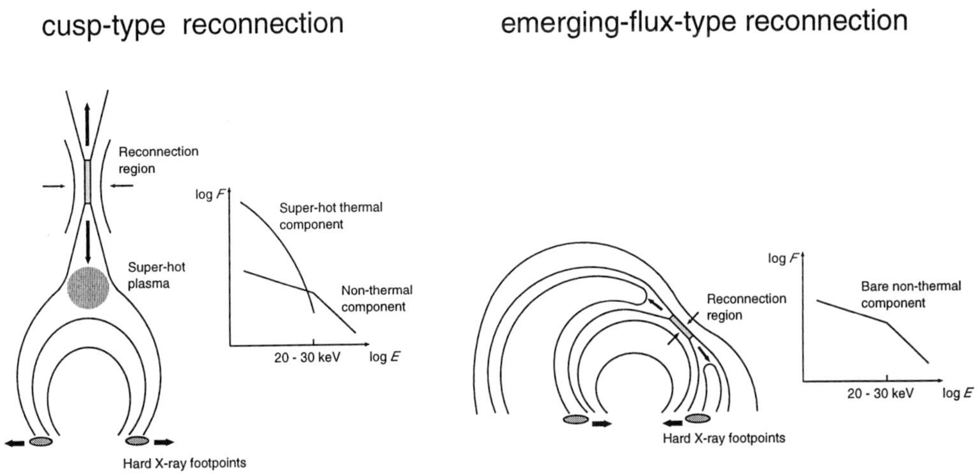

Fig. 3. Relationship between magnetic field configuration and thermal/non-thermal hard X-ray production in solar flares. Flares caused by cusp-type reconnection produce super-hot plasmas at the looptop while little such plasma is produced in flares caused by the emerging-flux-type reconnection. (after Sakao *et al.* 1998)

even a decrease. This tendency is caused by the difference in the magnetic configuration between both groups. A super-hot plasma is created near the looptop portion by the downward flow caused by cusp-type magnetic reconnection high in the corona. On the other hand, in the emerging-flux type reconnection such a high-temperature plasma is not created because there are no targets for the fast reconnection flow. Figure 3 shows a simple rendition of this interpretation.

Tsuneta *et al.* (1997) reported a high temperature source above the soft X-ray loop of the impulsive flare on January 13 1992. This source coincides in position with the above-the-looptop hard X-ray source and continues to sit above the soft X-ray loop throughout the flare. The single high-temperature source in the initial phase evolves into two high-temperature ridge structures in the peak and decay phases, and the compact hard X-ray source appears to be located in between the high-temperature ridges. They conclude that the high temperature region observed with SXT is heated by slow shocks associated with magnetic reconnection and that the above-the-looptop hard X-ray source is heated by the fast shock owing to the collision of the supersonic downward outflow with the reconnected magnetic fields. However, from numerical simulations Fletcher & Martens (1998) have shown that the above-the-looptop emission and its spectral and spatial structure, including the ridges, can also be generated by high energy electrons and protons that are bottled up in the cusp by the magnetic mirror effect.

ACKNOWLEDGEMENTS

We would like to express our sincere thanks to ISAS, NASA, SERC, and the *Yohkoh* team for their continuous and valuable support of the mission.

REFERENCES

Aschwanden, M.J., H.S. Hudson, T. Kosugi, and R.A. Schwartz, *ApJ*, **464**, 985 (1996*a*).

Aschwanden, M.J., M.J. Meredith, H.S. Hudson, T. Kosugi, and R.A. Schwartz, *ApJ*, **468**, 398 (1996*b*).

Aschwanden, M.J., T. Kosugi, H.S. Hudson, M. . Meredith, and R.A. Schwartz, *ApJ*, **470**, 1198 (1996*c*).

Brown, J.C., *Solar Phys.*, **18**, 489 (1971).

Duijveman, A., P. Hoyng, and M.E. Machado, *Solar Phys.*, **81**, 137 (1982).

Fletcher, L., and Martens, P.C.H., *ApJ*, **505**, 418 (1998).

Hanaoka, Y., *PASP*, **51**, 483 (1999).

Hudson, H.S., T. Kosugi, N.V. Nitta, and M. Shimojo, *ApJ Letters*, **561**, L211 (2001).

Hoyng, P., M.E. Machado, A. Duijveman, A. Boelee, C. de Jager, R. Fryer, M. Galama, R. Hoekstra, J. Imhof, H. Lafleur, H.V.A.M. Maseland, W.A. Mels, A. Schadee, J. Schrijver, G.M. Simnett, Z. Svestka, H.F. van Beek, W. van Tend, J.J M. van der Laan, P. van Rens, F. Werkhoven, A.P. Willmore, J.W.G. Wilson, and W. Zandee, *ApJ Letters*, **244**, L153 (1981).

Jain, R., H. Dave, K.S.B. Manian, A.B. Shah, N.M. Vadher, V.M. Shah, S.L. Kayasth, V.D. Patel, and M.R. Deshpande, in *Probing the Sun with High Resolution*, in press (2002).

Kosugi, T., T. Sakao, S. Masuda, H. Hara, T. Shimizu, and H.S. Hudson, in *New Look at the Sun with Emphasis on Advanced Observations of Coronal Dynamics and Flares, Proc. of Kofu Symp.*, eds. S. Enome and T. Hirayama, p. 127 (1994).

Kosugi, T., K. Makishima, T. Murakami, T. Sakao, T. Dotani, M. Inda, K. Kai, S. Masuda, H. Nakajima, Y. Ogawara, M. Sawa, and K. Shibasaki, *Solar Phys.*, **136**, 17 (1991).

Lin, R.P., and the *HESSI* Team, in *High Energy Solar Physics: Anticipating HESSI*, ASP Conference Series 206, 1 (2000).

Lin, R.P., R.A. Schwartz, R.M. Pelling, and K.C. Hurley, *ApJ Letters*, **251**, L109 (1981).

Masuda, S., Ph. D. thesis, The university of Tokyo (1994).

Masuda, S., T. Kosugi, and H.S. Hudson, *Solar Phys.*, **204**, 55 (2001).

Masuda, S., T. Kosugi, T. Sakao, and J. Sato, in *Observational Plasma Astrophysics: Five Years of Yohkoh and Beyond*, eds. T. Watanabe, T. Kosugi, and A. C. Sterling, p. 259, Kluwer Academic Publisher, Dordrecht (1998).

Masuda, S., T. Kosugi, H. Hara, T. Sakao, K. Shibata, and S. Tsuneta, *PASP*, **47**, 677 (1995).

Masuda, S., T. Kosugi, H. Hara, S. Tsuneta, and Y. Ogawara, *Nature*, **371**, 495 (1994).

Miller, J.A., P.J. Cargill, A.G. Emslie, G.D. Holman, B.R. Dennis, T.N. LaRosa, R.M. Winglee, S.G. Benka, and S. Tsuneta, *JGR*, **102**, 14631 (1997).

Nishio, M., K. Yaji, T. Kosugi, H. Nakajima, and T. Sakurai, *ApJ*, **489**, 976 (1997).

Ogawara, Y., T. Takano, T. Kato, T. Kosugi, S. Tsuneta, T. Watanabe, I. Kondo, and Y. Uchida, *Solar Phys.*, **136**, 1 (1991).

Sakao, T., Ph. D. thesis, The university of Tokyo (1994).

Sakao, T., T. Kosugi, and S. Masuda, in *Observational Plasma Astrophysics: Five Years of Yohkoh and Beyond*, eds. T. Watanabe, T. Kosugi, and A. C. Sterling, p. 273, Kluwer Academic Publisher, Dordrecht (1998).

Sakao, T., T. Kosugi, S. Masuda, K. Yaji, M. Inda-Koide, and K. Makishima, in *New Look at the Sun with Emphasis on Advanced Observations of Coronal Dynamics and Flares*, eds. S. Enome and T. Hirayama, p. 169 (1994).

Sakao, T., T. Kosugi, S. Masuda, M. Inda, K. Makishima, R.C. Canfield, H.S. Hudson, T.R. Metcalf, J.-P. Wuelser, L.W. Acton, and Y. Ogawara, *PASP*, **44**, L83 (1992).

Sato, J., *ApJ Letters*, **558**, L137 (2001).

Sato, J., T. Kosugi, and K. Makishima, *PASP*, **51**, 127 (1999).

Shibata, K., S. Masuda, H. Hara, T. Yokoyama, S. Tsuneta, T. Kosugi, and Y. Ogawara, *ApJ Letters*, **451**, L83 (1995).

Takakura, T., M. Inda, K. Makishima, T. Kosugi, T. Sakao, S. Masuda, and T. Sakurai, *PASP*, **45**, 737 (1993).

Takakura, T., K. Tanaka, and E. Hiei, *Adv. Space Res.*, **4**, 143 (1984).

Tanaka, K., *PASP*, **39**, 1 (1987).

Tsuneta, S., S. Masuda, T. Kosugi, and J. Sato, *ApJ*, **478**, 787 (1997).

Tsuneta, S., H. Hara, T. Shimizu, L.W. Acton, K.T. Strong, H.S. Hudson, and Y. Ogawara, *PASP*, **44**, L63 (1992).

Tsuneta, S., L. Acton, M. Bruner, J. Lemen, W. Brown, R. Caravalho, R. Catura, S. Freeland, B. Jurcevich, M. Morrison, Y. Ogawara, T. Hirayama, and J. Owens, *Solar Phys.*, **136**, 37 (1991).

Tsuneta, S., T. Takakura, N. Nitta, K. Ohki, K. Tanaka, K. Makishima, T. Murakami, M. Oda, and Y. Ogawara, *ApJ*, **280**, 887 (1984).

Uchida, Y., M.S. Wheatland, R. Haga, I. Yoshitake, and D. Melrose, *Solar Phys.*, **202**, 117 (2001).

Wang, H., D.E. Gary, H. Zirin, T. Kosugi, R.A. Schwartz, and G. Linford, *ApJ Letters*, **444**, L115 (1995).

LOOPTOP AND FOOTPOINT IMPULSIVE HARD X-RAYS AND STOCHASTIC ELECTRON ACCELERATION IN FLARES

V. Petrosian

Center for Space Science and Astrophysics, Stanford University, Stanford, CA 94305, USA

ABSTRACT

The discovery of hard X-rays from tops of flaring loops by the HXT of *Yohkoh* represents a significant progress in the understanding of solar flares. This report describes the properties of 20 limb flares observed by *Yohkoh* from October 1991 to August 1998, 15 of which show detectable impulsive looptop emission. Considering the finite dynamic range (about a decade) of the detection it can be concluded that looptop emission is a common feature of all flares. The light curves and images of a representative flare are presented and the statistical properties of the footpoint and looptop fluxes and spectral indices are summarized. The importance of these observations, and those expected from *RHESSI* with its superior angular, spectral, and temporal resolution, in constraining the acceleration models and parameters is discussed briefly.

INTRODUCTION

The most significant discovery of the HXT instrument on board the *Yohkoh* satellite has been the detection of hard X-ray emission from the top of solar flare loops as well as their footpoints. The first so-called "Masuda" flare is that of January 13, 1992 (Masuda *et al.* 1994; see also Alexander & Metcalf 1997), which is clearly delineated by a soft X-ray (thermal) loop, and shows three compact hard X-ray sources, two located at the footpoints (FPs) and a third near the loop top (LT). Several other such sources are described in Masuda's thesis (1994). As pointed out by Masuda *et al.* (1994), these observations lend support to theories that place the location of flare energy release high up in the corona. The power law hard X-ray spectra of the LT sources indicate that electron acceleration is indeed occurring at or near these locations. The exact mechanism of the acceleration is a matter of considerable debate. In previous works (see Petrosian 1994 and 1996) we have argued that among the three proposed particle acceleration mechanisms (electric fields, shocks, and plasma turbulence or waves) the stochastic acceleration of ambient plasma particles by plasma waves provides the most natural mechanism and can explain the observed spectral features of flares (Park, Petrosian, & Schwartz 1997; hereafter **PPS**). In two recent works (Petrosian & Donaghy, 1999 and 2000; **PD**) we demonstrated that the observed characteristics of the Masuda flares can be used to constrain the model parameters. In order to gain a clearer picture of the frequency of occurrence of LT sources and the relative values of the fluxes and spectral indices of the FP and LT sources, we (Petrosian, Donaghy, & McTiernan 2002; **PDM**) have expanded and extended Masuda's analysis. In the next two sections I first summarize the results of this work and then comment on their consequence for the acceleration mechanism.

DATA ANALYSIS AND RESULTS

We have used the *Yohkoh* HXT Image Catalogue (Sato *et al.* 1988) to search for flare candidates for detection of LT emission. We have used Masuda's (1994) selection criteria (heliocentric longitude > 80 degrees, peak count rate > 10 counts per sec per subcollimator in the $\sim 33 - 53$ keV range, i.e. the M2 channel). We found 20 such events from 10/91 through 8/98, of which 11 were selected by Masuda for the period of 10/91 to 9/93. Observations of two events are interrupted by spacecraft night. Of the remaining 18 events, 15 show detectable impulsive looptop emission. As described below, considering that the finite dynamic range (about a decade) of the detection introduces a strong bias against observing comparatively weak looptop sources, one can conclude that LT emission is a common feature of all flares. An interesting aside, is that of the 9 new events, 3 appear to be examples of interacting loop structures with multiple LT and FP sources, of the type analyzed by Aschwanden *et al.* (1999). It is surprising that none of the 11 Masuda events are in this category.

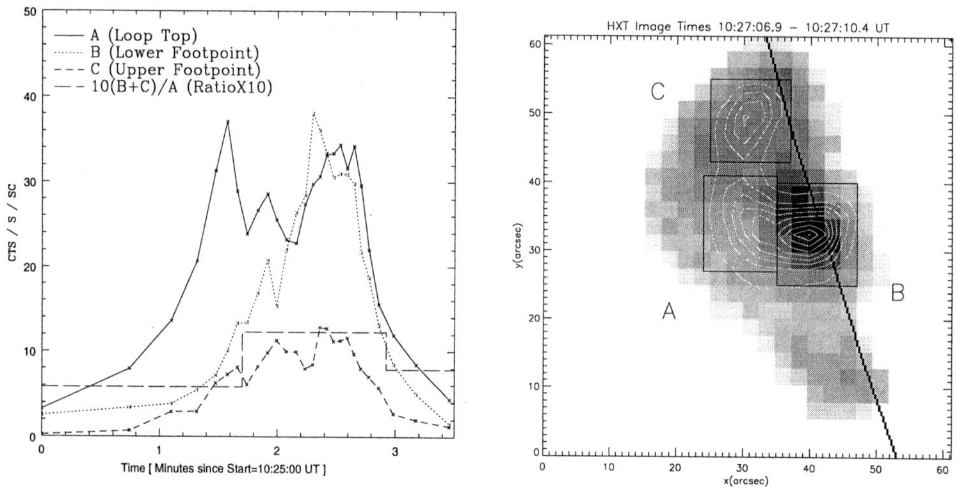

Fig. 1. Images (*Right Panel*) and light curves (*Left Panel*) for the December 18 1991 flare. The contours and the gray scale show the HXT (channel M2; $\sim 33 - 53$ keV) and SXT images of the loop, respectively, for the specified time. The diagonal line shows the location of the solar limb. The brightest contour and the contour separations are $B_{max} = 8.1$ and $\Delta B = 0.73$ counts/pixel with 2.5 sq. arc second size pixels. The light curves of the the LT and FP sources refer to the counts integrated over regions shown on the right panel. The dashed histogram shows the average of the ratio of counts $\mathcal{R} = FPs/LT$ (multiplied by 10) for three time intervals.

We have constructed HXT images and investigated their evolution throughout all these flares using the *Yohkoh* spectral and spatial analyses software packages. In a few cases we have also used the Alexander & Metcalf (1997) "pixon" method of image reconstruction. From the investigations of these images we have determined the locations of LT and FP sources and produced separate light curves for the well defined sources. Figure 1 shows an example of a simple loop with an intermediate strength LT source and Figure 2 shows a flare with a more complex morphology consisting of two loops with different but related temporal evolution.

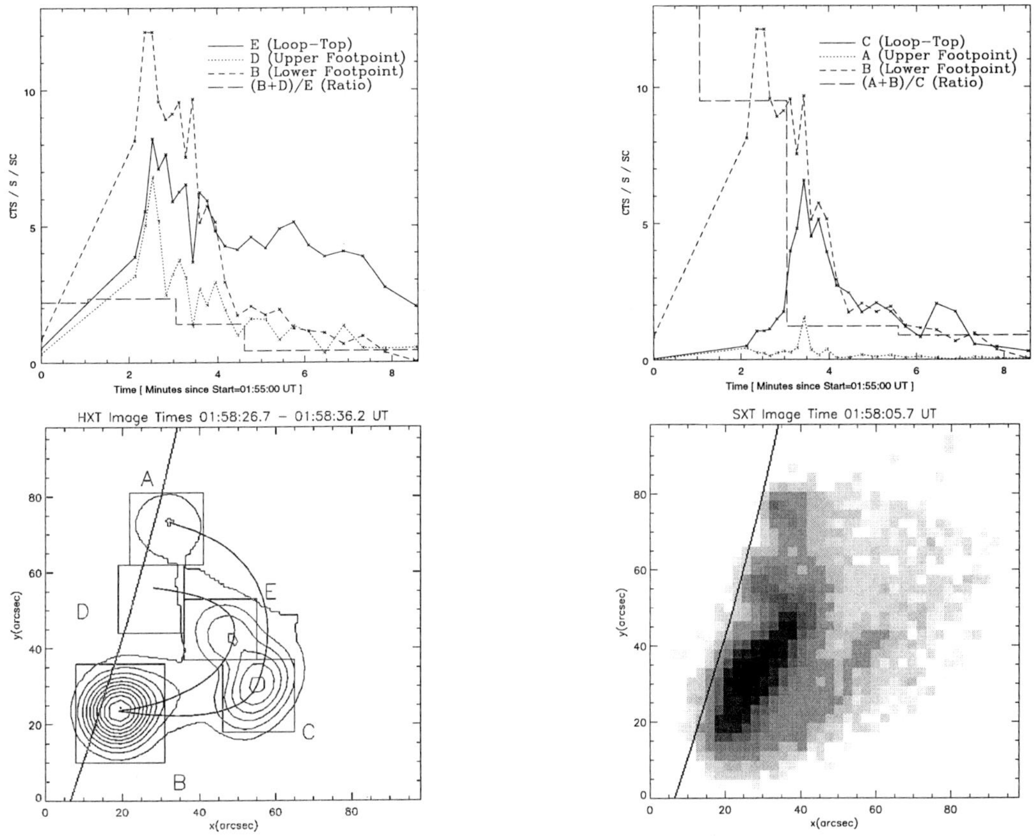

Fig. 2. Same as Figure 1 for the August 18, 1998 flare. The upper left and right panel light curves represent the southern (AEB) and the northern (BDE) loops, respectively. Note that for the LT source D we plot counts divided by 3. In the HXT image (lower left panel) $B_{max} = 14.8$, $\Delta B = 0.82$ counts/pixel, the digonal line shows the limb location, and the two arcs sketch the presumed loop outlines. The SXT image shown on the lower right panel was taken nearly two minutes after the HXT image.

We determine the relative fluxes of the LT and FP sources and obtain rough measures of some of the spectral characteristics (e.g. power-law indices). Figure 3 shows the M1 channel ($\sim 23 - 33$ keV) counts of the FPs vs. LT sources for all flares (left panel) and the distributions of the count ratio $\mathcal{R} = FPs/LT$ (right panel). We use a representative time period around an impulsive peak and avoid the later stages (the third periods of the histograms shown along the light curves) which can be contaminated by thermal emission. Note that some flares (those connected by dashed lines) contribute more than one data point.

Analysis of these results lead to the following very important conclusions (see also **PDM**).

• The LT hard X-ray emission seems to be a common characteristic of the impulsive phase of solar flares, appearing in some form in 15 of the 18 selected flares. The absence of LT emission in the remaining cases (those indicated by the horizontal arrows in Figure 3) is most likely due to the finite dynamic range of the imaging technique which is about 10. The hatched diagonal regions show this range. Flares outside the area between these two bounds will have either a too weak a FP or LT source to be detected by HXT. From

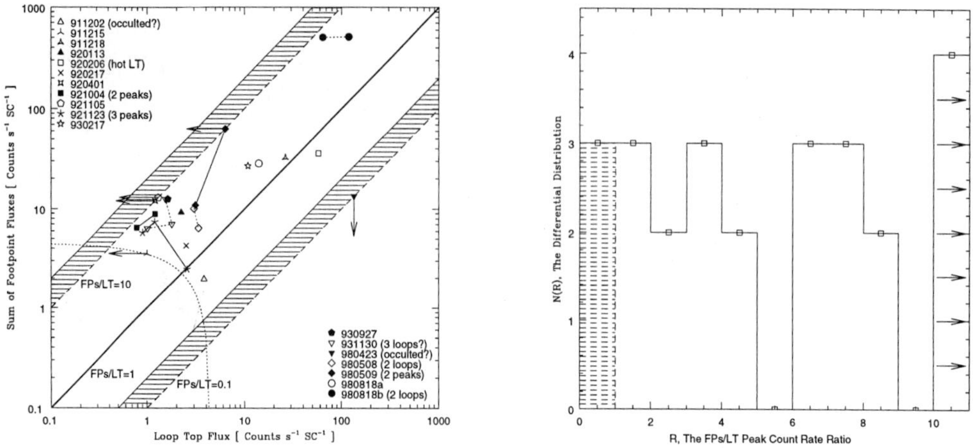

Fig. 3. (*Left Panel*): Counts from two FPs vs LT counts, in the M1 channel. The diagonal lines show lines of constant ratio ($\mathcal{R} = FPs/LT$) and represent detection thresholds arising from the finite dynamic range of the instrument which is about 10. Flares with undetected LT source are denoted by an arrow placed on the upper bound of detection of $\mathcal{R} = 10$. The dotted curve shows the event selection threshold of 10 counts at the M2 channel. (*Right Panel*): The differential distribution of the ratio $\mathcal{R} = FPs/LT$ of all flares. The arrows indicate ratios greater than the dynamic range. Some of the flares in the shaded area with $\mathcal{R} < 1$ may be occulted or be dominated by a superhot thermal component.

this we conclude that *LT emission is present in all flares.* However there are very few flares with $\mathcal{R} < 1$ and there are indications that the three such cases seen in Figure 3 are either partially occulted or are dominated by a superhot LT source. Thus one may conclude that, in general, the ratio $\mathcal{R}$ has a relatively flat distribution between 1 and 10, with few cases outside this range. A larger sample with a wider dynamic range will be required for a better determination of this distribution.

Figure 4 shows the distribution of the overall spectral index (left panel) and the distribution of the difference between low ($\sim 13 - 28$ keV, L and M1 channels) and high ($\sim 28 - 53$ keV, M1 and M2 channels) energy indices. Clearly with only a four channel data one must be cautious in the interpretation of these histograms. Nevertheless, some significant conclusions can be drawn from these results as well.

• The overall distribution of the power-law spectral index γ rises rapidly above 2, peaks around 4 and then declines gradually thereafter. This is similar to previous determinations of this distributions from HXRBS on board the *Solar Maximum Mission* (see e.g. McTiernan & Petrosian 1991), but contains a few more steep spectra, specially for LT sources. This difference could be due to thermal contamination and/or because HXT is sensitive to lower photon energies than HXRBS. On the average, the spectral index of LT sources is larger (i.e. spectra are steeper) than that of the FP sources by one unit; $\bar{\gamma}_{LT} = 6.2 \pm 1.5, \bar{\gamma}_{FP} = 4.9 \pm 1.5$. The physics of the acceleration process must certainly play a role here.

• The spectra tend to steepen at higher energies (spectral index γ increases by 1 to 2), especially for sources with $\gamma < 6$, for which the thermal contribution should be the lowest. This is the opposite of what is observed at higher energies, where spectra tend to flatten above 100's of keV (McTiernan & Petrosian 1991). The directivity of the X-ray emission and the albedo effect for the limb flares under consideration could play some role here, especially for the FP sources.

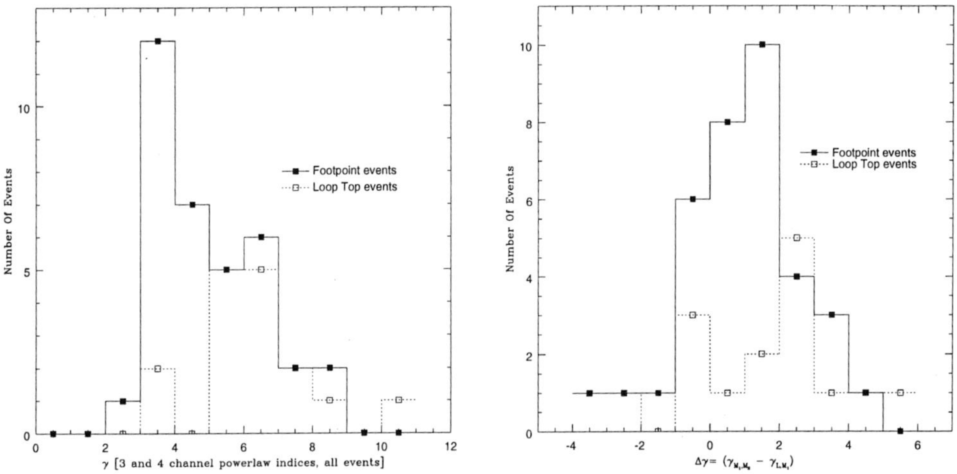

Fig. 4. The distribution of the overall spectral power-law index γ (*Left Panel*), and the distribution of $\Delta\gamma = \gamma_{M1,M2} - \gamma_{L,M1}$ (*Right Panel*). The solid histograms and filled points represent the FP sources and the dotted histogram and open points represent the LT sources.

• Finally, we note that solar flares occur in many different morphologies, the most common being a simple flaring loop with one LT and two FP sources. However, interacting loop models and even more complicated structures are frequently observed. There is a hint that the frequency of occurrence of complex morphologies may be different for the declining and growing phases of the solar cycle.

THEORETICAL IMPLICATIONS

The above results can be used to constrain the model parameters describing the plasma in the acceleration site and those describing the acceleration mechanism. These parameters define several important timescales: The acceleration time scale is related to the energy diffusion coefficient D_{EE} as $\tau_{\rm ac} \sim E^2/D_{EE}$. The mean scattering time is inversely proportional to the pitch angle diffusion coefficient, $\tau_{\rm sc} \sim 1/D_{\mu\mu}$. The time for a particle with velocity v to cross the acceleration site of size L is $\tau_{\rm tr} \sim L/v$; these two timescales determine the escape time from the acceleration site (for $\tau_{\rm sc} < \tau_{\rm tr}$, $T_{\rm esc} \sim \tau_{\rm tr}^2/\tau_{\rm sc}$, otherwise $T_{\rm esc} \sim \tau_{\rm tr}$). Finally the energy loss timescale for an electron of energy E is $\tau_{\rm L} = E/\dot{E}_{\rm L}$, which for the non relativistic electrons under consideration here is dominated by the Coulomb losses, $\tau_{\rm Coul} = vE/(4\pi r_0^2 {\rm ln}\Lambda n m c^4)$, where $4\pi r_0^2 {\rm ln}\Lambda = 2 \times 10^{-23}$ cm^2 and m is the mass of the electron. The values of these time scales depend on the plasma density n, magnetic field B, plasma turbulence energy density $w_{\rm turb}$ and size L, and their variations with energy depend on these parameters and the spectrum of the turbulence (for details see **PPS** and **PD** and reference cited there).

For example, if $T_{\rm esc}$ is large the accelerated electron spend a long time in the acceleration site or at the loop top giving rise to a strong LT source. Inversely, a weak LT source is expected for a short $\tau_{\rm Coul}$. Very roughly, the ratio of the FP to LT emission is expected to vary as $\mathcal{R} = J_{FPs}/J_{LT} \sim \tau_{\rm Coul}/T_{\rm esc}$, where the J's refer to the expected bremsstrahlung fluxes. Furthermore, the spectral shape of these fluxes are also related to the above mentioned parameters. It is clear, for example, that if the acceleration time is short compared to the escape time, then more electrons get to higher energies resulting in a flat accelerated electron spectrum and LT hard X-rays. For a power law accelerated spectrum, $f(E) \propto E^{-\delta}$, the LT (*thin target*) hard X-ray spectrum at photon energy k is $J_{LT} \propto k^{-\delta-1/2}$. On the other hand, the spectrum of the electrons that escape the acceleration region and reach the footpoints is $f(E)/T_{\rm esc}(E) \propto E^{-\delta-s'}$, assuming that for the small energy range of the HXT we can use the approximation $T_{\rm esc}(E) \propto E^{s'}$. These electrons will emit a *thick target* spectrum at the footpoints with $J_{FPs} \propto k^{-\delta-s'+1}$. Thus, for LT spectra that are steeper than the FPs spectra we require $s' < 3/2$. When the escape is determined by the traverse time $s' = -1/2$ and this condition is satisfied. And when scattering dominates, this requires $\tau_{\rm sc} \propto E^{s}$, with $s > -5/2$. The energy dependence of $\tau_{\rm sc}$ depends on the characteristics of the turbulence. In general, one expects a positive value for s, and even for for a very steep spectrum of the turbulence one has $s > -1$ (see e.g. Pryadko & Petrosian 1997).

However, it should be noted that these relations are very approximate and valid only for a limited energy range and very steep electron spectra; they break down completely for $\delta < 2.5$. Nevertheless, this exercise demonstrates that using the observed values of the spectral indices and FP and LT counts we can determine the plasma and acceleration characteristics. As shown in **PD** the values of the parameters such derived from the *Yohkoh* high spatial resolution data are very reasonable, and agree with those derived by **PPS** from fits to large dynamic range overall spectra. It is clear then that more refined and simultaneous observations of the flare characteristics can yield important information about the the acceleration mechanism, the energy release, and the evolution of solar flares. We eagerly anticipate the increased spectral, temporal and spatial resolution possible with the instruments of the *RHESSI* satellite.

ACKNOWLEDGEMENTS

This work is supported in parts by NASA grants NAG-5-7144-0002 and NAG5-8600-0001. This paper was completed during the authors stay at the Institute for Theoretical Physics at UC Santa Barbara, which is supported in part by the National Science Foundation Under Grant No. PHY99-07949.

REFERENCES

Alexander, D., and Metcalf, T.R., *ApJ*, **489**, 442 (1997).

Aschwanden *et al.*, *ApJ*, **526**, 1026 (1999).

Masuda, S., Ph.D. Thesis, The University of Tokyo (1994).

Masuda, S. *et al.*, *Nature*, 371, 455 (1994).

McTiernan, J.M., and Petrosian, V., *ApJ*, **379**, 381 (1991).

Park, B.T., Petrosian, V., and Schwartz, R.A., *ApJ*, **489**, 358 (1997) [**PPS**].

Petrosian, V., in *High-energy Solar Phenomena*, edsJ.M. Ryan and W.T. Vestrand, AIP ConfProc. **294**, 162 (1994).

Petrosian, V., in *High Energy Solar Physics*, eds. R. Ramaty, N. Mandzhavidze, and X-M. Hua, AIP Conf. Proc. **374**, 445 (1994).

Petrosian, V., and Donaghy, T.Q., *ApJ*, **527**, 945 (1999) [**PD**].

Petrosian, V., and Donaghy, T.Q., in *High Energy Solar Physics-Anticipating RHESSI*, eds. R. Ramaty and N. Mandzhavidze, ASP Conf. Series, **206**, 215 (2000) [**PD**].

Petrosian, V., Donaghy, T.Q., and McTiernan, J.M., *ApJ*, **569**, 459 (2002) [**PDM**].

Pryadko, J.M., and Petrosian, V., *ApJ*, **482**, 774.

Sato, J., Sawa, M., Masuda, S., Sakao, T., Kosugi, T., and Sekiguchi, H., The *Yohkoh* HXT Image Catalogue, Noboyama Radio Observatory Publication (1998).

SOFT X-RAY HIGH-TEMPERATURE REGIONS ABOVE SOLAR FLARE LOOPS

S. Akiyama[1,2] and H. Hara[3]

[1]*Institute for Computational Sciences and Informatics, George Mason University, Fairfax, VA 22030, USA*
[2]*Space Science Division, Naval Research Lab., Code 7670A, Washington DC 20375-5352, USA*
[3]*National Astronomical Observatory, 2-21-1 Osawa, Mitaka, Tokyo 181-8588, Japan*

ABSTRACT

We analyze data from 141 solar flares that were observed with the *Yohkoh* soft and hard X-ray telescopes to investigate the general characteristics of the soft X-ray high-temperature region above soft X-ray flare loops. We performed a careful analyses in order to obtain accurate temperature maps of the high-temperature regions, and found high-temperature regions of 15–35 million degrees above soft X-ray flare loops for 64 flares out of the 141 events. The region appears at the hard X-ray impulsive phase and its temperature reaches a maximum before the time of the soft X-ray peak. The volume emission measure of the region is 10^{47-48} cm^3 on the average, which is about ten times smaller than that of the 10 million degree flare loop below it. The high-temperature regions tend to move outwards at a speed of 10–20 km s^{-1}, which is faster than the rising speed of the flare loops of 5–10 km s^{-1}.

INTRODUCTION

Masuda *et al.* (1994) discovered hard X-ray sources above soft X-ray loops, which suggested that a reconnection process might take place in impulsive flares as well as in LDE flares. According to Masuda *et al.* (1994), the loop-top hard X-ray sources had effective temperatures of 100–150 MK and a total emission measure (EM) of 10^{44} cm^{-3} under the assumption of a thermal (superhot) source. Moreover, the areas above the flare loop showed higher temperatures than other regions in the main loops. The temperatures and the total EMs in these areas were about 30 MK and 10^{47-48} cm^{-3}, respectively. Tsuneta *et al.* (1997) investigated the high temperature (HT) region of the January 13, 1992 flare and reported following characteristics. (1) The temperature of the hot source (15–20 MK) was 1.5 times higher than that of flare loop (9–12 MK). (2) The total EM of the hot source (2×10^{48} cm^{-3}), was an order of magnitude less than that of the flare loop (10^{47} cm^{-3}). (3) A single HT region appeared before the soft X-ray peak and evolved into two HT sources during the peak and decay phases. Based on these observations the authors suggested that the HT region was heated by a slow shock associated with magnetic reconnection. Similar HT regions above flare loops have been observed in 16 events out of 33 with the soft X-ray telescope (SXT) and the Bragg Crystal Spectrometer (BCS) aboard *Yohkoh* (Doschek 1999). Warren *et al.* (1999) reported hot plasma (15–20 MK) at the top of an arcade by comparing *Transition Region and Coronal Explorer* (*TRACE*) and SXT images.

Until recently the physics of the HT regions was not well understood. First, it was difficult to make accurate temperature maps for faint structures in SXT images. Second, a statistical approach for the HT region was not thoroughly done, with discussion of their physical parameters and their relationship to the soft X-ray loop. Due to limitations of space, we refer for a detailed description of how accurate temperature maps are made to Akiyama (2001). In this paper we focus on the physical parameters of the HT region and examine the characteristics of HT regions above flare loops.

OBSERVATION AND ANALYSIS

The observations were made with SXT on *Yohkoh*. Only partial-frame images (PFIs) in flare mode were used, which were taken in high temporal and spatial resolution. For this study flares are selected from the *Yohkoh* HXT image catalog, edited by Sato *et al.* (1998), that satisfy the following conditions: (1) Flares are observed by both the Al 11.6 and the Be 119 filters, which is the most temperature-sensitive filter combination for flaring plasma. (2) The number of full-resolution PFI images obtained from the two filters is larger than 30, so that the temporal evolution of the HT region can be examined. (3) Near-limb flares located at heliocentric longitudes exceeding ± 60 degrees are chosen to study the vertical structures above the soft X-ray loops. A total of 141 flares was selected with the above-mentioned criteria from the period October 1, 1991 to August 31, 1998.

Generally speaking the temperatures obtained from SXT images by a filter ratio method are less reliable in the pixels with low intensity. Therefore we use only pixels in the images with intensities larger than 1/50 of the maximum intensity and with a Poisson noise error less than 10%. Successive images, usually taken in two minute intervals, are coaligned and summed to improve the photon statistics. Only scattered light structures from the central part of the point spread function (PSF) are considered in the analysis, applying the PSF calculated from the procedure "SXT_PSF" in the *Yohkoh* software.

An HT region is defined as an area which satisfies the following criteria: (1) The temperature of the region should be more than 1.5 times the average temperature of the main flare loop. In this paper, the flare loop is defined as the area where the emission measure in each pixel is more than 30% of the peak emission measure. (2) The region should have more than two contiguous pixels, which have temperatures satisfying the first criterion. (3) The region should be continuously observed in more than three maps. To find the HT regions we made movies of temperature maps which consisted of running-mean difference images integrated for about two minutes. We found 64 flares with HT regions above the flare loop tops. Next, the size of the HT regions is defined as the area where the temperatures are more than 80% of the maximum temperature in the selected region. We then use this region to obtain the physical parameters of the HT region. Figure 1 shows an example of a flare loop and HT region. The white line and the black line in the left panel correspond to the flare loop region and the HT region.

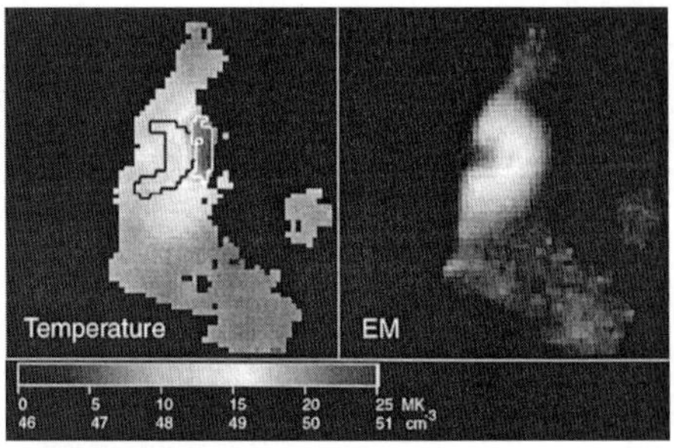

Fig. 1. HT region and flare loop. The left and the right panel show the temperature and the EM maps, respectively. In the left panel, the HT region and the flare loop are shown by a white line and a black line respectively.

RESULTS AND DISCUSSION

We study the average temperature, the average volume emission measures (VEM), the size of the flare loops and the HT regions, to investigate the relationship between the flare and the HT region. Each quantity is obtained at the time closest to the observed peak in the hard X-ray telescope (HXT) M1 band. In addition, since both the flare loops and the HT regions rise with time, we examine the rise speeds. Figure 2 shows the relation between the flare loops and the HT regions. As for the temperature, we can clearly see from Figure 2a that there is a correlation between the HT regions and the flare loops. The average quantities for the flare loops and the HT regions are 11 MK and 19 MK, respectively. The average VEMs of the flare loops and the HT regions are $10^{49.9}$ cm^{-3} and $10^{48.6}$ cm^{-3}, respectively. The VEM of the flare loop is about one order larger than that of the HT region. We do not see a correlation for the VEMs. The average sizes of the flare loops and the HT regions are 2.0×10^8 km^2 and 1.5×10^8 km^2, respectively. We see from Figure 2c that there is a correlation in size. Figure 2d shows the relation between the average rise speed of the HT regions and the flare loops. Apparently the rise speed of the HT regions is faster than that of the flare loops. The averages for the flare loops and the HT regions are 6.0 km s^{-1} and 15.6 km s^{-1}, respectively. Furthermore, as Tsuneta *et al.* (1997) point out, the HT region sometimes separates into two HT ridge structures during the peak and decay phases. A total of 19 out of 64 HT regions show the characteristic that the HT region separates into a few patches.

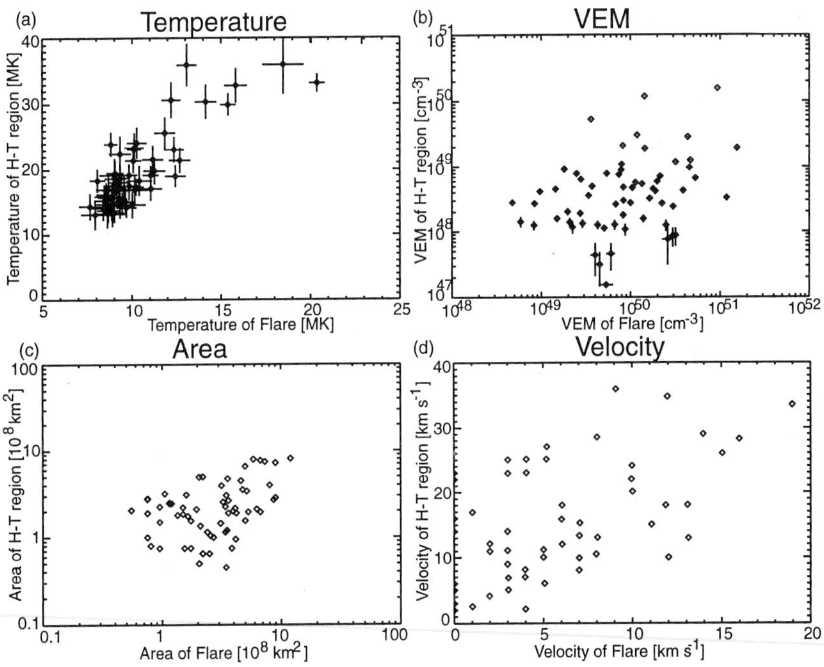

Fig. 2. The panels show the relation between the flare loops and the high temperature regions.

Figure 3 shows an example of the temporal variations of the high-temperature regions and the flare loops for the January 13, 1992 flare. Figure 3a shows time variations of the X-ray flare intensity obtained from Be 119 filter images of SXT (dot-dashed line) and the HXT M1-band (solid line). The other three panels show the time variation of the HT region (asterisks) and the flare loop (dashed line) in temperature, emission measure, and size, respectively. The following common features can be seen in these figures: the temperatures of the

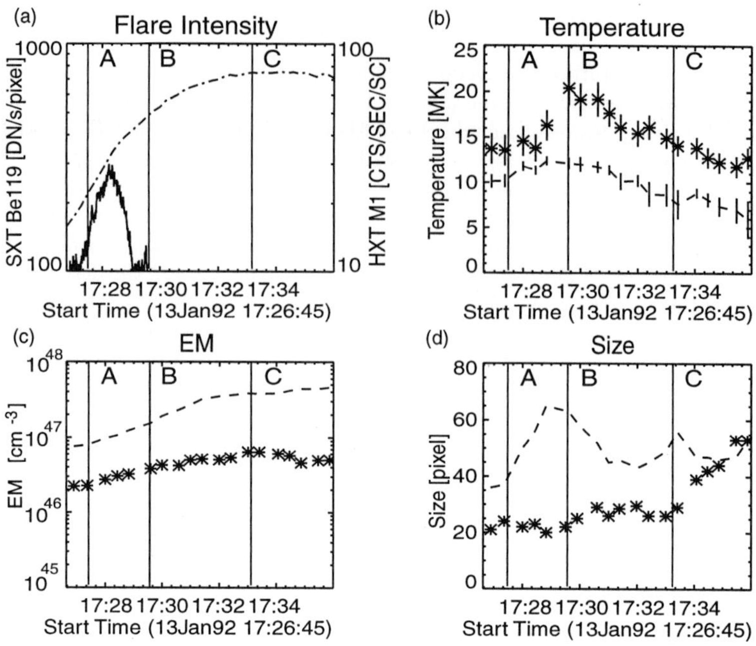

Fig. 3. Temporal variations of the flare loop (dashed line) and the HT region (asterisks). Time profiles of the soft and hard X-ray intensities (a), temperature (b), emission measure (c), and size (d) are plotted. The lines A, B, and C show the time when the intensity of the HXT M1-band begins to increase, the time of the maximum temperature of the HT region, and the time of the maximum intensity in soft X-rays, respectively.

HT regions rapidly increase during the hard X-ray impulsive phase, and the temperature of the HT region reaches its maximum between the peaks of the hard and the soft X-ray emission. This can be clearly seen in Figure 3a and Figure 3b. We see from Figure 3c that the VEM of the HT region is about one order of magnitude smaller than that of the flare loop at the coincident time of line B. It appears that the low density plasma of the HT region is heated earlier than that of the flare loop, which consists of evaporated chromospheric plasma, and we conclude that the HT region could be formed by directly heated coronal plasma, prior to evaporation. Many *Yohkoh* observations have contributed to the scenario that magnetic reconnection contributes to energy release in flares. Assuming that the reconnection region is located above the flare loop, one possibility is that the heating mechanism of the HT region could be closely related to the magnetic reconnection process.

REFERENCES

Akiyama, S., PhD Thesis., The Graduate University for Advanced Studies (2001).

Doschek, G.A., *ApJ*, **527**, 426 (1999).

Masuda, S., T. Kosugi, H. Hara, S. Tsuneta, and Y. Ogawara, *Nature*, **371**, 495 (1994).

Sato, J. *et al.*, The *Yohkoh* HXT Image Catalog (1998).

Tsuneta, S., S. Masuda, T. Kosugi, and J. Sato, *ApJ*, **478**, 787 (1997).

Warren, H. P. *et al.*, *ApJ Letters*, **527**, L121 (1999).

SCIENTIFIC RESULTS FROM RHESSI – A PREVIEW

A. G. Emslie

Department of Physics, UAH, Huntsville, AL 35899, USA

ABSTRACT

The *RHESSI* spacecraft was successfully launched on February 5, 2002. The instrumentation on board represents a quantum advance in our ability to analyze physical processes on spatial and temporal scales of physical interest, and to perform true gamma-ray spectroscopy on an astrophysical source.

INTRODUCTION

After several setbacks in its schedule, the *Reuven Ramaty High Energy Solar Spectroscopic Imager* (*RHESSI*)[1] mission was finally launched into orbit by a Pegasus rocket on February 5, 2002. After a week or so spent preparing the spacecraft, it made its first observations on February 12, 2002 and became fully operational soon afterwards. It is still too early to discuss actual flare events observed by *RHESSI*. Furthermore, significant developments in the data analysis software, taking into account the actual on-orbit response of the instrumentation, must still be performed before a true in-depth analysis of the data is possible. Nevertheless, the data so far obtained show that the scientific promise from *RHESSI* is excellent. In this "forward-looking review," I will discuss some of the exciting scientific investigations that *RHESSI* will enable us to pursue.

RHESSI CAPABILITIES

The core instrumentation of *RHESSI* (Lin *et al.* 1998) consists of two parts: (i) cooled Germanium detectors that provide exquisite spectral resolution of a few keV across the energy range from a few keV to several MeV, and (ii) matching pairs of absorbing grids that rotate with the spacecraft and so provide imaging information through the rotating modulation collimator technique (see below). Together, this combination of spectral and spatial information will provide us with data that are not only a quantum leap above that from previous instruments such as the *Yohkoh* HXT, but is also, for the first time, commensurate with spatial, spectral and temporal scales of physical interest.

The spectral resolution of the *RHESSI* Germanium detectors is sufficiently good to permit two fundamentally new investigations. First is the determination of bremsstrahlung continuum spectra with an energy resolution fine enough to define the spectral shape without the need for fitting of mathematical forms. (Spectra from low-resolution scintillators must be determined using an iterative procedure wherein the points corresponding to the count rates in each bin are used to determine a spectral form, which is in turn used to determine the weighted energy in each bin and so the position of the points used.) Subtle changes in the spectral shape, over both space and time, can thus be determined and used to constrain the physics affecting the electrons that produce the observed bremsstrahlung. Second, the width of the *RHESSI* energy channels is lower than

[1] *HESSI* was renamed *RHESSI* in March 2002, in honor of the late NASA scientist who pioneered the fields of solar-flare physics, γ-ray astronomy, and cosmic ray research.

the FWHM of all significant gamma-ray lines in the solar spectrum (with the exception of the very narrow deuterium production line at 2.223 MeV); this permits determination of shifts and shapes of these lines, with the associated wealth of true *spectroscopic* information contained therein.

In addition, *RHESSI* offers us the opportunity to perform imaging spectroscopy on cosmic sources that appear in the several-degree field of view of the instrument. These include the Crab Nebula and serendipitous gamma-ray bursts.

Imaging

Spatial information on *RHESSI* is obtained through the interpretation of the temporal modulation of the source signal as the various pairs of absorbing grids rotate across it. The basic principle is straightforward (Hurford 2001): a point source at polar coordinates (θ, ϕ) from the spacecraft rotation axis will, as the spacecraft rotates, appear to trace out a circle across the collimated grids in the spacecraft frame. This creates a temporally modulated signal as the source executes a projection of this circular path (i.e. simple harmonic motion) perpendicular to the orientation of the grids. The number of modulations per rotation cycle is equal to the number of grid slats traversed, and so provides the value of radial coordinate θ, while the phase of the modulation pattern provides the azimuthal coordinate ϕ. Each point source location (θ, ϕ) thus has its characteristic "basis" temporal modulation pattern. Identification of the strength of each of these "basis functions" in the measured signal provides the intensity of the source at each point in the plane and hence a map of the source. Equivalently, this can be viewed as each grid providing a circular trace of the source in the (u, v) Fourier plane. Inversion of these components can then be used to construct an image, using various techniques with time-honored use throughout the radio astronomy community.

The spatial resolution of this technique is the ratio of the pitch of the absorbing grids to the distance between the grid planes at the front and rear of the telescope, respectively. The finest-pitch grids are also thin (to provide a sufficiently large field of view) and so are effectively transparent to higher energy radiation. Similarly, the thicker grids do provide source modulation (and hence spatial information) at higher energies, but these grids must have larger pitches (to provide the necessary field of view) and so cruder spatial resolution. The spatial resolution is therefore a function of energy, but is as fine as 2.3 arc seconds for the thinnest grid pair. (Indeed, if we exploit higher harmonics of the [triangular] rotating modulation collimator signal, then the resolution can be pushed down to $1/n$ of this fundamental resolution value, where n is the harmonic number. The strength of the associated signal, decreases, however, like $1/n^2$.) The thinnest grids therefore effectively image, with resolution of order 2000 km or better, the deka-keV hard X-rays that are a diagnostic of electrons of similar energy, electrons that have long been known to carry a large fraction of the total energy released in the flare. With typical flare loops extending over at least 10,000 km, *RHESSI* thus has the capability of resolving the hard X-ray emission spectrum at different locations along the electrons' paths; the high spectral resolution noted above means that quite subtle differences in the spectra at adjacent locations in the structure will be resolvable and accurately determined.

The *temporal* resolution of *RHESSI* is limited mostly by photon flux issues. Nevertheless, with a half-rotation every 2 seconds, *RHESSI* has the capability of sampling the entire set of Fourier components accessible to it, and so constructing a full image, on this timescale. It is also noteworthy that rapid modulation through a grid pair is evident on timescales much smaller than the full 2-second half-rotation timescale, so that partial information on the source structure and its evolution is available on timescales as low as tens of milliseconds. (Again, it should be stressed that the quality of such data will be dependent on the number of photons collected in this time interval and hence on the intensity of the source. Also, the information will not permit the construction of a true "image" of the source; comparison of such short time duration observations with models is best done in terms of Fourier components – see below.)

Gamma-Ray Spectroscopy

As mentioned above, the energy bandwidth of the *RHESSI* Germanium detectors is lower than the FWHM of all significant gamma-ray lines in the solar spectrum (with the exception of the narrow deuterium production line at 2.223 MeV). This permits determination of shifts and shapes of these lines, so providing information on not only the energy spectrum, but also the angular distribution, of the energetic electrons and ions that produce the observed emission. Synthetic gamma-ray line profiles for different species, and distributions (angular and energy), of exciting particles have been derived (Werntz *et al.* 1990), for comparison with the *RHESSI* data.

Recent analysis by Ramaty *et al.* (1995) of the ^{20}Ne line at 1.634 MeV (a line with a relatively low excitation threshold, and also one which has been hitherto very difficult to resolve) has shown that the accelerated proton spectrum in flares may be significantly steeper at low energies than previously thought. Because of the low intensity of this line, its separation from nearby lines and continua is problematic, and indeed was only accomplished by summing *Solar Maximum Mission* spectra from a number of different flares (Share & Murphy 1995). *RHESSI* will clearly resolve the ^{20}Ne line in individual events, and thereby permit a more refined estimate of the energy content in protons and ions, to be compared with the energy content of the electrons.

Hard X-Ray Spectroscopy

The source-integrated hard X-ray flux $I(\epsilon)$ (cm^{-2} s^{-1} keV^{-1} at the Earth) is given by (e.g. Brown 1971, Tandberg-Hanssen & Emslie 1986)

$$I(\epsilon) = \frac{A}{4\pi R^2} \int_\epsilon^\infty \int_\epsilon^{E_o} F_o(E_o) \frac{\sigma(\epsilon, E)}{|dE/dN|} dE\, dE_o, \tag{1}$$

where A is the flare area, R the Earth-Sun distance, $F_o(E_o)$ is the injected differential electron flux (cm^{-2} s^{-1} keV^{-1}), $\sigma(\epsilon, E)$ is the bremsstrahlung cross-section (cm^2 keV^{-1}) and dE/dN is the energy loss rate per unit column density $N = \int n\, dz$. Reversing the order of integration in (1) yields the equivalent, and useful, result

$$I(\epsilon) = \frac{A}{4\pi R^2} \int_\epsilon^\infty \frac{\sigma(\epsilon, E)}{|dE/dN|} G(E)\, dE, \tag{2}$$

where $G(E) = \int_E^\infty F(E_o)\, dE_o$.

Traditionally, Eq. (2) has been used in the context of a collisionally-dominated *cold target* energy loss process, for which $dE/dN = -2\pi e^4 \Lambda / E$, where e is the electronic charge and Λ is the Coulomb logarithm (e.g. Emslie 1978). With an ambient medium that can be considered "cold" in the sense that the ambient electron thermal velocity is much smaller than that of the energetic electrons, and *a constant value of* Λ *throughout the flare region*, Eq. (2) can be inverted rather straightforwardly (Brown 1971) to infer $G(E)$ and hence the injected electron spectrum $F_o(E_o)$ for a given spatially-integrated hard X-ray spectrum $I(\epsilon)$. This inversion process requires knowledge not only of $I(\epsilon)$ but also its first and second derivatives (e.g. Brown 1971) and so it cannot be meaningfully applied to low-resolution scintillator spectra, for which calculation of the required derivatives is subject to large uncertainties. On the other hand, *RHESSI's* excellent spectral resolution permits this inversion procedure (which has been in existence for over 30 years) to finally be carried out in practice.

However, an important factor relevant to a realistic solar atmosphere target significantly affects this simple picture. The value of the Coulomb logarithm Λ, and hence the electron energy loss rate dE/dN, depends significantly on the ionization state of the target – specifically it is a factor of ~ 3 higher in the (ionized) corona than in the (near-neutral) chromosphere (e.g. Brown 1972, Emslie 1978). Therefore, higher energy electrons, with their greater penetration depth, "see" a trajectory-averaged energy loss rate that is lower than that they would have in a target that was fully ionized throughout. Hence spectral inversions that assume a uniform, ionized target substantially overestimate the number of high-energy electrons required to produce a given hard X-ray yield.

Brown *et al.* (1998) have analyzed this problem in some detail. They showed that the true injected electron spectrum $F_o(E_o)$ is related to that inferred from a uniform, ionized, target inversion, $F_o^*(E_o)$, through the relation

$$f(\eta) + \nu f(\eta + 1) = f^*(\eta). \tag{3}$$

Here $\eta = (E_o^2/4\pi e^4 \Lambda_{\rm cor} N_{\rm cor})$ (where $\Lambda_{\rm cor}$ and $N_{\rm cor}$ are the Coulomb logarithm and column density of the coronal portion of the flare) is a change of variable for E_o (such that $f(\eta)\,d\eta = F_o(E_o)\,dE_o$) and $\nu = (\Lambda_{\rm cor}/\Lambda_{\rm chrom}) - 1 \simeq 2$ reflects the different energy loss rates in the coronal and chromospheric regions of the flare. [For a wholly ionized target, ν would equal zero and $f(\eta)$ would be the same as $f^*(\eta)$.] To apply equation (3) to data, we first invert the observed hard X-ray spectrum to obtain the equivalent fully-ionized target spectrum $f^*(\eta)$ and then use (3) to solve for the actual injected spectrum $f(\eta)$.

Although the solution of equation (3) for $f(\eta)$ for a given $f^*(\eta)$ is nonunique (even for perfect data), Brown *et al.* (1998) showed that a valid (and, in some sense, most "straightforward") solution of (3) is

$$f(\eta) = \mathcal{L}\left\{\frac{\mathcal{L}^{-1}\left[f^*(\eta); s\right]}{1 + \nu e^{-s}}; \eta\right\}, \tag{4}$$

where $\mathcal{L}$ and $\mathcal{L}^{-1}$ are the Laplace transform operator, and its inverse, respectively. Application of this equation requires a *functional form* of $f^*(\eta)$ (and not just a handful of data points). The excellent (~ 2 keV) spectral resolution of the *RHESSI* detectors will provide accurate empirical forms for $f^*(\eta)$, enabling equation (4) to determine $f(\eta)$ and so the true injected spectrum $F_o(E_o)$ (with, however, the nonuniqueness caveats mentioned by Brown *et al.* 1998).

Imaging Spectroscopy

The true power of *RHESSI*, of course, lies not within its imaging, nor its spectroscopy, but rather in the combination of the two. Every half-rotation (~ 2 seconds), *RHESSI* samples a large number of points in the spatial Fourier plane, each at an energy resolution of a few keV. This permits the measurement of accurate hard X-ray spectral forms at different locations within the target, locations sufficiently close that subtle changes in the hard X-ray spectra between the two locations can be measured and used to infer the physical processes affecting the bremsstrahlung-producing electrons themselves.

Emslie (1981, 1986) has provided explicit expressions for the predicted variation of hard X-ray intensity, and electron flux, respectively, throughout a target in which the dominant energy loss process is Coulomb collisions with ambient electrons (and hydrogen atoms). This analysis has recently been extended (Emslie, Barrett, & Brown 2001) to describe the *inversion* of spatially-resolved hard X-ray data in order to determine *empirically* the form of the electron energy redistribution rate as a function of energy and so constrain the physical processes operating.

Consider a set of observations of the hard X-ray spectrum $I(\epsilon, z)$ at a (one-dimensional) set of positions z, oriented along the electron travel path. If the spatial resolution limit is significantly smaller than the characteristic distance over which energy loss occurs, then the hard X-ray emission $I(\epsilon, z)$ from a given "pixel" can be treated as thin-target, and straightforward inversion of the bremsstrahlung spectrum (e.g. Brown 1971, Johns & Lin 1992) gives the local electron flux spectrum $F(E, z)$ at that point in the flare.

Now, continuity of electron flux demands that

$$\frac{\partial F(E,z)}{\partial z} = -\frac{\partial}{\partial E}\left(F(E,z)\frac{dE}{dz}\right), \tag{5}$$

where dE/dz is the effective energy loss (taking into account all processes, including escape and energy redistribution within the target). Straightforward inversion of this equation yields

$$\frac{dE}{dz} = -\frac{1}{F(E,z)}\int_{E^*}^{E}\frac{\partial F(E,z)}{\partial z}\,dE, \tag{6}$$

where E^* is such that $dE/dz = 0$ for $E = E^*$ (e.g. $E^* = \infty$ for Coulomb collisions). Eq. (6) gives the energy loss (or gain) rate as a function of energy E in terms of the measured variation of $F(E, z)$ with position z.

Emslie, Barrett, & Brown (2001) used the $F(E, z)$ recovered from synthetic (noisy) hard X-ray count spectra $C(\epsilon, z)$ to obtain $\partial F(E, z)/\partial z$ as a function of energy at each position z and, using Eq. (6), the empirical forms for the electron energy change rate dE/dz. The results from these simulations agreed well, up to an energy of about 40 keV, with the *ab initio* energy loss rate used to generate the synthetic spectra $C(\epsilon, z)$. Above this energy, the effects of both photon noise rendered the results unreliable. More sophisticated spectral inversion techniques, such as those using Tikhonov regularization methods (e.g. Piana 1994, Piana & Brown 1998), have the potential to extend the useful applicability of the technique to higher energies.

Emslie, Barrett, & Brown (2001) showed that this technique could generate energy loss profiles $dE/dz(E)$ over the range 10 – 40 keV that are sufficient to discriminate between values of the energy loss index α (in the expression $dE/dz \sim E^{-\alpha}$) that differ by as little as ~ 0.5.

An ability to discern the empirical form of the energy loss rate to this level of accuracy can indeed be used to discriminate between candidate acceleration and transport models. For example:

- collisions result in a dE/dz which is negative, with $\alpha = 1$. If the empirical energy loss rate is found to be indeed consistent with $\alpha = 1$, then the magnitude of dE/dz can be used as a diagnostic of the ambient target density n.

- energy losses due to driving of a beam-neutralizing return current against ohmic losses (e.g. Knight & Sturrock 1977, Emslie 1980, Brown & Bingham 1984, van den Oord 1990) depend only on the global decelerating electric field (proportional to the total beam flux at a given depth) and are thus negative and independent of E (i.e. $\alpha = 0$). The magnitude of dE/dz in this case yields diagnostics on the ambient resistivity of the plasma.

- acceleration through resonant interaction with a low-amplitude spectrum of MHD waves (e.g. transit-time acceleration – Miller, LaRosa, & Moore 1996) produces a dE/dz which is proportional to p/v, where p is the particle momentum and v the velocity. In the non-relativistic regime appropriate to the bulk of hard X-ray production, this implies a dE/dz that is *positive* and independent of E ($\alpha = 0$ again; cf. the return current Ohmic loss case above).

• energy changes due to collective wave-particle effects result in the formation of an energy-space plateau in a region that originally was driven to have a positive slope (e.g. through the energy-dependent action of Coulomb collisions; Emslie & Smith 1984). Haydock *et al.* (2001) have introduced a parametric representation of the energy flux spectrum $F(E, z)$ as a function of position in the source, produced by a combination of Coulomb collisions and wave-particle relaxation, and applied Eq. (6) to these forms in order to yield an effective energy loss rate for the combined processes. The resulting dE/dz profiles are sufficiently different to permit discrimination between models with Coulomb collisions only and those which also have wave-particle processes operating. This is especially significant in light of Haydock *et al.*'s result that the *spatially integrated* hard X-ray spectrum is the same shape, whether or not wave-particle interactions are present.

Imaging in Fourier Space

Although the *RHESSI* instrument takes ~ 2 seconds for a half-rotation, and hence a complete survey of the Fourier components of the source structure, the basic information on the spatial structure of the source is contained in the temporal modulation of the detector signals as the grid slit/slat grid structure passes in front of the detector. Hence, *some* spatial information is present on timescales down to one slit→slit transition time, which can be as low as ~ 50 ms for the finest grids (Lin *et al.* 1998).

The appropriate methodology to address such spatial information is fundamentally different than the techniques currently being proposed to address full (> 2 s) datasets. Rather than inverting the temporal modulation patterns to obtain a "picture" of the source, for comparison with the predictions of various models, it may be more revealing to compare structures in "Fourier space," i.e., forward-transforming the model predictions into spatial Fourier information, which is then directly compared with the Fourier components deduced from the raw data.

To appreciate the power of this method, note that it is somewhat analogous to that used in the 1970s for determination of temperature structure in the solar atmosphere, using as basic data a set of intensities in different EUV lines. What was instead done then was to invert the line intensity data to derive an *empirical* Differential Emission Measure (DEM $= n^2 \frac{ds}{dT}$, where n is density, s position and T temperature) distribution, to be compared with *theoretical* DEMs derived from model atmosphere structures. The raw EUV data consisted of a set of line intensities, while the raw *RHESSI* data consists of a set of time-tagged photon detections. The comparison between models and observation is made at the "intermediate" level of the DEM (EUV data) or Fourier components (*RHESSI*). Inversion of a very short time interval of raw *RHESSI* data to obtain (perhaps limited) Fourier information on the source structure can still be meaningfully compared with the full predictions of model structure in Fourier space.

We illustrate the proposed methodology with a simple example. Suppose that during the impulsive phase of a flare the hard X-ray emitting structure is essentially a linear structure (e.g. along a magnetic field line), the extent of which increases linearly with time (possibly a result of an Alfvénic disturbance in the acceleration region, or of the motion of ion-acoustic fronts [Brown, Melrose & Spicer 1979] that confine the X-ray source). Suppose also that at that time a set of *RHESSI* grids is fortuitously oriented with the slits perpendicular to the length of the source, so that information on its extent can be obtained on a timescale as short as a slit→slit passage. (Note that such a configuration is not really that fortuitous - as long as the slits are not oriented *along* the source, the temporal modulation information that we seek will be present to some extent. Note also that the exact slit orientation will be *known* at the time of the observation.)

Let the intensity of the source at position x and time t be represented by the profile

$$I(x,t) = \begin{cases} I_o, & -Vt < x < Vt \\ 0, & \text{otherwise,} \end{cases} \tag{7}$$

where V is the characteristic velocity of the expanding source. Taking the spatial Fourier transform of the source structure gives

$$\hat{I}(k,t) = \int_{-\infty}^{\infty} e^{ikx} I(x,t)\, dx = \frac{2I_o}{k} \sin kVt. \tag{8}$$

Early in the lifetime of the source, we expect to see the greatest modulation in the finest (high k) grids, while later the greatest modulation will be in somewhat coarser (lower k) grids. We therefore compare the Fourier transforms represented by two grids of different k, say k_o and βk_o (where for *RHESSI* $\beta \simeq [\sqrt{3}]^n$, n reflecting both the difference in grid number and/or the harmonic (e.g. 1st vs. 3rd) of the triangular grid shadow function used). This allows us to construct the ratio

$$R(t) \equiv \frac{\hat{I}(\beta k_o)}{\hat{I}(k_o)} = \frac{1}{\beta}\frac{\sin \beta\theta}{\sin\theta}, \tag{9}$$

where $\theta = k_o Vt$. At $t = \theta = 0$, $R = 1$. As t increases, R decreases to zero at $\theta = \pi/\beta$ $(t = \pi/\beta k_o V)$, and then rises to (formally) infinity at $\theta = \pi$ $(t = \pi/k_o V)$. Since we know β and k_o from the properties of the instrument, measurement of the ratio of the modulation amplitudes in two different detectors as a function of time t provides a measure of the velocity V.

CONCLUSIONS

RHESSI is capable of providing a wealth of information on the physical processes behind particle acceleration and propagation in solar flares. The true features revealed by the data are, of course, yet to be realized, and I am sure that, just like the surprising "loop-top" hard X-ray sources revealed by the *Yohkoh* HXT, there will be many features that will require us to rethink many of our ideas on particle acceleration and energy transport in flares. At the time this talk was given, *RHESSI* was the subject of yet another launch delay. By the time of writing, however, *RHESSI* was fully operational in a nominal orbit. Given that the Sun will cooperate by producing an ample supply of hard X-ray and gamma-ray flares (a not-too-optimistic hope, based on the history of solar activity in the late stages of previous solar maxima), we look forward to many years of fruitful scientific enquiry.

ACKNOWLEDGEMENTS

I thank the whole *RHESSI* team for their patient education of the author on many matters relating to the instrument and its capabilities, and a host of other colleagues for stimulating discussions on the issues mentioned herein.

REFERENCES

Brown, J.C., *Solar Phys.*, **18**, 489 (1971).

Brown, J.C., *Solar Phys.*, **26**, 441 (1972).

Brown, J.C., and Bingham, R.R., *A&A*, **131**, L11 (1984).

Brown, J.C., McArthur, G.K., Barrett, R.K., McIntosh, S.W., and Emslie, A.G., *Solar Phys.*, **179**, 379 (1998).

Brown, J.C., Melrose, D.B. and Spicer, D.S., *ApJ*, **228**, 592 (1979).

Emslie, A.G., *ApJ*, **224**, 241 (1978).

Emslie, A.G., *ApJ*, **235**, 1055 (1980).

Emslie, A.G., *ApJ*, **249**, 817 (1981).

Emslie, A.G., in *Solar Flares and Coronal Physics Using POF as a Research Tool*, NASA CP-2421, p. 132 (1986).

Emslie, A.G., Barrett, R.K. and Brown, J.C., *ApJ*, **557**, 921 (2001).

Emslie, A.G., and Smith, D.F., *ApJ*, **279**, 882 (1984).

Haydock, E., Brown, J.C., Conway, A.J., and Emslie, A.G., *Solar Phys.*, in press (2002).

Hurford, G.D., personal communication (2001).

Johns, C., and Lin, R.P., *Solar Phys.*, **137**, 121 (1992) (see also Erratum in *Solar Phys.*, **142**, 219 [1992]).

Knight, J.W., and Sturrock, P.A., *ApJ*, **218**, 306 (1977).

Lin, R.P., *et al. SPIE*, **3442**, 2L (1998).

Miller, J.A., LaRosa, T.N., and Moore, R.L., *ApJ*, **461**, 445 (1996).

Piana, M., *A&A*, **288**, 949 (1994).

Piana, M., and Brown, J.C. *A&A Suppl.*, **132**, 291 (1988).

Ramaty, R., Mandzhavidze, N., Kozlovsky, B., and Murphy, R. J., *ApJ*, **455**, L193 (1995).

Share, G.H., and Murphy, R.J., *ApJ*, **452**, 933 (1995).

Tandberg-Hanssen, E.A., and Emslie, A.G., *The Physics of Solar Flares*, Cambridge University Press (1986).

van den Oord, G.H.J., *A&A*, **234**, 996 (1990).

Werntz, C., Kim, Y. E., and Lang, F. L., *ApJ Suppl.*, **73**, 349 (1990).

A RAPIDLY MOVING HARD X-RAY SOURCE IN A CME

H. S. Hudson

Space Sciences Laboratory, University of California, Berkeley, CA 94720, USA

ABSTRACT

A major flare occurring about two days beyond the solar west limb resulted in the *Yohkoh* detection of a hard X-ray source high in the corona (above about 8×10^4 km). The simultaneous observation of microwaves from this source allowed a determination of the plasma density of about 4×10^9 cm^{-3}. The surprisingly high intensity of hard X-ray radiation from this event implies a non-thermal tail fraction greater than 0.2% above 20 keV in the electron distribution function. The simultaneous occurrence of a fast coronal mass ejection (CME) and major solar particle event suggests that non-thermal electrons may play a role in the energetics of CMEs.

INTRODUCTION

The solar corona contains many kinds of non-thermal activity readily observed at metric and decimetric wavelengths. These phenomena typically require the presence of non-thermal electrons with energies ranging from a few keV up through the relativistic domain. Nevertheless it has proven extraordinarily difficult to study these phenomena using hard X-rays emitted by bremsstrahlung, owing to the inefficiency of this process and the limitations of existing hard X-ray instruments.

Limb occultation has been about the only practical method for observing coronal hard X-ray sources with non-imaging instruments. An over-the-limb distance of some 10° suffices in practice to eliminate the foot-point sources (see Tomczak 2001). Even for an imaging instrument such as *Yohkoh* HXT, the limitation on image dynamic range means that the bright footpoint sources still can obscure the fainter coronal ones. Early examples of non-imaging coronal hard X-ray sources include the events reported by Frost and Dennis (1971), Hudson (1978), and Hudson *et al.* (1982). In the *Yohkoh* decade we have not had comparable observations in spite of the availability of hard X-ray imaging. The Masuda source (Masuda *et al.* 1994) provides a well-known exception, but it has a closer relationship to the impulsive-phase (footpoint) sources rather than to the inherently coronal sources cited above.

OBSERVATIONS

In this paper we provide further analysis of the event of April 18, 2001 (Hudson *et al.* 2001), which revealed a limb-occulted hard X-ray source high in the corona. The event apparently took place in NOAA active region 9415, which had a history of high activity before rotating past the west limb. A remote over-the-limb location matches the appearance of an extremely large hard X-ray event detected by *Ulysses* from a more favorable heliolongitude (K. Hurley and J. McTiernan, personal communication 2001), and from the markedly non-Neupert relationship between soft and hard X-rays (Hudson *et al.* 2001). In the interpretation below we assume an over-the-limb rotation of 26.7°, corresponding to a line-of-sight altitude of 8.8×10^4 km.

The *GOES* soft X-rays showed a C2.2 long-decay event, whereas the HXT hard X-rays suggested an event an order of magnitude more energetic. The timing of hard and soft X-ray fluxes did not have the usual Neupert signature (Dennis & Zarro 1993). The hard X-ray burst lasted for about 70 s at half maximum and had a photon number spectral index of about 4, with a slight hardening after the time of maximum. As discussed by Hudson *et al.* (2001), the hard X-ray source matched the location of a relatively compact microwave source, both of which moved outwards at a projected velocity of about 930 km s^{-1}.

Figure 1 shows these motions, along with a later snapshot of the CME. The CME motion was even faster, and the image shown reveals a compact component that would normally suggest a filament eruption. According to the CUA catalog of CMEs (see `http://cdaw.gsfc.nasa.gov/CME_list/`), this was one of the fastest CMEs observed by LASCO. We therefore cannot match specific features easily, but it seems safe to conclude that the hard X-ray sources were embedded in the CME structure.

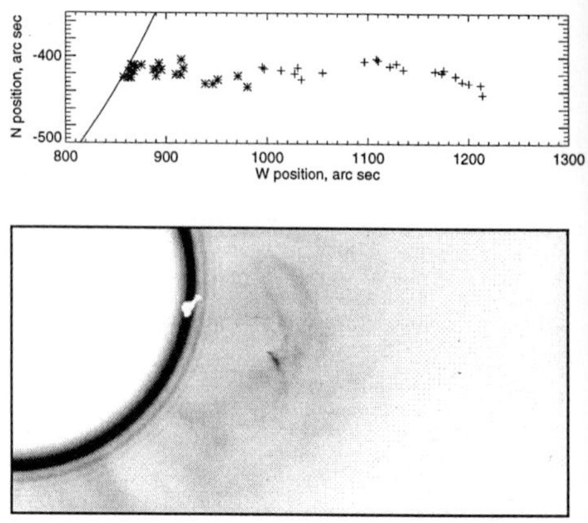

Fig. 1. Upper: Motion of the hard X-ray (∗) and microwave (+) source locations, covering the time range 02:14:30 UT to 02:17:36 UT. Lower: snapshot of the CME at 02:30:04 UT, at a larger scale as observed by the LASCO C2 coronagraph (negative image). Note the bright inclusion within the CME near frame center.

ENERGETICS

Figure 2 shows the hard X-ray spectrum at the time of peak hard X-ray flux, 02:14:45 UT). The spectral fit, 30.2 $(E/20keV)^{-3.71}$ ph $(\mathrm{cm}^2\ \mathrm{sec\ keV})^{-1}$, is not close to E^{-2}, so it differs substantially from coronal hard X-ray spectra seen with *OSO-5*, *OSO-7*, and in the *Yohkoh*/HXT observation of the flare of April 23, 1998 (Sato 2001).

The hard X-rays imply the presence of a certain number of fast electrons at energies greater than the observed X-ray energy. Such electrons carry substantial total energy in the impulsive phase, as shown by Kane & Donnelly (1971; see also Lin & Hudson 1976). Here we assess the energy contained in fast electrons, which by the circumstances do not radiate by thick-target processes from the usual footpoint sources.

From the bremsstrahlung cross-section one obtains a direct estimate of the instantaneous total electron energy, if one knows the target density (see Table). This estimate (the "thin-target approximation") represents a lower limit to the total electron energy, as long as the target does not have a lumpy distribution. This situation requires that a non-unity filling factor not confuse the density estimate, a situation we discuss below. In this paper we use the Bethe-Heitler cross-section for convenience (Brown 1971, Hudson *et al.* 1978). Because the main hard X-ray source matches the position of a microwave source with a thermal (free-free) spectrum, Hudson *et al.* (2001) could infer a source density of 4×10^9 cm^{-3}. Using this and the observed source area gives an estimate of the electron distribution in the background plasma. We thus find that the non-thermal tail population (> 20 keV) represents about 0.2% of the total electron population. This straightforward estimate provides a good basis for the conclusions discussed below, but we discuss possible complications here for the sake of thoroughness.

In this event we observe a location above the limb, with circumstances strongly suggesting that the electrons occur in an expanding loop geometry. In such a situation we would expect that electrons would scatter into the loss cone and precipitate, losing the bulk of their energy at the out-of-sight footpoints. In any case the thin-target approximation gives an estimate of the instantaneous number of electrons N_{20} above 20 keV. Depending upon the unknown speed of the escape process, the residence time of an electron might be of the order of one second, in which case the total number of electrons and the total non-thermal electron energy might be much larger. An upper limit comes from assuming that the electrons stop in the corona via collisions there, as in a thick target analysis, although at the inferred density one would not expect this. Here the stopping time would have to be shorter than the event duration (see Alexander & Metcalf 1999). We can thus use the thin-target and thick-target estimates to bound N_{20}: $1.3 \times 10^{36} < N_{20} < 2.6 \times 10^{37}$.

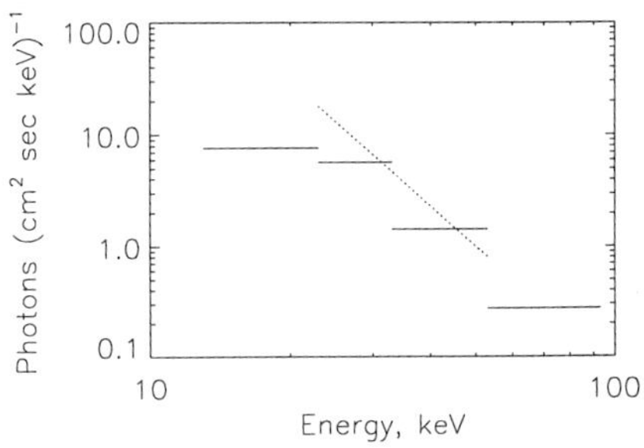

Fig. 2. Hard X-ray spectrum from *Yohkoh* HXT at the time of maximum hard X-ray flux, 02:14:45 UT. The dashed line shows the spectral fit to the count ratio of the two bands covering 23-53 keV, photons (cm^2 s keV)$^{-1}$. Lines show the four HXT bands as counts (cm^2 s keV)$^{-1}$ for the nominal energy ranges.

Table 1. Energetics of the April 18, 2001 event (reference time 02:14:45 UT)

Observed X-ray spectrum	$30.2\ (E/20keV)^{-3.71}$ ph (cm^2 sec keV)$^{-1}$
Compact source density	4×10^9 cm^{-3}
Compact source volume (02:16 UT)	1×10^{28} cm^3
Thin-target number	$1.3 \times 10^{36} > 20$ keV
Thin-target energy	6×10^{28} ergs > 20 keV
Thick-target number	$2.6 \times 10^{37} > 20$ keV

Finally, we consider the significance of a possible spatially-extended component in this event. The one-dimensional images shown by Hudson *et al.* (2001) suggest the presence of such a source, appearing initially at about 02:14:30 UT along with the moving compact source. If real, this source may also imply a large energy content, since the density in the larger-scale corona is lower. It would also imply an even larger tail population.

CONCLUSIONS

The main purpose of this paper has been to add more detail about the estimate of energetics in this event. In the simplest and most likely interpretation, we find that the tail population of the coronal electron distribution (above 20 keV) in the moving source amounts to some 0.2% of the background electron population. We do not know the temperature of the background distribution, but let us conservatively put it at a coronal temperature of 3×10^6 K. The tail energy and non-thermal pressure of the moving source thus are, at a lower limit, comparable to those of the bulk plasma, and probably much greater. The total electron energy, also shown in Table 1 for the thin-target case, amounts to about 6×10^{28} ergs, comparable to the total energy represented by the soft X-ray source. From this event we therefore conclude that the coronal non-thermal electron population can dominate the energetics, as it typically also does during the impulsive phase at lower altitudes. Observations of other related sources will show whether this conclusion can be generalized.

ACKNOWLEDGEMENTS

This work was supported under NASA contract NAS 8–37334. *Yohkoh* is a project of ISAS, Japan. I thank L. Fletcher, M. Wheatland, and the referee for comments. We have made use of the CUA catalog of CMEs, which is generated and maintained by the Center for Solar Physics and Space Weather, The Catholic University of America, in cooperation with the Naval Research Laboratory and NASA. *SOHO* is a project of international cooperation between ESA and NASA.

REFERENCES

Brown, J.C., The deduction of energy spectra of non-thermal electrons in flares from the observed dynamic spectra of hard X-ray bursts, *Solar Physics*, **18**, 489 (1971).

Cliver, E. W., B. R. Dennis, A. L. Kiplinger, S. R. Kane, D. F. Neidig, N. R. Sheeley, Jr., and M. J. Koomen, Solar gradual hard X-ray bursts and associated phenomena, *ApJ* **305**, 920, (1986).

Dennis, B.R., and D.M. Zarro, The Neupert effect - What can it tell us about the impulsive and gradual phases of solar flares? *Solar Physics* **146**, 177 (1993).

Frost, K., and B.R. Dennis, Evidence from hard X-rays for two-stage particle acceleration in a solar flare, *ApJ* **165**, 655 (1971).

Hudson, H.S., A purely coronal hard X-ray event, *ApJ* **224**, 235 (1978).

Hudson, H.S., Canfield, R.C., and Kane, S.R., Indirect estimation of energy deposition by non-thermal electrons in solar flares, *Solar Physics* **60**, 137 (1978).

Hudson, H.S., R.P. Lin, and R.T. Stewart, Second-stage Acceleration in a Limb-occulted Flare, *Solar Physics* **75**, 245 (1982).

Hudson, H.S., T. Kosugi, N.V. Nitta, & M. Shimojo, Hard X-radiation from a fast coronal ejection, *ApJ Letters* **561**, L211 (2001).

Lin, R.P., & H.S. Hudson, Non-thermal processes in large solar flares, *Solar Physics* **50**, 153 (1976).

Metcalf, T.R., and D. Alexander, Coronal trapping of energetic flare particles: *Yohkoh*/HXT observations, *ApJ* **522**, 1108 (1999).

Masuda, S., T. Kosugi, H. Hara, S., Tsuneta, and Y. Ogawara, A loop-top hard X-ray source in a compact solar flare as evidence for magnetic reconnection, *Nature* **371**, 495 (1994).

Kane, S.R., and R.F. Donnelly, Impulsive hard X-ray and ultraviolet emission during solar flares, *ApJ* **164**, 151 (1971).

Sato, J., Observation of the coronal hard X-ray sources of the 1998 April 23 flare, *ApJ Letters* **558**, L137 (2001).

Tomczak, M., The analysis of hard X-ray radiation of flares with occulted footpoints, *A&A* **366**, 294 (2001).

A SIMPLE ESTIMATE FOR THE ENERGIES OF ELECTRONS ACCELERATED IN FLARE CURRENT SHEETS ON THE SUN

Y. E. Litvinenko

Institute for the Study of Earth, Oceans, and Space, University of New Hampshire, Durham, NH 03824-3525, USA

ABSTRACT

A major result from *Yohkoh* is the frequency with which hard X-ray sources occur above solar flare loops. Observations suggest that electron acceleration in flares occurs in the magnetic reconnection region above the loops. Unfortunately, models for particle acceleration in reconnecting current sheets predict electron energy gains in terms of the reconnection electric field and the thickness of the sheet, both of which are extremely difficult to measure. It can be shown, however, that application of Ohm's law in a turbulent current sheet, combined with energy and Maxwell's equations, leads to a formula for the electron energy gain in terms of the flare power output, the magnetic field strength, the plasma density and temperature in the sheet, and its area. Typical flare parameters correspond to electron energies between a few tens of keV and a few MeV.

INTRODUCTION

Although observations of nonthermal hard X-ray emission away from the limb during solar flares have been known for some time, it is a major result from *Yohkoh* that hard X-ray sources in impulsive flares are frequently located in the cusp region above soft X-ray flare loops (Masuda *et al.* 1994). This result and further studies strongly suggest that particle acceleration in impulsive flares occurs in the magnetic cusp region above the loops (e.g. Aschwanden 1998, Metcalf & Alexander 1999). The most promising geometry for flare energy release and particle acceleration, which is in agreement with *Yohkoh* observations, is that of a large-scale reconnecting current sheet in the cusp region. The observed hard X-ray emission is generated via bremsstrahlung by accelerated electrons, and the most direct way of accelerating electrons to tens and hundreds of keV is by a strong (super-Dreicer) electric field in the current sheet.

Particle acceleration in model current sheets has been extensively studied (e.g. Martens 1988, Litvinenko 1996). The problem of charged particle motion in a current sheet is greatly simplified by the fact that typical acceleration length and time scales under solar flare conditions turn out to be very small compared with typical global parameters. This is why it is relatively easy to derive electron energies in terms of local quantities such as the reconnection electric field and the thickness of the sheet. This theoretical advantage, however, makes it very difficult to test the model predictions using the observational data. This note gives a simple example that shows how the electron acceleration model can be extended to express the predicted energies in terms of observable quantities.

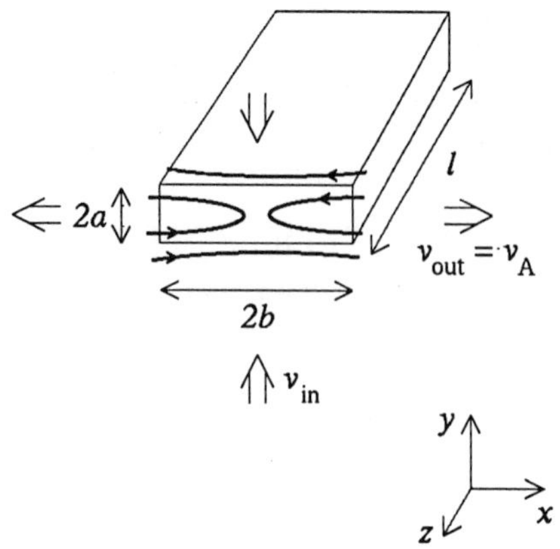

Fig. 1. Projection of the magnetic field in the reconnecting current sheet (length l, thickness $2a \ll l$, width $2b \approx l$) on the xy plane. The electric field E and the longitudinal magnetic field $B_{\|}$ are along the z-axis.

ELECTRON ACCELERATION MODEL

The usual approach in the study of charged particle orbits in a current sheet is to approximate the electric and magnetic fields in the sheet (located, say, at $y = 0$) by the first nonzero terms in the Taylor expansion:

$$\mathbf{B} = -(y/a)B_0\hat{\mathbf{x}} - B_\perp\hat{\mathbf{y}} + B_{\|}\hat{\mathbf{z}}, \tag{1}$$

where B_0 is the reconnecting magnetic field component, and a is the half-thickness of the sheet (Figure 1). The reconnection electric field in the sheet is $\mathbf{E} = E\hat{\mathbf{z}}$. Both E and the nonreconnecting component of magnetic field $B_{\|}$ may be assumed constant. Although $B_\perp = B_\perp(x)$ vanishes at the center of the sheet, a fixed nonzero value of $B_\perp$ can also be assumed on a given particle orbit, and the effect of its variation can be ignored for electron energies below a few MeV (Litvinenko 2000). In general $B_{\|} \neq 0$, and the fact that $\mathbf{E} \cdot \mathbf{B} \neq 0$ indicates the presence of a significant resistive term in Ohm's law, required for magnetic reconnection:

$$\mathbf{R} = \mathbf{E} + \mathbf{v} \times \mathbf{B}/c, \tag{2}$$

where $\mathbf{v}$ is the reconnection flow velocity.

Particle orbits in current sheets with $B_\perp \neq 0$ are very complex in general, but the situation is simpler in some limiting cases. Since the magnetic field in the solar corona is known to have a significant axial component along the coronal loops, the limit of a strong longitudinal field $B_{\|}$ on the order of the reconnecting field B_0 should be appropriate for flaring current sheets. A sufficiently large $B_{\|}$ magnetizes electrons and makes them follow the field lines. Integrating the magnetic field line equations defines the acceleration length l_{acc} as the displacement δz along the electric field, which corresponds to $|\delta y| = a$ when the magnetized electrons initially inside the sheet escape it along the field lines (see Litvinenko 1996, for a detailed discussion of particle orbits in this case). The typical energy gain in the test-particle approximation

$$\mathcal{E} = eEl_{\mathrm{acc}} \approx eEa\,(B_{\|}/B_\perp) \tag{3}$$

appears to be able to explain electron acceleration in solar flares.

The approach summarized above is theoretically very attractive. It is not, however, quite satisfactory from the observational standpoint because both the reconnection electric field E and the half-thickness of the sheet a are local parameters that are extremely difficult to measure (e.g. Foukal & Hinata 1991). This difficulty motivates the following three-step argument that leads to a simple formula for $\mathcal{E}$ in terms of global parameters, which employs neither of the two quantities. The idea is to eliminate E and a by using equations that describe the current sheet structure.

First, the Maxwell equation integrated across the sheet allows one to express the sheet thickness in terms of the reconnecting magnetic field component B_0 and the average electric current density j in the sheet:

$$B_0/a \approx 4\pi j/c. \tag{4}$$

Second, the reconnection electric field E is proportional to the Poynting energy flux into the sheet:

$$P/S \approx cEB_0/2\pi. \tag{5}$$

Here P is the flare energy release rate, and l and $S \approx l^2$ are the length and area of the sheet. Note that the energy equation defines only the product of E and B_0 rather than either of these quantities separately. It is this product though that determines the energy gain $\mathcal{E}$ after the first step is performed.

Third, the most critical assumption is required to calculate the electric current density j in the sheet. Fast magnetic reconnection corresponds to highly super-Dreicer electric fields and large electric current densities, exceeding the threshold for a current-driven instability. The resulting turbulence determines the anomalous resistive term $\mathbf{R}$ in Eq. 2 and limits the current density. Marginal stability arguments and numerous laboratory investigations (de Kluiver, Perepelkin, & Hirose 1991) indicate that the anomalous, enhanced resistivity $\eta_{\rm a}$ is proportional to the local electric field:

$$\eta_{\rm a}/\eta_{\rm c} \approx 10\, E/E_{\rm D} \tag{6}$$

for $E > E_{\rm D}$. Here $E_{\rm D}$ is the Dreicer electric field:

$$E_{\rm D} = 4\pi n e^3 \ln \Lambda/(kT), \tag{7}$$

n and T are the density and temperature in the sheet, and $\eta_{\rm c}[{\rm s}^{-1}] \approx 10^{-7} T^{-3/2}[{\rm K}]$ is the classical electric resistivity. An important consequence of the empirical Eq. 6 is that the electric current density is independent of the electric field:

$$j \approx J = 0.1\, E_{\rm D}/\eta_{\rm c} = {\rm const.} \tag{8}$$

Successive application of Eq. 4, 5, and 8 to Eq. 3 leads to the sought-after result for the electron energy gain in a flaring turbulent current sheet:

$$\mathcal{E} \approx \frac{e}{2} \frac{B_{\parallel}}{B_{\perp}} \frac{P}{JS}, \tag{9}$$

where $S \approx l^2$ and $J \sim nT^{1/2}$. It is worth noting that the predicted nonthermal energy gain for electrons is smaller for higher plasma temperatures in the sheet, $\mathcal{E} \sim T^{-1/2}$.

To summarize, although the electron energy gain in a turbulent flaring current sheet is defined by the local parameters E and a that are very difficult to measure, electron energies can be estimated using the following observable (at least indirectly) quantities: the total flare power output, two nonreconnecting magnetic field components, the plasma density and temperature in the sheet, and its area.

DISCUSSION

To demonstrate that the suggested argument leads to a reasonable energy range for flare-accelerated electrons, consider the following typical parameters of impulsive flares: $P \approx 10^{28}$ erg s^{-1}, $l \approx 10^9$ cm, $n \approx 10^9$ cm^{-3}, $T \approx 10^7$ K. Now Eq. 8 and 9 reduce to $J \approx 3 \cdot 10^8$ cgs and

$$\mathcal{E} \approx 5\,(B_{||}/B_{\perp}) \text{ keV}. \tag{10}$$

The nonreconnecting magnetic field component $B_{||}$ parallel to the sheet corresponds to the axial field in coronal loops, on the order of a hundred Gauss. The field component $B_{\perp}$ perpendicular to the sheet is much smaller than the reconnecting component B_0, roughly on the order of a Gauss (Martens 1988, Litvinenko & Craig 2000). Observationally, the ratio $B_{\perp}/B_0$ is defined by the ratio of the reconnection inflow speed and the outflow Alfvén speed (e.g. Yokoyama *et al.* 2001), which also leads to the ratio of the nonreconnecting components $B_{||}/B_{\perp} \approx 10^1 - 10^3$. The resulting estimate for the range of particle energies $\mathcal{E} \approx 50$ keV-5 MeV supports the viewpoint that the hard X-ray and gamma-ray emissions observed in impulsive solar flares are generated by electrons accelerated in magnetic reconnection regions above the soft X-ray flare loops.

ACKNOWLEDGEMENTS

This work was supported by NSF grant ATM-9813933 and NASA grant NAG5-7792.

REFERENCES

Aschwanden, M. J., in *Observational Plasma Astrophysics: Five Years of Yohkoh and Beyond*, eds. T. Watanabe *et al.*, p. 285, Kluwer, Dordrecht (1998).

de Kluiver, H., Perepelkin, N. F., & Hirose, A., *Phys. Rep.*, **199**, 281 (1991).

Foukal, P., & Hinata, S., *Solar Phys.*, **132**, 307 (1991).

Litvinenko, Y. E., *Astrophys. J.*, **462**, 997 (1996).

Litvinenko, Y. E., *Solar Phys.*, **194**, 327 (2000).

Litvinenko, Y. E., & Craig, I. J. D., *Astrophys. J.*, **544**, 1101 (2000).

Martens, P. C. H., *Astrophys. J.*, **330**, L131 (1988).

Masuda, S., Kosugi, T., Hara, H., Tsuneta, S., & Ogawara, Y., *Nature*, **371**, 495 (1994).

Metcalf, T. R., & Alexander, D., *Astrophys. J.*, **522**, 1108 (1999).

Yokoyama, T., Akita, K., Morimoto, T., Inoue, K., & Newmark, J., *Astrophys. J.*, **546**, L69 (2001).

HEAVY ION ACCELERATION IN SOLAR FLARES

J. A. Miller

University of Alabama in Huntsville, Department of Physics, Huntsville, AL 35899, USA

ABSTRACT

Stochastic acceleration by cascading MHD turbulence is able to account for a wealth of impulsive solar flare energetic particle properties. Here we show that it can also account for the interplanetary heavy ion distributions from the August 7 1999 flare, observed by the *Advanced Composition Explorer.*

INTRODUCTION

The acceleration of particles to high energies is an ubiquitous phenomenon at sites throughout the Universe. Impulsive solar flares offer one of the most impressive examples anywhere, releasing up to 10^{32} ergs of energy over timescales of several tens of seconds to several tens of minutes. Much of this energy is spent on accelerating ambient electrons and ions to suprathermal and even relativistic energies, which then either remain trapped at the Sun or escape into interplanetary space. The radiations from the trapped particles (Ramaty & Murphy 1987) are important signatures of the acceleration mechanism. However, the particles that escape into space (Reames 1990) and subsequently are detected there are perhaps the most direct (assuming there is no appreciable further acceleration away from the Sun) and important diagnostic, and are refered to as solar energetic particles (SEPs).

The observed SEP abundances from impulsive flares are illustrated in Table 1, which shows specifically the abundance ratios for particles $\gtrsim 1\,\mathrm{MeV\,nucleon^{-1}}$ (data from the "Great Debate" (1995) and Reames *et al.* references therein, Share & Murphy 1997, Reames 2001). (The ions labeled Kr and Xe are actually groups of ions with mean mass numbers of 85 and 128, respectively.) The abundances for the gradual events are essentially the same as those in the ambient corona, and this is the basic evidence for the shock accelerated nature of these particles. However, the abundances for the impulsive events show dramatic enhancements in several of the ratios. Namely, (1) ^{3}He is enhanced by a factor of ≈ 2000 relative to ^{4}He; (2) ^{4}He is enhanced by a factor of ≈ 5 relative to H (or H is suppressed by a factor of ≈ 5 relative to ^{4}He); (3) Ne, Mg, Si, Fe, and especially Kr and Xe are enhanced by progressively increasing factors relative to O; and (4) C, N, O, and ^{4}He are not enhanced relative to one another. Also shown is the charge-to-mass ratio Q/A for the first ion (the "numerator") in the ratio, for a temperature of 3×10^6 K. At this temperature, O is fully ionized and has a Q/A of 0.5, like ^{4}He, C, and N. Note that Q/A is also equal to that ion's cyclotron frequency in units of the H cyclotron frequency Ω_{H}.

One trend is immediately apparent: below the ^{4}He, C, N, and O cyclotron frequency of $0.5\Omega_{\mathrm{H}}$, the abundance enhancement of an energetic ion species is inversely proportional to its cyclotron frequency. Note that ^{4}He is also enhanced relative to H, but the degree of this enhancement is somewhat larger than the trend of the heavier ions would suggest. Another curious point is the magnitude of the ^{3}He enhancement, which is even more curious due to the location of its cyclotron frequency between that of H and ^{4}He. If one assumes an acceleration mechanism that has an efficiency that increases with increasing Q/A, H should be enhanced relative to ^{4}He, which is not the case; on the other hand, if the efficiency increases with decreasing Q/A,

Table 1. Ion Abundance Ratios

Ratio	Q/A	Impulsive Flares	Gradual Flares (Corona)
$H/^4He$	1	$\approx$2 ($\times$5 decrease)	$\approx$10
$^3He/^4He$	0.67	$\sim$1 ($\times$2000 increase)	$\sim$ 0.0005
$^4He/O$	0.5	$\approx$46	$\approx$55
C/O	0.5	$\approx$0.436	$\approx$0.471
N/O	0.5	$\approx$0.153	$\approx$0.128
Ne/O	0.42	$\approx$0.416 ($\times$2.8 increase)	$\approx$0.151
Mg/O	0.42	$\approx$0.413 ($\times$2.0 increase)	$\approx$0.203
Si/O	0.42	$\approx$0.405 ($\times$2.6 increase)	$\approx$0.155
Fe/O	0.26	$\approx$1.234 ($\times$8.0 increase)	$\approx$0.155
Kr/O	0.13	($\times$100 increase)	
Xe/O	0.11	($\times$1000 increase)	

then ^{4}He should be enhanced relative to ^{3}He, which is also clearly not the case. A second point regarding ^{3}He is that its energy distribution is often quite different from that of the other ions in interplanetary space (e.g. Mason *et al.* 2002), which are in turn quite similar. It should be emphasized that these enhancements are all valuable diagnostics of the impulsive flare acceleration mechanism.

Of course, the Q/A values for the ions depend upon the plasma temperature, and the above discussion is predicated upon a temperature of 3×10^6 K. However, it has been argued (Reames, Meyer, & von Rosenvinge 1994) that the initial plasma temperature must in fact be 3×10^6 K; otherwise, the Q/A values become either (1) interlaced in a manner which precludes any pattern at all, or (2) equal, in which case there is no known mechanism that can preferentially accelerate one ion over another.

The simplest overall explanation for these observations is that ^{3}He is subjected to an additional acceleration, from which the other ions are immune. The only known way this could occur is if ^{3}He were stochastically accelerated by cyclotron resonance with waves that are excited in a narrow frequency range just around the ^{3}He cyclotron frequency. This basic idea was originally proposed by Fisk (1978), and subsequent models just differ in the specific wave mode that is employed (e.g., cf. Fisk with Temerin & Roth 1992, and Miller & Viñas 1993). However, regardless of the specific model, ^{3}He should be considered essentially as proof that stochastic acceleration at least operates in impulsive flares.

THE MODEL

Given that stochastic acceleration is known to at least occur in impulsive flares, we advocate it as the mechanism whereby *all* ions and electrons are energized from thermal to relativistic energies. In our model, we suppose that everything (including ^{3}He) is stochastically accelerated by broad-band MHD shear Alfvén and fast mode waves, but that ^{3}He is also subjected to some extra acceleration via one of the models referenced above. This extra stochastic acceleration is assumed to be responsible for the spectral difference in the ^{3}He energy distribution as well as its huge enhancement, and will not be discussed further here.

Stochastic acceleration relies upon wave-particle resonance, and occurs when the condition $x \equiv \omega - k_{\|}v_{\|} - \ell|\Omega|/\gamma = 0$ is satisfied. Here, $v_{\|}$, γ, and Ω are the particle's parallel speed (with respect to the ambient magnetic field $\vec{B}_0$), Lorentz factor, and cyclotron frequency; while ω and $k_{\|}$ are the wave frequency and parallel wavenumber. The quantity x is the frequency mismatch parameter. If the harmonic number $\ell \neq 0$, its sign depends upon the sense of rotation of the wave electric field and the particle in the plasma frame: if both rotate in the same sense (right or left handed) relative to $\vec{B}_0$, then $\ell > 0$; if not, then $\ell < 0$. When $x = 0$, the frequency of rotation of the wave electric field is an integer multiple of the frequency of gyration of the particle in its guiding center frame, and the sense of rotation is the same. The particle thus sees an

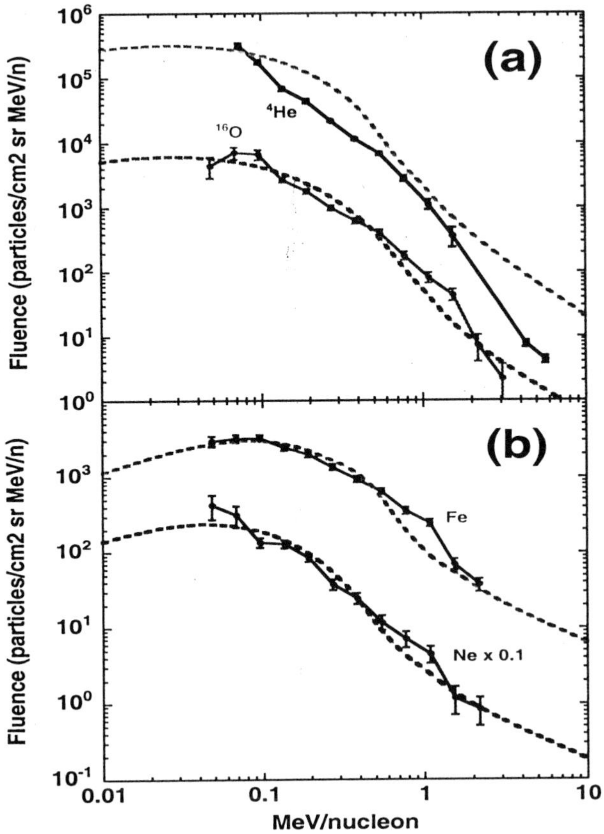

Fig. 1. Model fits (*dotted lines*) to the interplanetary ^{4}He, O, Fe, and Ne distributions from the August 7 1999 flare (Mason *et al.* 2002). Data (*connected dots with error bars*) is obtained from ULEIS and SIS on the *Advanced Composition Explorer*. The model distributions are from the cascading turbulence simulation, with an injection of $51\,\mathrm{ergs\,cm^{-3}\,s^{-1}}$ of turbulence into a region of size $10^8\,\mathrm{cm}$; the magnetic field is $100\,\mathrm{G}$ and the plasma density is $10^{10}\,\mathrm{cm^{-3}}$.

electric field for a sustained length of time and will be either strongly accelerated or decelerated, depending upon the relative phase of the field and the gyromotion. This is called cyclotron or gyroresonance. If $\ell = 0$, there is matching between the parallel motion of the particle and the wave parallel electric or magnetic field. This is called Landau or Cherenkov resonance.

The Alfvén waves posses a transverse left-hand polarized electric field, which can resonate with the ions via the $\ell = +1$ (or cyclotron) resonance. We see from the resonance condition and the Alfvén wave dispersion relation that in order for a wave to cyclotron resonate with a low-energy ion (say near the thermal speed), ω must be near Ω. On the other hand, as the ion gains energy, ω can become much less than Ω. Hence, a broad-band spectrum of Alfvén waves extending up to Ω can stochastically cyclotron accelerate ions from the tail of the thermal distribution to high energies (see also Barbosa 1979, Eichler 1979). On the other hand, Alfvén waves are not able to accelerate electrons from the background.

The fast mode waves posses a compressive magnetic field, which can couple with either ions or electrons via the $\ell = 0$ (or Landau) resonance. We see from the resonance condition and the fast mode dispersion relation that (1) a wave of any ω will only Landau resonate with a particle having a speed v greater than the Alfvén speed v_A, and (2) as the energy increases the wave propagation angle must approach 90°. Hence, fast mode waves having a distribution of propagation directions can resonate with particles from Alfvénic to relativistic energies, and this leads specifically to stochastic transit-time acceleration (Miller 1997). Now, initially in a flare plasma of temperature $T \approx 3 \times 10^6$ K, only electrons will be present in large numbers above v_A, and so these particles will be preferentially accelerated by the fast mode waves. However, should the ions become super Alfvénic (e.g. as a result of cyclotron acceleration by Alfvén waves), then they too will be transit-time accelerated. Lastly, even though fast mode waves of any ω can resonate with super Alfvénic particles, the acceleration rate is proportional to their wavenumber or frequency.

When a particle is in resonance with a single small-amplitude wave, $v_{||}$ executes approximate simple harmonic motion about that parallel velocity which exactly satisfies the resonance condition (Karimabadi *et al.* 1992). There is no energy gain on average. The frequency ω_b of oscillation is proportional to the square root of the wave amplitude, and if $|x| \leq 2\omega_b$ the particle and wave effectively are in resonance. Hence, the exact resonance condition $x = 0$ does not have to be satisfied in order for a strong wave-particle interaction to occur.

This brings us to resonance overlap, which is what yields large average energy gains. To understand overlap, consider two neighboring waves, i and $i+1$, where $i+1$ will resonate with a particle of higher energy than i will. A particle initially resonant with wave i will periodically gain and lose a small amount of $v_{||}$. If the gain at some time is large enough to allow it to satisfy $|x| \leq 2\omega_{b,i+1}$, where $\omega_{b,i+1}$ is the bounce frequency for wave $i+1$, then the particle will resonate with that wave next. After "jumping" from one wave to the next in this manner, the particle will have achieved a net gain in energy. If other waves are present that will resonate with even higher energy particles, the particle will continue jumping from resonance to resonance and achieve a maximum energy corresponding to the last resonance present. If the wave spectrum is discrete, then the spacing of waves is critical; however, if the spectrum is continuous (as is almost certainly the case in flares, and is the case in our model below), then resonance overlap will automatically occur. Of course, the particle can also move down the resonance ladder, but over long timescales there is a net gain in energy and stochastic acceleration is the result.

An important aspect of stochastic acceleration in general is that it is not directed, as with DC electric fields (e.g. Miller *et al.* 1997). This allows cospatial return currents to form, which draw particles up from the denser and cooler chromosphere, ensure charge neutrality, and provide the replenishment for the acceleration region that is necessary in order to sustain the huge fluxes of hard X-rays and gamma rays that also accompany these events (see Miller *et al.* 1997). We incorporate this into our model, which consists of just a few elements:

1. (Assumption) During the primary flare energy release phase, we suppose that long wavelength, low-amplitude ($\delta B/B \ll 1$) MHD Alfvén and fast mode waves are excited.
2. (Some fact, some assumption) These waves then cascade in a Kolmogorov-like fashion to smaller wavelengths (e.g. Verma *et al.* 1996), forming a power-law spectral density.
3. (Fact) When the mean wavenumber of the fast mode waves has increased sufficiently, the transit-time acceleration rate for super Alfvénic electrons can overcome Coulomb energy losses, and these electrons are accelerated out of the thermal distribution and to relativistic energies (Miller, LaRosa, & Moore 1996). As the Alfvén waves cascade to higher wavenumbers, they can cyclotron resonate with progressively lower energy ions. Eventually, they will resonate with ions in the tail of the thermal distribution, which will then be accelerated to relativistic energies as well (Miller & Roberts 1995).
4. (Fact) When the ions become superAlfvénic (above $\approx 1\,\mathrm{MeV\ nucleon^{-1}}$), they too can suffer transit-time acceleration by the fast mode waves and will receive an extra acceleration "kick."

Regarding item 3, the Alfvén waves will encounter Fe first, since it has the lowest gyrofrequency. (We do not discuss Kr and Xe here since these have not been incorporated into our simulation yet, but the same general scenario should apply to them as well.) Iron will be strongly accelerated but is not abundant enough to damp the waves. Thus, some wave energy will cascade to higher frequencies where it encounters Ne, Mg, and Si. The same way, these ions suffer strong acceleration, but the wave dissipation is not complete. Some wave energy then cascades to reach ^{4}He, C, N, and O. Thus, iron will resonate with the most powerful waves; Ne, Mg, and Si will resonate with waves having less power; and ^{4}He, C, N, and O will resonate with even less powerful waves. Hence, Fe should be enhanced more than Ne, Mg, and Si relative to ^{4}He, C, N, and O. Since ^{4}He, C, N, and O all have the same cyclotron frequency, they should not be enhanced relative to each other. In this way, cascading turbulence is qualitatively able to account for the observed abundances at, and below, the ^{4}He cyclotron frequency Ω_4.

We investigate this process quantitatively with a self-consistent quasilinear simulation, in which the two wave species are evolved with nonlinear diffusion equations and each particle species (electrons plus all the ion species) is described by a Fokker-Planck equation in energy space. We take into account the nonlinear cascading of the turbulence, its damping on the particles, the accompanying acceleration of the particles, their escape from the acceleration region, Coulomb losses, and replenishment by a cospatial return current. The simulation is also basically a two-parameter model, the two parameters being the turbulence injection rate and the acceleration region length. Of course, there are other parameters that must be specified, but variances in these have the same effect as varying either one (or both) of the above two quantities. We assume that the particles suffer no further acceleration once they escape from the flare site, and that interplanetary transport does not further alter the energy distributions.

RESULTS AND CONCLUSIONS

We cannot consider all the model results here, but do note that it is capable of accounting, simultaneously, for typical ion and electron fluxes, maximum energies, and acceleration timescales (Miller *et al.* 1997), as well as the energy integrated abundance ratios given in Table 1. We focus here on just one aspect of these results: namely, the agreement between the theoretical ion distributions and those observed in space. In Figure 1, we show the calculated distributions of ^{4}He, O, Fe, and Ne obtained from our cascading turbulence model, along with the observed distributions. It is important to remember that the relative normalizations of the model spectra are fixed by the simulation, as are the spectral shapes. All that is varied in order to fit the data is the absolute normalization of one distribution. This is excellent agreement for the cascading turbulence model, which, in addition to being able to account for the bulk properties of the energetic particles, is now able to explain as well the detailed ion distributions observed in space.

We conclude by discussing the one ion that has been left out thus far: namely, H. It is interesting that the most abundant ion may have the most convoluted acceleration. In the present version of our simulation, we assume that the Alfvén wave dispersion relation was not affected by ions heavier than H. That is, while these heavy ions were able to damp the cascading waves, they did not affect the dispersion properties of the plasma; in this case, the dispersion relation continued up to the H cyclotron frequency, and H could be directly accelerated out of the background by the cascading waves. This was a simplifying assumption, since the more realistic treatment (below) is quite involved. At any rate, we obtain a suppression of H acceleration due to wave absorption by ^{4}He that is qualitatively in agreement with the observations.

In reality, the plasma dispersion relation is not this simple, since ^{4}He is 10% as numerous as H and will significantly alter the dielectric tensor of the plasma. (All other heavy ions, which are much less numerous, can still be treated as test particles [i.e., neglected] as far as dispersion is concerned.) Specifically, as the wave frequency ω increases from the MHD regime, the Alfvén branch of the dispersion relation has a resonance, or "stops" at $\Omega_4 = 0.5\Omega_{\rm H}$. It begins again at the cutoff frequency of about $0.6\Omega_{\rm H}$ and continues up to $\Omega_{\rm H}$, where there is another resonance that finally ends the Alfvén branch altogether; the waves on this section of the Alfvén branch are commonly called $\rm H^+$ electromagnetic ion cyclotron (EMIC) waves.

Since the cascading waves on the Alfvén branch are generated at very low frequencies, they will now cascade only up to Ω_4, since this is where the dispersion relation terminates. There will consequently be no wave energy cascading up to the H cyclotron frequency Ω_H now. Protons will thus not be stochastically accelerated directly by the cascading Alfvén waves. However, fast mode waves, which accompany the Alfvén waves in the turbulent cascade, have a different polarization, do not interact with the background thermal ions, and are not totally damped by the background electrons until they reach frequencies somewhat above Ω_H. The initial situation which should develop is thus one in which there is a near equipartition between the two wave species below Ω_4 but only fast mode waves above Ω_4. As a result of mode-mode conversion, there will be a flow of energy from the fast mode branch to the EMIC branch, where these waves will then cyclotron resonate with and accelerate protons. This scenario qualitatively explains the enhanced ^{4}He/H ratio: ^{4}He is directly accelerated by the cascading waves, but H is accelerated only after an intervening nonlinear process occurs, which should decrease its overall efficiency. This qualitative picture needs further investigation, but roughly accounts for H acceleration in a realistic solar flare plasma.

ACKNOWLEDGEMENTS

This work was supported by NASA grants NAG5-8480 and NAG5-4608.

REFERENCES

Barbosa, D.D., *ApJ*, **233**, 383 (1979).

Eichler, D., *ApJ*, **229**, 413 (1979).

Karimabadi, H. *et al.*, *JGR*, **97**, 13853 (1992).

Fisk, L.A., *ApJ*,**224**, 1048 (1978).

Great Debates in Sp. Phys. 1995 (articles by Miller, Hudson, and Reames), *Eos* Trans. AGU, 76(41), 401 (1995).

Mason, G.M. *et al.*, *ApJ*, in press (2002).

Miller, J.A., *ApJ*, **491**, 939 (1997).

Miller, J.A., and Viñas, A.F., *ApJ*, **412**, 386 (1993).

Miller, J A., and Roberts, D.A., *ApJ*, 452, 912 (1995).

Miller, J A., LaRosa, T.N., and Moore, R.L., *ApJ*, **461**, 445 (1996).

Miller, J.A. *et al.*, *JGR*, **102**, 14631 (1997).

Ramaty, R., and Murphy, R.J. 1987, *Space Sci. Rev.*, **45**, 213 (1987).

Reames, D.V., *ApJS*, **73**, 235 (1990).

Reames, D.V., *ApJ Letters*, **540**, L111 (2001).

Reames, D.V., Meyer, J.P., and von Rosenvinge, T.T., *ApJS*, **90**, 649 (1994).

Share, G.H., and Murphy, R.J., *ApJ*, **485**, 409 (1997).

Temerin, M., and Roth, I., *ApJ Letters*, **391**, L105 (1992).

Verma, M.K., *et al.*, *JGR*, **101**, 21619 (1996).

THE INTENSE GAMMA-RAY FLARE ON 1997 NOVEMBER 6

M. Yoshimori[1], H. Ogawa[1], H. Hirayama[1], G. H. Share[2], and R. J. Murphy[2]

[1]*Rikkyo University, Tokyo 171-8501, Japan*
[2]*Naval Research Laboratory, Washington, D.C., 20375, USA*

ABSTRACT

An X9.4/2B flare at 11:52 UT on 1997 November 6 is the most intense gamma-ray event *Yohkoh* has observed so far. Gamma-ray lines resulting from nuclear de-excitation, neutron-capture and positron annihilation and >10 MeV gamma-rays were observed. In addition, the Oriented Scintillation Spectrometer Experiment on *Compton Gamma Ray Observatory* (OSSE/*CGRO*) detected flare-associated neutrons after the strong gamma-ray emission was over, suggesting the possibility of extended high-energy proton production. We discuss the accelerated-proton spectrum, photospheric ^{3}He abundance, plasma temperature at the positron annihilation site and production processes of >10 MeV gamma-rays from *Yohkoh* and OSSE observations.

INTRODUCTION

An intense flare (X9.4/2B, S18E64) occurred at 11:52 UT on 1997 November 6. A variety of observations of X-rays, gamma-rays, neutrons (Yoshimori *et al.*, 2002) and solar energetic particles (Mason *et al.* 1999, Falcone *et al.* 2000, Cliver *et al.* 2001) were reported. This flare is the best example for study of characteristics of high-energy phenomena such as particle acceleration and subsequent X-ray, gamma-ray and neutron production during a large solar flare.

OBSERVATIONS

The gamma-ray emission was intense over a wide energy band and lasted for 4 minutes. The background-subtracted gamma-ray count spectrum in 11:52:48-12:01:52 UT is shown in Figure 1 (Yoshimori *et al.*, 1999). The neutron-capture line at 2.22 MeV, de-excitation lines of C(4.44 MeV) and O(6.13 MeV) and a complex of nuclear lines in 1-2 MeV are superimposed on the continuum. Time profiles at 20-33 and 53-72 MeV and the count spectrum above 10 MeV are shown in Figures 2 and 3, respectively. Although the count rates gradually increased with time after the impulsive peak, it is due to a temporal variation in the background level. The gamma-ray count spectrum exhibits a broad excess above 40 MeV and is not fitted by a single power-law. The hard X-ray count rate was saturated during the peak phase of the flare and the hard X-ray spectrum was analyzed in the decay phase (11:55:30-12:00:00 UT). The positron annihilation line at 511 keV is shown in Figure 4. Taking into account the instrumental resolution of a NaI scintillator (FWHM is 48 keV at 511 keV), the intrinsic line width is estimated to be less than 16 keV(FWHM). The OSSE/*CGRO* missed the gamma-ray flare but detected neutrons between 12:08 and 12:28 UT.

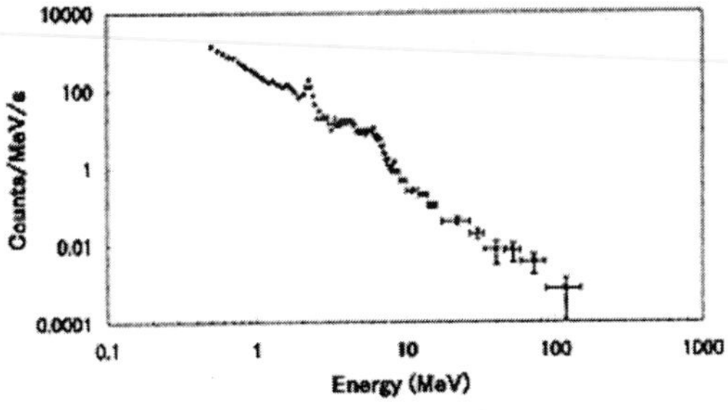

Fig. 1. Background-subtracted gamma-ray count spectrum (Yoshimori *et al* 1999).

DISCUSSION

We derive the accelerated-proton spectrum at 10-100 MeV from a ratio of the neutron-capture to C and O line fluxes. The power-law spectral index ranges from 3.0±0.5. A total number of accelerated protons above 10 MeV is calculated to be 9 x 10^{33} from the fluences of C and O lines.

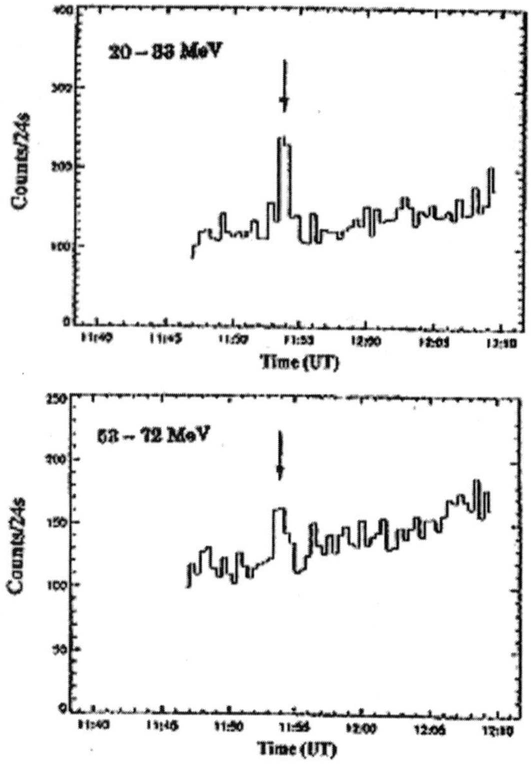

Fig. 2. Count rate time profiles at 20-33 MeV and at 53-72 MeV.

The photospheric ^{3}He to H abundance ratio is derived from a comparison of the time profiles between the 2.22 MeV and de-excitation lines. The decay time of 2.22 MeV line is 72±11 s which is longer than that of the de-excitation lines. It is due to behavior of neutrons in the photosphere. After the neutrons are thermalized by elastic scattering with protons, the following three processes take place: (1) radiative capture by proton to emit the 2.22 MeV line, (2) non-radiative capture by ^{3}He and (3) beta-decay (decay time is 918 s). Assuming the photospheric proton density is 1.2 x 10^{17}cm^{-3}, we determine the ^{3}He to H abundance ratio is (2.3±1.4) x 10^{-5}. It is in agreement with the previous values (Hua and Lingenfelter 1987, Murphy *et al.* 1997).

Three processes are thought to contribute to the production of >10 MeV gamma-rays: (1) bremsstrahlung of primary >10 MeV electrons, (2) neutral pion-decay and (3) bremsstrahlung of positrons from positive pion-decay.

In Figure 3 we show the three components which were calculated by Ramaty and Murphy (1987). The continuum below 30 MeV is fitted by the power-law of E^{-3} which is dominated by primary electron bremsstrahlung. On the other hand, the broad excess above 40 MeV is likely due to neutral pion-decay gamma-rays which peak at 70 MeV. The *Yohkoh* observation indicates that both electrons and protons were efficiently accelerated to energies above 10 and 300 MeV, respectively, in the peak phase of the flare.

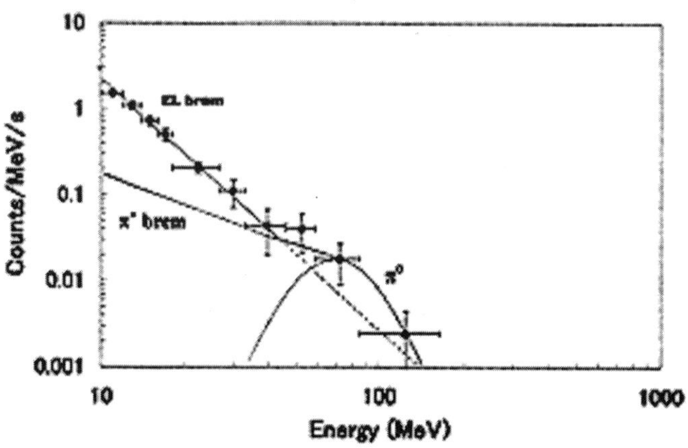

Fig. 3. >10 MeV gamma-ray count spectrum.

The intrinsic width of the 511 keV line depends on the temperature of the positron annihilation site. The FWHM of the line is expressed by 1.1 x ($T/10^4$) keV, where T is the temperature. The temperature is derived to be <2.1 MK from the observed intrinsic width. Share *et al.* (1996) obtained the temperature of 0.2-10 MK from

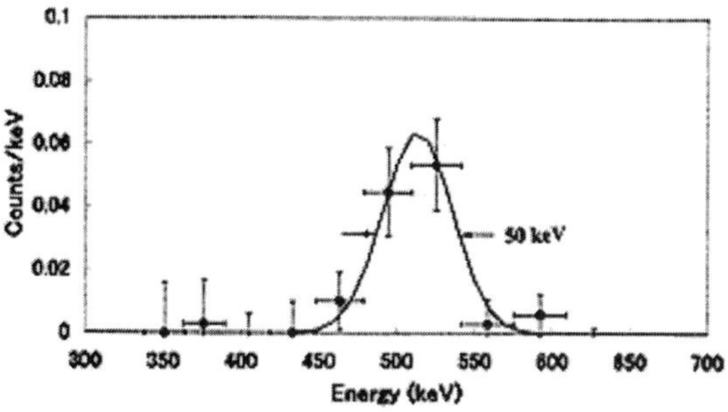

Fig. 4. Positron annihilation line spectrum.

measurements of the width of 511 keV line from seven solar flares. We compare the temporal variations between the 511 keV and de-excitation C and O lines. Figure 5 shows that the decay time of the 511 keV line is longer than that of the de-excitation line, suggesting the possibility that positrons are emitted by long-life positive beta-decay nuclei of ^{13}N and ^{15}O of which half-lives are 10 and 2 minutes, respectively.

The neutron arrival time profile provides information on the neutron production time profile and accelerated-proton spectrum on the Sun. The OSSE observations show that the neutron count rate gradually decreased with time, as shown in Figure 6. The arrival neutron time profile is calculated from the algorithm developed by Murphy *et al.* (1999). In order to explain the OSSE neutron time profile, we require that (1) the neutrons are produced in the peak (11:52:48-11:56:40 UT) and extended (11:56:40-12:06:00 UT) phases and (2) the proton spectrum hardens in the extended phase (spectral index is 3.0 in the peak phase and 2.5 in the extended phase).

A powerful instrument with gamma-ray imaging and spectroscopic capabilities is needed to advance the understanding of high-energy phenomena on the Sun. The *RHESSI* observation will provide us much more detailed information on particle acceleration and subsequent gamma-ray and neutron production.

ACKNOWLEDGEMENTS

This work has been supported by a Grant-in-Aid for Scientific Research of the Ministry of Education, Science and Culture C-1640290.

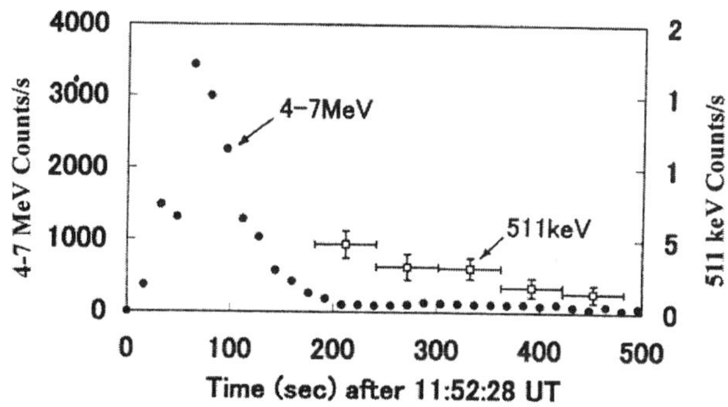

Fig. 5. Time profiles of the 511 keV line and 4-7 MeV gamma-rays

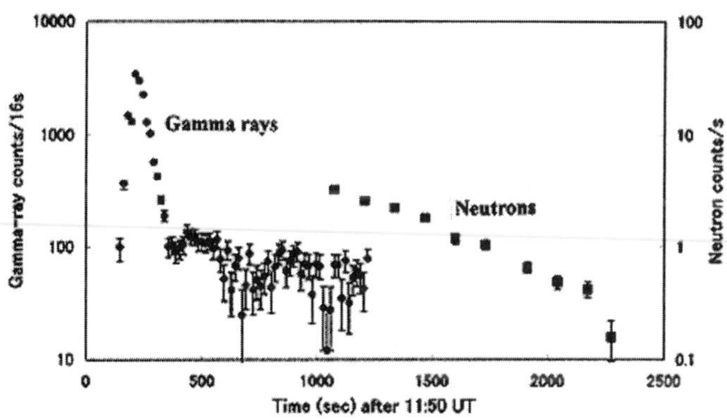

Fig. 6. Time profiles of 4-7 MeV gamma-rays and >16 MeV neutrons

REFERENCES

Cliver, E.W. *et al.*, *27th Intern. Cosmic Ray Conf.*, **8** , 3277 (2001).

Falcone, A.D. *et al.*, *in Acceleration and Transport of Energetic Particles Observed in the Heliosphere*, ed. R.A. Mewaldt *et al.*, p.193, AIP Conf. Proc. No.528, New York, NY (2000).

Hua, X-,M. and R.E. Lingenfelter, ApJ. **319**, 519 (1987).

Mason, G.M. *et al.*, Geophys. Rev. Lett. **26**, 141 (1999).

Murphy, R.J. *et al.*, ApJ. **490**, 883 (1997).

Murphy, R.J. *et al.*, ApJ. **510**, 1011 (1999).

Ramaty, R. and R.J. Murphy, Space Sci. Rev. **45**, 213 (1987).

Share, G.H. *et al.*, *in High Energy Solar Physics*, ed. R. Ramaty *et al.*, p.162, AIP Conf. Proc. No.374, New York, NY (1996).

Yoshimori, M. *et al.*, *26th Intern. Cosmic Ray Conf.*, **6**, 30 (1999).

Yoshimori, M. *et al.*, *Adv. Space Res.*, in press (2002).

HIGH-ENERGY MEASUREMENTS OF THE 1991 NOVEMBER 15 SOLAR FLARE

J. M. Ryan[1], M. Arndt[1], K. Bennett[3], A. Connors[1], H. Debrunner[4], J. Lockwood[1], M. McConnell[1], G. Rank[2], V. Schönfelder[2], R. Suleiman[1], O. Williams[3], C. Winkler[3], and C. A. Young[1]

[1]*Space Science Center, University of New Hampshire, Durham, NH 0382, USA*
[2]*Max-Planck Institut für Extraterrestrische Physik, Postfach 1312, 85741 Garching, Germany*
[3]*Astrophysics Division, ESTEC, Noordwijk, The Netherlands*
[4]*Physikalisches Institut, Bern, Switzerland*

ABSTRACT

We report γ-ray observations of the 1991 November 15 solar flare. Although the event was not as well measured as some with the *Compton Gamma Ray Observatory (CGRO)* because the axis of the *Compton* spacecraft was far from the solar direction, the data are comprehensive and span a wide range of energies with good sensitivity and resolution. Of particular interest are the observations that the energetic proton spectrum was harder during the impulsive phase, the transport of protons in the corona was diffusive rather than adiabatic, the abundance of ^{3}He may be higher than is normally believed and there is not enough energy in the energetic electrons to power the white light emission, although proton heating is still possible.

INTRODUCTION

The X1.5 solar flare of 15 November 1991 was one of the few solar flares detected and measured using the *Compton* Observatory while simultaneously being observed by instruments on the *Yohkoh* spacecraft. The soft X-ray flux peaked at the X1 level at 22:38 UT. One of the difficulties in interpreting high-energy photon data from a flare is the general absence of supporting or context data from other instruments, either space borne or ground based. These ancillary data are vital for understanding the processes of particle acceleration and transport in solar flares. For example, these context data provide the morphology and spatial dimensions associated with the flaring region. They also provide measures of other important quantities, such as soft X-ray power, white light images and energetic electron behavior. The *Yohkoh* observations and various ground based measurements of the 15 November 1991 solar flare constitute a rich source of information about the event that we can use to interpret the hard X-ray and γ-ray data. In this paper we report measurements of the flare radiation using data from BATSE, COMPTEL and EGRET instruments on the *Compton* spacecraft that span the energy range of 20 keV to 10 MeV. In later publications we will extend that range to 50 MeV. The data that we have used from *Yohkoh* include published results from the white light imager, hard X-ray telescope (HXT) and the Gamma Ray Spectrometer.

THE HIGH-ENERGY BEHAVIOR OF THE FLARE

The hard X-ray and γ-ray behavior of the flare is illustrated in Figure 1. Intensity-time profiles are shown in the 18-30 keV (X-ray) and 0.6-10 MeV (γ-ray) ranges. These data are most sensitive to the primary electrons (BATSE data) and protons (COMPTEL data), accelerated in the flare. Both electron bremsstrahlung and nuclear γ-ray emission contribute to the count rate in the 0.6-10 MeV channel with the bremsstrahlung dominating at the lowest part of the band (<1 MeV) and the nuclear component dominating above 1.2 MeV. We note that the hard X-rays have three significant spikes in intensity during the impulsive phase. The impulsive phase is followed by a gradual phase of approximately twice the duration of the impulsive phase, but the intensity is reduced and there is less temporal structure. We call this gradual phase the post-impulsive phase. Both phases are preceded by a precursor phase and followed by a decay phase, neither of which is addressed in this paper. We note that the γ-ray spikes, although less resolved than those in hard X-rays, are delayed with respect to the hard X-ray spikes by at least one time bin, in this case, 6 s. This γ-ray spike delay was observed early in the *Solar*

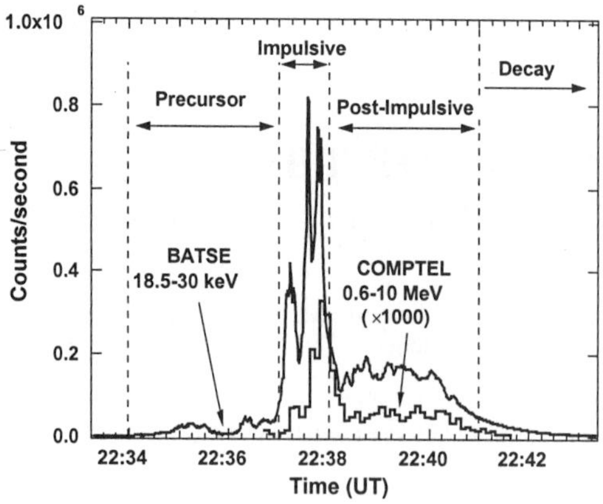

Fig. 1. Intensity-time profiles for X and nuclear γ rays.

Maximum Mission and regularly thereafter (Forrest & Chupp 1983). Again, the delay seen here is over the broad range of 0.6 to 10 MeV, i.e., both relativistic-electron bremsstrahlung and nuclear γ-rays. Further analysis is planned to study the delay *strictly* due to the nuclear component. The hard X-ray emission during this period is largely confined to two intense footpoints that, except for a perspective offset, coincide with the white-light bright points (Sakao *et al.* 1992).

The BATSE hard X-ray data (Schwartz, private communication) was fit to a double power-law spectrum. The best fit requires a spectral break, being harder at lower energies and breaking into a softer spectrum. The fit is consistent with the results obtained by other researchers, except those from the *Ulysses* spacecraft. The *Ulysses* data were obtained at a heliocentric latitude of 80°. The *Ulysses* spectrum is significantly harder (Kane *et al.* 1998). Among those data with similar heliocentric viewing perspectives, all spectra require a break in the neighborhood of 170 keV. The BATSE spectrum has a power-law index above 170 keV of –3.6.

The γ-ray data from COMPTEL were obtained from the Burst Spectrometer, a detector that recorded spectra from transients in the range of 0.6 to 10 MeV. This detector was omni directional, employed an on-board ^{60}Co source for calibration and had a relatively unconstrained view of the Sun on 15 November 1991. The main COMPTEL instrument axis was far removed from the solar direction so that good COMPTEL telescope data do not exist for this flare. In figures 2 and 3 we show deconvolved γ-ray spectra from the impulsive and post-impulsive phases. Most prominent is the deuterium formation line at 2.223 MeV indicating the presence of free neutrons produced by nuclear interactions of accelerated and ambient protons and ions. Other lines are present in both spectra, the strongest and clearest being between 1 and 2 MeV from de-excitations of ^{24}Mg, ^{20}Ne and ^{28}Si. The lines in these spectra do not occur at their proper energies. This is due to a correctable pulse-height analyzer calibration error in the Burst Spectrometer electronics. The calibration still places the 2.223 MeV line at its proper energy but other lines, both at higher and lower energies, are systematically at lower energies. Another artifact in these spectra occurs at 3.0 MeV arising from a digital electronics problem. Aside from these issues, we see that, when compared to the intensity at 2.223 MeV, there is a lower intensity of γ-rays above 2.223 MeV in the post-impulsive phase. Also plotted in these spectra are the extensions of the BATSE hard X-ray spectra obtained from the corresponding time intervals. The bremsstrahlung continuum converges nicely with the γ-ray data at ~700 keV for both phases.

The γ-ray photon spectra were produced using a maximum-entropy deconvolution of the COMPTEL count-rate spectra (background plus flare) using a background estimate from orbits approximately 24 hours earlier and later than the flare. Backgrounds offset by 24 hours faithfully represent the same orientation and background environments of the spacecraft provided the spacecraft is not in a region of steep background gradients. The background was removed in the final step after the deconvolution process meaning that the Gaussian-statistics requirement of the maximum-entropy process was satisfied.

DISCUSSION

We first note a few features of the high-energy emission from the 15 November 1991 event that have been observed in other flares. The first is that the ratio of the hard X-ray to γ-ray intensities varies between spikes. In fact, the first spike in hard X rays is difficult to identify in the γ-ray data. Second, as mentioned above, the γ-ray spikes are generally delayed with respect to those in hard X rays. Third, the hard X-ray spectrum is power-law in shape. Finally, we note that the γ-ray spectrum drops precipitously at ~9 MeV, the highest energy of solar γ-ray lines. With the present state of analysis, we can only place an upper limit on the primary bremsstrahlung emission above 10 MeV. The bremsstrahlung spectrum above 170 keV with a power-law index of –3.6 implies a primary electron spectrum of $E^{-4.6}$ under the assumption that the emission is thick target in nature. This assumption is supported by the hard X-ray images of HXT showing that most of the emission comes from the footpoints of a magnetic loop.

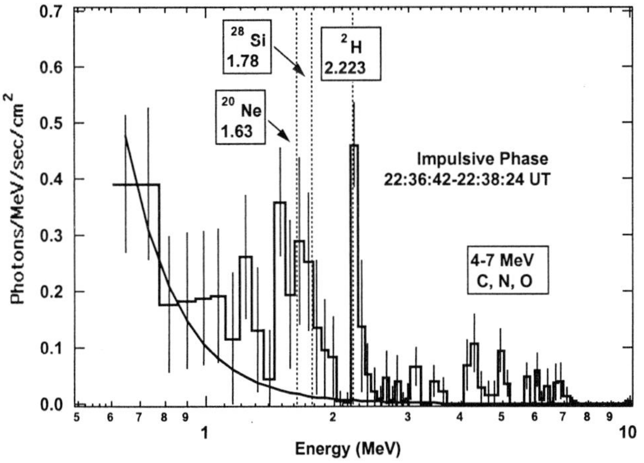

Fig.2. Impulsive phase γ spectrum.

The flare-integrated γ-ray spectrum is consistent with a proton power-law spectrum with an index between –4 and –5, similar to that of the electrons. The spectrum is also consistent with a K_2 Bessel function (approximately exponential) with an αT parameter (Murphy *et al.* 1987) between 0.008 and 0.015. The proton spectrum was inferred using the ratios of the integrated intensities of the 2.223 MeV line from free neutron capture and the 4.43 MeV line from carbon. Similar values were obtained using the ratios of the integrated intensities of the 2.223 MeV line and the integrated flux between 4 and 7 MeV (both broad and narrow lines with negligible bremsstrahlung continuum). These proton spectral shapes are consistent with those obtained by Kawabata *et al.* (1994) and Yoshimori *et al.* (1994) using these and other line ratios. Although compromised by the delayed nature of the 2.223 MeV emission, the relatively lower amplitude of the spectrum above 2.223 MeV in the post impulsive phase implies a softer proton/ion spectrum than that for the impulsive phase. This must be confirmed after correcting for the delayed behavior of the 2.223 MeV line emission. Although the data are still being analyzed to quantify the spectral softening, the same effect can be seen in the noticeable lack of structure (sharp lines) above 2.223 MeV in the post-impulsive phase. The C and O lines have higher thresholds than those of Mg, Ne and Si.

We can compute the energy embodied in the accelerated electrons and protons. We used the bremsstrahlung spectrum above 170 keV since we eventually wish to compare our energy estimates to those of Sakao *et al.* (1992) for the white-light emission. Electrons below ~170 keV cannot reach the photosphere from the corona. Above 170 keV the energy of the $E^{-4.6}$ electron spectrum is ~10^{28} ergs. This value is somewhat lower than one might first expect from a large flare, but our high value for the lower integration limit yields 3.8×10^{-3} of the energy from an integration starting at the conventional 20 keV. Thus, extending the spectrum downward to 20 keV with a harder spectrum represents a large amount energy in the form of electrons as we would expect from a large flare. We take the proton spectrum to be power-law in shape with an index of –4.5. Integrating the proton energy above 30 MeV (approximate threshold for production of most free neutrons) we obtained an energy of 5×10^{27} ergs. Extrapolating the spectrum down to 1 MeV, one obtains a proton energy of ~10^{30} ergs. Hudson *et al.* (1992) estimated the white-light energy to be ~10^{30} ergs *Yohkoh* data. Although the protons contain 10^{30} ergs above 1 MeV and the electrons contain a similar amount above 20 keV, most of those protons and electrons cannot reach the photosphere from the corona. Potential agreement can be found if the proton spectrum is as soft as the data allow and we integrate upwards from 10 MeV. It is difficult to reconcile the white light emission with the small amount of energy in electrons above 170 keV.

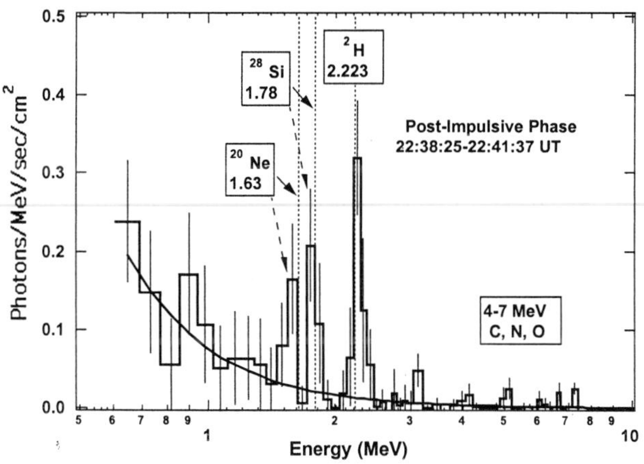

Fig. 3. Post-impulsive phase γ spectrum.

The advantage of having an image of the energetic particle precipitation pattern becomes clear when one wishes to study the transport of particles in the solar environment. If we take the hard X-ray image of the electron impact areas to also represent the areas of proton or ion precipitation we can then interpret the delay of the γ-ray spikes (produced by protons—the dominant contributor to the 0.6-10 MeV rate) with respect to those of the hard X-rays (produced by electrons). The separation of the white-light and hard X-ray footpoints is approximately 1.5×10^9 cm. We take this value as the characteristic scale of the loop overlying those footpoints. Over this distance, assuming that the protons and electrons were accelerated simultaneously and produce photons at the same altitude, the protons impact the chromosphere/photosphere ~6 s after the electrons. The electrons (typically 50 keV) travel significantly faster than the ~20 MeV (where the nuclear lines are mostly being produced) protons responsible for the γ-rays. Dividing the characteristic loop dimension by the γ-ray delay yields a "velocity" of $\sim2\times10^8$cm-s^{-1}, whereas the protons themselves move with a much greater speed of $\sim5\times10^9$cm-s^{-1}. This disparity between the delay and the proton velocity strongly suggests that the transport of the protons is diffusive in nature. This conclusion is still valid if one uses the velocity *difference* between the electrons and the protons. This contrasts with descriptions of adiabatic proton transport in loops often used by other researchers. The adiabatic motion of protons is inconsistent with the observed delays.

The decay of the neutron capture line at 2.223 MeV is fixed by the depth of the neutron penetration (equivalently the average neutron energy) into the photosphere (local density of hydrogen) and the abundance of ^{3}He. Thermal neutrons are captured on both H and ^{3}He, but only the proton captures reveal themselves radiatively. The intrinsic capture time is obscured because of concurrent production of neutrons. However, the standard method for extracting that information relies on modeling the data with production, traced by nuclear-line emission in the 4-7 MeV range, and a single intrinsic capture time. Performing that exercise on the 15 November 1991 flare data, we obtained a capture time of 26 (+20,-15) s, a lower figure than is typically reported, implying a ^{3}He/H ratio of $(2.0\pm1.5)\times10^{-5}$. Since we see no evidence of protons above 50 MeV in the γ-ray data (no deep penetration of fast neutrons), we tentatively ascribe the short capture time to a high abundance of ^{3}He. However, because of the large uncertainty in our ^{3}He estimate, our number is not in serious disagreement with other estimates (Young *et al.* 2001). New *RHESSI* results might provide improved data on this phenomenon.

CONCLUSIONS

The X1.5 solar flare of 15 November 1991 is *in toto* one of the best-observed γ-ray flares of the last solar cycle. Indications of diffusive proton transport in the corona and higher than "normal" ^{3}He abundances deserve further attention and will be addressed in subsequent publications.

ACKNOWLEDGEMENTS

We acknowledge Drs. G. J. Fishman and David Bertsch of the *CGRO* project. The COMPTEL project is supported by NASA under contract NAS5-26645, by the Deutsche Agentur fur Raumfahrtgelenheiten (DARA) under grant 50 QV90968 and by the Netherlands Organization for Scientific Research NWO.

REFERENCES

Forrest, D. J., & Chupp, E. L., Simultaneous Acceleration of Electrons and Ions in Solar Flares, *Nature*, **305**, 291 (1983).

Hudson, H. S., Acton, L. W., Hirayama, T., & Uchida, Y., White-Light Flares Observed by *Yohkoh*, *Publ. Astron. Soc. Japan*, **44**, L77 (1992).

Kane, A. R., Hurley, K., McTiernan, J. M., Boer, M., Niel, M. *et al.*, Stereoscopic Observations of Solar Hard X-Ray Flares Made by *Ulysses* and *Yohkoh*, *Astrophys. J.*, **500**, 1003 (1998).

Kawabata, K., Yoshimori, M., Suga, K., Morimoto, K., Hiraoka, T. *et al.*, Positron Annihilation Radiation from the 1991 November 15 Flare, *Astrophys. J. (Suppl. Ser.)*, **90**, 701 (1994).

Murphy, R. J., Dormer, C. D., & Ramaty, R., High-Energy Processes in Solar Flares, *Astrophys. J. (Suppl. Ser.)*, **63**, 721 (1987).

Sakao, T., Kosygin, T., Masuda, S., Inda, M., Makishima, K. *et al.*, Hard X-Ray Imaging Observations by *Yohkoh* of the 1991 November 15 Solar Flares, *Publ. Astron. Soc. Japan*, **44**, L83 (1992).

C. A. Young, K. Bennett, A. Connors, R. Diehl, M. McConnell *et al.*, Energetic proton spectra in the 11 June 1991 solar flare, *AIP Conference Proceedings, GAMMA 2001: Gamma-Ray Astrophysics 2001*, Eds. Ritz, S., Gehrels. N., and Shrader, C.R, **587**, pp. 623-627 (2001).

Yoshimori, M., Suga, K., Morimoto, K., Hiraoka, T., Sato, J. *et al.*, Gamma-Ray Spectral Observations with *Yohkoh*, *Astrophys. J. (Suppl. Ser.)*, **90**, 639 (1994).

RADIO SHOCKS FROM RECONNECTION OUTFLOW JET ?—NEW OBSERVATIONS

H. Aurass[1], M. Karlicky[2], B. J. Thompson[3], and B. Vršnak[4]

[1]*Astrophysikalisches Institut Potsdam, An der Sternwarte 16,D-14482 Potsdam, Germany*
[2]*Astronomical Institute, CZ-251 65 Ondřejov, Czech Republic*
[3]*NASA Goddard Space Flight Center, code 682.3, Greenbelt, MD-20771, USA*
[4]*Hvar Observatory, University of Zagreb, Kačicéva 26, HR-10000 Zagreb, Croatia*

ABSTRACT

We point out possible radio signatures of reconnection outflow termination shocks during the impulsive flare phase. The standing shock signature is identified between 350 and 800 MHz during the four minutes of most energetic hard X-ray emission. *SOHO*-EIT images in the corresponding time interval reveal a flare loop, the growth of a hot cusp-shaped arcade, and colder post-flare loops underneath.

INTRODUCTION

Models of dynamic (two-ribbon, arcade) flares involve the formation of a system of standing slow and possibly also fast mode shock waves associated with the fast reconnection process below the erupting filament. We have sketched the plasma-magnetic field configuration in Figure 1a. The inflowing plasma (arrows pointing toward the diffusion region, DR) is squeezed between the merging field line systems and is ejected at high velocity along the thin current sheet. Two pairs of slow mode standing MHD shocks (SMSS) are formed extending from the DR and separating the inflow and outflow plasma regions. The SMSSs are sites of strong electric currents (e.g. Sato & Hayashi 1979) with a current density comparable to the DR conditions. They are a potential source of nonthermal particles. These particles can be trapped between the SMSSs. As sketched in Figure 1a both reconnection jets are directed toward an obstacle (EP and PFL in Figure 1a). If the outflow jet is supermagnetosonic then most probably above the postflare loops a standing fast mode MHD shock will be formed. Such shocks can under certain circumstances excite a non-drifting type II burst.

Aurass, Vršnak & Mann (2001, Paper I henceforth) recognized for the first time the radio signature of a fast mode outflow termination shock in a dynamic radio burst spectrogram. It started almost 1 hour after the impulsive phase and lasted for more than 30 min. Simultaneous imaging observations showed a postflare loop arcade with a bright soft X-ray cusp. In Paper I favorable circumstances for the radio detection of a termination shock in the reconnection outflow were emphasized to be *a comparatively large height of the diffusion region, a low plasma to magnetic pressure ratio β upstream of the slow shocks, and a small angle between the reconnecting field lines.* We searched for nondrifting radio features with the typical **H**erring**B**one fine structure (**HB**, see Nelson & Melrose 1985) in the same frequency range as discussed in Paper I but during the impulsive flare phase. As an unique indicator of this time interval we used data of the new Czech hard X-ray spectrometer (HXRS) aboard the US MTI satellite (Farnik *et al.* 2001). The information about the geometry of coronal plasma structures is attributed in the given event by the Extreme Ultraviolet Telescope aboard the *Solar Heliospheric Observatory* (*SOHO*-EIT, Delaboudinière *et al.* 1995). A rough

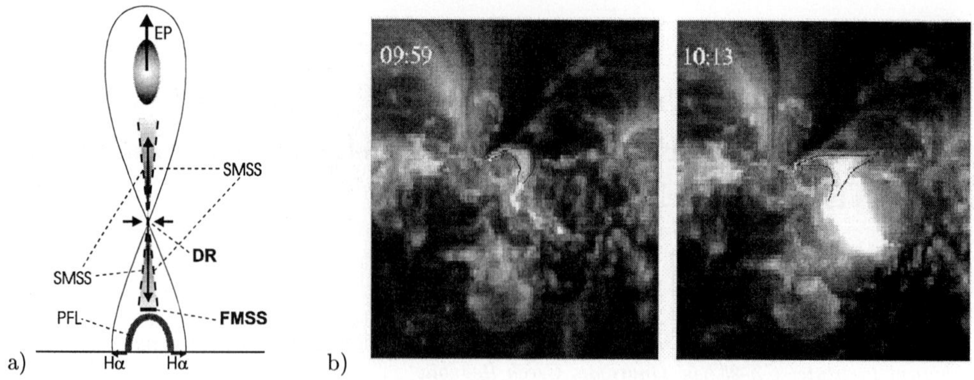

Fig. 1. a) Flare plasma–magnetic field configuration. DR–**D**iffusion **R**egion; EP–**E**rupting **P**rominence; SMSS–**S**low **M**ode **S**tanding **S**hocks; thick arrows–hot outflowing jets; FMSS–**F**ast **M**ode **S**tanding **S**hock; PFL–**P**ost**F**lare **L**oops; Hα–flare ribbons. b) 29 March 2001, AR 9393, *SOHO*-EIT images (195 Å–1.6·10^6 K) showing cusp evolution—the part of a) underneath **DR**—during the impulsive phase.

inspection of Nançay Radio Heliograph data[1] confirmed that the radio sources are situated in and around AR 9393. We give a brief description and discussion of this set of observations.

THE 29 MARCH 2001 EVENT - OBSERVATIONS

On 29 March 2001 there was a 1N, X1.7 flare reported in AR 9393 (N16W12) between 09:55 and 11:25 UT (NOAA Solar Geophysical Data, March 2001). *SOHO*-EIT images reveal that the preflare activation started not later than 09:24 UT. During the event also neighboring active regions are involved (NOAA AR 9401, 9394, 9395, 9405). Figure 1b shows the two images which are available at the beginning and near the end of the impulsive flare phase. In the 195 Å emission it is nicely seen how a previously visible flare loop starts to deform to a more and more height-extended cusp-shaped structure. Due to the favorable perspective we see directly the structure as expected from Figure 1a, the range below DR. During the flare, the two southward situated active regions are disturbed by the arriving flare wave at 10:13 UT (not shown in our Figure 1b). Simultaneously, a preflare noise storm radio continuum, and most of the radio flare components are suppressed (see Chertok *et al.* 2001) as recorded in the 40–800 MHz radio spectral data of Astrophysical Institute Potsdam (AIP). In Figure 2a we show single frequency cuts through the radio spectrum at 327 MHz (flare burst) and 70 MHz (continuum with flare-induced depression). In addition, Figure 2a gives the HXR flux of the flare in three channels between 12.6 and 147.2 keV. The most powerful energy release is detected in two enhancements between 10:04 and 10:12 UT with a minimum at 10:08 UT. Further, the 3 GHz radio flux (AO Ondrejov) is overplotted.

The coronagraph aboard *SOHO* reveals a halo CME ejected with 942 $\mathrm{km\,s^{-1}}$ and a (linearly) backwards extrapolated starting time of about 09:40 UT[2]. As expected (Aurass *et al.* 1999), in the estimated onset time interval (09:43 UT) some drift bursts are observed between 300 and 400 MHz.

In the radio spectrum, Figure 2b, three slowly drifting bursts are evident between 400 and 70 MHz. The **F**undamental-**H**armonic (F-H) mode distinction of these bursts is difficult due to off-scale flux at lower frequencies and ongoing noise storm continuum. From 300 to 100 MHz, there is a faint type II burst without F-H-pattern between 09:58 and 10:02 UT yielding a speed of 980 $\mathrm{km\,s^{-1}}$ if we assume F-mode emission and

[1] `ftp://mesola.obspm.fr/jrh_film/`
[2] `http://cdaw.gsfc.nasa.gov/CME_list/index.html`

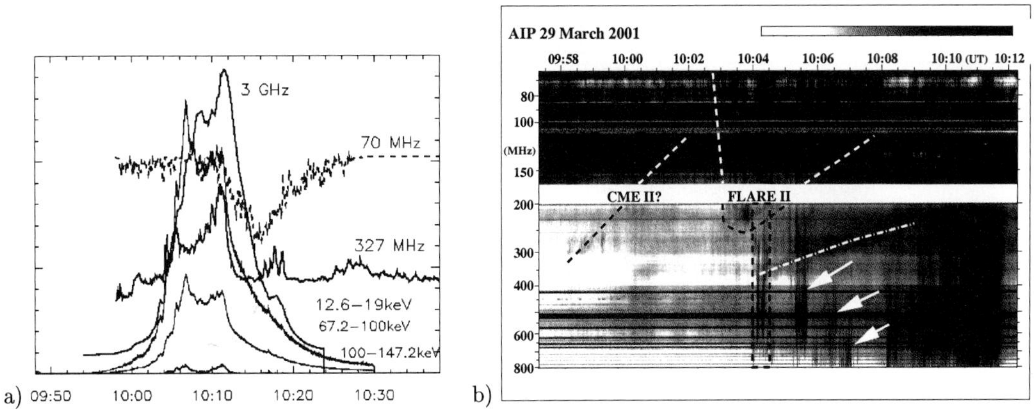

Fig. 2. Hard X-ray spectrometer (AO Ondřejov) and radio data (AI Potsdam, AO Ondřejov) of 29 March 2001. a) Three HXRS channels together with 3 GHz, 327 and 70 MHz plotted in arbitrary units. b) Dynamic radio spectrogram. Horizontal stripes result from strong interference. The black dashed box is enlarged in Figure 3a. Arrows point at other "HB" patches (see text). White dash-dotted line – the slowest drifting feature.

a one-fold Newkirk density model. This means we find a reasonable density model transforming the burst drift rate into the independently determined CME speed. If this burst is a CME signature the delay between first faint drift bursts and the proper CME-lift-off is not unusual (Aurass *et al.* 1999).

At 10:03–$>$10:07 UT, we recognize F and H mode signatures of a "conventional" type II burst between 250 and 70 MHz. During the type II emission the noise storm starts to get depressed. Eventually, at 10:04 UT a slowly drifting and narrowband enhancement appears (190–260 km s^{-1} with the Newkirk model; white dash-dotted in Figure 2b). This feature resembles to the sawtooth pattern discovered by Klassen *et al.* (2001). From *SOHO*-EIT images we estimate an EIT wave speed of about 200 km s^{-1} possibly corresponding to the slowest drifting radio feature. Furthermore, as shown in Figure 2b, there are patches of drift bursts between 250–800 MHz. At about 10:04:15 UT, one patch covers the starting point of the sawtooth-like slow drift burst. An F-H relation can not be recognized in these patches. But some of them seem to tend more towards lower, others more towards higher frequencies in the spectrum. An enlargement of one of these patches (Figure 3a) discloses that this can be HB emission. HB fine structure of type II bursts is due to nonthermal electrons energized at coronal shock waves. The HB patches are clearly not connected with the above mentioned low frequency ("conventional") type II burst. Because the patches itself do not show a well defined drift rate *we argue for radio emission from one (or more) standing fast mode magnetosonic shocks* such a sketched in Figure 1a as FMSS. In Figure 3b we have plotted smoothed cuts through the HB patches at some representative frequencies. Superposed are three channels of the HXRS data. The figure demonstrates that the HB patches appear *simultaneously* with the HXR emission in the (most energetic) 100–147 keV channel of HXRS.

CONCLUSIONS

We demonstrated the presence of patches of radio fine structures reminiscent of herringbones – a typical shock-excited radio fine structure – during the main maximum of energy release (shown by HXRS data) of a dynamic flare. This emission is observed in the decimeter range (here: 250–$>$800 MHz). As in Paper I we did not find a fundamental-harmonic mode pattern. We argue that these patches are excited by the perpendicular standing fast mode shocks sketched in Figure 1a. The simultaneous cusp formation and evolution (shown by *SOHO*-EIT data) supports our argument. Thus we extended the work started in

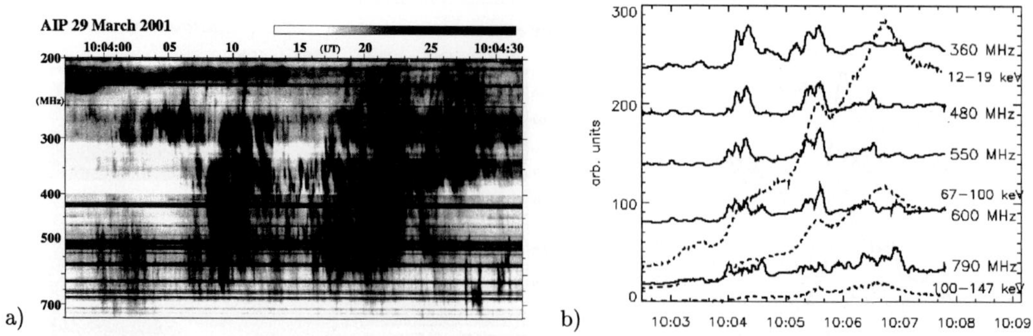

Fig. 3. Decimetric HB emission. a) Enlargement of the dashed box in Figure 2b. Notice drift rates of both signs. b) Selected single frequency flux of smoothed HB patches (baseline shifted for clarity). Superposed (dashed) are 3 HXRS channels (compare with Figure 2a).

Paper I by a well observed case of a probable decimetric standing shock signatures *during* the impulsive flare phase. The presented result confirms the assumption of Paper I that standing shock signatures in dynamic spectra are not rare but mostly not recognized as shock-driven emission due to insufficient sensitivity, time and frequency resolution of most sweep spectrometers, and due to previously missing supplementary observations in other spectral ranges.

ACKNOWLEDGEMENTS

HXRS is a joint endeavor by AICAS of the Czech Republic and SEC of NOAA, USA. The CME catalog is generated and maintained by the Center for Solar Physics and Space Weather[3]. *SOHO* is a project of international cooperation between ESA and NASA.

REFERENCES

Aurass, H., A. Vourlidas, M.D. Andrews, B.J. Thompson, R.H. Howard, & G. Mann, Nonthermal Radio Signatures of Coronal Disturbances with and without CMEs, *ApJ*, **511**, 451 (1999).

Aurass, H., B. Vršnak, & G. Mann, (Paper I), Shock-excited Radio Burst from Reconnection Outflow Jet? *A&A*, **384**, 273 (2002).

Chertok, I.M., S. Kahler, H. Aurass, & A.A. Gnezdilov, Sharp Decreases of Solar Metric Radio Storm Emission, *Solar Phys.*, **202**, 337 (2001).

Delaboudinière, J.P., *et al.*, EIT: EUV Imaging Telescope for SOHO, *Solar Phys.*, **162**, 291 (1995).

Farnik, F., H. Garcia, & M. Karlicky, A New Solar Hard X-ray Spectrometer, *Solar Phys.*, **201**, 357 (2001).

Klassen, A., H. Aurass, & G. Mann, Sawtooth Oscillations in Solar Flare Radio Emission, *A&A*, **370**, L41 (2001).

Nelson, G.S. & D.B. Melrose, Type II Bursts, in *Solar Radiophysics*, eds. D.J. McLean, & N.R. Labrum, Cambridge Univ. Press, Cambridge, 333 (1985).

Sato, T. & T. Hayashi, Externally Driven Magnetic Reconnection, *Phys. Fluids*, **22**, 1189 (1979).

[3]The Catholic University of America in cooperation with the Naval Research Laboratory and NASA.

THEORETICAL MODEL IMAGES AND SPECTRA FOR COMPARISON WITH RHESSI AND MICROWAVE OBSERVATIONS OF SOLAR FLARES

G. D. Holman[1], L. Sui[2], J. M. McTiernan[3], and V. Petrosian[4]

[1]*NASA's Goddard Space Flight Center, Code 682, Greenbelt, MD 20771, USA*
[2]*Catholic University of America, 200 Hannan Hall, 620 Michigan Avenue, Washington, DC 20064, USA*
[3]*University of California at Berkeley, Space Science Laboratory, Berkeley, CA 94720, USA*
[4]*Stanford University, Astronomy Program, Varian Bldg. 302C, Stanford, CA 94305-4060, USA*

ABSTRACT

We have computed bremsstrahlung and gyrosynchrotron images and spectra from a model flare loop. Electrons with a power-law energy distribution are continuously injected at the top of a semicircular magnetic loop. The Fokker-Planck equation is integrated to obtain the steady-state electron distribution throughout the loop. Coulomb scattering and energy losses and magnetic mirroring are included in the model. The resulting electron distributions are used to compute the radiative emissions. Sample images and spectra are presented here.

We are developing these models for the interpretation of *RHESSI* x-ray/gamma-ray data and coordinated microwave observations. The Fokker-Planck and radiation codes are available on the Web at: `http://hesperia.gsfc.nasa.gov/hessi/modelware.htm` .

THE MODEL

The model flare loop is semicircular with a radius of 17 arcseconds. The loop has a circular cross section with a radius of 2 arcseconds. The magnetic field strength at the top of the loop is 200 gauss, increasing linearly to 600 gauss at the footpoints. The temperature and density in the coronal part of the loop are taken to be 2×10^7 K and 2×10^{11} cm^{-3}, respectively. The density is much higher at the footpoints, simulating the solar chromosphere.

Energetic electrons are continuously injected at the top of the loop. Their density distribution is a power law in energy between 20 keV and 100 MeV and varies as E^{-5}. The injected electrons are isotropic in pitch angle and have a total density of 2×10^8 cm^{-3}.

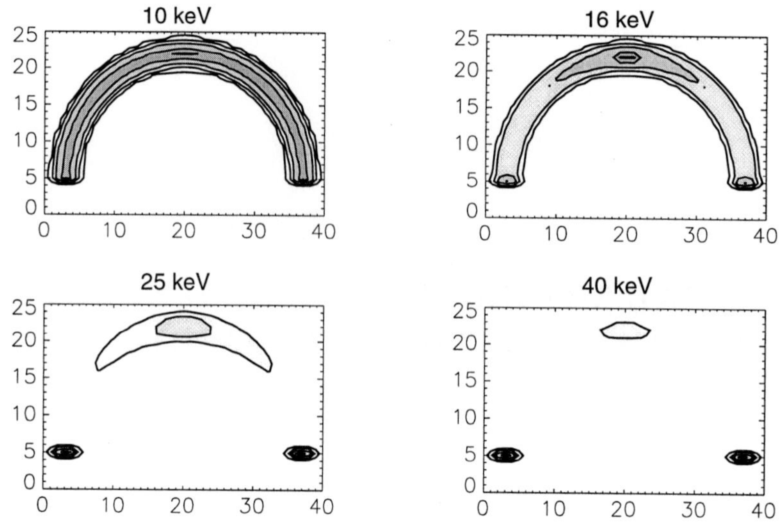

Fig. 1. Computed flare images at four x-ray energies. Contour levels are 0.1, 0.3, 0.6, 0.9, and 0.99 of the peak flux in each map. Axis labels are in arcseconds.

SAMPLE IMAGES AND SPECTRA

The observer's lines of sight are taken to be perpendicular to the plane of the loop. The x-ray image (contour map, Figure 1) at 10 keV is dominated by bremsstrahlung from the thermal plasma. At 16 keV and 25 keV looptop emission from the lower energy nonthermal electrons is apparent. The looptop emission becomes weaker with increasing photon energy and is almost negligible relative to the footpoint emission at 40 keV. This trend results from the steeper spectrum of the emission from the top of the loop relative to the thick-target footpoint emission (Figure 2). The spectra show a flattening below 20 keV because of the low-energy cutoff in the electron distribution. The spectra are flatter above a few hundred keV because of relativistic effects. A slight steepening of the footpoint spectrum can be seen above 10 MeV because of the cutoff in the electron distribution at 100 MeV.

Gyrosynchrotron results are not shown here because of space limitations.

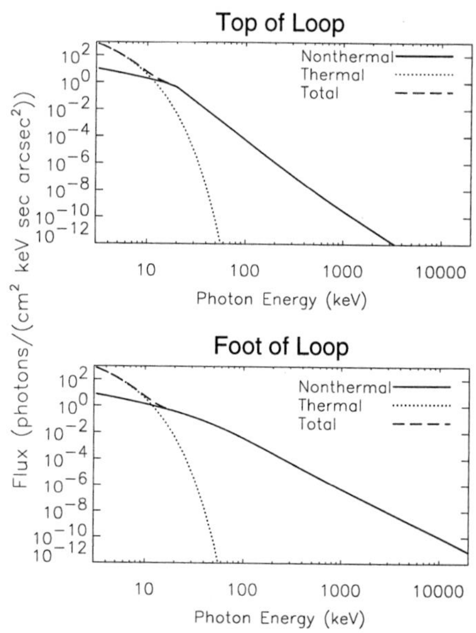

Fig. 2. Computed spatially-resolved spectra for a location at the top of the flare loop and a location at a footpoint.

ACKNOWLEDGEMENTS

This work is supported in part by the NASA Sun-Earth Connection Program.

HARD X-RAY OBSERVATIONS OF HIGH CORONAL REGIONS IN SOLAR FLARES

J. Sato

Physics Department, Montana State University, P.O. Box 173840, Bozeman, MT 59717, USA

ABSTRACT

Using Hard X-ray data obtained from the Hard X-ray Telescope (HXT) onboard *Yohkoh* I have studied coronal sources in three limb flares, including one particularly interesting event on April 23 1998. The results for the April flare show: (1) the existence of an extended source located in the high corona (~5000 km), and (2) dominant thermal and other sources do not come from the same looptop region. Preliminary results for the other two flares are also discussed.

INTRODUCTION

A solar flare is one of the most energetic phenomena in the solar corona and the energetic process is still mysterious. Since electrons are accelerated and energized in the corona, the study of coronal hard X-ray emission may give us information on the acceleration process. Using the Hard X-ray Telescope (HXT) on board the *Yohkoh*, the regions around soft X-ray looptops have been studied and the existence of hard X-ray sources has been revealed. Recently Sato (2001) and Hudson *et al.* (2001) showed the existence of a coronal source located high in the corona. In this paper I mainly show characteristics of hard X-ray sources observed in the high corona.

OBSERVATIONS OF THE APRIL 23 1998 FLARE

The April 23 1998 flare was a *GOES* X1.2 event that occurred beyond the limb. The estimated occultation height is $\sim 1.7 \times 10^4$ km. Figure 1 shows hard X-ray images in the L (grayscale) and M2 (contour) energy bands. Increasing time is indicated by "a" to "h" (see more detail in Sato 2001). In the L-band the brightest area is located initally in the south. Then a strong L-band source appears at lower altitude and moves northward (Figure 1(b)-(g)). In the M2-band, two bright sources (S1 and S2) appear in the southern region. After that, in the northern region the brightest source (N) is observed coincident with the L-band source, which is usually considered as thermal emission. Near the peak of the hard X-rays, an extended structure (E) is clearly seen in the higher corona with an estimated altitude of $\sim 5 \times 10^4$ km. Since the S1, S2, and E sources are not consistent with a thermal source (N), these are nonthermal sources in M2 band energy range (a detailed analysis is done in Sato & Hanaoka (1999) and Sato (2001)).

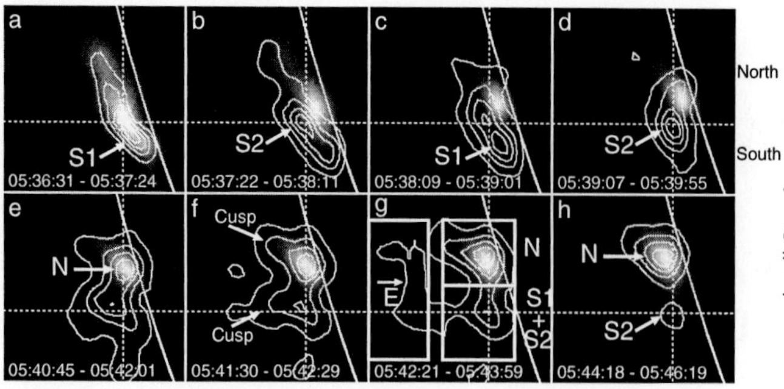

Fig. 1. *Yohkoh*/HXT image in the L (13–23 keV, gray scale) and M2 (33–53 keV, contour) energy bands. See Sato (2001) for a detailed analysis.

Table 1. High and low ( ) coronal hard X-ray emissions (cts/s) from fanbeam data

Date	98/04/23	98/05/09	98/08/18
L	22(1618)	7(991)	42(3454)
M1	9(234)	3(193)	24(716)
M2	4(19)	0(14)	17(102)
H	6(10)	0(6)	16(31)

DETECTION OF HIGH CORONAL HARD X-RAY EMISSION

In order to find other events showing hard X-ray emission in high corona, we have used fanbeam data of the HXT. Fanbeam data give us photon counts from a limited spatial area without imaging (see Sato 2001). In this case, we selected two fanbeam elements covering low (around the looptop) and high (~30 arcsec above the looptop) coronal regions respectively. Table 1 shows the observed hard X-ray photon counts (background is subtracted) in low and high coronal regions for the three limb flares. From Table 1 it is clear that the high coronal emission is very weak compared to the low coronal emission. Also, as observing energy increases, the ratio high corona/low corona becomes high. This means that the hard X-ray emission in the high corona has a hard spectrum.

DISCUSSION

Through an analysis of April 23 1998 flare and other flares, we found some interesting characteristics of hard X-ray sources in the low and high corona. Especially identification of the high coronal source is very important since the source is not well known. The high coronal emission derived from fanbeam data is very weak compared with the low coronal emission and therefore it is consistent that we cannot see the L-band source due to the limited dynamic range of the HXT (~10:1). On the other hand, in the higher energy ranges (M2 and H) there is a high possibility of seeing the source if the source size is not too large. However, the source has only been identified in the April 23 1998 flare. Therefore the source may have a large-sized structure like in April 23 1998 flare. If so, *RHESSI* will be very useful for more detailed study due to its high dynamic range (~100:1).

REFERENCES

Sato, J., Observation of the Coronal Hard X-ray Sources of the 1998 April 23 Flare, *ApJ*, **558**, 137 (2001).

Sato, J., and Hanaoka, Y., Observation of the Looptop Source of the 1998 April 23 Flare, *Solar Physics with Radio Observation*, eds. T. Bastian, N. Gopalswamy, and K. Shibasaki, p. 349, (1999).

Hudson, H.S., T. Kosugi, N.V. Nitta, and M. Shimojo, Hard X-radiation from a Fast Coronal Ejection, *ApJ*, **561**, 211 (2001).

MODELING OF X-RAY SOURCE OCCULTATION BY THE SOLAR DISK

J. Sylwester and B. Sylwester

Space Research Centre, Polish Academy of Sciences, Kopernika 11, 51-622 Wroclaw, Poland

EXTENDED ABSTRACT

The aim of the present research is to investigate the occultation of solar X-ray sources near and behind the edge of solar disk. Understanding the occultation details allows for precise coalignment of deconvolved SXT images with sub-arcsec resolution in cases where the limb position can be used as a reference, i.e. for flares occulted by the limb. Accurate image coalignment (down to 0.1 arcsec) is necessary in order to determine realistic maps of plasma temperature using the filter ratio technique. Present results indicate also that for a proper analysis of time history and apparent spectral evolution of sources located near the edge of solar disk, the spectral influence of the occultation needs to be taken into account.

In our analysis, we assume that the base of the solar atmosphere corresponds to one of the plane-parallel VAL models (Vernazza, Avrett, & Loeser 1981). We also assume that the X-ray source being occulted is point-like and that its spectrum comes from optically thin plasma of temperature T. We model the occultation as seen by SXT and HXT onboard *Yohkoh* in all energy channels using the effective areas $\eta(\lambda)$ from the respective SolarSoft files. In the solar context the X-ray optical thickness stems from two physical processes, absorption due to the presence of 'metals' (σ_a), and Thomson scattering (σ_T) on free electrons (Klein-Nishina at higher energies). Using the VAL-F model $N_{\rm H}(r)$ and $N_{\rm e}(r)$ dependencies we have calculated the energy (wavelength) dependent optical thickness along the line-of-sight l $[\tau(h,\lambda) = 2\int_0^\infty [\sigma_a(\lambda)\, N_{\rm H}(h,l) + \sigma_{\rm T}(\lambda)\, N_{\rm e}(h,l)]\, dl]$ (see the sketch in Figure 1a for definitions). We have calculated the spectra transmitted at a given height $I(h,\lambda)$ and the total signal $I(h)/I_0 = exp(-\tau)$ expected to be measured within individual SXT and HXT bands. In Figure 1b we show the results obtained for three SXT filters and all HXT bands for a source with $T = 15$ MK. Inspection of Figure 1b reveals that:

1. The exact position of the "occulting" limb is slightly different for individual SXT filters and HXT bands. Among the SXT filters, the Be119 limb position is 0.3 arcsec below its location for the "softer" channels. The most "unusual" $I(h)/I_0$ behavior is observed for Al12. This is related to a pronounced "two-band" character of the $\eta(\lambda)$ dependence for the Al12 filter.
2. The position of the "occulting" limb in HXT bands is about one arcsec below the respective SXT positions. However, the position of the "occulting" limb is still a few hundred km (on the Sun) above the optical limb.
3. The spectra observed become harder and harder as the source comes closer to the optical limb. This spectral hardening is related with a steep dependence of σ_a on the energy ($\sigma_a \sim E^{-3}$; cf. Somov 1975).

In accordance with (1), when coaligning Al12 with Be119 deconvolved images using the limb position as reference, the mentioned correction should be used. This correction amounts to 0.1 arcsec for cooler and 0.3 arcsec for hotter sources. From (1) and (3) it follows that the temperature of the source (as derived from the Be119/Al12 ratio) will increase as the source comes closer to the limb. For a source with temperature $T = 15$ MK the filter ratio temperature would be 30 MK at the height where the Al12 intensity falls to half the I_0 value. This effect is expected to produce an artificial high temperature edge close to the occulting

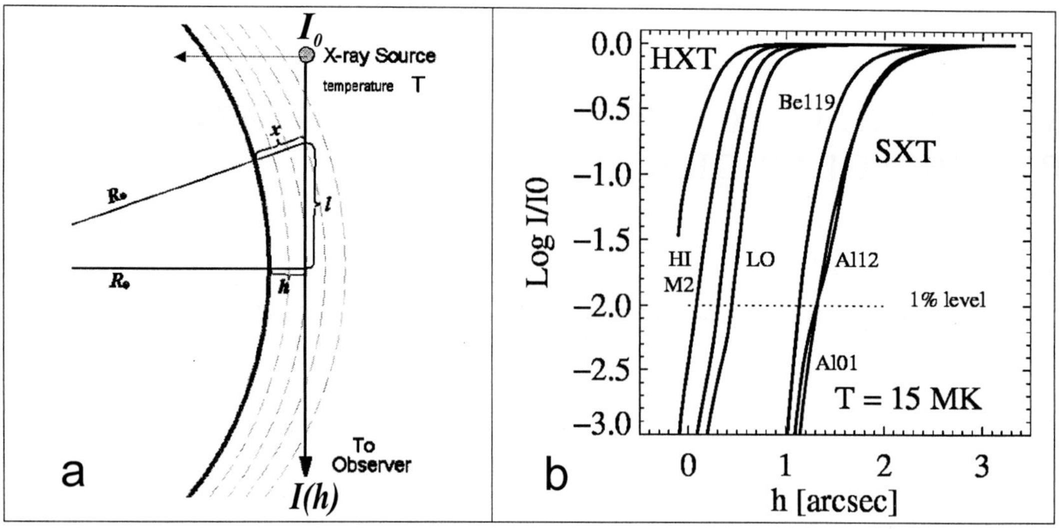

Fig. 1. a) Sketch of the physical situation considered (as seen from above the ecliptic plane): the point-like X-ray flare kernel of temperature T is observed near the limb, at an apparent distance h from optical edge of the disk, being partially occulted by the lower layers of solar atmosphere (not to scale). b) The height dependence of intensity $(I(h)/I_0)$ due to absorption of X-rays from the same source by the solar atmosphere for HI, M1, M2 and L bands of HXT and Be119, Al12, and Al01 filters of SXT.

limb. The width of this edge is 0.2 - 0.4 arcsec (less than 1/5 of an SXT pixel), and therefore would only show up in temperature maps derived from deconvolved images. From the results of occultation modeling it is possible to predict the shapes of light curves for sources evanescent behind the limb. In SXT observations it is seen that bright looptop kernels usually rise with a typical velocity of ~10 km/s. If such a rising source (assumed point-like) appears from behind the limb, we predict a specific pattern of light-curves for different bands of SXT and HXT. This pattern follows that presented in Figure 1b, if we apply a constant scaling factor of ~ 75 s/arcsec to the horizontal axis. As one may expect the first signal to increase is that of the hardest HXT channels, followed by the SXT ones after ~ 70 s. The shape of the increase of the lightcurve is similar for all bands (except Al12). Detailed modeling of the light-curves for real sources (somewhat extended and non-isothermal) emerging from behind the limb may allow for a diagnostic for both the source dynamics and the temperature structure. The results obtained here are independent of the atmospheric model considered, i.e. the location of the occulting edge is within 0.1 arcsec for any of the six solar atmosphere models considered in VAL.

ACKNOWLEDGEMENTS

This work has been supported by Polish Committee for Scientific Research Grant 2.P03D.024.17. The authors would like to thank the unknown referee for comments which led to substantial improvement of the present contribution.

REFERENCES

Somov, B.V., *Solar Phys.*, **42**, 235 (1975).

Vernazza, J.E., E.H. Avrett, and R. Loeser, *ApJ. Supp.*, **45**, 635 (1981), [VAL].

MONITORING THE CHANDRA X-RAY OBSERVATORY RADIATION ENVIRONMENT: CORRELATIONS BETWEEN GOES-8 AND CHANDRA/EPHIN DURING DOY 89 - 106, 2001

S. N. Virani[1], R. A. Cameron[1], P. P. Plucinsky[1], R. Mueller-Mellin[2], and S. L. O'Dell[3]

[1]*Harvard-Smithsonian Center for Astrophysics, USA*
[2]*University of Kiel, Germany*
[3]*NASA Marshall Space Flight Center, USA*

SUMMARY

The time period between DOY 89 (March 30) and DOY 106 (April 16), 2001 will likely be remembered as one of the most active time spans in this solar cycle. During this period of activity the Sun unleashed many M and X-class solar flares. Two of these, an X20 and an X17, were amongst the largest solar flares recorded in the last 10 years. Indeed the sunspot group responsible for this activity, AR 9393, will also likely be recorded as one of the most active sunspot groups of this solar cycle.

The *Chandra X-ray Observatory* (*CXO*), NASA's latest "Great Observatory", was launched on July 23 1999. The highly elliptical orbit of the *CXO* (perigee $\approx$ 10,000 km, apogee $\approx$ 140,000 km, 28.5 degree initial inclination) takes the satellite outside of the Earth's magnetosphere during a large fraction of the year. In this location the *CXO* is directly exposed to the particles released during solar flares and CMEs, resulting in high background and possible damage to the science instruments.

In this paper we present data from the EPHIN radiation monitor on-board the *CXO* and the SEM instrument on-board the *GOES*-8 satellite (in geostationary orbit). Not unexpectedly, considering their magnetic rigidities, we find strong correlations between the EPHIN P4 (5–8.3 MeV) and the *GOES*-8 P2 (4–9 MeV) proton channels, and also between the EPHIN P41 (41–53 MeV) and the *GOES*-8 P5 (40–80 MeV) proton channels. Since data from EPHIN are only available when a communication link has been established (nominally once every 8 hours), these correlations allow the Science Operations Team of the *CXO* to better gauge the radiation environment of the *CXO* and to take preventive measures, such as suspending observations and protecting the science instruments when necessary.

ACKNOWLEDGEMENTS

SNV, RAC, and PPP acknowledge support for this research from NASA contract NAS8-39073.

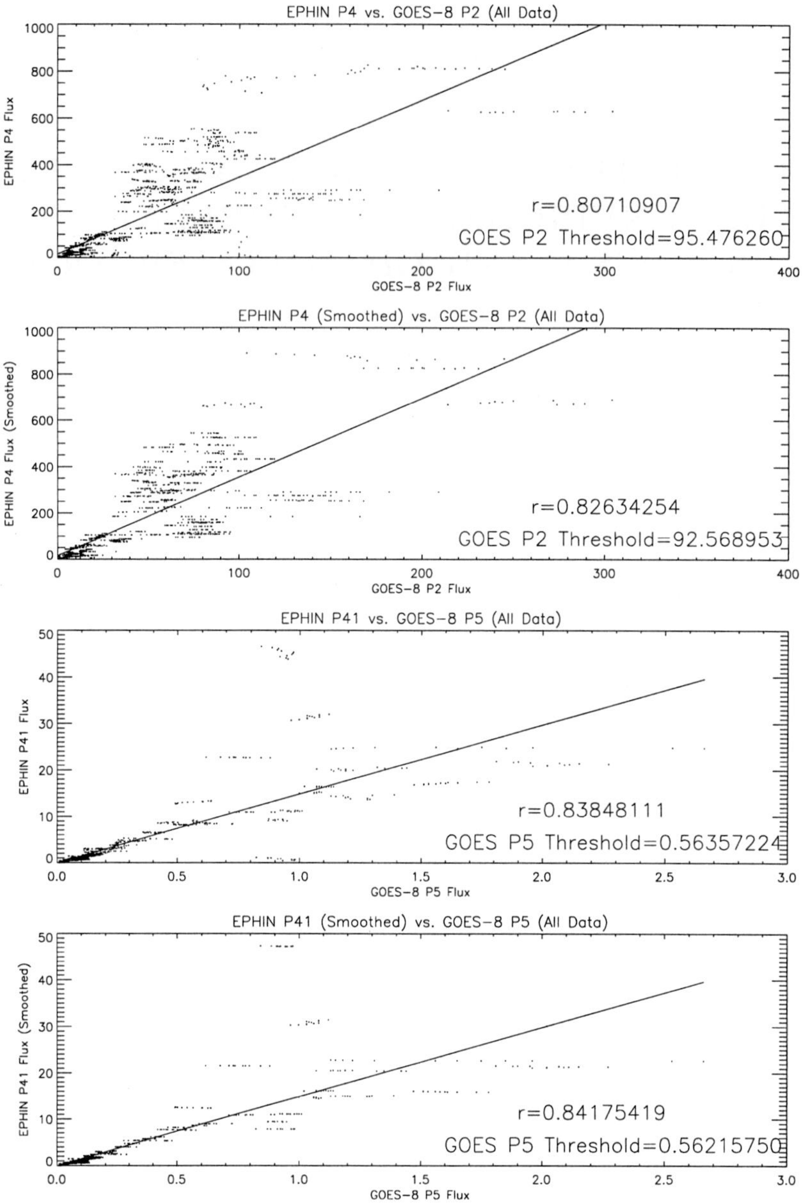

Fig. 1. *Top 2 Plots*: EPHIN P4 proton (5-8.3 MeV) data versus *GOES*-8 P2 proton (4-9 MeV) data. *Bottom 2 Plots*: EPHIN P41 proton (41-53 MeV) data versus *GOES*-8 P5 proton (40-80 MeV) data. The second plot in each grouping presents the correlation once the EPHIN proton data set has been smoothed (width = 5 EPHIN samples). Also included is the linear Pearson correlation coefficient, r. Fluxes are in units of protons/(cm^2 s sr MeV).

Section XIII.

Analysis Tools

NEW INTERFACES TO THE YOHKOH ARCHIVE AT MONTANA STATE UNIVERSITY

A. R. Davey and J. Sato

Montana State University, P.O. Box 173840, Bozeman, MT 59717, USA,

ABSTRACT

MSU recently brought online 2.1 Tb of storage with a primary aim of providing space for work on the new *Yohkoh* movie. In the process we have also been able to create another complete archive of *Yohkoh* data. We are investigating ways to make these data available to the community that seek to maximize the scientific potential and make it easier for users to identify data of interest. In the first stage of the work, we will be providing a web based interface to the *Yohkoh* HXT Flare Image Archive work of Jun Sato. We will then look to expand this interface to include fully integrated SXT/HXT data. It is intended that this and future work will be part of the *Yohkoh* Galileo Project and of MSU's involvement in the *Virtual Solar Observatory*.

YOHKOH HXT FLARE IMAGE ARCHIVE

Recently Jun Sato has been working on building a *Yohkoh* HXT Flare Image Archive which is an expansion on the previous *Yohkoh* HXT Image Catalogue (Sato *et al.* 1998). This work has involved for example, converting HXT data to FITS format, overlaying HXT data on SXT images and tying in *GOES* and Hα data. Whereas the image catalogue has one image per HXT band per flare, we aim to produce images at the beginning, peak, and decay phases of the flare. If capacity permits we would like to include more images and possibly movies too. We are now working on abstracting the information contained in this work into a relational database. This is the open source database, *MySQL*. Using a database provides several advantages when interrogating data. We can provide complex, unconstrained searches in data space, which may be especially powerful for example when performing data queries which span time scales such as a solar cycle. Use of judicious indexing and data organization provides us with the ability to search large amounts of data quickly. By providing an intuitive web interface, we hope to make HXT data more accessible to the community and provide better tie-ins to SXT, *GOES*, and Hα data.

FUTURE WORK

The work detailed represents not only what we hope will be a useful tool for the solar community but also a learning tool for future work. We intend to expand on the work above to create a full *Yohkoh* mission knowledge database involving all instruments, which we aim to be part of the collaboration on the *Yohkoh* Galileo project and MSU's involvement in the *Virtual Solar Observatory*, which would provide further integration with other solar missions. The possibility of integrating refereed journals with links to papers and corresponding data sets is one avenue we would like to explore, provided copyright issues can be resolved. This would also aid in one aspect of work that we would like to pursue which would be scientific keyword searches. If sufficient insight into the data is obtained, we should be able to perform keyword searches, such as "arcade downflows" and identify datasets of interest. There also exist a number of studies which individuals have performed as part of their scientific research. We anticipate providing a framework by which these studies might be incorporated into the wider database.

REFERENCES

Sato, J., Sawa, M., Masuda, S., Sakao, T., Kosugi, T. and Sekiguchi, H. *The Yohkoh HXT Image Catalogue*, Nobeyama Radio Observatory/NAO, Japan (1998).

BLIND DECONVOLUTION OF THE SXT PSF CORE PART

S. Gburek[1], J. Sylwester[1], and P. C. H. Martens[2]

[1]*Space Research Centre, Polish Academy of Sciences, Solar Physics Division, Kopernika 11, 51-622 Wroclaw, Poland*
[2]*Montana State University, P.O. Box 173840, Bozeman, MT 59717, USA*

ABSTRACT

We explore a blind iterative deconvolution algorithm for the determination of the core part of the point spread function of the soft X-ray telescope aboard the *Yohkoh* spacecraft. The algorithm has been adapted and modified to deal with the in-flight recorded X-ray images of solar flares. Particular care has been taken to achieve good data selection and initial conditions in order to improve the algorithm performance and convergence. We show an example of a deconvolved point spread function core profile and compare it with ground calibration data.

THE ALGORITHM AND DATA SELECTION

We have adapted the blind iterative deconvolution (BID) algorithm (Ayers & Dainty 1988) to deal with data from Soft X-ray Telescope (SXT) aboard the *Yohkoh* satellite. The result obtained in the Ayers & Dainty paper on synthetic data showed that the algorithm is capable of giving good restorations for both the deconvolved image and the point spread function (PSF). Our, independent, tests revealed that the performance and speed of the BID algorithm depend on the initial guess for the shape of the PSF and the quality of data. Therefore, we took special care in data selection and processing. We have chosen full resolution SXT images taken in the Al12 filter for our first PSF restorations. From analysis of compact flare kernel images we came to the conclusion that a good guess for the PSF can be provided directly from images of X-ray compact structures observed by SXT. Trial deconvolutions showed also that the use of compact sources accelerates the BID code convergence. From compact structures that we had found earlier in a search through the entire mission-long database of SXT full resolution frames (Gburek & Sylwester 2002), we selected the data for the year 2000, a period relatively short in comparison with the duration of the *Yohkoh* mission, but long enough to ensure good coverage over the CCD. For comparison of our PSF BID restorations with SXT ground calibration data (Martens, Acton, & Lemen 1995) we have chosen the *bx02_apr23* series of microfocus source images taken in Al-K line, which lies near maximum of the SXT effective area curve for the Al12 filter. The deconvolutions of the SXT PSF were performed as follows. First, for a given calibration image, we took a sequence of Al12 compact flare images from the same area on the CCD. Then, the initial guess for the PSF core was made with the steepest descent method; a term by Sylwester and Gburek. In short, this method takes an image sequence, normalizes each image to $[0,1]$ and co-registers them. Then it searches the entire sequence for the lowest signal at any pixel position. The minimum values for each pixel are collected in a new array of the original image size from the sequence. The authors have checked that good estimates of PSF core can be obtained by this method. The final PSF estimate was then deconvolved from the most compact flare image of the sequence and compared to the chosen calibration beam image. The results are discussed below.

RESULTS AND CONCLUDING REMARKS

Tests of BID on real SXT data have revealed expected and desirable features: in the deconvolved image there is an increase in signal range and a separation of nearby sources. The structures in the "clean" images are much sharper. Sharpening is also seen in the restored PSF. No significant deformations of the image and PSF, which may come from noise or method artifacts, were detected during the tests. The deconvolved PSF is still slightly more fuzzy than the calibration beam profile in Al-K line (see Figure 1). Because the observed solar radiation is not monochromatic, one would expect some broadening of the in-orbit PSF with respect to the on ground calibrated in monochromatic spectral lines. In conclusion, an improvement of the morphological and photometric properties of the SXT images can be obtained with the BID method described above. First results show that blind deconvolution is capable of determining the shape of the SXT PSF core in-orbit. We find close agreement with the ground calibrated PSF.

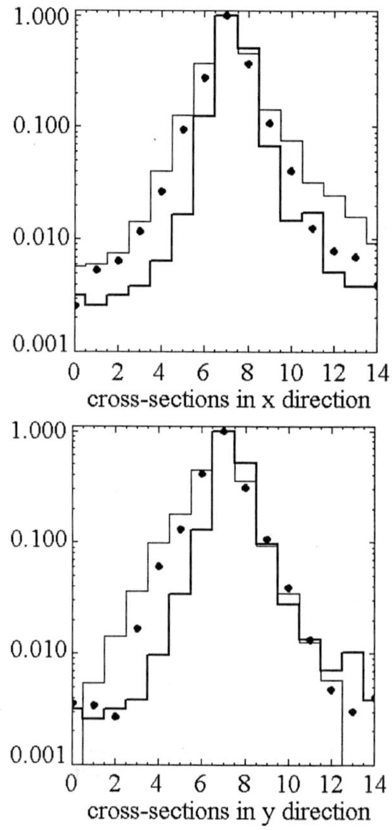

Fig. 1. A result of blind iterative deconvolution for the SXT PSF core and its comparison to ground calibration data is shown in cross-sections in the plots above. Thin solid line: cross-sections of the initial PSF guess constructed by the steepest descent method from a flare image sequence; black dots: cross- sections of PSF obtained from BID; thick solid line: cross-sections of the SXT PSF from ground calibrations.

ACKNOWLEDGEMENTS

This contribution has been supported by Polish KBN grant 2.P03D.024.17 and a grant from the *Yohkoh* 10th Anniversary Meeting organization.

REFERENCES

Ayers G. R., Dainty J. C., Iterative blind deconvolution method and its applications, in *Optics Letters*, **13**, no. 7, (1988).

Gburek S., Sylwester J., Search for compact X-ray sources in SXT observations, *Solar Phys.*, in press, (2002).

Martens P., Acton L., Lemen J., The point spread function of the soft X-ray telescope aboard *Yohkoh*, *Solar Phys.* **157**, 141, (1995).

THE TEMPERATURE ANALYSIS OF YOHKOH/SXT DATA USING THE CHIANTI SPECTRAL DATABASE

M. Shimojo[1], H. Hara[2], and R. Kano[2]

[1]*Nobeyama Radio Observatory, Nobeyama, Minamimaki, Nagano, 384-1305, Japan*
[2]*National Astronomicat Observatory, Osawa, Mitaka, Tokyo, 181-8588, Japan*

ABSTRACT

The choice of atomic database for X-ray emission lines is very important for the temperature analysis of *Yohkoh*/SXT imagery since the temperature of the solar plasma is derived from a model X-ray spectrum based on the database. Recently Dere *et al.* (2001) recalculated the atomic database for X-ray emission lines and released a new version of the CHIANTI package (Version 3.03). We calculate the temperature responses of the SXT analysis filters using the new CHIANTI package and compare the results with temperature responses derived from the Mewe database, which has been the standard database for *Yohkoh*/SXT until now. We find that when applied to SXT data the Mewe database yields higher temperatures than the CHIANTI database. In particular, a temperature analysis using the Al.1/Al12 filter combination of the SXT instrument is very sensitive to the differences between the two databases, and we find that the resulting temperatures from the two databases can differ by as much as 30 to40% at 4-5 MK. Since the Al.1/Al12 filter combination is typically used for determining temperatures of active region and flare plasmas from SXT data, it is likely that previous analyses of SXT data have over-estimated the thermal energy, flux due to thermal conduction, and flux due to radiative losses for active regions and flares.

TEMPERATURE RESPONSES

Figure 1 shows CCD count rates as a function of temperature for the 'Al.1', 'Al12', and 'Be' SXT analysis filters ('Temperature Responses'). The dotted lines are the standard temperature responses (Tsuneta *et al.* 1991) based on the Mewe database (Mewe *et al*,1985, 1986) using the abundance model of Meyer (1985) and the ionization equilibrium model of Arnaud & Rothenflug (1985). The solid, dashed, and dash-dotted lines are the temperature responses[1] based on the CHIANTI database using different abundances (solid: Feldman, dashed: Waljeski, dash-dotted: Meyer) and the ionization equilibrium model of Mazzotta *et al.*(1998). The figure indicates that the CHIANTI-derived temperature responses are more sensitive to $\sim$ 10 MK plasma than the Mewe-derived responses, and that the peak temperatures based on CHIANTI are higher than for Mewe. The differences are due in part to differences in the abundance models used and in part to differences in the ionization equilibrium models used, which determine the peak temperature of the ionized Fe fraction. The abundance of Fe as given by both the Feldman and Waljeski coronal abundance models (Feldman 1992, Waljeski *et al.*1994) is larger than the value given by the Meyer abundance model. On the other hand, the recent ionization equilibrium model shifts the peak temperature of the ionized Fe fraction to a higher value than previous models.

[1]Available at `ftp://solar.nro.nao.ac.jp/pub/user/shimojo/SXT_resp/`.

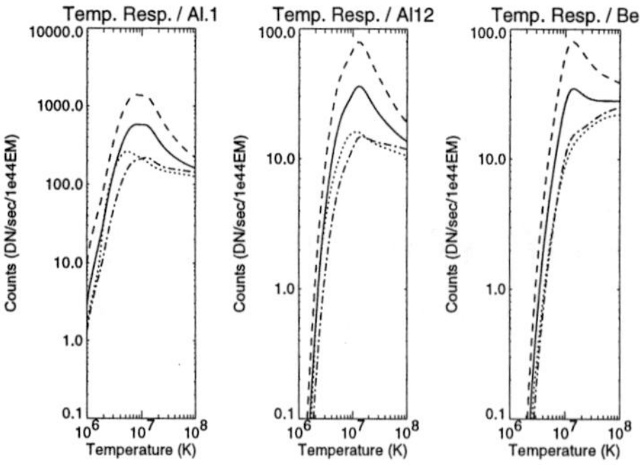

Fig. 1. Temperature responses of *Yohkoh*/SXT 1st column:Al.1, 2nd column:Al12, 3rd column: Be.

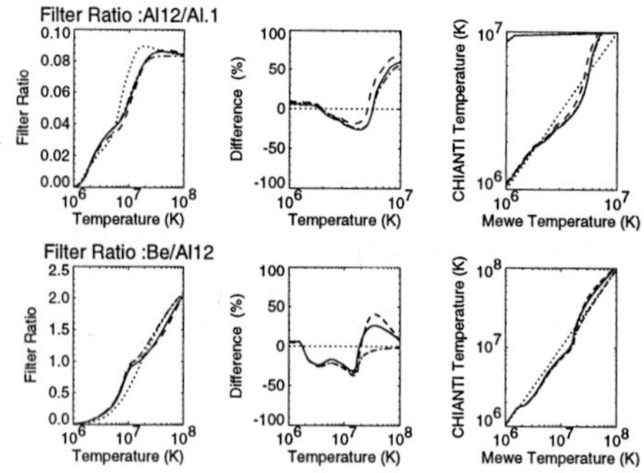

Fig. 2. Al12/Al.1 and Be/Al12 Filter Ratios of *Yohkoh*/SXT 1st column:filter ratio, 2nd column: The difference between Mewe and CHIANTI, 3rd column: Mewe vs. CHIANTI.

FILTER RATIOS

The filter ratios determined from the temperature responses are the most important functions for temperature analyses of SXT data. Figure 2 shows the filter ratios of the usual filter combinations and the differences between the temperature analyses based on Mewe and CHIANTI (dotted: Mewe, solid: Feldman, dashed: Waljeski, and dot-dashed: Meyer, CHIANTI). The plots of the Al.1/Al12 (first row) and the Al12/Be (second row) filter combinations illustrate the large differences between Mewe and CHIANTI. In particular, in the 4-5 MK range, the Mewe-derived Al.1/Al12 ratio differs from the CHIANTI-derived ratio by 30-40 %. Similarly, in the vicinity of 20 MK the Mewe-derived Be/Al12 ratio differs from the CHIANTI- derived ratio by 20-40 %. For example, a plasma temperature of 4 MK as derived using from the Al.1/Al12 filter ratio using the Mewe database is reduced to 2 MK if the CHIANTI database is used. Hence it is likely that previous analyses of SXT data have over-estimated the thermal energy, flux due to thermal conduction, and flux due to radiative losses for active regions and flares.

REFERENCES

Arnaud, M., and Rothenflug, R., An updated evaluation of recombination and ionization rates, *ApJ. Supplement*, **60**, 425 (1985)

Dere, K.P., Landi, E., Young, P.R., and Del Zanna, G., CHIANTI-An Atomic Database for Emission Lines. IV. Extension to X-Ray Wavelengths, *ApJ. Supplement*, **134**, 331 (2001)

Feldman, U., Elemental abundances in the upper solar atmosphere *Physica Scripta*, **46**, 202 (1992)

Mazzotta, P., Mazzitelli, G., Colafrancesco, S., and Vittorio, N., Ionization balance for optically thin plasmas: Rate coefficients for all atoms and ions of the elements H to NI *A&AS*, **133**, 403 (1998)

Mewe, R., Gronenschild, E.H.B.M., and van den Oord, G.H.J., Calculated X-radiation from optically thin plasmas. V, *A&AS*, **62**, 197 (1985)

Mewe, R., Lemen, J.R., and van den Oord, G.H.J., Calculated X-radiation from optically thin plasmas. VI, *A&AS*, **65**, 511 (1986)

Meyer, J.-P., Solar-stellar outer atmospheres and energetic particles, and galactic cosmic rays, *ApJ. Supplement*, **57**, 173 (1985)

Waljeski, K., *et al.*, The composition of a coronal active region *ApJ.*, **429**, 909 (1994)

THE POINT SPREAD FUNCTION OF THE YOHKOH SOFT X-RAY TELESCOPE

J. Shin and T. Sakurai

Department of Astronomical Science, The Graduate University for Advanced Studies, National Astronomical Observatory, Osawa 2-21-1, Mitaka, Tokyo 181-8588, Japan

ABSTRACT

We have analyzed the pre-launch calibration data for characterizing the point spread function (PSF) of the *Yohkoh* Soft X-Ray Telescope (SXT). Our study shows that both the undersampling effect and the noise in the data should be considered very carefully. The full width at half maximum (FWHM) of the SXT PSF is found to be nearly constant (about one pixel size) over the central area of the CCD where the solar disk is located. The similarity of the results obtained from different wavelengths implies that the contributions from scattering are negligible in the core part of the PSF.

FITTING PROCEDURE

Due to the finite width of the SXT PSF a certain amount of blurring is inherent in the observed images. Though there have been many attempts to determine the SXT PSF using various methods, we believe that the results were not satisfactory in describing the pattern of blurring. Especially we found it crucial that the effects of undersampling and noise in the data are treated properly. For finding the best fit solution for the PSF from experimental data using the χ^2 minimization method, the consideration of the non-photon noise component (mostly the readout noise), $I_{\rm np}$, is important along with the Poisson fluctuation by photon noise. Thus the total variance should be expressed by the combination of these noises. To consider them together we convert the photon count to the number of detected electrons, F, which is proportional to the number of incident X-ray photons ($F = kI_{\rm p}$). The variance of F can be described as $\sigma_{\rm F}^2 = k^2\sigma_{I_{\rm p}}^2 = k^2 I_{\rm p} = kF$, assuming Poisson statistics. Consequently, χ^2 will have an expression of $\chi^2 = \sum_{i,j}[F_{\rm obs}(i,j) - F_{\rm cal}(i,j)]^2/[kF_{\rm cal}(i,j) + \sigma_{I_{\rm np}}^2]$, where $F_{\rm obs}(i,j)$ and $F_{\rm cal}(i,j)$ are the electron number of the test data and the value of the analytical function at a pixel (i,j), respectively. The best fit solution of the PSF will be determined by minimizing this value. The rms of non-photon noise was obtained from the dark region of each data, and found almost independent of the position on the CCD.

It is known from previous studies that the FWHM of the SXT PSF is about one pixel size ($\sim$2.46 arcsec), which implies that the test data are possibly undersampled. This undersampling effect will influence the shape of the measured PSF, especially near the core. What is more, the real peak of the PSF is not always located exactly at the center of the pixel having maximum value. Thus most of the undersampled PSFs show asymmetric distributions. For these reasons it is necessary to consider the undersampling effect when determining the original shape of the PSF. We first construct a PSF with pixels much smaller than the SXT CCD pixel size. And then we integrate the small pixels until the original CCD pixel size is obtained. Finally χ^2 is calculated by comparing this undersampled model PSF with the test image of 21×21 pixels.

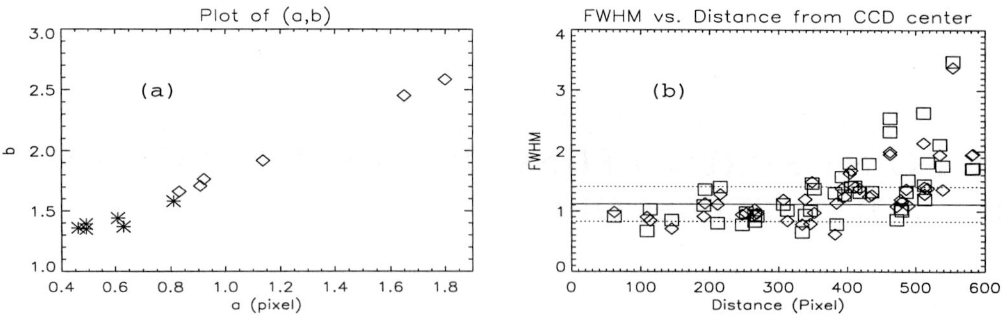

Fig. 1. (a) Comparison of the results for six PSFs obtained by Martens *et al.* (1995) (diamonds) and by this study (asterisks). (b) Positional variation of the FWHM of the best fit Moffat function for the data sets 'bx21_apr24' (rectangles) and 'bx42_apr25' (diamonds) obtained using C-K line (44.7Å). The solid and the dotted lines represent the mean and the standard deviation of the results within 450 pixel distance, respectively.

The mathematical form of the SXT PSF can be characterized by the elliptical Moffat function (Martens, Acton, & Lemen 1995). For finding the best fit solution, however, we must consider a total of 7 unknown parameters plus the undersampling effect simultaneously. Since the PSF is very sharp and the dynamic range of the SXT CCD is comparatively narrow, only a few pixels near the peak are available for the actual calculation. Therefore it is necessary to reduce the number of parameters in the equation to increase the accuracy of the result for each parameter. The contours of the data show that the PSFs near the central area of the CCD are almost rotationally symmetric and the ellipticity is seen only near the edges of the CCD. For this reason we assume in our calculation a rotationally symmetric Moffat function $\sim [1 + (r/a)^2]^{-b}$.

Figure 1a shows the values of the Moffat coefficients a and b for six PSFs obtained at the same location on the CCD. Compared to the results of Martens *et al.* (1995), ours show a much reduced dispersion since we have considered the undersampling effect very carefully in evaluating the best functional form of the fit. The dispersion that still remains in our result may be due to noise in the data.

POSITIONAL VARIATION OF THE PSF

Previous studies on the SXT PSF have suggested that its shape may vary with the position on the focal plane. It is very important to understand this spatial variation of the PSF because coronal activity happens on the limb as well as on the disk center. Four sets of data obtained at different wavelengths are used in our study. Each data set contains a total of 49 calibration data measured at different locations on the focal plane. Figure 1b shows the positional variation of the FWHM of the best fit Moffat function. We can see that the FWHM stays nearly constant within the error bound up to about 450 pixel distance from CCD center. The increase of the FWHM beyond this distance might be related to the ellipticity of the PSF at the edges and/or to the low signal to noise ratio. As a consequence we believe that the degree of blurring is almost the same over the central area of the CCD where the solar disk is located. Also, the contributions from scattering are negligible in the core part of the PSF, because the results obtained from different wavelengths (not shown here) are almost identical within the error range.

REFERENCES

Martens, P.C.H., L.W. Acton, and J.R. Lemen, The Point Spread Function of the Soft X-Ray Telescope aboard Yohkoh, *Solar Phys.*, **157**, 141 (1995).

AN EFFICIENT AND VERSATILE VIDEO SERVER SYSTEM FOR STUDYING THE YOHKOH MISSION ARCHIVE

G. L. Slater[1] and J. Bartus[2,3]

[1]*Lockheed Martin Solar and Astrophysics Laboratory, Bldg 252, 3251 Hanover St. Palo Alto, CA 94303 USA*
[2]*Solar Physics Research Corporation, 4720 Calle Desecada, Tucson, AZ 85718, USA*
[3]*Institute of Space & Astronautical Science, 3-1-1 Yoshinodai, Sagamihara, Kanagawa 229-8510, Japan*

ABSTRACT

We present a fast, compact, notebook-based video server system for viewing the entire *Yohkoh* SXT mission archive of full frame images taken with the Al-Mg filter. The system may be developed further to handle multiple, coaligned image archives from different satellites and observatories and is thus a prototype for an efficient 'visual exploration' tool for solar physics.

INTRODUCTION

The entire archive of images taken with *Yohkoh's* Soft X-ray Telescope exceeds 5 million frames. Now that the complete archive is kept online at several institutions, it is relatively easy to search the archive by time, pointing, filter combinations, resolution, etc. However, there is still a lack of powerful tools for visually exploring the archive. In this poster we present an initial step toward developing such an 'archive exploration' tool. It is a laptop video server containing the mission-long archive of all full disk SXT images taken with the Al-Mg filter. Using a Matrox RT2500 real time video editing card and a user interface provided by Adobe Premier, the user can visually explore the entire database from end to end easily and efficiently, with total control over position, slew rate, and slew direction. The interface we present here is a relatively simple example of a tool that could in the future be adapted to much larger archives, which might include multiple coaligned image databases. With the certainty of ever larger image archives looming ominously in the future, such tools will assume an increasingly crucial role in efforts to fully exploit the torrent of data that gushes from the current generation of ground and space based telescopes and which threatens to inundate the scientific community.

DATA PREPARATION PROCEDURE

All images were prepared with the latest SXT calibration software which includes background subtraction, leak subtraction, desaturation, exposure compositing, normalization, point spread function deconvolution, de-spiking, and registration. Images are converted from native SXT format (SSC/SSS) first to GIF format (in order to be accessible to the Adobe/Matrox software), and then to a special Matrox AVI MPEG2 format which the RT2500 card is able to decode on the fly, thus enabling the streaming of images from hard disk. The AVI files for the entire mission long movie are then viewable with a variety of interfaces. We are currently using one from Adobe Premier.

VIDEO SERVER SYSTEM

We initially made use of a desktop PC system running Windows 2000. The system had dual 866 Mz Intel PIII processors, 1 GB of SDRAM memory, and a RAID 5 array of Seagate SCSI Ultra 160 disk drives for the video disk system. After much frustration in attempting make this system work with the Matrox RT2500 video card, we finally settled on a more modest system consisting of a single PIII processor, 256 Mb RAM, and an ATA 100 IDE drive for the video disk. This configuration performed adequately. For the purposes of this presentation we decided to develop a third system, based upon an IBM ThinkPad laptop computer. In order to attach the Matrox card to this system, we had to attach the laptop to an 'expander box' which had a single PCI slot. Unfortunately, the Matrox RT2500 card was too long to house internally in the expander box. It was necessary to construct an external housing for the laptop/expander box combination in order for the Matrox card to be attached externally. In addition, a second disk drive for use as the video server disk was attached via firewire connection to the Matrox card itself. This is the configuration that was displayed at *Yohkoh* 10 conference, and which proved adequate for driving the Matrox card.

USER INTERFACE

The user interface for the video server system consists of a simple, typical movie player, in this case provided by Adobe Premier. The key feature of this player is the slide bar which can be used to slew rapidly across the archive and position the player at arbitrary individual frames of the movie. This high degree of positioning resolution is provided by the Matrox RT2500 card.

YOHKOH DVD

As a by-product of the *Yohkoh* SXT full mission movie, we have produced a preliminary version of a DVD containing the archive.

CONCLUSIONS

We have produced an efficient, compact video server for displaying the *Yohkoh* SXT archive with a convenient user interface. We are encouraged by its performance and consider it to be a prototype for a more sophisticated, general image data exploration tool for multiple archives.

Section XIV.

Future Observing

A HIGH-SPEED Hα CAMERA FOR SOLAR FLARE OBSERVATIONS

Y. Hanaoka, M. Noguchi, K. Ichimoto, and T. Sakurai

National Astronomical Observatory of Japan, 2-21-1 Osawa, Mitaka, Tokyo 181-8588, Japan

ABSTRACT

We have developed a new digital imaging system for the Hα imager of the Solar Flare Telescope at Mitaka, National Astronomical Observatory of Japan (NAOJ), for high-cadence observations of solar flares. It covers a 6.4 × 4.9 arcmin field of view with 660 × 494 pixels, and the best time resolution is 1/30 sec (depending on the field of view). We started regular observations with this system in July 2001, and have detected 156 flares up to the end of 2001. We hope to collaborate with *RHESSI* to observe energetic phenomena in solar flares.

INTRODUCTION

To resolve individual spikes (elementary bursts) of impulsive solar flares, a time resolution of less than 1 sec and a spatial resolution of about 1 arcsec are required. It is difficult to realize such high temporal and high spatial resolution simultaneously using hard X-ray instruments such as HXT on *Yohkoh* and *RHESSI*. On the other hand it is easy to obtain one-arcsecond resolution images at video rates of Hα flare observations. Therefore Hα observations are useful in the study of impulsive flares and various attempts to develop high-speed Hα camera systems have been made (e.g. Kiplinger 1989, Wülser & Martin 1989, Wang *et al.* 2000). Observations with such high resolution produce a huge amount of data, but recent advances in computer technology enable us to handle this vast amount with a small computer.

We have developed a high-speed Hα camera, which constitutes a new digital imaging system for the Hα imager of the Solar Flare Telescope at Mitaka, NAOJ (Sakurai *et al.* 1995). The high-speed Hα camera is based on a real-time image processing system. We started regular observations with this camera in July 2001 and collaborated with *Yohkoh* for half a year. After *RHESSI*'s launch we expect to collaborate with *RHESSI*.

DESCRIPTION OF THE SYSTEM

The technical specifications and the operation of the high-speed Hα camera are as follows:

- Telescope: a 15cm refractor + Zeiss Lyot filter, passband 0.25 Å
- Digital System: camera TAKEX FC-300 (1/3 inch), CCD 660 × 494 pixels, 10 bit A/D, 30 frames/sec, interface card Graphin IPM-8540D, and a personal computer with WindowsNT operating system

Fig. 1. An X1.6 flare on October 19, 2001 observed with the High-speed Hα camera.

Operation of the High-Speed Hα Camera:

- 'Quick look' images (every 30 sec) and high-cadence images from the digital camera are recorded on a hard drive during the daily observation cycle.
- In baseline operations full-frame high cadence images are recorded every 0.5 sec. The maximum rate is more than 5 frames/sec for full-frame images (the field of view $6.4' \times 4.9'$) and 30 frames/sec for partial frame images. The daily amount of raw data with a time resolution of 0.5 sec is about 40 GB.
- After the daily observations an automatic flare search is conducted.
- The high-cadence data for the time periods of the detected flares are automatically selected and recorded on CD-R's.

EXAMPLE OF OBSERVED FLARES

During the 2001 July–December period, 156 flares have been detected. The biggest one is an X1.6 flare on 2001 October 19. Figure 1 shows Hα images of the flare.

Daily Hα images and movies, lists of observation time periods and detected flares, and quick-look pictures of the flares are available on the internet. Visit `http://solarwww.mtk.nao.ac.jp/en/database.html` , and follow the link there.

REFERENCES

Kiplinger, A.L., B.R. Dennis, & L.E. Orwig, A high-speed digital camera system for the observation of rapid Hα fluctuations in solar flares, in *Max '91 Workshop 2: Developments in Observations and Theory for Solar Cycle 22*, p.346 (SEE N90-12459 03-92) (1989).

Sakurai, T. et al., Solar flare telescope at Mitaka, *Publ. Astron. Soc. Japan*, **47**, 81 (1995).

Wang, H., J. Qiu, C. Denker, T. Spirock, H. Chen, & P.R. Goode, High-Cadence Observations of an Impulsive Flare, *Astrophys. J.* **542**, 1080 (2000).

Wülser, J.-P. & H. Martin, High time resolution observations of Hα line profiles during the impulsive phase of a solar flare, *Astrophys. J.* **341**, 1088 (1989).

BALLOON-BORNE HARD X–RAY SPECTROMETER FOR FLARE OBSERVATIONS

K. Kobayashi[1], S. Tsuneta[2], T. Tamura[2], K. Kumagai[2], Y. Katsukawa[1], S. Kubo[2], T. Yamagami[3], and Y. Saitoh[3]

[1]*University of Tokyo, 2-21-1 Osawa, Mitaka, Tokyo 181-8588, Japan*
[2]*National Astronomical Observatory of Japan, 2-21-1 Osawa, Mitaka, Tokyo 181-8588, Japan*
[3]*Institute of Space and Astronautical Science, 3-1-1 Yoshinodai, Sagamihara, Kanagawa 229-8510, Japan*

ABSTRACT

We have developed a balloon-borne hard X-ray detector system for observing high resolution hard X-ray spectra of solar flares. The instrument consists of 16 cadmium telluride detectors, each $10 \times 10 \times 0.5$ mm in size. It has a 3 keV energy resolution over the energy range of 15–100 keV. The first flight of this instrument took place from Sanriku, Japan on August 29, 2001. While no large flares occurred during the 3 hours of level flight, detector performance was verified, and one possible microflare detected.

INTRODUCTION

High resolution hard X-ray spectra are essential for understanding particle acceleration in solar flares. The transition between the non-thermal and thermal electrons, typically between 20 and 60 keV, is of particular interest. We have developed a balloon-borne instrument for observing high-resolution flare spectra. The first flight took place on August 29, 2001. While no flare was observed on this flight, severals aspects of the instrument design were successfully demonstrated.

INSTRUMENT DESIGN

The instrument consists of a pressurized detector enclosure mounted above the gondola, an electronics module inside the gondola, and shields for passively cooling the detector enclosure. The detector enclosure contains sixteen $10 \times 10 \times 0.5$ mm CdTe detectors as well as preamplifiers and a high voltage battery for detector bias. We chose CdTe detectors with Indium electrodes which act as Schottky barriers and dramatically reduce the leakage current, allowing a high bias voltage of 275 V. These detectors were recently developed by Takahashi *et al.* (1998) and manufactured by Acrorad Co. The detector enclosure was pressurized to prevent coronal discharge, and a CFRP/Rohacell composite window is used. Detector quantum efficiency and resolution were measured on the ground, and all 16 detectors showed a FWHM resolution of 3 keV or better.

The detectors must operate below 0° C for optimal performance. The instrument is designed to maximize passive cooling. Aluminum coated Polyimide films are placed around the detector enclosure to block sunlight and infrared radiation from the ground while maximizing the view of sky. The detector enclosure and collimator assembly are covered by Ag-Teflon, a high-emissivity low-absorptivity material that acts as a radiator.

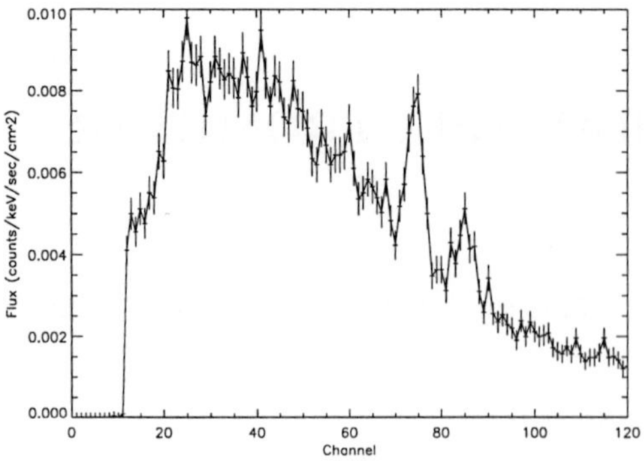

Fig. 1. Summed background spectrum.

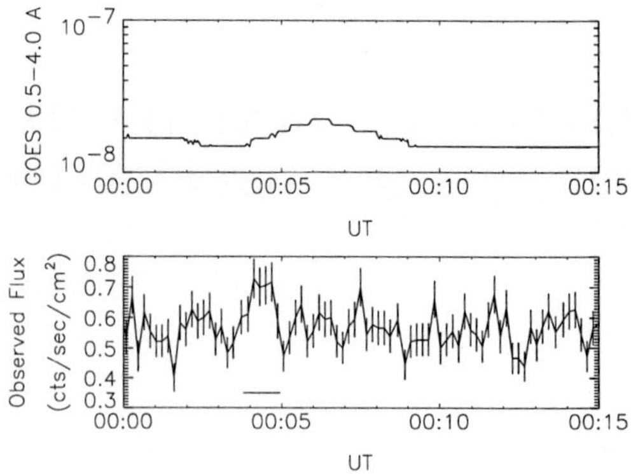

Fig. 2. Top: *GOES* plot from 00:00 to 00:15 UT, showing a microflare starting 00:04. Bottom: our corresponding data.

The gondola attitude is controlled using the input from a wide angle sun sensor. A second 2-D sun position sensor was used to measure the attitude. This sensor uses a pinhole and position sensitive detector, and has a resolution of 1 degree.

FLIGHT RESULTS

The instrument was launched successfully at 6:30 JST August 29, 2001 from the Sanriku Balloon Center in northern Japan. It reached the 41km target altitude at 9:00. Observation was stopped at 9:45 due to battery problems. The instrument was successfully recovered.

The thermal design of the detector system proved successful. The detector enclosure temperature stabilized at $-13°$ C, far better than the required 0 degreesC temperature.

The background spectrum summed over the 45 minutes of level flight is shown in Figure 1. The two peaks represent emission lines from the lead shielding; we chose not to use a graded-Z shield so that the lead lines can be used to calibrate the detector gain.

Figure 2 shows the light curve of our measured count rate along with the *GOES* X-ray flux near the beginning of our observation. The *GOES* plot clearly shows a microflare occurring at 00:05 UT, and our instrument shows an increased rate from 00:04 to 00:05 UT that corresponds in time to the leading edge of the *GOES* X-ray brightening. We believe that this represents the successful detection of a microflare. However, due to the few counts in the increase we were unable to quantitatively analyze this signal.

REFERENCES

Takahashi, T., Hirose, K., Matsumoto, C., Takizawa, K., Ohno, R., Ozaki, T., Mori, K., Tommita, Y., *Proc. SPIE*, **3446**, 29 (1998).

PROSPECTS FOR HARD X-RAY SOLAR FLARE POLARIMETRY WITH RHESSI

M. L. McConnell[1], D. M. Smith[2], A. G. Emslie[3], R. P. Lin[2], and J. M. Ryan[1]

[1]*Space Science Center, University of New Hampshire, Durham, NH 03824*
[2]*Space Sciences Laboratory, University of California, Berkeley, CA 94720, USA*
[3]*Dept. of Physics, University of Alabama, Huntsville, AL 35899*

ABSTRACT

Designed primarily as a hard X-Ray imager and spectrometer the *Reuven Ramaty High Energy Solar Spectroscopic Imager (RHESSI)* is also capable of measuring the polarization of hard X-Rays (20–100 keV) from solar flares. These studies will provide the capability to probe the geometry of the acceleration process. Although not originally designed to study hard X-Ray polarization of solar flares, it was realized during the development of *RHESSI* that the essential ingredients for measuring the polarization, namely, an array of detectors in a rotating spacecraft, were already present. All that was needed was the addition of a strategically placed cylinder of Be in the cryostat to Compton scatter the hard X-Rays (20–100 keV) into the rear segments of the adjacent Ge detectors, since the direction of the scattering depends on the polarization of the incoming photon. Monte Carlo simulations indicate that a 20–100 keV polarization sensitivity of less than a few percent can be achieved for X-class flares, by comparing the counting rates of these rear segments.

RHESSI AS A POLARIMETER

The capability for doing polarimetry arises from the inclusion of a small unobstructed Be scattering element (3 cm in diameter by 3.5 cm long) that is located within the detector cryostat, near the center of the Ge detector array (Figure 1). Directly in front of the cryostat is a graded-Z shield that is designed to absorb a large fraction of the flux below 100 keV, flux that tends to dominate the flare event. Openings in this shield provide an unattenuated path for low energy photons from the Sun to reach the front surface of the cryostat directly in front of each Ge detector and directly in front of the Be scattering block. Thinned windows in the cryostat are designed to maximize the transmission of low energy solar photons to each Ge detector and to the Be scattering block. The Ge detectors are segmented, with both a front and rear active volume. Low energy photons (below about 100 keV) can reach a rear segment of a Ge detector only indirectly, by scattering. Some fraction of the photons which reach the Be block are Compton scattered into the rear segments of adjacent Ge detectors, four of which have an unobstructed view of the Be block. The angular distribution of the scattered photons (in which is embedded the polarization signature) is sampled by the adjacent Ge detectors. The spacecraft rotation serves to increase the angular sampling frequency and to reduce the effects of any systematic variations in the intrinsic detection efficiencies.

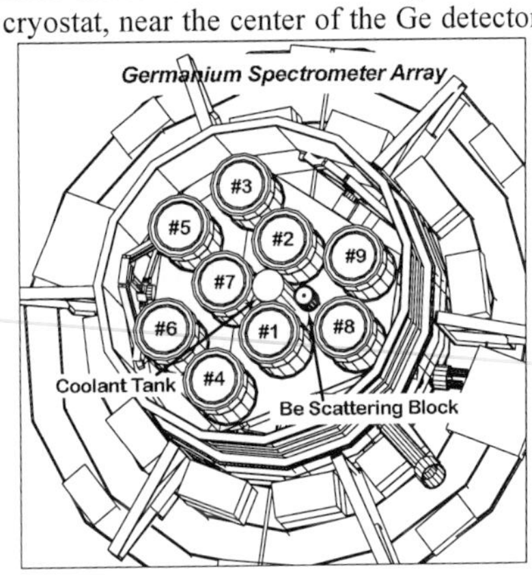

Fig.1. A view of the spectrometer array from the *RHESSI* mass model, showing the location of the Be scattering block.

For performing polarimetry response simulations, we have used a modified version of GEANT3 along with a RHESSI mass model that includes not only the *RHESSI* scientific instrumentation, but also the spacecraft support structure. In our analysis of the simulation data, we use events only from the four Ge detectors closest to the Be. We have found that the Ge detectors that are further from the Be do not provide a polarization signature with sufficient signal-to-noise to be useful in polarization studies.

We characterize the polarization response of *RHESSI* using two parameters: 1) the effective area, which represents the effective area for events satisfying the necessary criteria (single energy deposit in rear segments of the selected Ge detectors); and 2) the polarization modulation factor, a quantity ranging between 0 and 1, that is a measure of the quality of the polarization signature (e.g., McConnell *et al.* 1999). Of particular importance here is the significant impact of spacecraft scattered photons, which is most easily seen in terms of the effective area (Figure 2). Above about 50 keV, scattered photons become important. At energies near 100 keV and above, scattering completely dominates the response. The effects of scattering are also seen in the plot of modulation factor versus energy (Figure 3). Since the scattered component carries with it no polarization signature, the modulation factor decreases at energies above ~30 keV, where scattering becomes important.

It is difficult to define a 'typical' solar flare to use as a baseline for estimating polarization sensitivities. The X-Ray classification depends only on the peak X-Ray flux. The polarization sensitivity for a given class of flare will also depend on the specific spectral shape and on the duration of the event. We have used durations ranging from 20 seconds up to 1000 seconds, with an average spectrum corresponding to that given by Chanan, Emslie, & Novick (1988) for an X2 class flare. The background is estimated using data from the Ge detector of the Wind/TGRS experiment. We have found that *RHESSI* will have sufficient polarization sensitivity to measure the polarization of X-class flares down to a level below 10% and, in some cases, below 1%. This level of sensitivity will be useful in constraining various models that have been published in the literature, some of which predict polarization levels as high as 20 or 30% (e.g., Leach & Petrosian 1983). Our sensitivity estimates have not yet considered the effects of albedo flux scattered from the Earth's atmosphere. Although the level of the albedo flux may be significant, its modulation by the spacecraft rotation (with a different modulation pattern than that of the polarization signal) will help us to distinguish the albedo component from the direct source flux.

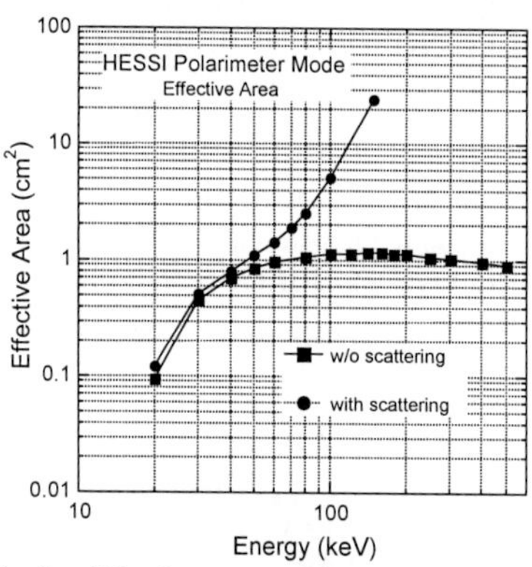

Fig. 2. Effective area of the polarization mode both with and without scattering.

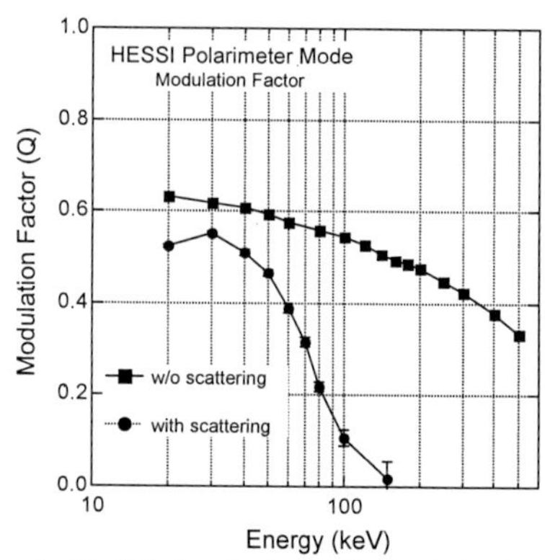

Fig. 3. Modulation factor of the polarization mode both with and without scattering.

ACKNOWLEDGEMENTS

This work is currently supported by NASA grant NAG5-10203.

REFERENCES

Chanan, G., Emslie, A.G., & Novick, R., *Solar Phys.*, **118**, 309 (1988).
Leach, J. & Petrosian, V., *Ap. J.*, **269**, 715 (1983).
McConnell, M.L., Macri, J.R., McClish, M., & Ryan, J., *Proc. SPIE*, **3764**, 70 (1999).

LIST OF ACRONYMS

The editors have not insisted that the following acronyms be spelled out in each paper:

Satellite	*Instrument*	*Meaning*
Yohkoh		"Sunbeam"
	SXT	Soft X-ray Telescope
	HXT	Hard X-ray Telescope
	BCS	Bent Crystal Spectrometer
	WBS	Wide Band Spectrometer
TRACE		Transition Region and Coronal Explorer
SOHO		SOlar and Heliospheric Observatory
	EIT	Extreme-ultraviolet Imaging Telescope
	CDS	Coronal Diagnostic Spectrometer
	SUMER	Solar Ultraviolet Measurements of Emitted Radiation
	LASCO	Large Angle Spectroscopic COronagraph
	UVCS	Ultra-Violet Coronal Spectrometer
GOES		Geostationary Operational Environmental Satellite

LIST OF PARTICIPANTS

Yohkoh 10th Anniversary Meeting
Kona, Hawaii -- January, 2002

Name	*Institution*	*Country*
Acton, Loren W.	Montana State University	USA
Akiyama, Sachiko	Naval Research Laboratory	USA
Altrock, Richard C.	Air Force Research Laboratory at Sacramento Peak	USA
Asai, Ayumi	Kwasan and Hida Observatories, Kyoto University	Japan
Aschwanden, Markus J.	Lockheed Martin Solar & Astrophysics Laboratory	USA
Aurass, Henry	Astrophysikalisches Institut Potsdam	Germany
Banerjee, Dipankar	Centre for Plasma Astrophysics	Belgium
Bartus, Janos	Solar Physics Research Corporation	Japan
Benevolenskaya, Elena E.	Stanford University	USA
Bookbinder, Jay A.	Smithsonian Astrophysical Observatory	USA
Brosius, Jeffrey W.	NASA Goddard Space Flight Center	USA
Bruner, Marilyn	Bermar Science & Technology LLC	USA
Canfield, Richard C.	Montana State University	USA
Cauffman, David P.	Lockheed Martin Advanced Technology Center	USA
Cirtain, Jonathan W.	Montana State University	USA
Correia, Emilia	Universidade Presbiteriana Mackenzie	Brazil
Cranmer, Steven R.	Smithsonian Astrophysical Observatory	USA
Davey, Alisdair R.	Montana State University	USA
DeForest, Craig E.	Southwest Research Institute	USA
DeLuca, Edward E.	Smithsonian Astrophysical Observatory	USA
Dobrzycka, Danuta	Harvard-Smithsonian Center for Astrophysics	USA
Doschek, George A.	Naval Research Laboratory	USA
Emslie, Gordon	University of Alabama in Huntsville	USA
Falconer, David A.	NASA Marshall Space Flight Center/University of Alabama in Huntsville	USA
Farnik, Frantisek	Astronomical Institute of the Academy of Sciences	Czech Republic
Fludra, Andrzej	Rutherford Appleton Laboratory	USA
Foley, Carl R.	Mullard Space Science Laboratory	Great Britain
Gburek, Szymon J.	Space Research Centre, Polish Academy of Sciences	Poland
Golub, Leon	Smithsonian Astrophysical Observatory	USA
Hagino, Masaoki	Meisei University	Japan
Hanaoka, Yoichiro	National Astronomical Observatory of Japan	Japan
Hansteen, Viggo H.	Institute of Theoretical Astrophysics	Norway
Hara, Hirohisa	National Astronomical Observatory of Japan	Japan
Harra, Louise K.	Mullard Space Science Laboratory	United Kingdom
Hassler, Donald M.	Southwest Research Institute	USA
Hirayama, Tadashi	Meisei University	Japan

Hirose, Shigenobu	Department of Physics, Science University of Tokyo	Japan
Holman, Gordon D.	NASA Goddard Space Flight Center	USA
Holmes, Charles P.	NASA Headquarters	USA
Hori, Kuniko	National Research Council/NASA Marshall Space Flight Center	USA
Hudson, Hugh S.	Space Sciences Laboratory, UC Berkeley	Japan
Ichimoto, Kiyoshi	National Astronomical Observatory of Japan	Japan
Isobe, Hiroaki	Kwasan and Hida Observatories, Kyoto University	Japan
Karlicky, Marian	Astronomical Institute	Czech Republic
Katsukawa, Yukio	University of Tokyo	Japan
Khan, Josef I.	Mullard Space Science Laboratory/Institute for Space and Astronautical Science	Japan
Kisich, Diane M.	Space Sciences Laboratory, UC Berkeley	USA
Kliem, Bernhard	Astrophysical Institute Potsdam	Germany
Klimchuk, James A.	Naval Research Lab	USA
Ko, Yuan-Kuen	Harvard-Smithsonian Center for Astrophysics	USA
Kobayashi, Ken	University of Tokyo	Japan
Kosugi, Takeo	Institute of Space and Astronautical Science	Japan
Kozu, Hiromichi	Kwasan Observatory, Kyoto University	Japan
Kundu, Mukul R.	University of Maryland, Department of Astronomy	USA
Kusano, Kanya	Hiroshima University	Japan
LaBonte, Barry J.	Johns Hopkins Applied Physics Laboratory	USA
Lang, James	Rutherford Appleton Laboratory	United Kingdom
Larson, Michelle B.	Space Sciences Laboratory, UC Berkeley	USA
Li, Jing	Institute for Astronomy/University of Hawaii	USA
Li, Youping	Purple Mountain Observatory	China
Litvinenko, Yuri	University of New Hampshire	USA
Madjarska, Maria S.	Armagh Observatory	Northern Ireland
Magara, Tetsuya	Montana State University	USA
Martens, Petrus C.	Montana State University	USA
Mason, Helen E.	University of Cambridge	United Kingdom
Masuda, Satoshi	Solar-Terrestrial Environment Lab, Nagoya University	Japan
Matsuzaki, Keiichi	Institute of Space and Astronautical Science	Japan
McKenzie, David E.	Montana State University	USA
McMullen, Rebecca A.	Montana State University	USA
Metcalf, Thomas R.	Lockheed Martin Solar & Astrophysics Laboratory	USA
Miller, James	University of Alabama in Huntsville	USA
Miyagoshi, Takehiro	National Astronomical Observatory of Japan	Japan
Moore, Ronald L.	NASA Marshall Space Flight Center	USA
Morimoto, Taro	Hida and Kwasan Observatories, Kyoto University	Japan
Morita, Satoshi	Science University of Tokyo	Japan
Nagata, Shin'ichi	Institute of Space and Astronautical Science	Japan
Nakagawa, Tomoko	Tohoku Institute of Technology	Japan
Narukage, Noriyuki	Kwasan and Hida Observatories, Kyoto University	Japan

Nightingale, Richard W.	Lockheed Martin Solar & Astrophysics Laboratory	USA
Nishida, Atsuhiro	Japan Society for the Promotion of Science	Japan
Nitta, Garry Y.	University of Hawaii	USA
Nitta, Nariaki V.	Lockheed Martin Solar & Astrophysics Laboratory	USA
Noonan, Elizabeth J.	Montana State University	USA
Ogawara, Yoshiaki	Institute of Space and Astronautical Science	Japan
Ohyama, Masamitsu	Shiga University	Japan
Owens, John H., Jr.	NASA Retired	USA
Panasenco, Olga A.	Institute of Nuclear Physics	Russia
Parnell, Clare E.	University of St. Andrews	Scotland
Petrosian, Vahe'	Stanford University	USA
Pevtsov, Alexei A.	National Solar Observatory	USA
Reeves, Katharine K.	Harvard-Smithsonian Center for Astrophysics	USA
Romashets, Eugene P.	Institute of Terrestrial Magnetism, Ionosphere and Radiowave Propagation (IZMIRAN)	Russia
Ryan, James M.	University of New Hampshire	USA
Saba, Julia L. R.	Lockheed Martin Solar & Astrophysics Laboratory	USA
Sakao, Taro	Institute of Space and Astronautical Science	Japan
Sakurai, Takashi	National Astronomical Observatory of Japan	Japan
Sato, Jun	Montana State University	USA
Sawant, Hanumant S.	Instituto Nacional De Pesquisas Espacias	Brazil
Schmahl, Edward J.	NASA Goddard Space Flight Center / University of Maryland	USA
Sersen, Michal	Comenius Univeristy	Slovakia
Shibata, Kazunari	Kwasan Observatory	Japan
Shimizu, Toshifumi	National Astronomical Observatory of Japan	Japan
Shimojo, Masumi	Nobeyama Radio Observatory/NAOJ	Japan
Shin, Junho	National Astronomical Observatory of Japan	Japan
Slater, Gregory L.	Lockheed Martin Solar & Astrophysics Laboratory	USA
Sterling, Alphonse C.	NASA Marshall Space Flight Center	USA
Sturrock, Peter A.	Stanford University	USA
Sui, Linhui	NASA Goddard Space Flight Center	USA
Sylwester, Barbara	Space Research Centre, Polish Academy of Sciences	Poland
Sylwester, Janusz	Space Research Centre, Polish Academy of Sciences	Poland
Takeda, Aki	Solar Physics Research Corporation	Japan
Takeuchi, Akitsugu	Yonago National College of Technology	Japan
Tanuma, Syuniti	Hida and Kwasan Observatories, Kyoto University	Japan
Tarbell, Theodore D.	Lockheed Martin Solar & Astrophysics Laboratory	USA
Tikhomolov, Evgeniy	TRIUMF, Canada's National Laboratory	Canada
Title, Alan M.	Lockheed Martin Solar & Astrophysics Laboratory	USA
Tsuneta, Saku	National Astronomical Observatory of Japan	Japan
Uchida, Yutaka	Physics Department, Science University of Tokyo	Japan
van Driel-Gesztelyi, Lidia	University College London	Great Britain
Vats, Hari Om	Physical Research Laboratory, Ahmedabad	India

Verma, V. K.	State Observatory, Naini Tal	India
Virani, Shanil N.	Harvard-Smithsonian Center for Astrophysics	USA
Warren, Harry P.	Harvard Smithsonian Center for Astrophysics	USA
Watanabe, Tetsuya	National Astronomical Observatory, Japan	Japan
Weber, Mark A.	Stanford University	USA
Wikstol, Oivind	Institute of Theoretical Astrophysics	Norway
Wills-Davey, Meredith J.	Montana State University	USA
Winter, Henry D.	Montana State University	USA
Yaji, Kentaro	Kawabe Cosmic Park	Japan
Yashiro, Seiji	Center for Solar Physics and Space Weather, The Catholic University of America	USA
Yokoyama, Takaaki	National Astronomical Observatory of Japan	Japan
Yoshimori, Masato M.	Rikkyo University	Japan
Yoshimura, Keiji	Institute of Space and Astronautical Science	Japan

AUTHOR INDEX

P

R

S

T

U

V

W

Y